International Human Resource Management and International Labour Law

A Human Resource Management Accouting Approach

by

Prof. Dr. habil. Wilhelm Schmeisser,

Univ.-Prof. Dr. Dieter Krimphove

and

Dipl.Kffr. (Univ.) Rebecca Popp

with the collaboration of

Kristin Kirchhoff, Martin Schuster, Edith Teschner, Susan Burghardt, Lydia Clausen, Alexa Hellweg, Daniela Liersch, David Luwisch, Bettina Pape, Cornelia Pape, Karin Peters, Katja Schneider, Anja Seifert, Daniela Woitok

Oldenbourg Verlag München

Bibliografische Information der Deutschen Nationalbibliothek

Die Deutsche Nationalbibliothek verzeichnet diese Publikation in der Deutschen Nationalbibliografie; detaillierte bibliografische Daten sind im Internet über http://dnb.d-nb.de abrufbar.

Rosenheimer Straße 145, D-81671 München
Telefon: (089) 45051-0
www.oldenbourg-verlag.de

Lektorat: Anne Lennartz
Herstellung: Tina Bonertz
Titelbild: thinkstockphotos.de
Einbandgestaltung: hauser lacour
Gesamtherstellung: Grafik & Druck GmbH, München

Dieses Papier ist alterungsbeständig nach DIN/ISO 9706.

ISBN 978-3-486-71649-8
eISBN 978-3-486-72119-5

Authors list

Wilhelm Schmeisser, Prof. Dr. habil., holds the chair of business administration. He is a director and speaker of the Kompetenzzentrum Internationale Innovations- und Mittelstandsforschung (Competence Centre for Innovation and SME Research), Berlin, as well as of the Forschungsstelle Europäisches Personalmanagement und Arbeitsrecht (Research Centre for European HR Management and Industrial Law, EPAR) at the University of Paderborn.

As its full professor Dr. jur. **Dieter Krimphove** teaches at the University of Paderborn European Business-Law, which emphasis European Labour-Law, European Banking-; Finance-Law as well as Competition-Law. At the University Paderborn he chairs also the "Institute of Harmonization of Law, Business Law and Finance". Concurrently he works as visiting professor of the Donau-Universität Krems. The European Commission granted Prof. Dr. Krimphove the "Jean-Monnet Chair 'ad personam'". His numerous professional publications and scientific papers made him known to a broad audience.

Rebecca Popp, Internationale Dipl.-Kffr. (Univ.), graduated in international business administration at the University of Nuremberg and is currently a freelance research assistant and a doctoral researcher with the Kompetenzzentrum Internationale Innovations- und Mittelstandsforschung, Berlin (www.mittelstandsforschung-berlin.de). She is the translator of the book.

Imperial Tobacco
REEMTSMA
Are well-known brands a big boost to your career?
ANNA-JULIA DLUGOSCH & JULIA MOHR
Junior Marketing Manager & Trainee Sales – Reemtsma
A person's first independent steps on their career path are often the most important. This is why we provide our employees with a great deal of scope from the very beginning – scope that they can fill with their own ideas. Moreover, with international brands such as JPS, Gaulouises, West, Davidoff and Drum, we also offer extensive experiences when it comes to business activities in our world wide locations and a manufacturing footprint in more than 50 factories.
Exciting products and excellent career opportunities make us an extremely attractive employer. So come and join us! We would very much look forward to helping you design your career. www.imperial-tobacco.com
GUIDED BY OUR VALUES.

Preface

On International Management and also on International Human Resource Managements there is from the point of view of behavioral science an almost unmanageable amount of literature. Common topics are personnel recruitment, staff selection, personnel development, international assessment centers and reintegration issues for delegated persons. It is almost impossible to develop and write a fundamentally innovative textbook in this genre. Looking at the daily business of business managers and lawyers who work in the field of international personnel management and international labor law of a company makes clear that their activities are not only coined by classical personnel functions, but also by international design of contract, international remuneration systems with stock-option programs, social, pension and other insurance issues for delegated persons, international human resource controlling and international human resource information systems, diversity management in Europe and world-wide as well as European Labor Law, if applicable supported by first country studies.

One will try in vain to find all those contents in one textbook on International Personnel Management. This is mainly due to formation problems of Human Resource staff that have not learned their competence field from a classical business managerial, especially finance-oriented and including legal point of view. In order to close this research and educational gap and to support students but also practitioners in this field with a textbook, this textbook was created.

The authors

Contents

List of Abbreviations

AGG	General Equal Treatment Act
AktG	German Stock Corporation Act
APB	Accounting Principles Board
Art.	article
BDSG	German Data Protection Act
BetrVG	Industrial Constitution Law
BGG	Act on the Equal Treatment of Persons with Disability
CFROI	Cash Flow Return on Investment
comp.	compare
DCF	Discounted Cash Flow
DCKG	German Corporate Governance Codex
DP	data processing
e.g.	exempli gratia, for example
EBIT	Earning before interest and taxes
EC	European Community
EC	Employee Capacity
ECHR	European Convention on Human Rights
ed.	editor, edition

EEC	European Economic Community
EStG	German General Fiscal Code
et seq.	and following
etc.	etcetera
EUR	Euro, Euros
EVA	Economic Value Added
FASB	Financial Accounting Standards Board
fig.	figure
GDP	Gross Domestic Product
GeWO	German Trade, Commerce and Industry Regulation Act
HGB	German Commercial Code
HR	Human Resources
IAS	International Accounting Standards
IASB	International Accounting Standards Board
IFRS	International Financial Reporting Standards
MIDP	Motor Industry Development Programme
mn	marginal note
N.N.	nomen nescio (name unknown, no name)
N°, n°	number
NUMSA	National Union of Metalworkers of South Africa
OECD	Organization for Economic Cooperation and Development

p.	page
ROCE	Return on Capital Employed
ROE	Return on Equity
SAR	Stock Appreciation Rights
sec.	section
SFAS	Statement of Financial Accounting Standards
SGB	German Social Code
subs.	subsection
tab.	table
TEC	Treaty Establishing the European Community
TSR	Total Shareholder Return
UDHR	Universal Declaration of Human Rights
US	United States of America
USD	US-Dollar
vol.	volume
VorstAG	Appropriateness of Management Remuneration Act
VorstOG	Executive Board Compensation Disclosure Act

I Finance oriented Human Resource Management for the support of globalization strategies of international enterprises

1 Introduction

"Globalization stands for the technical and economic change which has taken place in the last decades all around the world. The reduction of trade restrictions, regulations regarding service and capital movements as well as quickly diminishing transport and communication costs which are made possible by new technologies are causal for this development."[1]

A company expects higher profit chances, increase of sales in foreign markets and the achievement of competitive advantages through internationalization. Besides, diversification of risks and transfer of technology and know-how can be achieved. Sometimes, also public subsidies can be used on the basis of internationalization, when a subsidiary is established abroad and trade with a foreign partner is reinforced. A further important aim of internationalization is to react to exchange rates developments and other trade restrictions.

The internationalization of a company has effects on all areas and thus also on human resources with administrative, legal and finance-oriented characteristics and on behavioral personnel management. The tasks of international personnel management are much more comprehensive as the ones of national HR management, because an important task of international personnel management is to promote, organize and control foreign assignments.

Foreign assignments involve numerous measures, as for example a comprehensive remuneration management system with contractual problems with respect to foreign assignments regarding for example matters relevant for tax, pension and insurance law, the employment of staff and their support, personnel development and country specific

---

1 Regnet/Hofmann, 2000, p. 43, own translation.

characteristics, an international HR information and accounting system for a better diversification of corporate group activities etc.

2 Differentiation of international and national human resource management

Personnel management is the systematic analysis, evaluation and arrangement of all HR aspects in a company, including on a worldwide basis, e.g. by a German, international and thus European corporate group. This comprehends amongst other things the determination of needs, the analysis of inventories, labor displacement as well as personnel recruitment, selection, development and remuneration.

This is due to the fact that personnel management is the HR key to a sustainable success in competition of an international company, which means that it has to be regarded as one of the most important value drivers in value creation in finance-oriented human resource management and international, value-oriented business management.[2]

With the growing internationalization of a company also personnel management has to adjust to new circumstances. This does not only imply handling language barriers, but also taking cultural, political, economic and social hurdles, because these environmental factors and circumstances provide different stability for a foresighted personnel management with the instrument of country research.

"The most influential frame conditions of personnel management are

1. rise of national/ international competition (competitors)
2. quality and customer orientation
3. development of value of employees
4. globalization of competition
5. merging of economies
6. decrease in workforce supply
7. change of employment structures
8. ecological challenges."[3]

However, not every internationally active company has to have a special department for international personnel management. Depending on the kind and intensity of the company's activities abroad the HR management is adjusted to the cooperating foreign partner. In case of a cooperation, the influence on personnel management by the foreign cooperating partner

---

2 Comp. Schmeisser, 2008 as well as Schmeisser/Clausen, 2009

3 Wolf, 1994, p. 2, own translation.

is only limited or even inexistent. If a Joint Venture is concluded with a foreign partner, the national personnel management is only influenced in the case of staff exchange.[4] In case of a buy-out of a company abroad, however, the national company can determine the management of the foreign office via the corporate structure. The same happens in case of a foundation of a new subsidiary abroad.

In this case the International Human Resource Management is influenced by the local country cultures (comp. country research, e.g. South Africa) as well as by country specific personnel management systems. Also the strategy of internationalization and the management system of the company and the company culture play a role.

The international human resource management can be distinguished from the national human resource management by the following factors:

- bigger amount of activities,
- global perspective,
- consideration of privacy of employees,
- more risks in Corporate Governance and Compliance Management,
- consideration of different economic, legal and cultural conditions and factors and
- attempt of non-discrimination of different nationalities through diversity management.[5]

3 Problems and aims of international HR management

Task and aim of international personnel management is to quantitatively and qualitatively assure the right allocation of labor on a national and international level.

In order to fulfill this task cultural understanding as well as disposition and capacity of mobility of employees are promoted, e.g. with the help of business trips and shorter delegations abroad. Besides, the international personnel management supports a harmonious and constructive contact between the employees on a worldwide basis, in order to make a cross-national know-how transfer possible. Via the development of an open and tolerant company culture the adaptation of local branches of an international company to other countries' different economic and cultural frame conditions can be more efficiently achieved.

With regard to the tasks of international personnel management, one has to pay attention that different patterns of thinking and acting of country culture are taken into consideration in

4 Comp. Drumm, 2008, p. 625

5 Comp. Bartlett/Ghoshal, 1990, p. 99 et seq.

order to avoid misunderstandings and failures. Besides, the delegated employees are required to have the capacity to react flexibly to challenges.[6]

The international personnel management also has to consider social targets besides the HR-related support of the whole company when striving for economic and profit aims. The legal minimum requirements are different in every country (e.g. minimum wage). Furthermore, international personnel management contributes to the "creation of a uniform identity of the company at home and abroad by the unification of the frame of action of personnel management within the frontiers of governing law."[7]

4 Options of personnel policy in international personnel management

Depending on the degree of internationalization, the company has to adjust its strategy to new challenges. Each strategy offers advantages and disadvantages for the company and personnel management. "On the one hand it might create a competitive advantage for the company, if personnel tasks as selection, training or remuneration can be globally coordinated. On the other hand it can be problematic if specific requirements of the local branch are not sufficiently considered by adequate adjustments with regard to local or regional prerequisites like regulations of working hours, labor union requirements, salary structures or different labor mentalities of the other culture."[8] Four variants of international personnel management are differentiated which combine advantages and disadvantages differently.

Among these are:

- national,
- multinational,
- global and
- transnational model.

---

6 Comp. Regnet/Hofmann, 2000, p. 261

7 Drumm, 2008, p. 628, own translation.

8 Kumar/Wagner, 1998, p. 5, own translation.

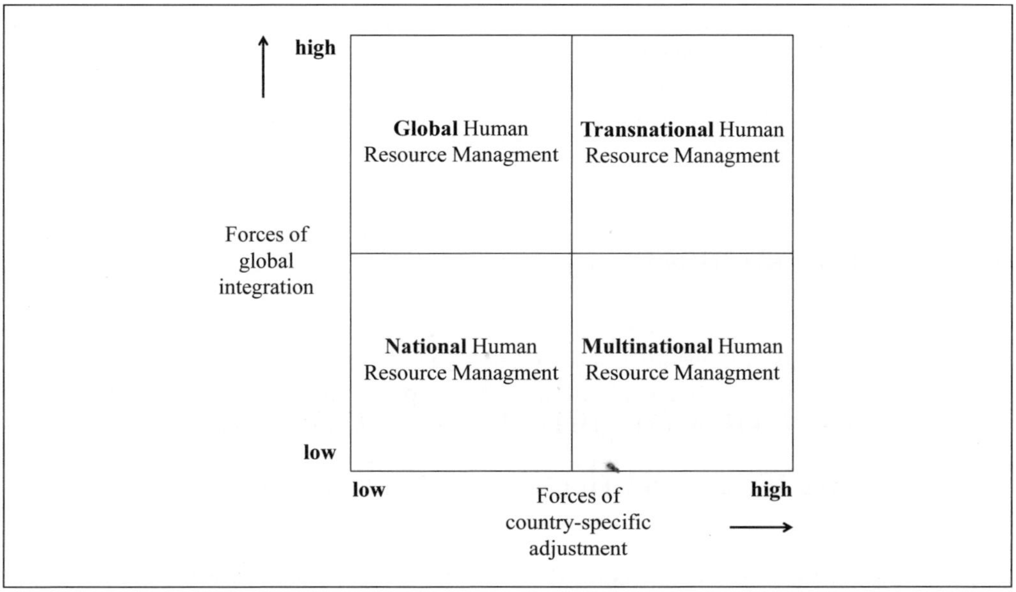

Fig. 4.1: Variants of international personnel management[9]

As can be seen in Fig. 4.1., the different models can be differentiated by the degree of global integration and local adjustment.

4.1 National model

The national model of international personnel management stands for little foreign activities. The small number of overseas branches, which often serve as pure sale base, does not yet require a global integration and local adjustment. Coordination of the foreign activities and of the business activities is undertaken in the main office, which transfers all important decisions to the branch offices. This model is often applied in the adjustment phase of the company to internationalization, during which personnel management can be arranged in a flexible way.

4.2 Multinational model

In the case of multinational personnel management the companies act more often in very different countries. Because of the strong differentiation of the individual countries it is

9 Comp. ibid., p. 5

necessary to adjust the personnel management to the frame conditions of the country in question. Due to the differentiated adjustment it becomes difficult to implement standardized actions in the corporation. Personnel related problems are solved independently and without connection to other corporate parts in the local offices.[10] Thus, central coordination in the main office is not necessary.

4.3 Global model

The global model of international personnel management is inappropriate for companies which are inexperienced in the area of international business. It is rather an adequate model for companies which act in many countries with many differences. "It is characterized by globally standardized personnel measures as selection criteria or motivation systems which can well be new to the host countries."[11] Therefore, experience is needed in order to deal with resistance and achieve advantages. Besides, cost advantages can be reached, because uniform measures are applied. On the other hand specific aspects of the host country are neglected.

4.4 Transnational model

The transnational model is applied in companies "which are acting in many different cultures, but which have enough experience and capacities in order to achieve worldwide coordination advantages despite of this."[12] It is possible, for example, to achieve cost advantages by coordinating trainings uniformly. Besides, the internationalization of the company is promoted because of the interchangeability of staff.

5 Internationalization strategies in personnel management

"Company aims, strategy and company culture have to be adjusted to the requirements of internationalization and globalization of the individual company, taking the company's historical development, the company structure and the national and international sector-

---

10 Comp. Kumar/Wagner, 1998, p. 6 et seq.

11 Ibid., p. 6, own translation.

12 Ibid., p. 6

specific conditions and development trends into consideration."[13] Because of growing internationalization of the company also personnel management has to be adjusted to new challenges. Different internationalization strategies provide the basis for this, as will now be described.

5.1 Basic strategies

Permutter's categorization is one of the basic strategies.[14] He distinguishes the ethnocentric, polycentric and geocentric strategy. The regiocentric strategy, which is a split form of the geocentric strategy, has been added recently.

5.1.1 Ethnocentric strategy

The ethnocentric strategy is characterized by the centralization of decisions in the head office, i.e. all decisions are taken in an authoritarian way in the head office and assigned to local offices. Besides, all controlling measures of the home country are also applied to the host country. In case of this strategy, the communication flows almost exclusively from the head company to the subsidiary. Key jobs are in most cases filled in with skilled labor and managers from the head company.[15] The aim of this strategy is to transfer proven concepts to the subsidiaries.

Cultural distance with leaders at the home office can be avoided by the appointment of personnel from the home office in leading positions. There might even be a personnel relation between them. The disadvantage is that higher costs arise because of the delegation of specialized and executive staff (travel expenses, costs of further trainings). Furthermore, the possibility of promotion of the employees of the local branch is limited, because the executive staff generally comes from the head office.[16]

5.1.2 Polycentric strategy

In case of the polycentric strategy the local branches are seen as independent units. Generally, decisions are taken on a local level, which leads to little flow of information and less communication intensity between the head office and the subsidiaries. Each subsidiary takes its own decisions and adapts the nationality and culture of the host country. The

---

13 Regnet/Hofmann, 2000, p. 40, own translation.

14 Comp. the Perlmutters E.R.R.G. categorization quoted from Perlitz ,1995, p.142 et seq.

15 Comp. Scholz, 2000, p. 96

16 Comp. Scholz, 2000, p. 97

organizations are different and independent from each other. Multinational companies chose this strategy for their personnel management.

The disadvantages of the ethnocentric strategy are equal to the advantages of the polycentric strategy. Only small remuneration costs result, because delegation costs do not apply and the employees of the subsidiary have enough possibilities of promotion, because the leading positions are filled in with employees of the individual subsidiary. Besides, communication problems of the employees of the subsidiary in the host country are avoided. On the other hand there are communication problems with the head office, because there is often a language barrier. Besides, conflicts between the ideals of the individual subsidiaries can arise. A further problem can be the recruitment of specialized and executive staff. As every country searches its own executive staff higher searching and placement costs can be the result.[17]

5.1.3 Geocentric strategy

The geocentric strategy aims at a uniform concept for all subsidiaries. Employees are motivated to reach international and national targets because of awards. The communication between the subsidiary and the head office is very intense in this case. In this case, the individual subsidiary identifies itself with the worldwide business, taking national interests into consideration. Worldwide cooperation between the head office and the individual subsidiaries is needed for decision taking.

The regiocentric strategy differentiates between regional characteristics. There is a close cooperation between the subsidiaries in the individual regions. On a regional level the individual units of the subsidiaries strongly depend from each other and adjust to the geographic conditions of the region.

A big advantage of this type of strategy is the fact that a uniform company culture is developed. However, also in this strategy very high costs and acceptance problems arise.

5.2 Culture strategies (Monoculture, multiculture, mixed culture)

The culture of a company has direct effects on its personnel management. The culture strategies determine the content of cultural transfer and lead to totally different measures.[18]

---

17 Comp. ibid., p. 97

18 Comp. Scholz, 2000, p. 99

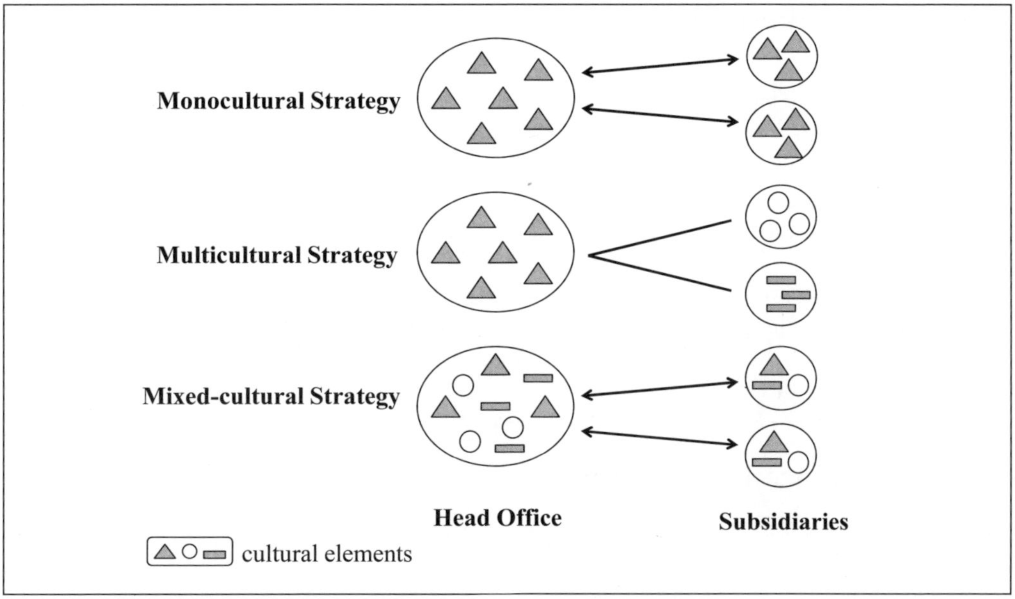

Fig. 5.1: Three cultural strategies[19]

In the case of the monocultural strategy, the company culture of the head company is transferred to the subsidiary abroad. The head office considers "its own company culture as superior to the foreign one and takes care that a company culture which is identical to the one of the head office is created in the foreign subsidiary."[20]

The multicultural strategy, in contrast, is characterized by the fact that the subsidiaries develop their own company cultures and adopt them to their individual country cultures. This means that it is possible that the subsidiaries have a company culture which is completely different from the one in the head office. However, generally, the main features of the head office are also recognizable in the company culture of thc subsidiary.

It is called strategy of mixed cultures when cultures of the subsidiary and the head company are mixed and a homogeneous company culture is created. Because of the influence of the country culture on the subsidiaries a culture results which matches at least to a certain degree each individual country culture.

The choice of the cultural strategy has direct effect on the individual tasks of human resource management. In the area of human resource development an appropriate choice of development measures has to be made in the case of the monocultural strategy. A job rotation is only possible between the head office and the subsidiary. In the case of the multicultural strategy, however, one has to keep the qualification level of employees all

---

19 Comp. ibid., p. 99

20 Comp. ibid., p. 98

around the world in mind. There are only few possibilities of rob rotation in case of this strategy, because every subsidiary has its own culture. The mixed-cultural strategy on the other hand is very appropriate for a systematic job rotation for the integration of employees across all companies.[21]

5.3 Decision strategies (central, local, federal)

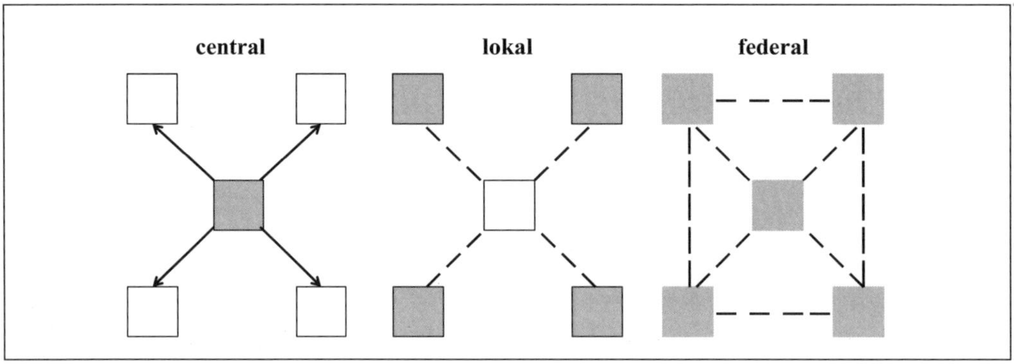

Fig. 5.2: Three decision strategies[22]

"In the case of the central structure there is a parent company where all important decisions are taken."[23] These decisions are then transferred to the subsidiary through exact hierarchies and the abandonment of autonomy.

The local structure is characterized by free and independent decisions. However, the parent company still provides guidelines which are binding for the subsidiary. "In extreme cases it is even possible that the subsidiary defines own targets which need only to be approved by the parent company."[24] Through individual decisions the subsidiaries negotiate their relationship themselves. The result is a partnership of convenience where important aspects are agreed upon.

21 Comp. Scholz, 2000, p. 100

22 Comp. ibid., p. 101

23 Comp. ibid., p. 100

24 Scholz, 2000, p. 101

5.4 Nine strategies of internationalization

Nine different internationalization strategies in personnel management are the result of these three cultural strategies and three decision strategies. These strategies connect the characteristics of cultural strategies with the organizational design of decision strategies.

culture strategy / decision strategy	monocultural strategy	multicultural strategy	mixed-cultural strategy
central	(1)	(2)	(3)
lokal	(4)	(5)	(6)
federal	(7)	(8)	(9)

Fig. 5.3: Internationalization strategies of personnel management[25]

Alternative 1 combines the monocultural strategy with the central decision strategy, which means that all decisions are taken within the head company and "from there jointly taken to the different subsidiaries."[26] Variant 2 is different: In this case a central structure exists, too, but in combination with the multicultural strategy, which accepts different company cultures. This internationalization strategy is applied when a company is active in different country cultures and wants to avoid problems of a inflexible culture.

---

25 Ibid., p. 102, own translation

26 Ibid., p. 101, own translation

"Only alternative 3 can be realized on a long term basis, i.e. the combination of the central structure with the strategy of mixed cultures. In this case there is a uniformly defined decision center in the corporation. However, through the multitude of merger processes a company culture is created which consists of the different partial cultures. […]."[27]

Alternative 4 is characterized by a local decision strategy. It is true that in case of the local solution, there are "several decision centers, but there is always one which disposes over some more power at least."[28] Exactly this characteristic can be observed in the fourth alternative, in which the company culture of the lead company is transferred to the subsidiaries. In the case of alternative 5 the company culture is mixed with country specific characteristics. Alternative 6 leads to the formation of a totally new company culture, as the subsidiaries contribute the respective cultural characteristics. Alternative 7 is a combination of the federal decision strategy and the monocultural strategy, which means that there is no single decision center and "one has to assume that a company disposes of a dominant company culture because of coincidence."[29] In alternative 8 "different decision strategies with different company cultures exist which jointly try to solve problems."[30] Alternative 9 makes the adaptation of company cultures possible.

6 Conclusion

"More and more companies will have to seriously think about how to reach their financial and performance oriented aims more effectively abroad."[31] In order to reach these aims it is also essential to dispose over a personnel management oriented towards the strategy of the company, because a good personnel management will only have success if the cultural, legal, technical, economic and ecological conditions of the host countries are analyzed and taken into consideration for the creation of personnel management strategies and measures."[32]

This leads to "… a change of the quantitative and qualitative labor supply with regard to regional and structural aspects:

- labor supply grows,
- labor supply develops differently from region to region due to different birth rates

---

27 Scholz, 2000, p. 102

28 Ibid., p. 102, own translation

29 Ibid., p. 103, own translation

30 Ibid., p. 103, own translation

31 Wolf, 1994, p. 6; own translation.

32 Wolf, 1994, p. 7; own translation.

- differing growth leads to different age structures of economically active people (industrial countries → higher, developing countries → younger) and
- worldwide adjustment of qualification (more students from developing countries, introduction of a dual education in China)"[33].

Literature

Bartlett, Ch. / **Ghoshal**, S.: Internationale Unternehmensführung. Frankfurt on Main/New York 1990.

Drumm, H.-J.: Personalwirtschaft. 6th edition, Springer Verlag, Berlin 2008.

Kumar, N. / **Wagner**, D.: Handbuch des Internationalen Personalmanagements. Verlag C. H. Beck, Munich 1998.

Regnet, E. / **Hofmann**, L. M. (ed.): Personalmanagement in Europa. Verlag für Angewandte Psychologie, Göttingen 2000.

Schmeisser, W.: Finanzorientierte Personalwirtschaft. Oldenbourg Wissenschaftsverlag, Munich 2008.

Schmeisser, W. / **Clausen**, L.: Controlling und Berliner Balanced Scorecard Ansatz. Oldenbourg Wissenschaftsverlag, Munich 2009.

Scholz, C.: Personalmanagement. 5th edition, Verlag Franz Vahlen, Munich 2000.

Stahl, G. K. / **Mayrhofer**, W. / **Kühlmann**, T. M. (ed.): Internationales Personalmanagement – neue Aufgaben, neue Lösungen. Rainer Hampp Verlag, Munich and Mering 2005.

Wolf, J.: Internationales Personalmanagement – Kontext, Koordination, Erfolg. Gabler Verlag, Wiesbaden 1994.

[33] Ibid., p. 5, own translation.

II International remuneration system

The task of international personnel management is to organize the company policy and, thus, the remuneration system in a uniform way for employees from different countries and with different cultural backgrounds.

The assignment of employees beyond the frontiers of the lead company's country is an important instrument, in order to cope with this task. These assignments can be limited or unlimited. Among the limited assignments are business trips, secondments, delegations, deployments and dislocations, which are different with regard to the duration of the employment abroad. In the case of unlimited assignments a local work contract exists. The employee is not contracted by the home company but completely integrated in the foreign enterprise.

If an employee is sent abroad for a medium term, this is referred to as secondment. If the stay abroad is shorter, it is called business trip, in the case of longer assignments the nomenclature would be dislocation.[34]

Also when an employee from a foreign company is contracted by the home country for a limited period of time, this is called a secondment.

The delegating company follows different aims with regard to international personnel management. These can be divided into three categories:[35] On the one hand know-how is transferred between the companies, on the other hand the company units at home and abroad are easier to coordinate and control. Finally the aspect of personnel development is an important aim of international assignments.

---

34 Comp. Haufe Personal Office Online, 2009

35 Comp. Schmeisser, 2008, p. 255

7 International delegation of executive managers – exemplified by BASF

Die BASF – The Chemical Company BASF is actively represented by over 300 subsidiaries and holdings on five continents.[36] The named continents, on which the company BASF is globally active, are Europe, North America and Central America (including the Caribbean), South America, Africa and Asia (Pacific Rim). The company BASF is present in over 100 important sites in 170 countries all over the world with its production and is, thus, close to the customer, ensuring a reliable provision of its products. Therefore, the company can be found in all economic regions with production and sales, because it wants to benefit from the opportunities of profitable growth all over the world. Europe is the home market of the BASF, but the dynamic regions North America, Asia and South America are becoming more and more important.[37]

At the end of the year 2007 the company BASF employed 95,175 employees all over the world, with about 1,000 executive managers. In the BASF executive managers belong to one of the four upper levels. There are regional differences with regard to the labor turnover rate, which was about 1.6 percent in 2007 on the worldwide average. Moreover, in 2007 2,160 trainees as well as 22,223 employees of other companies, who rendered technical and other services for the company in the different regions worked for the BASF. The following table shows the number of employees by region, in which the BASF is supra-regionally present (status: 31 December 2007).[38]

Regions	Employees	Percentage 2007	Percentage 2006
Europe	61,020	64.1	64.5
Of which Germany	46,890	49.3	49.7
North America	15,191	16.0	16.3
Asia, Pacific Rim	13,278	13.9	13.4
South America, Africa, Middle East	5,686	6.0	5.8
total	**95,175**	**100.0**	**100.0**

Table 7.1: Number of employees in the regions[39]

[36] Comp. Brinkkötter, 1997, p. 413

[37] Comp. BASF – The Chemical Company, Globale Präsenz, 2007

[38] Comp. BASF – The Chemical Company, Mitarbeiter in den Regionen, 2007

[39] Adapted from: BASF – The Chemical Company, Mitarbeiter in den Regionen, 2007

In 2007 the BASF spent 6,648 million Euros on personnel expenses and social contribution, which includes salaries, social security payments, old age contributions and social support. The amount has grown by 7.1 percent compared to the year 2006 (see table 7.1). The BASF offers a market-conform, attractive remuneration to its employees, as the remuneration is combined to social contributions, which are for example an additional health insurance, a company pension scheme and stock purchase programs.[40]

Remuneration components	m. EUR	Variation in percent
Wages and salaries	5,379	6.9
Social security taxes and old age provision and support	1.269	7,5
of which old-age provisions	359	25.1
Personnel expenses	**6,648**	**7.1**

Table 7.2: personnel expenses of BASF – The Chemical Company[41]

7.1 Importance of international deployment of staff

"Equal opportunities for all – regardless of their personnel characteristics."[42], also called "diversity", is valid for any BASF location around the world and is strongly fixed in the company's fundamental values and guidelines. This means that all employees have equal opportunities in the companies regardless of their sex, nationality, education or culture group. BASF believes that diversity creates open-mindedness, for cultural as well as demographic changes in the company as well as for customers' matters and desires. Also force of innovation is fostered this way. Different cultural backgrounds, individual perspectives, experiences, ways of thinking and personal approaches contribute to quicker and more creative reactions to challenges. Therefore, the worldwide best experts and managers are recruited and the integration of employees from other countries is supported with intercultural offers for further education. On the long run, the company BASF has the aim to raise the percentage of international staff among the executive managers as well as the percentage of women among them. In 2006 the international percentage of executive managers was about 30 percent, with a female percentage of 5.6 percent. Moreover, the company aims at rising the percentage of executive managers with international experiences even further. In 2006, the percentage of internationally experienced executive officers was

[40] Comp. BASF – The Chemical Company, Mitarbeiter in den Regionen, 2007

[41] Adapted from: BASF – The Chemical Company, Mitarbeiter in den Regionen, 2007

[42] BASF – The Chemical Company, Diversity, 2007

71 percent. In comparison, in 2003 it was only 64 percent.[43] This shows that deployments abroad are becoming more important. One has to take into consideration that the deployments abroad do not only take place from or to Germany, but also between subsidiaries, e.g. between Great Britain and Singapore or between Belgium and Brazil.[44]

7.2 Structural and contractual aspects

The BASF distinguishes between four types of international staff deployment depending on aim and duration, which is shown in the table below.[45] The examples are based on the typical employee who is delegated with his family and household from his home country to his workplace abroad.

Type	Aim	Duration
Informational delegation	Acquaintance with language and environment	5 weeks to 12 months
Deployment	Fulfillment of specific tasks	5 weeks to 24 months
Delegation	Know-how-transfer, development of executive manager	4 to 8 years
Transfer	Definitive assumption of a function	unlimited (after 8 years)

Table 7.3: Types of international deployment of staff of the BASF [46]

In practice the BASF – The Chemical Company uses three standardized contracts; the first contract – the so called delegation contract regulates the relation between the delegated person and the home company.

In the second contract the labor relations between the receiving company and the delegated person are set down. The third contract contains arrangements between the home company and the receiving company which are made for the delegation. In the case of a delegation contract the original labor contract of the delegated employee with the home company is not

43 Comp. BASF – The Chemical Company, Diversity, 2007

44 Comp. Brinkkötter, 1997, p. 414

45 Comp. ibid., p. 414

46 Comp. ibid., p. 414

annulled, but put at rest for the duration of the delegation. Thus, the expatriate has the right and the duty to return.[47]

7.3 Salary system

There are uniform guidelines within BASF – The Chemical Company for the determination and development of the earnings abroad, which precede the explanations on the salary system of executives for international delegations. These guidelines are handed to all delegated employees and have the following wording:[48]

- "BASF guarantees a certain living standard in the country of delegation to the delegated person, which is based on the net earnings of the employee in the home country and guarantees a delegation bonus on top.
- The delegated person is integrated into the salary structure of the country of delegation as far as possible, with the corresponding job market as basis. If the "net living standard" is not reached, necessary supplementary salary components are added.
- The company bears the risks with regard to changes of external conditions regarding the 'net living standard'
- This is guaranteed by
  - determination of earnings abroad (before leaving)
  - regulation of earnings (yearly revision/ adaption)
  - analytical follow-up (after the end of the year/ of the delegation) valid within BASF group.
- The personal situation of the employee is taken into consideration under individual aspects (e.g. family status, number of children, tax class, salary) as well as under global assumptions (domestic circumstances, attitudes of consumption) under protection of privacy
- The components of the salary system as well as the parameters of the frame conditions are transparent and explained. The sheets of calculation are provided to the delegated person.
- The exercised systems and parameters are uniformly valid for all delegated persons and are not subject to negotiations."[49]

7.3.1 Reference salary and other payments

The starting point for determining the earnings abroad is the reference salary, also called shadow salary, of the delegated person in the home country (comp. table 7.5).

---

[47] Comp. Brinkkötter, 1997, p. 415

[48] Comp. ibid., p. 429

[49] Comp. ibid., p. 429

Regular payments are added to the salary, as described in the following chapters. Variable payments, e.g. bonuses, are only added in the context of the analysis.[50] Taxes and social contributions are deducted from the reference earnings. The personal tax characteristics, e.g. tax class and number of children, of the income tax card are the basis for the tax calculation. The church tax paid before the delegation is not deducted, but only displayed for information, because this is a personal decision of the employee. This approach is not influenced by the German peculiarity that the church tax has to be deducted by the employer. The voluntary payment should not lead to a net cutting for the company and, thus, represent a disadvantage of expatriates. After this, the employee's contribution to employee's insurance, unemployment insurance, company old age provisions (pension fund) and general contributions to health and long-term care insurance are deducted. Moreover the generalized market rent of the hometown is deducted from the reference earnings after taxes. The resulting value is the basis for the "net living-standard" that has to be guaranteed in the country of delegation and is the reference figure for the increase for delegation, which is determined afterwards.[51]

Home country			
	Reference salary	EUR	177,777
+	Other earnings (fix)	EUR	0
+	Bonus (variable)	EUR	0
=	**Gross reference earnings**	EUR	**177,777**
–	Income tax	EUR	39,040
–	Solidarity surcharge (Germany)	EUR	2,928
=	**Reference earnings after taxes**	EUR	**135,809**
–	Employee's insurance	EUR	9,216
–	Unemployment insurance	EUR	3,120
–	Pension fund	EUR	3,556
–	Health insurance	EUR	4,284
–	Long-term care insurance	EUR	486
–	Market rent in home country (general)	EUR	23,400
=	**Available after rent**	EUR	**91,747**

50 Comp. Brinkkötter, 1997, p. 429, 431

51 Comp. ibid., p. 431 et seq.

	Buying power adjustment	EUR	
	Available before rent	EUR	115,147
	Of which share of consumption in country of delegation	70 percent	80,603
	Company car	yes	0
=	Share of consumption after rent	EUR	57,203
+/–	**Buying power adjustment**	25 percent	**14,301**
	Delegation bonus	EUR	**35,400**
	Net change compared to year before		
	Available after rent compared to year before	EUR	0
	Balance compared to year before	EUR	0

Table 7.4: Example of calculation home country[52]

7.3.2 Delegation and mobility bonus

In practice, the BASF – The Chemical Company concedes a financial bonus to an employee who is sent abroad by company order. This bonus is meant to encourage employees to go abroad, which is referred to as mobility incentive. Moreover, the motivation to live and work in another country shall be compensated with a difficulty bonus. The aspect of personal enrichment by gaining experience is not taken into consideration in the financial regulations.

In order to determine the amount of bonuses, one has to differentiate between the different countries of delegation and to create a ranking order. This means that a comparability of the countries has to be guaranteed, which is assessed on the basis of certain evaluation criteria. The company BASF evaluates countries in accordance with the following criteria:[53]

- Safety, domestic and external politics, court system
- Environment regarding health care and medical care
- Climatic disadvantages, acts of God
- Social and cultural isolation
- Quality of housing
- Infrastructure, public supply, availability of goods

---

52 Adapted from: Brinkkötter, 1997, p. 432

53 Comp. ibid., p. 415 et seq.

- Economic context, banking system
- Schools, possibilities of recreation and entertainment

The single criteria have to be attributed to the countries in question. They are determined on the basis of the circumstances in the country by international advisors and in accordance with a point system. This way, it can be guaranteed that well-founded values are compared.[54] This is an important aspect to be able to take the different circumstances into consideration in the case of international delegations. Delegations from a developing country to an industrial country entail not only better living and working conditions, but also a considerable adjustment and difficulties because of cultural differences. The balance of advantages and disadvantages with regard to the comparison of living and working conditions in the home country in comparison with those of the country of delegation lead to a certain value of complication. This value is different in every case, because every executive perceives and evaluates individual elements differently depending on age, family status and personal ideal. E.g. a family with children puts more emphasize on the school aspect than singles. Culture and sport enthusiast attach special importance to leisure time activities.[55]

Regardless of the emphasis on certain criteria, the general coherence of the individual attributions has to be taken into account. Therefore, the BASF uses a ten-stage system, which allows a necessary and sufficient differentiation. The countries of delegation presented in table 7.6. which result at the end of 1996 on the basis of objective criteria and after an examination of coherence, are differentiated in accordance with a ten-step system, with Germany as home country basis.

Differentiation	Countries of delegation
1	Austria, Switzerland
2	Belgium, Portugal, USA, Australia
3	Malaysia, Singapore, South Africa, Uruguay
4	Costa Rica, Hong Kong, Poland, Tokyo
5	Rio/Sao Paolo, Indonesia, Mexico City, Istanbul
6	Egypt, Bombay, Rumania, Beijing
7	Bogota, Mangalore, Petersburg, Nanjing
8	Banglad., Karachi, Minsk, Teheran
9	Ethiopia, Libya
10	Algeria, Nigeria

Table 7.5: Examples of the ten-step differentiation within the BASF[56]

54 Comp. Brinkkötter, 1997, p. 418

55 Comp. ibid., p. 418

56 Adapted from: Brinkkötter, 1997, p. 419

If, however, the delegated person comes from another country, as for example the USA, other classifications would result. This means that the degree of complication of going to e.g. Malaysia is different to the one presented above.[57]

For several years the aspect of differentiation with regard to family status has been discussed in many companies. The main argument of such considerations is that today often both partners are working and that the partner of the delegated person would give up his job or career. Several companies disapprove this differentiation, because the delegating company has a contractual relationship only with the expatriate and not with his husband or wife. The BASF, on the other hand, considers it appropriate to take the family status and size of family into consideration in many ways. This can be seen in the salary determination on net basis by determination of the corresponding tax class and deductions, in the regulations regarding homeward flights with tickets for all family members, in the allowances to rent and private health insurance and the refunding of school costs and language courses.

If however, the environment in the country of delegation changes, corresponding corrections should be made. In case of deterioration of the conditions in the country of delegation it makes sense to concede higher allowances on the key date also to employees who have already been sent abroad. Even if the employee has lived and worked under bad financial conditions before, the objective new evaluation of the country should lead to higher allowances out of fairness reasons. If the contrary is the case, i.e. if the situation of the conditions in the country of delegation improves, the new evaluation on the key date is valid for new delegates only. The situation improves if safety or supply aspects improve or a school which leads to a general qualification for university is built. The question is whether this is valid for already delegated employees. For reasons of protection of trust it seems justifiable to maintain the delegation bonus.

In Europe, also for executive managers an international labor market with more preparedness to move has developed because of strong economic interdependences. For that reason several companies have renewed and reduced their allowances within Europe. Since 1994 the BASF has paid a one time allowance, which is paid at the beginning of the delegation for four years instead of the yearly complication bonus. Married employees receive two net salaries and for every child coming to the country of delegation a quarter of the salary is paid.

The mobility bonus, which depends on the country of delegation and the reference salary, is higher for executive managers than for other employees, because they usually earn higher salaries.

The delegation bonus is the sum of mobility and complication bonus. Table 7.7 shows that when comparing different salaries and country hierarchies the delegated executive is relatively seen not paid too much despite the higher total bonus in absolute numbers.[58]

---

[57] Comp. ibid., p. 419

[58] Comp. ibid., p. 422

complication + mobility = delegation bonus

complication: South Africa	complication: India		mobility: South Africa	mobility: India		delegation bonus: South Africa	delegation bonus: India
12 T€	32 T€	+	5 T€	6 T€	=	17 T€	38 T€
12 T€	32 T€	+	5 T€	15 T€	=	17 T€	47 T€

Table 7.6: Examples for delegation bonuses of DASF[59]

7.3.3 Share of rent

With regard to the share of rent often a statistical experience-based rate is employed, as for examples 15 percent of the gross salary or 20 percent of the net salary. A further possibility is to use the actually paid rent. A few years ago the company BASF decided to use the approach of the so called market rent, which is kept general and graduated in lump sums. Based on practical experience, the market rent considers both factors, salary and the percentage of the apartment size (see table 7.8). The rent is based on market surveys of the inland home company and can be adapted in correspondence with market developments in appropriate time intervals. The market rent fixed when delegating the employee is generally only adapted after four years abroad in accordance with the valid parameters at this point in time.

Salary up to	1 person	2–3 persons	4 ≥ persons
80,000	600	900	1,100
100,000	700	1,000	1,200
120,000	800	1,100	1,300
140,000	950	1,250	1,450
160,000	1,100	1,400	1,600

Table 7.7:Excerpt of the BASF-market rent table (monthly rent)[60]

The market rent takes into account that a co-contribution for the rented apartment in the country of delegation has to be made and sometimes a housing allowance has to be granted by the company. The amount of housing allowances depends on a fixed rent limit, which is

59 Following: ibid., p. 422

60 Following: Brinkkötter, 1997, p. 423

calculated in collaboration with the local advisors. The amount which exceeds the limit as well as the ancillary rental receivables are paid by the delegated employee.[61]

Literature

BASF – The Chemical Company: Diversity, 2007, URL: http://www.corporate.basf.com/de/sustainability/mitarbeiter/diversity.htm?id=kJKV1C_XJbcp*OB Status: 08 July 2008.

BASF – The Chemical Company: Globále Präsenz, 2007, URL: http://corporate.basf.com/de/ueberuns/?id=f5PDIC_BUbcp3aA Status: 08 July 2008.

BASF – The Chemical Company: Mitarbeiter in den Regionen, 2007, URL: http://corporate.basf.com/de/ueberuns/mitarbeiter/?id=f5PDIC_BUbcp3aA Status: 08 July 2008.

Berndt, R. / **Altobelli**, C. F. / **Sander**, M.: Internationales Marketing Management. 3rd revised and extended edition, Springer Verlag, Berlin 2005.

Berufsgenossenschaft der Feinmechanik und Elektrotechnik (bgfe): EWR-Staaten, 2007, URL: http://www.bgfe.de/betrieb/bb_ewrstaaten.html Status: 03 July 2008.

Blom, H. / **Meier**, H.: Interkulturelles Management, Interkulturelle Kommunikation, Internationales Personalmanagement, Diversity-Ansätze im Unternehmen. 2nd edition, Verlag Neue Wirtschafts-Briefe, Herne, Berlin 2004.

Borgmann, B.: Die Entsendung von Arbeitnehmern in der Europäischen Gemeinschaft, Wechselwirkungen zwischen Kollisionsrecht, Grundfreiheiten und Spezialgesetzen. Peter Lang GmbH, Frankfurt on Main 2001.

Bornschein, T. / **Thomas**, M.: Leben und Arbeiten im Ausland. 1st edition, Interna, Bonn 2004.

Brinkkötter, H.-O.: Grundfragen aus der Entsendungspraxis der BASF mit besonderer Berücksichtigung von Oberen Führungskräften. in: Clermont, A. / Schmeisser, W. (ed.): l.c., 1997, S. 413–428.

Bröckermann, R. / **Pepels**, W. (ed.): Handbuch Recruitment, Die neuen Wege moderner Personalakquisation, Planung, Beschaffungswege, Auswahlverfahren. Beiträge aus Forschung und Praxis, Cornelsen Verlag, Berlin 2002.

Brüch, A.: Kulturelle Anpassung deutscher Unternehmensarbeiter bei Auslandsentsendungen, Eine empirische Studie in den USA, Kanada, Japan und Südkorea zu Kriterien und Einflussfaktoren erfolgreicher Aufenthalte von Fach- und Führungs-

---

61 Comp. ibid., p. 423

kräften. Peter Lang Europäischer Verlag der Wissenschaften, Frankfurt on Main 2001.

Bundesversicherungsanstalt für Angestellte: Beschäftigung im (vertragslosen) Ausland, Rechtsvorschriften, Erläuterungen und praktische Hinweise. 8th Edition, Bundesversicherungsanstalt für Angestellte, Berlin 1999.

Burghaus, A.: Auslandseinsatz von Mitarbeitern, Maßnahmen zur erfolgreichen Reintegration von Expatriates. VDM Verlag Dr. Müller, Saarbrucken 2006.

Clermont, A. / **Schmeisser**, W. (ed.): Internationales Personalmanagement. Verlag Franz Vahlen, Munich 1997.

Clermont, A. / **Schmeisser**, W. / **Krimphove**, D. (ed.): Personalführung und Organisation. Verlag Franz Vahlen, Munich 2000.

Deutschen Gesellschaft für Personalführung e. V. Düsseldorf: Der internationale Einsatz von Fach- und Führungskräften, Ein Ratgeber von Experten für die Praxis. 2nd revised and extended edition, Wirtschaftsverlag Bachem, Cologne 1995.

Deutsche Verbindungsstelle Krankenversicherung – Ausland (DVKA): Merkblätter für im Ausland Beschäftigte (Entsandte), 2007, URL:
http://www.dvka.de/oeffentlicheSeiten/Merkblaetter/Merkblaetter_Arbeiten.html
Status: 04 July 2008.

Deutsche Verbindungsstelle Krankenversicherung – Ausland (DVKA): Arbeiten im Ausland, 2007, URL:
http://www.dvka.de/oeffentlicheSeiten/Abkommenstaaten.htm
Status: 04 July 2008.

Dommermuth, T. / **Klinger**, M. (2001): Die optimale Gehaltsgestaltung. 1st edition, Rudolf Haufe Verlag, Munich 2001.

Förster, H. / **Heidenreich**, J. / **Heuser**, A.: Auslandsentsendung und Beschäftigung ausländischer Arbeitnehmer, Rechtliche Aspekte beim internationalen Mitarbeitereinsatz. Hermann Luchterhand Verlag, Neuwied, Kriftel 2002.

Gnann, T.: Arbeitsvertrag bei Auslandsentsendung. C. H. Beck'sche Verlagsbuchhandlung, Munich 1993.

Haufe Personal Office Online: Haufe Index. Version 13.4.22.0, 2009.

Heuser, A.: Die Entsendung deutscher Mitarbeiter ins Ausland. W. Bertelsmann Verlag, Bielefeld 2004.

Hild, B.: 30 Minuten für erfolgreiche Arbeit im Ausland. Gabal Verlag, Offenbach 2004.

Hofmann, K.-W. / **Nowak**, H. / **Rohrbach**, T.: Auslandsentsendung, Vorteile, Vorschriften und Gestaltungsmöglichkeiten der Entsendung im Arbeitsrecht, Steuerrecht und Sozialversicherungsrecht. Rudolf Haufe Verlag, Munich 2002.

Hogh, M.: Internationale Entsendung von Führungskräften. Erich Schmidt Verlag, Bielefeld 2000.

Hoppe, J.: Die Entsendung von Arbeitnehmern ins Ausland, Kollisionsrechtliche Probleme und internationale Zuständigkeit, Eine Untersuchung anhand praktischer Vertragsgestaltungen. Duncker und Humblot, Berlin 1998.

Jacobs, O. H.: Internationale Unternehmensbesteuerung, deutsche Investitionen im Ausland, ausländische Investitionen im Inland. 5th revised and extended ed., Stuttgart 2002.

Kose, W. / **Winckler** E. M. (ed.): Gesundheit und beruflicher Auslandsaufenthalt, Risiken, Vorsorge, Versorgung und Absicherung. Weißensee Verlag, Berlin 2003.

Kühlmann, T.: Auslandseinsatz von Mitarbeitern, Praxis der Personalpsychologie. Hogrefe-Verlag, Göttingen 2004.

Lang, J. M.: Moderne Entgeltsysteme, Leistungslohn bei Gruppenarbeit. 2nd revised edition, Deutscher Universitäts-Verlag, Wiesbaden 2001.

Laufenweiler, J.: Ausstrahlung im Arbeits- und Sozialrecht. Duncker & Humblot, Berlin 1999.

Marburger, H.: Auslandseinsatz von Arbeitnehmern. in: RdW, Schriftreihe: Das Recht der Wirtschaft, Richard Boorberg Verlag, Stuttgart 2004.

Mauer, R.: Personaleinsatz im Ausland, Personalmanagement, Arbeitsrecht, Sozialversicherungsrecht, Steuerrecht. Verlag C. H. Beck, Munich 2003.

ORC Worldwide GmbH (2007): ORC Global Workforce, Der Location Evaluation Report die ORC – Empfehlung für die Hardship Allowance, 2007, URL:
http://gwf.orcworldwide.de/ORC-GWF-Q1-2007-GERMANY.pdf
Status: 12 June2008.

Pellens, B. / **Crasselt**, N.: Bilanzierung von Stock-options. in: Der Betrieb, Wochenschrift für Betriebswirtschaft, Steuerrecht, Wirtschaftsrecht, Arbeitsrecht, Heft 05, 30 January 1998, Fachverlag der Verlagsgruppe Handelsblatt GmbH, Bochum 1998.

Pulte, P.: Arbeitsverträge bei Auslandseinsatz. 3rd revised Edition, Verlag Recht und Wirtschaft, Heidelberg 2004.

Pusch, M. / **Sauter**, N. / **Köhler**, P. / **Wallendorf**, A.: Handout Personalrekrutierung, Personalrekrutierung über Bewerberdaten. URL:
http://www.unierfurt.de/angewandte_mikrooekonomie/ss04personal/Handout-Personalrekrutierung.pdf
Status: 14 July 2008.

Schaub, G. / **Koch**, U. / **Neef**, K. / **Schrader**, P. / **Vogelsang**, H.:, 2008 Arbeitsrechtliches Formular- und Verfahrenshandbuch. 9th edition, C. H. Beck, Munich 2008.

Schmeisser, W. / **Brinkkötter**, H.-O. / **Krimphove**, D. (ed.): Internationales Entgeltmanagement. Rainer Hampp Verlag, Munich and Mering 2004.

Schmeisser, W.: Finanzorientierte Personalwirtschaft. Oldenbourg Wissenschaftsverlag, Munich 2008.

Söllner, A.: Einführung in das Internationale Management, Eine institutionenökonomische Perspektive. 1st edition, Gabler Verlag, Wiesbaden 2008.

Stahl, G.: Internationaler Einsatz von Führungskräften, Managementwissen für Studium und Praxis. R. Oldenbourg Wissenschaftsverlag, Munich 1998.

Swiss Life Deutschland: Goodbye Deutschland! Immer mehr Menschen wandern aus, Gastarbeiter und Expatriates, 2008, URL:
http://www.swisslife.de/slde/de/home/vorsorge/versicherungsthema/0801_ausland/0801_auswandern.html#aPar72744
Status: 14 July 2008.

Tung, R. L.: Selection and training procedures of U.S., European and Japanese multinationals. California Management Review, 1982.

Wirth, E.: Mitarbeiter im Auslandseinsatz, Planung und Gestaltung. Gabler Verlag, Wiesbaden 1992.

Annex

Annex I

Overview over delegation types

	Business trip	Secondment	Delegation	Relocation	Transfer
Characteristics of delegation types	Short term deployment abroad, timely limited journey for a clearly defined purpose with maintenance of the active core work place in the home company.	Short term to middle term delegation in connection with a timely limited project abroad, e.g. project management, market study.	Middle term delegation, duration of delegation can usually not be estimated beforehand	Middle- to long term delegation, fulfillment of a task in the foreign company, legal and factual integration in the foreign company	Long term and often final employment abroad with complete and generally unlimited integration into the foreign company
Duration	1 day to 3 months	3 months to 1 year	1 year to 3 years	2 to 5 years with option of prolongation	unlimited
Contractual construction	no further contract needed	Persistence of labor contract, additional contract for secondment possible	Persistence of labor contract, additional contract for delegation possible	Contract in home country is terminated or suspended, local labor contract with foreign company	Contract with home company is completely terminated, new contract with foreign company
Remuneration	Unchanged, compensation of travel expenses	Unchanged basic remuneration, compensation of business trip expenses + subsistence allowances	Unchanged basic remuneration, compensation of business trip expenses + subsistence allowances	Remuneration agreement with local company	Remuneration agreement with foreign company
Domicile	Maintenance of domicile and hotel or other accommodati on abroad	Maintenance of domicile in home country + 2nd domicile in country of secondment	Maintenance of domicile in home country + 2nd domicile in country of delegation	Transfer of residence	Abroad

Source: Following Mauer, 2003, pp. 3.

Annex II

Checklist for expatriates before travelling abroad

Checklist
General □ Terminate or suspend memberships □ Unregister with kindergarten/school □ Give notice of change of residence to child benefit fund □ Arrange mail forwarding (chargeable) □ Unsubscribe phone/newspaper □ Get certificates translated and certified □ Take address lists along □ Last will □ Book flight tickets in coordination with HR department **Departure** □ Personnel documents □ Visa, working and residence authorization □ Passport □ International Driving License **Medical Provisions** □ Consult GP and dentist □ If necessary, examination for fitness for tropics □ Vaccination certificate □ First aid kit

Household

□ Organize relocation

□ Have furniture stored

□ Clarify regulations on import and customs

□ Car

□ Unregister or re-register car

□ Import limitations

Insurances/Banks

□ Healthcare/Long-term care/Unemployment insurance

□ Old age provisions

□ Householder and liabilities insurance

□ Baggage insurance

□ Clarify tax questions

□ Examine standing orders with banks

□ Open bank account abroad

□ Authorizations for withdrawals from bank accounts

□ Credit cards

Apartment

□ Terminate leasing contract or rent out apartment

□ Register and unregister with authorities

□ Terminate contracts for electricity, gas, refuse disposal

Source: Adapted from Förster/Heidenreich/Heuser, 2002, pp. 9.

Annex III

Delegation agreement in the case of further existence of labor contract at home

(Detailed example)

The .... [name of the delegating home company],

hereafter called the COMPANY

and

Mr. / Mrs. / Ms. .... hereafter called the employee conclude the following

Delegation Agreement

Commencement and Duration of the Contract

The employee is employed in the [affiliate company/branch office/office] in .... as .... from .... until presumably.............. ....

A prolongation of the contract is possible with mutual agreement by both parties. During this time, the previous labor contract between the COMPANY and the employee persists. For the duration of the delegation abroad the following conditions apply:

Activities

………….. is disciplinary supervisor of the employee, but ……….. is his functional supervisor.

Residence and employment permit

The employee is responsible for the possibly necessary residence and employment permit. The COMPANY assists in this matter as far as necessary. The resulting costs will be borne by the company.

Remuneration

The employee receives

a) a monthly basic salary of EUR -----

b) a delegation allowance of EUR------

The basic salary is paid as fictive comparable home salary and is examined on an annual basis. It serves at the same time as assessment base for the benefits from the company pension scheme/contributions to the pension fund. The adaptation of the delegation allowance depends on the development of the living costs and the foreign exchange rate.

The employee receives furthermore:

a) a one-time clothing allowance for different climatic conditions amounting to EUR ……..

b) for the first two weeks of the delegation abroad a bridging allowance of EUR ……….. per day.

c) from the day of departure to commence the work to the day of departure of his/her husband/spouse/family, but not longer than for … months, a separation compensation of EUR …….

Holidays

For a delegation of two years the employee has the right to ……….. weeks of holidays, of which he/she can spend up to …….. weeks continuously at home. The holiday is included in the named duration of the contract.

The holiday at home starts on the day of departing from the country and ends with the day of reentering the country. During this time the employee is available for the COMPANY for business meetings, the time needed is not deducted from the available holidays. The duration of the absence shall not be longer than ….. weeks. The employee will coordinate the dates of the holidays at home in due time with his supervisor and the home main department. The holidays at home in the last year of work abroad have to be connected with the return travel due to the termination of the delegation abroad.

During the duration of the visit to the home country – including the absence due to business matters –delegation allowances continued to be are paid.

Accident insurance

For the duration of the delegation the COMPANY concludes for the employee a group accident insurance with the following amounts of coverage:

a) in the case of death EUR.....;

b) in the case of total disability EUR .....

In the case of partial disability the last sum is diminished in accordance with the degree of disability.

Travel and moving expenses

The COMPANY takes over

a) the travel costs [actual commuting expenses and airfares in economy class in accordance with the travel expense policy] for the employee and for the spouse and children up to the age of 18 for the departure journey, travelling to the other country and back to the home country for holidays and the journey back once the delegation is terminated.

b) the transportation costs for furniture including transportation insurance up to an appropriate amount which arise due to the dislocation of the employee, under the condition that the details of the dislocation and the return are coordinated with the .....[responsible specialized company department] in due time.

c) the costs of back transport in case of a life-threatening disease of the employee or their spouse or child.

The company grants an accommodation allowance in the amount of the equivalent of EUR .... per month.

As far as the employee has school-age children during his/her work abroad, the COMPANY pays a school fee in the amount of EUR ....

Final provisions

The employee has to abide by the tax law of .... [country of delegation] and to fulfill the duties arising from it.

Source: Adapted from Schaub/Koch/Neef/Schrader/Vogelsang, 2008

Annex IV

Labor contract abroad (detailed form)

The ………………………………… [name of the foreign company] with its head office

in ………………………… hereafter called the COMPANY,

and

Mr./Mrs./Ms. ……………………, living in …………………………………,

hereafter called the employee conclude the following

Labor Contract

Work and area of responsibility

The employee is employed as ……………………… in ……………………….

He has the following functions: ………………………………………………………………

The employee has to report to ……………… [e.g. Board of Directors].

The company retains the right to give the employee also another function corresponding to his/her professional qualification and his/her capacities and to change the supervision because of organizational reasons, without causing a unilateral termination of the contract or individual provisions.

Commencement, duration of contract, termination

The contract commences on ……………… /with the day of departure for the commencement of work at ……………………………………………………………… ………………………………

The contract is concluded for an indefinite period and can be terminated during the first ….. months/ with a one month's notice to the end of the month. After this period the reciprocal termination period is …. weeks/months to the end of a calendar month/a calendar year.

The labor contract ends without termination of the contract with the end of the month of the 65th birthday of the employee.

Remuneration

The employee receives as remuneration for his/her work:

a) a monthly basic salary of EUR ..... and

b) a delegation allowance of EUR .....,

payable at the end of the month. This covers the complete time of his/her work for the company.

Furthermore, the employee receives special allowances related to the calendar year/business year in accordance with the following principles:....................

This special allowance will be granted in the year of entry and departure on a pro rata temporis basis.

The above mentioned allowances will be paid in .... [corresponding currency] at the official conversion rate of ....[corresponding state bank] at the due date.

Taxes

The employee is responsible for the correct payment of taxes of his/her remuneration in the country of delegation in accordance with the tax laws.

Social security

During the delegation the legal provisions concerning social security in the home country continue to apply for the employee. The COMPANY will pay the employer contributions to the pension, health care, nursing and unemployment insurance.

Supplementary benefits, reimbursement of expenses

The COMPANY takes 50 percent/100 percent of the fees for a life insurance or the corresponding fees for a state old age insurance of the employee's choice over as well as for a health insurance, but up to 50 percent/100 percent of the maximum contribution rate of the social security of the home country of the employee.

The employee bears the total costs/pure rental costs for an appropriate housing up to an amount of .... percent of his/her monthly basic salary/up to the equivalent of EUR ...:. The

COMPANY takes the remaining housing costs/pure rental costs fully over up to the equivalent of the maximum of EUR ……….

The company refunds the expenses which the employee incurs because of the execution of his function, in accordance with national practice/with the travel expenses policy.

In case of termination of the labor contract and in case of life-threatening diseases and the return of the employee to ………..[home country] due to these circumstances, the COMPANY takes over the travel expenses of the employee including transportation costs for relocation of goods to an appropriate extent.

For the duration of the contract in ……….. the COMPANY offers the employee a car in accordance with the company car regulations of the company. The company car can also be privately used by the employee.

Insurance

The COMPANY insures the employee for the duration of the contract against accidents with the following coverage:

EUR ---------------------------- for the case of death

EUR ---------------------------- for the case of disability.

The insurance includes work as well as private accidents and has to be paid to the employee or his/her successor at law. It ends with the moment of termination of the contract.

Moreover, the COMPANY procures comprehensive company liability insurance for the totality of the employee's activities.

Holidays

<u>In the case of work in other European countries:</u>

The employee disposes over vacation for --------------- working days. The duration and the repartition have to be coordinated with …..[e.g. the responsible supervisor].

Generally, holidays shall be combined in one continuous period. Unused holiday days can only be transferred to the next calendar year if urgent company or personal reasons justify it. Remaining vacation days which are not used until the 31st of March of the next calendar year are forfeited without compensation.

<u>In the case of work in non-European countries:</u>

For every ... years of work the employee is entitled to .... weeks of vacation. The employee can spend a continuous period of up to .... weeks in Europe each year. The Europe holidays commence with the day of departure and end with the day of return. During this time, the employee is available to the COMPANY and their business partners for business meetings. The time needed is not deducted from the vacation.

The employee will coordinate the time of his/her holiday in Europe with his/her supervisor in due time. The vacation in Europe during the last working year has to be combined with the return journey due to the termination of the labor contract.

For the duration of the vacation in Europe, including the business related absence, the delegation allowances are continued to be paid to the employee.

Employee leave benefits

In the case of disease causing incapacity to work the company pays the monthly salary for the time of .... calendar days/weeks/months to the employee.

Final provisions

Claims from this labor contract have to be asserted within one month after receipt of the last payroll, otherwise they will be forfeited.

Taking into consideration the external position of the employee the contract is drafted in ..... and .... (e.g. Spanish) language. Both versions are legally binding; in the case of discrepancies between both versions only the .... text is decisive.

The labor contract concluded and designed with this provisions is subject to the law of ....., as far as the mandatory national law is not preceding.

Source: Adapted from Pulter, 2004, p. 13ff.

Annex V

Contents of EEC regulation N° 1408/71

Country of delegation as basic rule

Countries	Insurance branch	Regulation	Period	Form	responsible insurance
Belgium					compulsory health
Denmark		Delegation:			insurance in case of
Germany		art. 14 n°. 1a	for up to	E 101	private health
Finland		EEC regulation n°	12 months		insurance
France	health	1408/71			responsible old age
Greece	nursing				insurance carrier
Great Britain	old age				
Ireland	accident-				
Iceland	and	Prolongation	for		responsible carrier in
Italy	unemployme nt	of the delegation	up to a	E 102	the country
Liechtenstein	insurance	art. 14 n°. 1b	maximum of further		of delegation
Luxemburg		EEC regulation n°	12 months		
Netherlands		1408/71			
Norway					
Austria					
Portugal		Special		form-	
Sweden		agreement:	diverse*-	less	DVKA

Switzerland		Art. 17 EEC-regulation		appli-	
Spain		n° 1408/71		cation	
* Some examples: Spain/Netherlands/Belgium/Ireland 5 years, Italy 4 years, France 6 years (Comp. DVKA)					

Source: Adapted from Schmeisser, 2008, p. 274.

Annex VI

Countries with social security treaties

Country of contract	**Insurance Branch**					**Limitation in time**
	HI	NI	OAI	UEI	AI	delegation
Belgium	X	X	X	X	X	12 months
Denmark	X	X	X	X	X	12 months
Bosnia-Herzegovina	X	–	X	X	X	12 months
Bulgaria	X	X	X	X	X	12 months
Chile	–	–	X	X	–	36 months
China	–	–	X	X	–	48 months
France	X	X	X	X	X	12 months
Greece	X	X	X	X	X	12 months
Great Britain	X	X	X	X	X	12 months
Italy	X	X	X	X	X	12 months
Israel	X	–	X	–	X	unlimited
Japan	–	–	X	X	–	60 months
Canada	–	–	X	X	–	60 months
Korea	–	–	X	X	–	24 months
Croatia	X	X	X	X	X	24 months
Malta	X	X	X	X	X	12 months
Morocco	X	–	X	X	X	36 months
Macedonia	X	–	X	X	X	24 months
Poland	X	X	X	X	X	12 months
Portugal	X	X	X	X	X	12 months
Rumania	X	X	X	X	X	12 months
Slovakia	X	X	X	X	X	12 months
Turkey	X	–	X	X	X	unlimited

Tunisia	X	–	X	–	X	12 months
USA	–	–	X	–	–	60 months
Cyprus (Greek part)	X	X	X	X	X	12 months

HI=Health Insurance; NI = Nursing Insurance; OAI = Old Age Insurance; UEI = Unemployment Insurance, AI = Accident Insurance

Source: Adapted from Schmeisser, 2008, p. 277

III Stock-option programs as part of the remuneration management system

8 Basic considerations on economic philosophy and ethics

8.1 Shareholder-Value-Approach

In the beginning of the 90s a new management concept was discussed in Germany. This concept was published by Alfred Rappaport in his book "Creating Shareholder-Value" at the end of the 1980s.[62] The term "value" in this context means the company value in the sense of the market value of equity.

Rappaport provided with his concept a comprehensive illustration on value-oriented company management. The focus of this approach lies on the direction of the management towards the interests of the shareholders of a company, pursuing mainly the aim of increasing the company value. First German companies, as e.g. Daimler AG applied the shareholder value approach in the mid 1990s and integrated the concept in their company management.[63]

The company value can be calculated with different methods. In the context of the shareholder value analysis often the Discounted Cashflow Method is used. Put simply, the cashflow is calculated as difference between incoming and outgoing payments which a company has at its disposal for e.g. investments or dividend payments. For the determination of the current company value, however, not the future, but the current payment flows are relevant.[64]

---

62 Comp. Werder, 1998, p. 69–71

63 Comp. Fröndhoff, 2009

64 Comp. Friedrichsen, 2000, p. 17

In accordance with Rappaport the current company value can be determined with the help of present values of the future free cash flows. The free cash flows of a company are discounted over a certain period together with the residual value at the end of the period of prognosis. Four discounting the free cash flow the weighted average cost of capital is used.[65]

By discounting the estimated cash flows with the weighted average cost of capital (WACC) the total value of the company is obtained. By deducing the market value of the dept capital from the calculated total value of the company, the shareholder value can be determined. Value is added for the shareholders if the calculated yield is higher than the costs of capital.[66]

Therefore, the managers have to direct the company management to the interests of the shareholders if they do not want to lose them or want to win new investors for the company. The shareholders or potential investors will only invest into the company, if they receive an adequate yield for their capital. If they do not receive an adequate yield, they will sell their shares. This leads to falling market prices of the shares and potential investors will avoid an investment into the company. This can lead to difficulties of the company to raise capital. Because of these possible problems the managers have to direct the company management in such a way that a lasting increase of the yields on shares can be reached. The capital markets show positive reactions to the increase of yields on shares, which again leads to an increase of the price of shares on the middle or long run.[67] The shareholder value is influenced by so called value contributors. In accordance with Rappaport such value contributors are all values on which the company can have influence. The management can directly influence the value contributors trough operational business as well as through investment and financing decisions and thus have influence on the shareholder value.[68] Only those investment measures are carried out which achieve an appropriate yield. If a satisfactory yield cannot be reached, the measure has to be turned down. The focus of the management on high-yield business areas can lead to an optimization of the cash flows and thus, to an increase of enterprise value.[69]

In order to avoid losses of confidence on the shareholders' side it is also important for the company to pursue a shareholder-friendly information policy. An unclear communication could lead the shareholders or investors of the company to estimating the enterprise value too low. This erroneous estimation would have a negative impact on the market value of the shares.

The advantage of the shareholder-value-approach is the use of future-oriented payment flows and the consideration of the risk of costs of capital. Balance sheet related possibilities of

---

65 Comp. Kramarsch, 2004, p. 22

66 Comp. Schmeisser/Hahn/Schindler, 2004, p. 5

67 Comp. Friedrichsen, 2000, p. 18 et seq.

68 Comp. Achleitner/Wichels, 2002, p. 5

69 Comp. Schmeisser/Hahn/Schindler, 2004, p. 4

manipulation of the company profit are avoided in the application of the shareholder-value-approach.[70]

The critique of the shareholder value lies mainly in the unilateral approach to the interests of the investors (shareholders). According to the critics, stakeholders like customers, suppliers and employees remain disregarded. Rappaport counters:

> *"Not only the owner benefits from a value creating company, but it serves all other interested groups; on the other hand, all interested groups are endangered if the management does not achieve the creation of shareholder value."*[71]

Furthermore, Free Cash Flows used to calculate the shareholder-value are considered as easy to manipulate. For example, the reduction of staff or cost reductions in the area of research and development can be used to have a short-term positive influence on cash flows. Such a short-term course of action, however, is contradictory to the shareholder value concept that aims towards a lasting increase of the company value.

The reduction of work places can be necessary for the company in order to maintain or optimize its competitive capacity on a national and also international level. On the middle and long run this is good for the employees, because on a long-term basis this permits to maintain existing work places and to create new jobs.

German companies like the DAX-30 company RWE are aware of their responsibility towards the stakeholders and care for their interests in their company management.

Also within the context of the financial crisis, the shareholder-value-approach was reason for controversial debates. The shareholder-value-approach is accused of promoting a short term profit maximization and therefore, of being also responsible for the crisis. In accordance with Rappaport, however, the concept of the shareholder-value-approach is not directed towards short-term management decisions. Apart from the generation of increasing cash flows, a long term increase of the company value and an adequate risk assessment are the targets of his approach. According to Rappaport, there would not have been a crisis if these components had been taken into consideration by top managers.[72]

The management can be motivated to adhere to and execute the shareholder-value concept, by receiving apart from the common remuneration package profit oriented remuneration via value driven incentive systems as stock-option programs.[73]

---

70 Comp. ibid, p. 6

71 Comp. Rappaport, 1999, p. 8, own translation.

72 Comp. Rickens, 2009

73 Comp. Winter, 2000, p.7

8.2 Principal-agent theory

The execution of the shareholder-value-concept in the company can lead to conflicts between the shareholders and the management. This can be attributed to the fact that shareholders and management partially pursue different aims. Already in the 18th century the economic theorist and business ethicist Adam Smith recognized this problem. He wrote:[74]

> *"The directors of such companies, however, being the managers rather of other people's money than of their own, it cannot well be expected that they should watch over it with the same anxious vigilance with which the partners in a private copartnery frequently watch over their own. Like the stewards of a rich man, they are apt to consider attention to small matters as not for their master's honor, and very easily give themselves a dispensation from having it. Negligence and profusion, therefore, must always prevail, more or less, in the management of the affairs of such a company."*

The divergences in interest between the investors in form of the shareholders and the employed managers are described as principle-agent-theory in business sciences. In this theory the shareholders are called principals and the managers are called agents. The shareholders entrust the managers with the direction of the company. The shareholders do not need to be active themselves in the company in which they invest.[75]
The main interest of the shareholders is a profitable capital investment. The shareholders pursue exclusively the monetary goal of high increase in value of the invested capital.[76] Furthermore, the shareholders are more adventurous than the managers. This is expressed by the fact that financing and investment measures are chosen from which higher profit can be expected.[77]

The managers are interested in maximizing their income. Besides this monetary goal managers can also pursue non-monetary goals. These include for example the expansion of the personal sphere of control or the improvement of one's own reputation.[78] The willingness to take risks of the managers is generally weaker than the one of the shareholders. This is due to the fact that their workplace depends on the economic development of the company. Managers are rather reluctant to take risky actions.[79]

---

74 Smith, 1879, p. 254f, own translation.

75 Comp. Engelsing, 2001, p. 26

76 Comp. Engelsing, 2001, p. 27

77 Comp. Friedrichsen, 2002, p. 23

78 Comp. Engelsing, 2001, p. 27

79 Comp. Friedrichsen, 2002, p. 23

The different goals of both parties cause goal conflicts if the agents do not act in the principals' interest, but pursue mainly their own goals. This means they act in an opportunist way. The resulting conflict of goals can be enforced by informational asymmetries.

The informational asymmetry describes the uneven repartition of information between two contracting partners. The agent has an informational advantage over the principal, because of his position in the company. Because of this unilateral advantage, the agent is in the position to take decisions in his own favor and to the disadvantage of the principal.

There are different types of informational asymmetries like hidden action and hidden information. It is typical for these forms of information asymmetries that the informational advantage of the agent occurs only after the conclusion of the contract.

Hidden action describes the fact that the principal can observe the behavior of the agent only with difficulties or with high costs.[80] Because the agent has the choice between different alternative actions and the principal has the observation problem, the agent can take the decision himself how much effort he puts into the execution of the chosen action. The principal will only know the final result.[81] However, he does not know how to assess the final result, because he does not know whether the result is due to external influences and how far the agent has influenced the result with his own actions.

Hidden information describes the problem that the principal can observe the actions of the agent and is aware of the options of the agents. The agent can simulate wrong facts with regard to his characteristics and capacities before contract conclusion which the principal knows only after the contract has come to effect. Thus, there is the risk of the agent using existing informational asymmetries to his own advantage.

If the investors do not insist on corresponding retaliatory actions, the conflict causes that the shareholder investment can not evolve in the best way and that the manager uses his leeway for action to the company's and the principal's disadvantage.

In order to mitigate these asymmetries the principal has two possible options: [82]

1. There is the possibility to build information and controlling systems or to improve them. They are meant to give the principal a better insight into the agents' behavior.
2. The second possibility is the output oriented control, in which the agents are supervised by the principal with respect to the desired and future results of their actions.

One of the controlling instruments which can be used for the purpose of encouraging the agents to behave as far as possible in the principals' interest is stock-option programs. With this solution, the remuneration of the managers (agents) is linked to the increase of company

---

80 Comp. Dietz, 2004, p. 27

81 Comp. Dietz, 2004, p. 27

82 Comp. Engelsing, 2001, p. 29

value. Only because of the increase of the market price of the shares and, thus, of the company, the manager get paid for their work.

8.3 Goals of stock-option programs

8.3.1 Minimization of the principal-agent problem

The increase of the market price of the shares and thus, of the company-value is the preferred goal of the shareholders. The problem is that managers do not only exclusively pursue this target but also own opportunistic goals. By connecting the remuneration with stock-option programs the remuneration of the manager is connected to the development of the market price of the shares. Thereby, the managers are encouraged to direct the company in the shareholders' interest, because only by increasing the market value of the share the manager can obtain an additional financial advantage.

The increase of the company value and of the shareholder benefits is expressed in an increase of the stock quotation, from which the investors can finally also gain a financial advantage.[83]

A further reason which is named in the context of the use of stock-option programs is the so called risk-taking-effect.[84] The presumption is that managers act generally in a risk-averse way. This means that investment decisions which would have positive capital values are possibly not taken because the risk is too high for the managers. However, the investment would have been in the shareholders' interest. By the use of stock-option the right incentives can be given so that the manager's willingness to take risks is increased.[85]

As the assessment base for stock-option programs is the shareholder value, the manager is more likely to take such investments into consideration which contribute to an increase of the enterprise value.[86] This way, an adjustment of interests between shareholders and managers can be achieved. Thus, both stakeholders have the same aim, i.e. the increase of the company value.

A further adjustment of aims can be reached between managers and investors if a personal investment e.g. in the form of shares is required from the managers.[87]

---

[83] Comp. Engelsing, 2001, p. 29

[84] Comp. Schmeisser/Hahn/Schindler, 2004, p. 42

[85] Comp. Friedrichsen, 2002, p. 27

[86] Comp. Engelsing, 2001, p. 30

[87] Comp. ibid, p. 30

8.3.2 Personnel commitment and recruitment of top managers

Companies compete more and more for internationally mobile managers. The globalization has reduced national barriers and enforced the international competition for top managers. A competitive remuneration system, as provided by the stock-option program, is therefore necessary to tie the own top managers to the company and to protect them from labor piracy by competitors.

The existence of stock-option programs is internationally very different. In Canada and the US the use of stock-options is at 85 and 66 percent.[88] The take-over of managers of these countries can be simplified by using stock-option programs as remuneration instrument. The managers of these countries know the profit potential of stock-options, because they have been used in these countries for a long time as remuneration component. Therefore, it is the task of personnel management in German companies to create an efficient remuneration system in order to gain these top managers' interest for German companies acting on an international basis.

A further important aspect is that stock-option programs can contribute to staff commitment if they are structured correspondingly. It is important for this to use long-term stock-options with lockup periods and forfeiture clauses. The lockup period means that managers can exercise their options only after a certain time. If the manager leaves the company before the end of the lockup period, he forfeits the right to exchange his options against company shares. The stock-option cannot be handed on to others as this is mostly excluded by contract. Thus, the managers are encouraged to remain in the company, because the financial losses can be considerable in the case of a positive economic development of the company.

In the context of company mergers the different remuneration forms of the merging companies have to be equally taken into consideration. German companies cannot exclude themselves from internationally usual remuneration practices, but have to adjust themselves, so that in this context stock-option programs for the remuneration of managers are used.

8.3.3 Improvement of the company's liquidity

The exclusive remuneration of managers with traditional remuneration components, as basic salary or success related bonuses causes always a loss of liquidity or a reduction of the company cash flow.[89] This problem can be avoided by using stock-options.

If the stock-option program is exercised via an approved or conditional capital increase, the liquidity of the company is not affected. When the capital of the company is increased the value of the company is divided into a bigger number of shares. It is true that this leads to a delusion of the shares of the existing shareholders, on the other side the increase in capital strengthens the equity capital base of the company.

---

88 Comp. Towers, 2009

89 Comp. Engelsing, 2001, p. 31

When the stock-option is exercised the nominal capital increases by the face amount of the new shares. At the same time the capital reserves rise by the difference between the new shares and the base price to be paid.[90]

Especially the lack of effect on company liquidity in the case of an increase of capital has to be positively mentioned, because German companies dispose over a lower equity-ratio in international comparison. By giving the managers of the company stock-options the company is able to pay an appropriate remuneration in international comparison and at the same time to preserve the company's liquidity.

8.4 Criticism of the use of stock-options

Stock-options for variable remuneration of managers also bear some disadvantages that will be explained below:

8.4.1 Manipulation of stock quotation

The manager shall be encouraged by the remuneration via a stock-option program to increase the shareholder value. The stock-options that the managers receive for their work are connected with the stock quotation.

The problem is that the stock quotation of a share can be influenced by more aspects than merely the shareholder value-oriented decisions of the managers, as for example through company external factors. E.g. the stock quotation can be influenced by the up and downward trends on international markets as well as by the development of interest rates and currency exchange rates. Thus, the stock quotation can rise higher than the performance of the management would justify. A manager who exercises his real stock-options at this moment receives shares at reduced prices or if he exercises virtual stock-option he receives especially high capital gains for which he has not been responsible.[91]

This has lead to considerations on excess remuneration of managers, discussing stock-options as disadvantageous approach and pointing out, that if employees or managers are responsible for losses, they should not be able to claim high compensations under labor law. A performance orientation, as provided by the Berlin Human Capital Assessment Model and long term bonuses have to be considered more often in the company's interest, as will be discussed later on.

A possible solution to solve the problem within this approach is to connect the basic price additionally to the sector or market index. Thus, not only the absolute increase of the market value of the share, but also the development of the market value in comparison with a sector

---

90 Comp. ibid, p. 31

91 Comp. Zitzewitz, 2003, p. 34

is decisive for the exercise of a stock-option. In this case, the owner of a stock-option can only exercise it if the company evolves economically better than the competitors.

A further disadvantage is that the capital market is very sensible to negative news from within the company. Thus, there is the risk of possibilities for manipulation, when the managers holding stock-options give certain information to the public. Rumors which are spread at the right moment can influence the stock quotations in one or the other direction. The result is that the basic price of stock-options which follows the stock quotation can go up or down.[92] The problem of insider knowledge of the managers can be reduced or contained by trading windows.

8.4.2 Changes in the capital structure of the company

A further disadvantage results in the case of use of stock-option programs through a conditional increase in capital. In this case it comes to a dilution effect of the voting right and reduction of the share in capital of the existing shareholders if there is an exclusion of the purchase right of the shareholders.[93] However, when the stock-option program is financed by the buyback of own shares, there is no dilution of the voting rights of the shareholders. On the other hand there is a leverage effect of the share in capital. The dilution is caused when the shares are handed down to the participants of the stock-option program below the cash value of the share. [94]

The question is whether these arguments disfavor the use of stock-options in companies. The resulting dilution effects remain generally acceptable and the increase of stock quotation and profit achieved with the help of stock-option programs compensate for this disadvantage.[95]

8.5 Stock-option programs in the context of modern remuneration systems

Previously it has been shown that the remuneration of the managers should be connected to their performance and success, in order to implement e.g. the shareholder value concept in the company.

This chapter shows how German companies have reacted to the changed requirements with regard to remuneration. In this context the changes of the remuneration structure and the

---

92 Comp. Friedrichsen, 2000, p. 53

93 Comp. ibid, p. 53

94 Comp. ibid, p. 232

95 Comp. Friedrichsen, 2000, p. 53

introduction of a value-oriented remuneration system in German companies are discussed first.

After this, the legal provisions concerning shares and company law which have to be respected in the remuneration of managers are discussed.

9 Changes in the remuneration structure in Germany

Up to only several years ago the design of the total remuneration was little sophisticated and it was mainly composed of small variable portion and a high basic salary. Depending on the hierarchic level and the sector of the upper management the amount of variable remuneration components corresponded to 10 to 20 percent of the total remuneration. In international comparison this amount of variable remuneration components of the total remuneration is very low. In a study from 1997 25 percent of the companies said to use revenue targets and performance related bonuses for the remuneration of the top managers. The bonuses generally include the participation of the members of the board of directors in the annual profits of the company. The basic salary of the management corresponded to up to 64 percent of the total remuneration.[96]

The following illustration shows the composition of the total remuneration in the mid 1990s and today.

---

96 Comp. Kramarsch, 2004, p. 4 et seq.

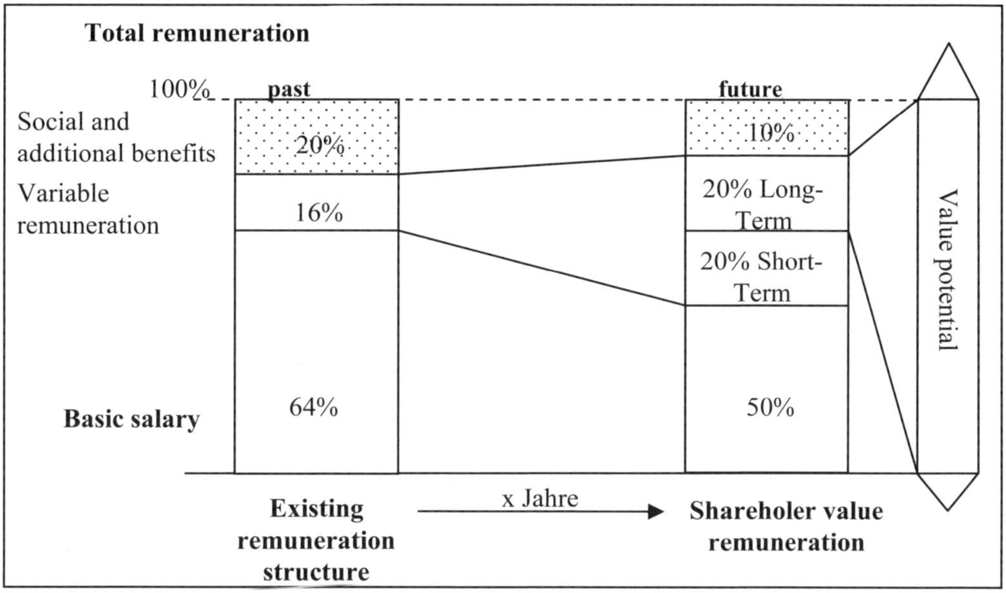

Fig. 9.1: Tendencies in the remuneration structure[97]

The formerly used remuneration structure with a high basic salary and a low share of bonuses in the total remuneration was criticized especially against the background of calls for a value-oriented direction of the company in the sense of the shareholder value approach. Also the commencing internationalization in connection with the increasing globalization and the competition for successful managers lead to a demand for a reformulation of the German remuneration system.[98]

It was criticized that the basic salary does not serve as appropriate remuneration for the performance of the top manages. Neither did this form of remuneration portray the success of the company in an adequate way. The lack of connection between the stock quotation and the bonuses granted to the top managers was also seen to be a problem. Bonuses are calculated with the help of key figures from previous business years, which caused that the development of the stock quotation was not taken into consideration for the remuneration. The result was that the management had no incentive to direct their decisions towards the increase of the stock quotation of the company.[99]

Furthermore, German accounting standards were criticized with regard to bonuses, because they permit assessment and accounting options and make a falsification of the accumulated equity possible. The danger was recognized that especially high bonuses could be achieved

---

97 Kramarsch, 2004, p. 5

98 Comp. Schmeisser, 2008, p. 68

99 Comp. ibid, p. 68 et seq.

when short-term yearly targets were pursued. This form of traditional remuneration system also affected the liquidity of the company to a great extent.[100]

These points of criticism show that a reform of the traditional remuneration system was necessary in Germany. In this context value-oriented remuneration systems were implemented in German companies to support the introduction of the shareholder value concept. Today it is merely unthinkable not to use them and they have become an important strategic instrument for the process of increasing company value.[101]

9.1 Demands for performance-oriented remuneration systems

If stock-option programs shall be successfully used, they have to fulfill certain requirements. This chapter describes these requirements.

Value-oriented remuneration systems have to be designed in such a way that the decisions of the management are controlled in the interest of the shareholders.[102] Incentives shall be created to encourage management to take decisions in the principal's interest in order to achieve an adjustment of the interests between shareholders and management.

In order to direct the behavior in the shareholders' interest, value-oriented remuneration systems have to be designed in such a way that permits to control them in a quick and performance-oriented way in order to guarantee the connection between performance and remuneration. Furthermore, it has to be taken care that the performance and remuneration are in brought into an appropriate relation.[103]

A successful incentive system has to fulfill the following requirements:

1. *Adequacy of target values:* The assessment base determines which factors permit to judge over the management's performance. It has to match the chosen aim to achieve adjusted interests between the shareholders and the management. The management shall only receive a higher remuneration if the desired targets of the shareholders are met or exceeded.[104]

---

100 Comp. Schmeisser, 2008, p. 69

101 Comp. ibid, p. 69

102 Comp. Laux, 1999, p. 12

103 Comp. Achleitner/Wichels, 2002, p. 7

104 Comp. Pellens/Crassselt/Rockholtz, 1998, p. 14

2. *Long-term orientation:* By the use of long-term success factors the danger of manipulation of the assessment base by short-term management decisions shall be compensated.[105]
3. *Transparency:* The design of the incentive system has to be comprehensible and clear for the management.[106]
4. *Acceptance and ease of communication:* The success of an implemented incentive system for remuneration depends decisively on whether it is accepted by the management. For this, it is important to communicate with the entitled persons and to show the reasons for and preconditions of the introduction of a value-oriented remuneration system.[107]
5. *Profitability:* From a shareholder point of view the costs of the incentive system shall not be higher than the benefit from it.[108]

The Boston Consulting Group has developed a performance-oriented remuneration system, which is presented in the following figure:

[105] Comp. Achleitner/Wichels, 2002, p. 10

[106] Comp. Pellens/Crasselt/Rockholtz, 1998, p. 14

[107] Comp. Achleitner/Wichels, 2002, p. 10

[108] Comp. Pellens/Crasselt/Rockholtz, 1998, p. 14

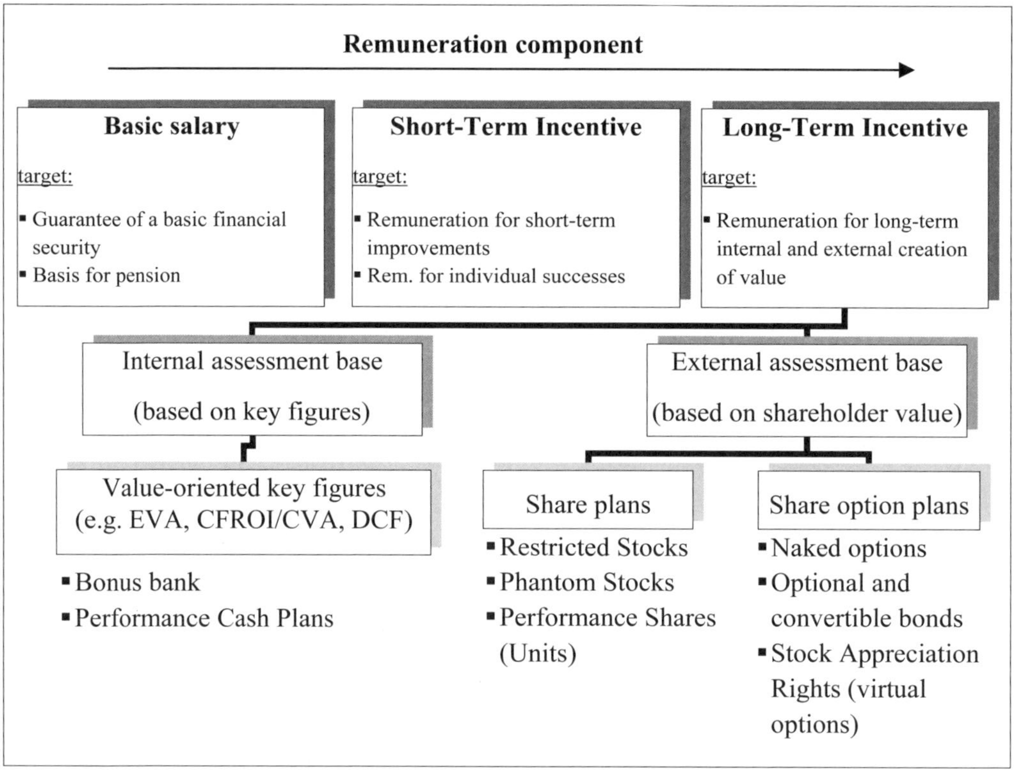

Fig. 9.2: Classifications of value-oriented remuneration systems[109]

The value-oriented remuneration system of the Boston Consulting Group is composed of three remuneration components, the basic salary and the variable components, which can be further divided into short-term and long-term remuneration components.

The basic salary is the basis for the management remuneration. Its amount is individually negotiated.[110] Usually, it is paid in 12 monthly installments. The amount of the basic salary is determined by the remuneration policy of the company. In practice, the basic salary serves above all to assure and preserve the management's living standard.[111] The basic salary is neither variable nor performance-related. Thus, the fix basic salary does not give any assistance in solving the principal-agent conflict.[112]

---

[109] Following: Schmeisser/Hahn/Schindler, 2004. p. 14

[110] Comp. Kramarsch/Becker, 2006, p. 24

[111] Comp. Lazar, 2007, p. 37

[112] Comp. Pape, 2004, p. 67

The short-term variable remuneration is also called short-term incentive. It consists for example of annual bonuses or royalties. Short-term incentives are generally based on strategic key figures of accountancy. The strategic incentives refer to the business year of the company and are mostly paid in a one-time payment when the target values are met at the end of the business year. The used key figures can be divided in traditional key figures of accountancy as profit or revenue and value-oriented key figures as the Cashflow Return on Investment (CFROI) or the Economic Value Added (EVA).[113]

The long-term incentives are long-term variable remuneration components, which are lastingly directed by the company profit.[114] They can be divided with the help of the underlying reference values, which are oriented towards the increase of company value. The increase of company value can be classified in internal and external reference values. Long-term incentives which are based on internal assessment values as benchmark for increase in value, try to calculate the economic value on the basis of company internal key figures. This value is than used as remuneration criterion. This happens under the condition that an increase of economic value induces rising stock quotations and thus, the increase of the value of the enterprise.[115]

Furthermore, long-term incentives, which are based on external assessment values, can be divided into stock and stock-option programs. Stock quotation oriented remuneration systems are oriented by the stock quotation and thus, directly by the development of the shareholder-value. The participation of the management in the company via stock and stock-options is a good method to harmonize the interests between the management and the shareholders and to maximize the shareholder capital which is embodied by shares.[116]

The German Corporate Governance Codex contains enough recommendations with regard to the composition of the total remuneration.

9.2 Company law aspects of remuneration

9.2.1 The German Corporate Governance Codex

The term Corporate Governance has received more attention in last years from the German public and was also correspondingly discussed in specialized literature. Corporate Governance includes especially basic rules for a good and responsible management and control of a company. These basic rules are written down in Germany in the German Corporate Governance Codex (GCGC).

---

[113] Comp. Schmeisser/Dittmann, 2004, p. 40

[114] Comp. Kramarsch, 2004, p. 35

[115] Comp. Achleitner/Wichels, 2002, p. 7 et seq.

[116] Comp. ibid, p. 9

This section deals at first with the development of the Corporate Governance Discussion and explains then the aims and the composition of the German Corporate Governance Codex (GCGC). After that the recommendations of the GCGC and its influence on the remuneration structure of directors in analyzed in detail.

Furthermore, the Codex contains recommendations for the design of stock-option programs and other variable long-term remuneration forms for management.

Development of the Corporate Governance Discussion

The Corporate Governance discussion started off in 1969 in the US. The American Institute of Law was engaged to conceive a draft of the "Federal Securities Code". After intense discussions at the end of the 1970s, 1992 the "Principles of Corporate Governance" were published.[117] In the context of the spectacular accounting scandals of the US companies WorldCom and Enron the call of shareholders and investors for more transparency grew louder. This lead the US government to pass the "Sarbanes-Oxley Act of 2002" on 30 July 2002. The Sarbanes-Oxley Act completes the regulations of the Securities and Exchange Act (SOX) of 1934. The SOX has prominently the goal to create more transparency in the management.[118] Also the confidence of the investors in the company's accounting should be re-established.

The passing of SOX did not only affect US companies but also German companies listed on the US stock market, as e.g. Allianz or Siemens. These companies are also subject to the regulations of the Sarbanes Oxley Act.

In Germany the Federal Government showed reactions after the Holzmann case with the establishment of a Government Commission "Corporate Governance – company management – company control – modernization of the stock corporation law." The Government Commission war installed with a letter by the former German Chancellor Gerhard Schröder on 29 May 2000. Theodor Baum was the head of the commission. The task of the Government commission was to show weaknesses of the German System of company management and control.[119] The German system of Corporate Governance was meant to be strengthened and weaknesses to be removed. The international developments with regard to company management were thereby also considered in Germany. Suggestions for reform should be elaborated in order to stay abreast of the new developments, so that the financial centre Germany could assert itself in international competition and survive in the system of international Corporate Governance.[120] The commission published its final report in July 2001 and recommended the formation of a new commission for the elaboration of the German Corporate Governance Codex.

---

[117] Comp. Grattenthaler, 2007, p. 172 et seq.

[118] Comp. Carl, 2005, mn. 1623 et seq.

[119] Comp. Bericht der Regierungskommission, 2001, p. A1

[120] Comp. ibid, p. A1

Further company crises as e.g. Flowtex[121] or Comroad[122] proved that there were deficits with regard to company management and control in Germany. The boards of directors of the companies have come under criticism, besides accountancy companies, because they did not detect the fraud. The trust of institutional investors, suppliers and other stakeholders in the company management was decisively weakened.

The newly set up commission under the direction of Dr. Gerhard Cromme, who was the chairman of the board of directors at that moment, commenced its work in September 2001. The commission was composed of academics and well-known business representatives.[123] The results of the Baum Commission provided the basis for the German Corporate Governance Codex. The finished version was published on 26 February 2002 and has been checked every year since then.

The German Corporate Governance Codex should represent a document created to re-establish and enforce the trust of the investors and of other stakeholders in the management of German enterprises. The German Corporate Governance Codex provides besides recommendations also ideas for business leaders for a good and responsible business management and control.

Targets and design of the German Corporate Governance Codex

The introduction of the German Corporate Governance shall especially deal with the following points of critique of the German corporate constitution:[124]

- Lack of orientation on shareholder interests,
- Dual corporate constitution with Executive Committee and Supervisory Board,
- Lack of transparency of German management,
- Lack of independence auf German Supervisory Board
- Limited independence of annual auditors

The Codex deals with points of critique and considers the shareholders' and institutional investors' demand for more transparency in the management in Germany. The preamble of the codex formulates the goals of the codex as follows:[125]

---

121 Comp. Heise, 2009, p. 7

122 Comp. N.N., 2002

123 Members of the commission are Cr. Gerhard Cromme, Dr. Gerhard Cromme, Dr. Paul Achleitner, Dr. Rolf-E. Breuer, Dr. Hans Friedrich Gelhausen, Ulrich Hocker, Max Dietrich Kley, Professor em. Dr. Dr. h.c. Marcus Lutter, Volker Potthoff, Heinz Putzhammer, Peer Michael Schatz, Christian Strenger, Pro et seq. Dr. Axel von Werder, Dr. Wendelin Wiedeking.

124 DCKG, Preamble, 2008

125 DCKG, comment by the commission head Gerhard Cromme, 2002

"This German Corporate Governance Codex (the 'Codex') shows essential legal provisions for management and control of German stock-listed companies (direction) and contains internationally and nationally recognized standards of good and responsible management. The Codex shall make the system of German Corporate Governance more transparent and comprehensible. Its aim is to boost the confidence of international and national investors, customers, employees and of the public in management and control of German stock listed companies."

The Codex contains so called "shall" recommendations and "should" or "can" proposals. Via the declaration of compliance in accordance with s. 161 German Stock Corporation Act (AktG) it is legally consolidated. Executive and Supervisory Board of listed companies have the duty to give annual information on whether the recommendations of the German Corporate Governance Codex were followed. In case of deviations from the codex it has to be explained which recommendations had not been followed and the reasons for the deviation should be given. This approach corresponds to the Anglo-Saxon principle: "Comply or Explain". The declaration of compliance has to be published on the homepage of the company in accordance with s. 161 subs. 2 German Stock Corporation Act.

If the company deviates from proposals, no explanation is necessary. In this case the deviations must not be disclosed. The Codex is mainly directed towards listed companies, but its application is also recommended for companies that are not listed on the stock market.

The codex is divided in seven parts:

- Preamble,
- Shareholders and General Assembly,
- Cooperation of Executive and Supervisory Board,
- Executive Board,
- Supervisory Board,
- Transparency,
- Accounting and Auditing.

The preamble contains the purpose of the Codex. The second and third section describes provisions for the General Assembly and the Shareholders' right. Besides the remuneration structure for managers, the individual tasks and responsibilities of Executive Board and Supervisory Board are described in the sections four and five. The sections six and seven contain regulations with regard to transparency as well as accountancy and auditing.

The financial crisis and the upcoming criticism in print and other media of the remuneration of the managers have lead to enacting the Act on Appropriateness of Management Board Remuneration. Due to this, the government commission came together and duly adjusted the German Corporate Governance Codex. Innovations become only binding for companies when published in the German Federal Gazette. The currently valid version is now from 19 July 2009. The most important alterations are:

- New regulations concerning the composition and the transition of members of the Executive Board to the Supervisory Board[126],
- Regulations on the composition of the Audit Committee[127],
- Adjustment of remuneration structure and amount of directors[128],
- The principles of the social market economy have to be respected by the managers[129],
- Innovations with regard to the D&O insurances[130].

Additions and amendments which come up because of the design of the Corporate Governance Codex status as of 19 July 2009 are explained in more details in the next section. Furthermore, the recommendations of the Codex concerning the remuneration structure of the directors and their disclosure is underlined and explained in more detail.

GCGC-Recommendations for the remuneration of directors

The GCGC gives recommendations in section four as to how the remuneration of members of the executive board is to be construed. Section 4.2.3 of the Codex says:

- The total remuneration of directors shall comprehend fix and variable remuneration components. The variable remuneration components shall contain a yearly returning component, which is connected to the success of the company and comprehend components with long-term incentive effects and risk character.
- The GCGC recommends as components with long-term incentive effect and risk character above all shares, stock-options and comparable remuneration programs (e.g. phantom stocks) with long lock-up periods. For the remuneration components challenging, relevant reference parameters shall be used and a subsequent change of the performance targets shall not be possible. Furthermore, the Supervisory Board shall determine a profit cap.
- The chairman of the Supervisory Board shall inform the General Assembly on the remuneration structure of the Executive Board and about potential changes.

Sections 4.2.4 and 4.2.5 give information on the disclosure of manager salaries. Thereby, the demand by shareholders and other stakeholders for transparency in remuneration can be met. Section 4.2.4 recommends the disclosure of management remuneration as follows:

- The total remuneration of each member of the Executive Board has to be broken down in success independent and success related remuneration components with long-term incentive effect.

---

126 DCGK, section 5.4.4, section 5.4.5 sentence 2, section 5.4.1, 2009

127 DCGK, section 5.3.2 sentence 3, 2009

128 DCGK, section 4.2.2, section 4.2.3 par. 1 und 2, 2009

129 DCGK, preamble par. 2, 2009

130 DCGK, section 3.8 par. 2, 2009

- The remuneration components have to be disclosed individually, unless the General Assembly with three quarter majority decides otherwise.

The Codex recommends in section 4.2.5 where the disclosure of manager salaries has to be made:[131]

- The disclosure of manager remuneration shall be made in a remuneration report which forms part of the Corporate Governance report. It has to be taken care to formulate the disclosure in a comprehensible way.
- Furthermore, the parameters of the design of stock-option programs or other remuneration systems shall be explained by long-term incentive effects. The values for each remuneration component equally have to be published.

Currently, 29 of the 30 DAX groups comply with the requirement to disclose the management salaries. In 2008, Merck was the only company to deviate from the GCGC. Thereby, the companies make their contribution to more transparency and make it much easier for shareholders and other stakeholders to gain an insight into the remuneration policy of the company. The shareholders and investors can make a better assessment of how their capital was used for the remuneration of managers.

The government committee, which was set up by the Federal Ministry of Justice, has adapted the German Corporate Governance Codex with regard to the financial crisis and the following criticism of remuneration of Board members. Now, the recommendations with regard to remuneration of members of the Executive Board are more closely explained. Section 4.2.3 of the Codex has been elaborated and contains the following additions:[132]

- The remuneration of the members of the Executive Board has to be designed in such a way that it serves a permanent company development.
- Positive and negative developments shall be considered in the design of the variable remuneration components.
- All remuneration components have to be appropriate and designed in such a way that they do not tempt the board member to take inappropriate risks.
- For example, share based or key figure based remuneration elements can be used as variable remuneration components.

The new version of the Codex takes the changed economic situation into consideration. Besides, the Codex tries via its addendum to achieve that the remuneration is adapted to a long-term company development.

---

131 DCGK, section 4.2.5, 2008

132 DCGK, section 4.2.3, 2009

9.2.2 VorstOG – Executive Board Compensation Disclosure Act

Despite the introduction of the German Corporate Governance Codex the institutions of stock-listed companies had refrained for a long time from individually publishing their compensation. This was discovered in a survey of the consulting firm Towers Perrin from the year 2008. While in 2002 only five DAX 30 companies disclosed the salaries of their directors, 18 companies followed the recommendations of the GCGC in the business year 2004.

It could not be determined by shareholders and other stakeholders how the executive board compensation was composed. This meant that the shareholders could not understand of which remuneration components the total compensation of the Executive Board members was composed.

With the Executive Board Compensation Disclosure Act the DAX-30 companies were obliged to individually disclose the Compensation of the Executive Board from 2006 on. This meant that it was not sufficient anymore to only publish the total remuneration paid. They had to distinguish between success independent and success related components as well as components with long-term incentive effects (e.g. stock-options). Since then, it has been legally required to describe the total compensation in detail with individual components in the business report.[133] In sec. 285 Sentence 1 n° 9 German Commercial Code (HGB) the Disclosure Duties are fixed as follows:

> *"The total compensation granted for the activities in a business year (salary, profit participations, subscription rights, options and other stock market based remunerations, expense compensations, insurance remunerations, commissions and other compensations of all kinds). This total compensation also includes forms of compensation, which are not disbursed but transformed in claims of other kind or lead to the increase of other claims.*
>
> *Beside the compensation for the business year, further compensations have to be disclosed which have been granted in the business year, but not been published in any annual accounts. Options and other stock market based remunerations have to be directly disclosed with the number and the value at the point of time when they are granted. Further developments of value which result from a change of conditions for exercising these rights have to be taken into consideration.*
>
> *In case of a stock listed incorporated company the compensation of each individual board member with the name of the member has to be individually disclosed and divided into success independent and success related components as well as components with long-term incentive effect."*

[133] Sec. 285 sentence 1 n° 9 HGB, sec. 314 subs. 1 n° 6 HGB

If the company does not comply with these duties of disclosure, this can lead to a monetary fine of up to EUR 50,000.[134] However, the companies can be freed from the duty of disclosure of compensation. This happens in the General Assembly, where a three-quarter majority has to give their consent to the renunciation to disclosure. This legislative regulation can be criticized as inadequate, because family owners could for example hold the majority of the company, so that they could avoid the detailed disclosure of the remuneration. A control by the investors with regard to the remuneration policy of the company can therefore not take place.

The only DAX-30 company which does not individually disclose the compensation of the Executive Board in the annual report is Merck KGaA.

In this case the general partner – the E. Merck OHG – has the control of personnel over the personally reliable members of the management. For this reason the regulations of the VorstOG does not have to be applied by the company.

9.3 Remuneration aspects under stock corporation law

The introduction of stock-option programs and other long-term variable remuneration components for members of the Executive Board lead to renewed criticism of adequacy of their compensation. With regard to the adequacy of remuneration of the Executive Board there are further legal regulations in Germany which shall be introduced in detail in the followings sections.

9.3.1 Regulations in accordance with sec. 87 Companies Act (AktG)

The remuneration of members of the Executive Board of German incorporated companies is not only regulated by the GCGC but has also to be assessed in accordance with the rules of sec. 87 Companies Act. The regulations of sec. 87 AktG have their origin in the German Stock Companies Act of 1937 and shall harmonize the basic principles of correct remuneration. The introduction of sec. 87 Companies Act was meant to protect incorporated companies and their shareholders from excessively high compensations for board members.[135]

---

[134] Sec. 334 subs. 3 HGB

[135] Comp. Grattenthaler, 2007, p. 323 et seq.

Sec. 87 AktG states that the total remuneration of the individual members of the Executive board shall be in an appropriate relation to their tasks and to the company's situation.[136]

Provisions with regard to how far the remuneration of the members of the Executive Board is appropriate are only generally regulated in sec. 87 AktG. The Supervisory Board has to check the total remuneration of the Executive Board for adequacy. According to law the following components have to be checked for adequacy:

- the salary,
- possible profit participation,
- expense compensations,
- insurance remuneration,
- commissions,
- other compensations of all kind.

Furthermore it is only regulated in sec. 87 subs. 2 AktG that a reduction of the remuneration of a member of the Executive Board is possible when the company's situation has essentially deteriorated and the company's survival is endangered.

The adequacy of the remuneration of the Executive Board came again into focus of public discussions with the financial crisis. For example, the members of the Executive Board of the Commerzbank had to accept a limitation of their salaries. This was a precondition so that the predicated bank could get money from the bank rescue fund Soffin. The amount of the remuneration of the members of the executive board should not be higher than EUR 500,000.[137] In the previous few years the members of the executive board had still received amounts of up to EUR 5.2 million.

What can be criticized with regard to sec. 87 German Stock Companies Act is the fact is that it is hardly put into practice. Therefore, it is often called a "Dead-Letter-Law".[138]

Because of the discussed points of critique the Federal Government of Germany has passed the new Appropriateness of Management Remuneration Act (VorstAG) in order to reduce the weaknesses of sec. 87 German Stock Companies Act. This act will be discussed in the following section:

9.3.2 Appropriateness of Executive Board Remuneration Act (VorstAG)

On 18 June 2009 the Federal Government of Germany passed the Appropriateness of Executive Board Remuneration Act (VorstAG). This Act was meant to create more

---

[136] Comp. Klahold, 1999, p. 46

[137] Comp. N.N., 2008

[138] Comp. Boecken/Düwell/Hümmerich, 2007, p. 119

incentives for companies to design the remuneration of their Executive Boards in the context of sustainable enterprise development. With the introduction of this Act, the adaptations of the GCGC became necessary.

For big income earners amongst the top managers harder regulations were elaborated. Above all, stock-options are concerned by the new regulations. The new regulation now requires a lock-up period for stock-options of four years instead of the formerly usual two years (sec. 193 subs 2 n° 4 AktG new version). Only after the end of this period managers can exercise their options.

A further point deals with the determination of manager remuneration. Up to then the Supervisory Board had the possibility to delegate decisions on the remuneration of the Executive Board to a Committee. Now, this possibility does not exist any more. The remunerations of the Executive Board can now only be decided by the Supervisory Board of a Company. The existing sec. 107 subs. 3 sentence 3 AktG has been changed in this respect. Furthermore, the liability of the members of the Supervisory Board with regard to the remuneration of the Executive Board has been intensified. If the Supervisory Board does not determine an adequate remuneration of the Executive Board, it becomes liable for damages against the company.

Additionally, there are amendments for members of the Executive Board who want to change to the audit committee of the incorporated company. For them a "cooling-off period" applies, which means that persons who have been members of the Executive Board during the last years, may not change into the Committee (sec. 107 subs. 3 sentence 4 et seq. AktG new version).

Furthermore, the law stipulates a cooling-off period if a member of the Executive Board wants to change into the Supervisory Board. In controversial points – as the limitation of manager salaries – it was not possible to come to an agreement. Also the Treasury Secretary of the US is not in favor of a legal limitation of manager salaries. In an interview he said: "I don't think our government should set caps on compensation."[139] Instead he supports the "Say-on-Pay-concept".[140] The "Say-on-Pay-concept" deals with the co-determination right of the shareholders of a company in the manager remuneration. In the US, this form of right to say becomes more and more popular. Since the end of January 2009 more than 70 institutional investors, church groups and foundations have required from more than 100 incorporated companies to give shareholders the right to co-determine the remuneration of the Executive Board in future.[141]

---

[139] Comp. Christie/Homan, 2009

[140] Comp. ibid

[141] Comp. N.N., 2009

10 Categorization of Stock-option Programs

10.1 Overview

Stock-option programs can be counted among long-term incentives. Long-term incentives can be subdivided into programs with stock market based and key figure based management remuneration. The stock market based management remuneration can be further divided in real and virtual equity instruments.[142] After that it can be further distinguished between systems with an option based or a share-based character.

For the remuneration with real equity instruments the remuneration is paid via shares (restricted stocks), stock-options or performance-oriented stock allocations (performance shares). The remuneration via stock-options is possible via the integration of the option into convertible bonds or exclusively via "naked warrants".[143]

Real equity instruments are executed with the help of equity components in form of shares or subscription rights with regard to shares. The recipients are put in the position of a shareholder or are given the right to come into this position at a later point in time when receiving stock-options or performance shares.

In the case of virtual equity instruments, on the other hand, money is paid. Virtual equity instruments imitate the mode of operation of real option or share programs. This means that the assessment of the profit is also determined by the stock quotation. Amongst often used virtual equity instruments are Stock Appreciation Rights. They do not work like real stock-options. The profit is not paid in shares but in cash.

Stock market based remuneration in form of real or virtual equity instruments is generally only exercised in stock listed companies. The reason is that this form of long-term incentives is governed by the stock quotation of the shares of the company.[144] Figure 10.1 summarizes these considerations.

Companies that are not stock-listed, on the other hand, need key figure based instruments. Key figure based programs are also used by stock listed companies.

Key figure based instruments can be divided into phantom plans and other cash-based long term remunerations. The focus of the following section lies on the share-based remunerations, especially on stock-options. Phantom stocks are included in the deliberations, because their design is quite similar to the one of the share-based management remuneration.

---

142 Comp. Kramarsch, 2004, p. 129

143 Comp. Dietz, 2004, p. 45

144 Comp. Kramarsch, 2004, p. 129 et seq.

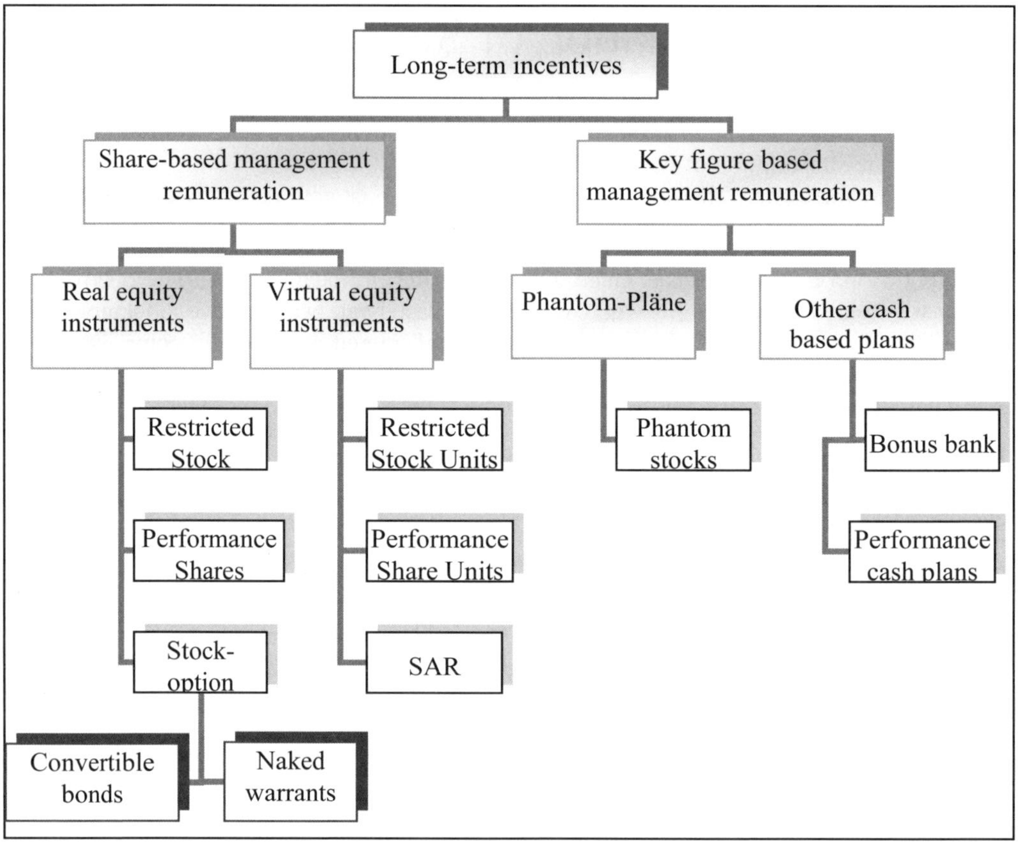

Fig. 10.1: Categorization of stock-option programs[145]

10.1.1 Convertible bonds and warrant bonds

The convertible bond certifies besides the rights of normal bonds the right to convert the bond after a certain period into stock. This means that it is composed of two components: the bond and the right to conversion.[146]

The bond is subject to interests. The interest generally goes by the long-term Federal Government bond. The conversion right makes it possible for the owner of the bond to convert the bonds into company stocks after expiration of a certain period. The conversion ratio is fixed in the terms of issue at the beginning. The price of conversion which the owner of a convertible bond has to pay, results from the difference between conversion price and

145 Following: Kramarsch, 2004, p. 130

146 Comp. Kramarsch, 2004, p. 54

nominal value. If the owner of a bond does not use his right to conversion, the bond will be paid back to him.[147] If he uses his right to conversion, the owner becomes a shareholder and the bond expires.

A warrant bond is similar to a convertible bond. The owner of the warrant bond is also entitled to the subscription of company stocks. The warrant bond, however, does not expire when used by the owner. The warrant bond continues to exist until repaid by the company.[148]

The provision of convertible bonds in the context of the introduction of a stock-option program was very popular in Germany up to the introduction of the Control and Transparency in Enterprises Act, as it was the only possible way to give stock-options to employees and managers. Today, this solution is considered as relatively complicated and legally too complex.[149] With the introduction of the Control and Transparency in Enterprises Act it has become possible to issue naked warrants to employees and managers.

When issuing convertible option and options, stock corporation regulations have to be complied with as will be discussed further on.

For the incorporated company giving out options or convertible options in the context of a stock-option plan is a way to gain foreign capital. The dilution of capital of the existing shareholders is the disadvantage. An advantage is that when the right to conversion is used, foreign capital is converted into equity and the repayment of the option is not necessary. On the other hand, one has to have in mind that the purpose of stock-option programs is the increase of enterprise value and of the stock quotation.[150]

A consent or authorization decision of the General Assembly is necessary to issue convertible bonds or warrant bonds.

If warrants or convertible bonds are given out, it is necessary to have a resolution of approval or an enabling resolution of the General Assembly in accordance with sec. 221 subs. 1 sentence 1 AktG. For the resolution at least a three-quarters majority of the authorized capital represented at the moment of resolution is necessary. The approval or authorization is valid for no more than a period of five years. The executive board and the chairman of the supervisory board have to leave the resolution and a report on issue of convertible bonds with the commercial registry in charge. The report has to contain essential information on the bond conditions and information on the reason for exemption from subscription rights (sec. 221. subs. 4 sentence 2 in connection with sec. 186 subs. 4 sentence 2 AktG).[151]

---

[147] Comp. Kramarsch, 2004, p. 54

[148] Comp. Schmeisser/Hahn/Schindler, 2004, p. 67

[149] Comp. ibid, p. 69

[150] Comp. ibid, p. 67

[151] Comp. Schmeisser/Hahn/Schindler, 2004, p. 68

10.1.2 Naked warrants

Naked warrants correspond to the general character of options. Stock-options give the right to:

- A purchase (call) or sale (put),
- of a fixed number of shares (base value),
- at a fixed price (base price),
- at a fixed moment (European option) or
- within a fixed period of time (American Option).

There are two kinds of options, the purchase option (call option) and the sales option (put option).[152] The call option gives the right to buy the share at a fixed price. The put option gives the right to sell the share at a fixed price.

In case of call options the purchaser participates in price advances, if the stock quotation is higher than the fixed purchase price of the option. The owner of the option can purchase the share from the issuing party at a lower price than the current stock quotation. The difference between the stock quotation and the previously fixed purchase price is the profit. The exact opposite is the put option. The owner of a put option makes profit, if the fixed purchase price is higher than the current stock quotation. The share can be purchased by the option holder for a lower price at the stock market and be sold for the sales price which lies above the current stock quotation. The difference is the option profit.[153]

The following figure shows the difference between call and put options.

[152] The option types „short“ and „long“ will not be described in more details, because they are not important in practice

[153] Comp. Kramarsch, 2004, p. 135 et seq.

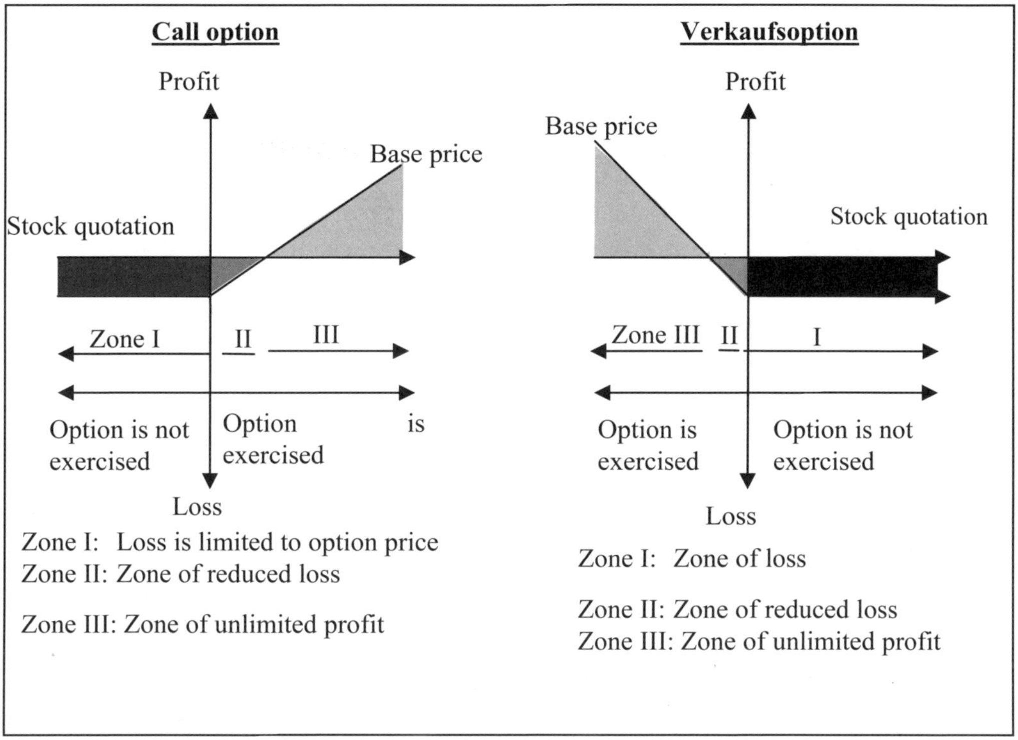

Fig. 10.2: Functionality of a stock-option[154]

Stock-options are differentiated in European and American options. In stock-option programs exclusively American options are applied. In Germany, the execution of stock-options is limited to certain periods because of insider dealing provisions of the German Securities Trades Act. Furthermore, the stock-option programs are subject to stock-options regulations, according to which it is only possible after expiration of a certain time to exercise the option.

Stock-options are generally granted without payment of an option premium to the manager. The participant in a stock-option program, however, does not receive the stock-option fully free of charges. Firstly, he has already "paid" the stock-option with his work.[155] Secondly, the risk for the recipients of a naked warrant is higher than in case of a convertible option. In case of a convertible option the participant gets his full investment with full interests back even in case of negative price developments of the stocks. Naked options, on the other hand, are forfeited without value, if the manager does not achieve to push the stock quotation towards the fixed performance target during its runtime period.[156]

---

[154] Following: Herrling/Schmidt/Zahner, 2008, p. 155

[155] Comp. Dietz, 2004, p. 47

[156] Comp. Schmeisser/Hahn/Schindler, 2004, p. 69

10.1.3 Stock Appreciation Rights

Stock Appreciation Rights (SAR) is a further form of value-oriented remuneration. They can be considered as adapted form of stock-options, because they imitate them – however virtually. They are also called value appreciation rights, which give the managers the right to participate in the increase in value of the company. The stock quotation serves as value for the calculation of the increase in value. The beneficiaries receive a payment at a fixed point in time, which corresponds to the difference between the stock quotation and the base price at the moment of exercise.

Stock Appreciation Rights are not financed by issuing shares. Therefore, they do not induce a dilution of the shares of the existing shareholders. Also, the administrative workload for realizing a virtual stock-option program is smaller.

However, issuing stock appreciation rights also brings some disadvantages. The payment of the virtual stock-option program affects the company's liquidity.[157]

10.2 Excursion: incentive programs

Besides stock-option programs there is a variety of other long-term incentive systems of share-based remuneration programs which are used by companies to compensate their managers in a performance related way. Because of the increasing use of those incentive programs, these are explained in more detail in the following section.

10.2.1 Phantom shares

Phantom shares are often used when corporate law forbids issuing shares. For example, companies which are not incorporated companies or partnerships limited by shares cannot offer stock-option programs for their managers. It can be still interesting for those companies to direct the remuneration of their managers towards the long-term growth of the company. Phantom shares are one possible way to do so.

Starting point of this participation model is that the managers do no receive real shares, but a virtual number of shares. These shares are a fictive part of the company value. The value of the company does not have to be calculated via the stock quotation, but can be determined via a procedure chosen by the company.[158] After expiration of the program the recipient is paid the sum. Furthermore, the plan can integrate the payment of a bonus to the beneficiary.

---

[157] Comp. Kramarsch/Becker, 2006, p. 49

[158] Comp. Kramarsch/Becker, 2006, p. 50

An advantage of the remuneration via phantom shares is the flexible design. Legal provisions, which would apply in the case of stock-options, do not have to be considered.

A disadvantage is the outflow of company liquidity at the moment of outpayment. As long as no personal investment is foreseen, the manager bears no financial risk.

10.2.2 Restricted Stocks/Restricted Stock Units

Restricted stocks are a real equity instrument which makes it possible for the manager to fully acquire shares. The attribution of shares to the beneficiaries underlies certain time restrictions. These restrictions are time limitations of disposal, as e.g. certain lock-up and holding periods, which usually consist of three years.[159]

The use of performance targets to measure management performance is not common in this case. At the end of the runtime of the stock program the participant either gets the right to purchase a certain number of shares at favorable conditions or the shares are attributed. During the holding period the participant has the voting right and the right to payment of dividends.[160]

There is also the possibility of the virtual design of restricted stocks. In this case, they are called restricted stock units. The participants do not receive an attribution of shares at the end of the runtime, but a cash settlement.[161]

As in the design of restricted stock plans no performance hurdles are used, they are no alternative to the success oriented stock-option programs. They are rather a sensible addition to a long-term remuneration.[162]

10.2.3 Performance Shares/Units

In the case performance shares the participant is attributed with shares for which he generally does not have to pay. The condition for the attribution of shares is that certain performance goals as EBIT or return on capital are achieved. There is also the possibility of using a virtual design. This form of share program is referred to as performance share units. At the end of the waiting period the participant receives a cash settlement. The amount of the settlement is calculated from the number of attributed performance shares multiplied with the stock quotation at the end of the waiting period.[163]

---

159 Comp. Kramarsch/Schelter, 2006, p. 1

160 Comp. Kessler/Babel, 2003, p. 41

161 Comp. Meiffert, 2008, p. 192

162 Comp. Kramarsch/Schelter, 2006, p. 1

163 Comp. Meiffert, 2008, p. 192

The following Fig. illustrates the shortly presented explanation graphically via the example of the performance share plans of the RWE group.

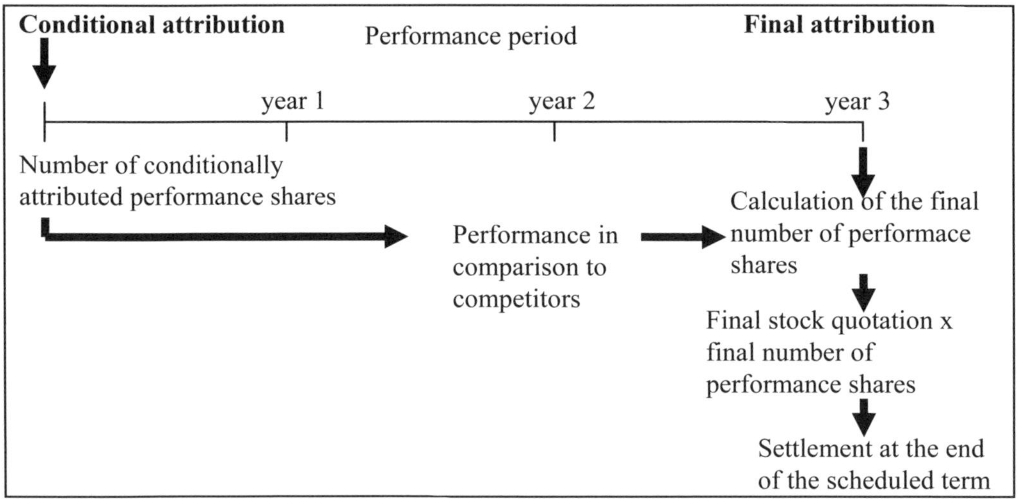

Fig. 10.3: Performance Share Plan des RWE group[164]

The performance share plan "Beat" of the RWE group has a term of three years. It is virtually designed and the participant does not receive shares but a cash settlement. During the runtime of the plan the participant receives a conditional attribution of performance shares. The final number of performance shares the participant receives is only calculated at the end of the performance period. The amount of settlement results from the multiplication of the final quotation of the RWE share with the final number of performance shares. RWE's condition for the outpayment is that the TSR of the RWE share is increased by at least 25 percent as measured by the benchmark index Dow Jones Stoxx Utilities. If the performance target is not met there is no outpayment.

10.2.4 Performance Cash-Plan

The performance cash-plan is referred to as long-term success bonus. In this case, the beneficiary receives the promise for a cash payment which is directed by the long-term company success. Generally a three-year performance period is fixed. After expiration of the period certain performance targets will have to be fulfilled so that there is an outpayment to the beneficiary. The performance targets can be based on internal or capital market oriented key figures.[165]

---

164 RWE AG, 2005, p. 4

165 Comp. Filbert/Kramarsch, 2008, p. 9

11 Criteria for the arrangement of stock-options programs

In order to fulfill the purpose of incentive effects and the real implementation of the shareholder-value-thought and the acceptance by the shareholders, shareholder option programs have to fulfill certain criteria with regard to their arrangement. Incorrectly designed stock-option programs result in high costs without bringing the desired benefit.[166] The most important parameters for the arrangement of stock-option programs are presented in the following sections.

The design of the stock-option programs for employees is realized by the executive board. If the beneficiary is part of the executive board, the supervisory board (sec. 87 AktG) takes the design over.

11.1 Circle of beneficiaries

Before a stock-option program can be installed or introduced in a company, the circle of beneficiaries, i.e. the employees to whom the real or virtual stock-options shall be offered, shall be determined. The determination of the circle of beneficiaries is closely connected with the targets pursued through the stock-option program.

Stock-option programs pursue mainly the aim of creating incentives in order to increase the company value in the sense of the shareholders and to reduce the divergence of interests between principal and agent. Only the top management has the total responsibility for the company and is in the situation to directly influence the company value with its decisions.[167] For this reason, the circle of participants in the stock-option program has to be limited to the top management.

If stock-options are given to the employees, anyway, who have little responsibility or little or no influence on the stock quotation, there is the danger that stock-options do not create additional incentives. The incentive effect of stock-option programs is then noticeably reduced. For this reason stock-option programs are generally limited to the top management level.[168]

---

[166] Comp. Winter, 2001, p. 85 et seq.

[167] Comp. Klemund, 1999, p. 70

[168] Comp. Weber, 2002, p. 31 et seq.

11.2 Determination of the base price

The base price is the price to be paid for which the owner of stock-option when exercising the option can purchase company shares. The difference between the base price and the stock quotation at the moment of exercising the option is the profit of the option which the owner of the option can realize through the exercise of the stock-option.[169] This means that the option profit depends largely on the base price. The realized profit is the higher the lower the base price is.[170]

There are two options for the determination of the base price. One possibility is to fix the base price at the moment of granting the stock-option, so that it remains unchanged during the complete runtime of the option. The other possibility is to fix it during the runtime and to align it with the stock quotation at this moment or at the moment of exercise of the option.[171] The determination of the base price during the runtime of the stock-option is mostly chosen, if a relative performance target is agreed on in the stock-option program, in which the development of the stock quotation is connected to a reference index. In most stock-option programs, however, the base price is determined at the moment of grant.[172] Hereby, the base price can have three different forms. The base price can be equal to the current stock quotation (at-the-money-option), lower (in-the-money-option) or higher (out-of-the-money-option). Granting stock-options with a base price below the current stock quotation is not acceptable for the shareholders because the beneficiaries would already benefit at the moment of grant of the option. From the perspective of the existing shareholders this leads to an immediate destruction of shareholder-value and is opposed to the aims of the stock-option program.[173] This is one of the reasons why in practice mostly a base price is chosen which corresponds to the current stock quotation or approximates the current stock quotation.[174]

For the determination of the base price either rates at a certain reporting date or average rates can be used. If the rate at a reporting date is chosen, the base price corresponds to the stock quotation of a trading day. In the case of average rates the base price corresponds to the average value of several trading days. In practice average rates have to be preferred to reporting day rates, because random price fluctuations can be eliminated this way.[175]

---

169 Comp. Engelsing, 2001, p. 54

170 Comp. Friedrichsen, 2000, p. 169

171 Comp. Friedrichsen, 2000, p. 171 et seq.

172 Comp. Klahold, 1999, p. 31

173 Comp. Achleitner/Wichels, 2002, S 18

174 Comp. Kramarsch, 2004, p. 136 et seq.

175 Comp. Baums, 1997, p. 3

11.3 Performance targets

Performance targets, also called exercise hurdles, determine which aims have to be reached after expiration of a lock-up period and permit the owner of the option to convert the stock-option into shares or to receive cash from the exercise of his virtual option. It the stock-option program is financed via a conditional capital increase or the retirement of own shares, the determination of a performance target is obligatory in accordance with sec 193 subs. 2 n° 4 German Stock Corporation Act (AktG). Additionally, the German Corporate Governance Codex requires stock-option programs to be connected with challenging, relevant reference parameters.[176]

The dilution effect for existing shareholders shall be compensated by the use of performance targets. The dilution effect is reduced, because when performance targets are reached, the company value is appreciated. There is the possibility in the stock-option program to combine one or several performance targets of which at least one has to be fulfilled for the exercise of the option. When a combination of several targets is used, it has to be determined whether both targets (and-rule) have to be reached or whether the achievement of one of both targets is sufficient (or-rule).

Depending on the form of the performance targets, these can be divided in absolute and relative performance targets. While absolute performance targets are connected with the absolute increase in stock quotation, in the case of relative performance targets, a comparison of the share with a sector or benchmark index is undertaken.

Analysts and shareholder representatives favor relative performance targets because from their point of view the danger of windfall profits can be avoided this way.[177] Furthermore, there is the possibility of using key figures from accountancy as targets.

The following figure shows possible forms of target design.

---

[176] Comp. DCGK, 2008, par. 4.2.3

[177] Comp. Kramarsch, 2004, p. 163

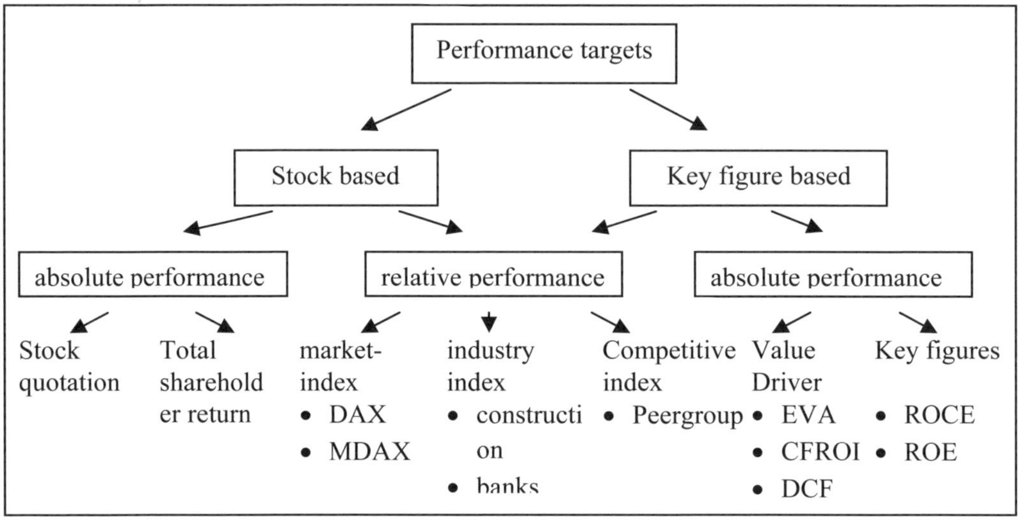

Fig. 11.1: Variants of performance targets[178]

Generally, the use of relative performance targets prevails on an international level. The design of performance targets is especially complex in Great Britain. In the US there is hardly any performance target determined in stock-option programs. If a performance target is determined anyway, mostly the total shareholder return in comparison to a sector index or a peer group is used.[179]

The following points describe the diverse possibilities to design success fees. In this context the different types of performance targets and the possible use in the management of stock-option programs is examined.

Absolute performance target

When determining an absolute performance target a minimum performance of the stock quotation is determined. This means that the stock quotation has to have increased by a certain, previously determined percentage within a certain period between the moment of granting and exercising the option. Mostly the base price is calculated at average rates and not at reporting date rates.[180]

[178] Ibid., p. 163

[179] Comp. Meiffert, 2008, p. 191

[180] Comp. Baums, 1997, p. 18

In practice an absolute rise in share price by at least eight percent is recommended.[181] If this minimum performance is not achieved, the owner of the option cannot exchange the option for company shares and the option forfeits.[182]

The following figure represents the model of an absolute performance target

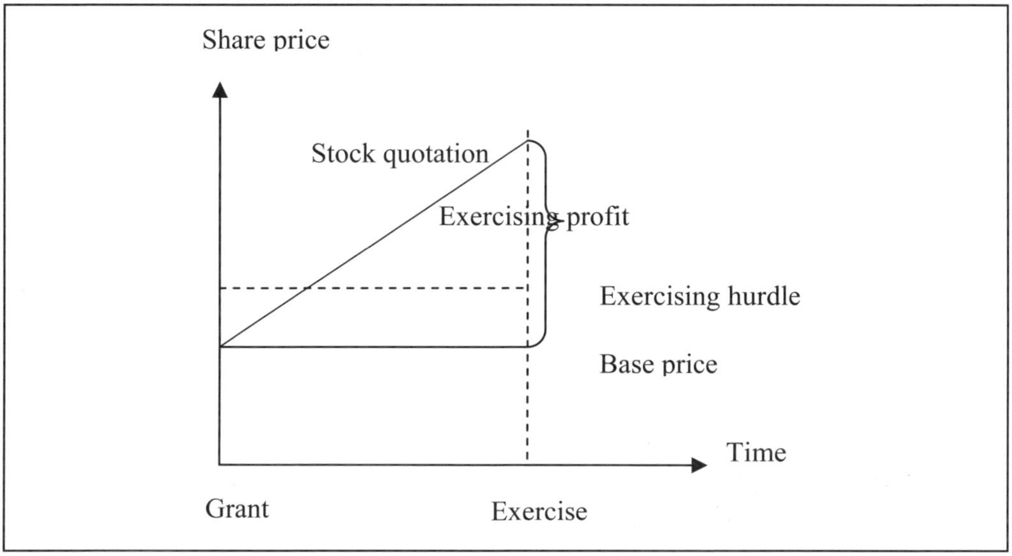

Fig. 11.2: Display of an absolute performance target[183]

The owner of the option has two possibilities in this case:

- He can exercise the stock-option, i.e. he purchases shares at the previously fixed base price and sells them at the stock market. The exercising profit results from the difference between the base price and the current share price.
- He can exercise a purchase option, i.e. he buys shares and puts them into his share deposit account, because he hopes that the prices would rise even further.

The exclusive use of absolute success targets as was common in the first stock-option programs applied by German companies is considered as disadvantageous.[184]

The most criticized aspect is the fact that the beneficiaries are remunerated for developments which are not connected to their performance.[185] This is the case, when the increase of the

[181] Comp. Union Investment, 2006, p. 4

[182] Comp. Schmeisser/Hahn/Schindler, 2004, p. 52 et seq.

[183] Following: Brühl, 2004, p. 465

[184] Comp. Weber, 2002, p. 38

[185] Comp. Baums, 1997, p. 12

stock quotation is caused by exogenous factors as e.g. interests, development of USD, current situation of stock market or development of regional markets. There is the danger that the owner of the stock-option might be remunerated for stock market developments which are unrelated to his performance. This way he can realize so called "windfall profits".[186]

In the adverse case there is the risk that a baisse at the stock market can cause that the stock-options of the beneficiary are forfeited even if the company development and his performance are positive. This can lead to a demotivation of the beneficiary. In specialist literature relative performance targets, which avoid "windfall profits", are favored.[187]

Relative performance target

In the case of relative success targets the relative development of share prices is used. Hereby, a comparison between the percentual development of prices of a share and the percentual development of the market or sector index or a peergroup is made.[188]

The exercise of a stock-option is only possible, if the development of the stock quotation of the own share exceeds the development of the sector or market index. In this case, a so called outperformance is given, in which the management has achieved an above-average performance and is therefore entitled to exercise the stock-option.[189]

For the indexation of the performance target, a national or international market index as e.g. DAX, MDAX or Dow Jones 30 can be considered at first. The disadvantage of the use of market indices is that the special characteristics of the sector in question are neglected. For this reason, special sector indices as e.g. the MSCI World Chemicals Index or the Dow Jones Euro Stoxx Bank Index seem more appropriate. The performance of the management can be better evaluated, because companies within a sector are generally facing the same external conditions.[190]

For especially highly diversified companies, the use of the sector index is not appropriate, as they can not be assigned to any sector and no comparable index exists.[191] In this case, an individual reference group is recommendable, a so called peer group. For each business area of the company, a group of comparable competitors is composed. By weighting of the individual sectors, an own reference index can be calculated.[192] The criteria for the

186 Comp. Klahold, 1999, p. 32 et seq.

187 Comp. Bösl, 2004, p. 110

188 Comp. ibid, p. 111

189 Comp. Friedrichsen, 2000, p. 151 et seq.

190 Comp. Rappaport, 1999, p. 137

191 The Deutsche Lufthansa and also Daimler AG have made an index of several competitors in their programs.

192 Comp. Friedrichsen, 2000, p. 154

composition of the peergroup should be transparent for shareholders and investors in order to avoid a possible manipulation of weighting of the peergroup created.[193]

The following figure shows the possible performance developments of stock quotation and reference index in the case of rising rates.

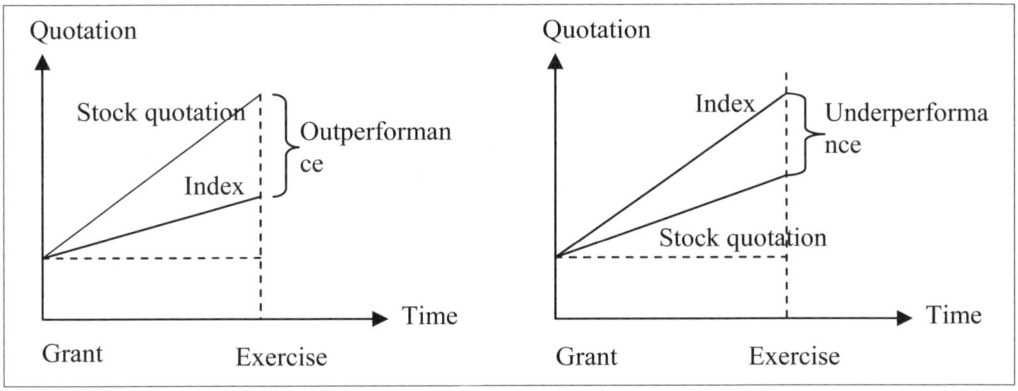

Fig. 11.3: Display of relative success fees in the case of rising prices[194]

The left graphic of Fig. 11.3 shows the exercising profit that the beneficiary can achieve if he exercises his stock-option. An exercise is possible when the stock quotation has a better performance than the reference index. In the right graphic the owner of the option cannot exercise the stock-option, because the stock quotation had a worse performance than the reference index. This is a case of underperformance of the manager.

When the relative performance target is connected to an index the stock-option can even be exercised by the beneficiary when the share prices are falling. This possibility of exercise of an option in times of a negative stock market situation makes it possible to adequately remunerate above average performance of the management, as long as the stock quotation of the company in question performs better than the market despite of the negative return.[195]

The following figure displays these deliberations:

[193] Comp. Heitzer/Klose/Steiner, 2000, p. 351

[194] Following: Brühl, 2004, p. 465

[195] Comp. Achleitner/Wichels, 2002, p. 17 et seq.

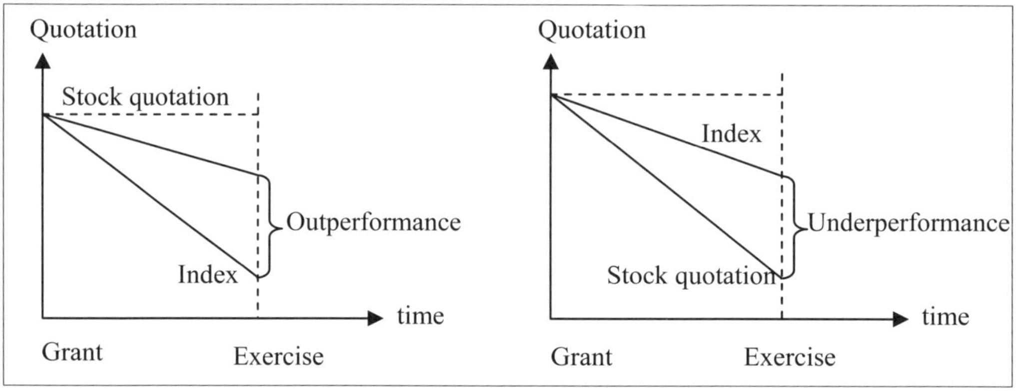

Fig. 11.4: Display of a relative performance target in the case of falling share prices[196]

The left graphic in Fig. 11.4. shows the maximum exercising profit that the beneficiary can obtain when exercising the option. An exercise of the stock-option is possible if the stock quotation has dropped less than the reference index. The right graphic in the Fig. shows that the share prices have dropped more than the reference index. In this case it is not possible to exercise the option.

Key figure based performance targets

The performance target does not necessarily have to be connected to the stock quotation. There is also the possibility to connect the performance target to accountancy key figures or value-oriented figures. For the determination of performance targets often key figures from accountancy, as e.g. the earning per share or the return on capital are used.[197]

However, the use of key figure related performance targets has to be considered as disadvantageous. The main point for critique is the mentioned danger of manipulation and the misuse of accountancy data. The connection of the performance target to the stock quotation has the advantage of being less susceptible to such manipulations by the management.

The most important counter argument for the use of key figures in a stock-option program consists in the fact that with turning away from the connection of the performance target with the stock-option means in a certain way also the renunciation of the shareholder-value-concept.[198] The use of the accountancy figures, like earnings per share, is meaningful for the shareholders, but is not connected with the long-term increase of the company value. The incentive effect of this type of program is not different from those of traditional remuneration

[196] Following: Brühl, 2004, p. 467

[197] For example the programs of Fresenius and Linde have a target per share and the Daimler AG usees the capital return as target.

[198] Comp. Kallmeyer, 1999, p. 98

models. The goal of the stock-option program to create a motivation for the beneficiary to increase the company value is not achieved.[199]

11.4 Time components

11.4.1 Chronology of a stock-option program

Granting and exercising stock-options is connected with different stages in time. This section gives a first overview over the different points in time, and the following sections will examine the different timely components in more detail.

The following fig. demonstrates the different timely components of a stock-option program.

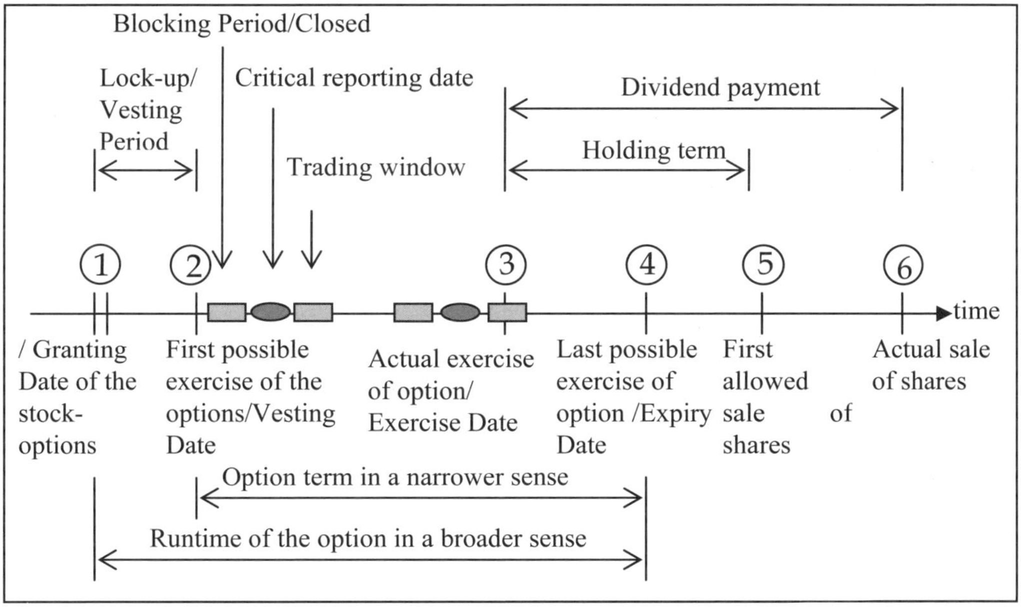

Fig. 11.5: Stages of a stock-option program[200]

An essential element in the design of the stock-option program is the total runtime also called option term. The runtime of the option determines the period of time from the first time concession to the last possible exercise of the stock-option. The length of the runtime has to

[199] Comp. Friedrichsen, 2000, p. 166 et seq.

[200] Comp. Schmeisser/Hahn/Schindler, 2004

be chosen in such a way that a long-term minimization of the principal-agent-conflict can be achieved.[201] In practice, a runtime of five to ten years is mostly chosen.

In point one of the figure the beneficiary is granted his stock-options. At this moment, the subscription rights are booked into his depot. In most cases, a lock-up period of the option program is agreed on, with the consequence that the beneficiary is only allowed to exercise the option after expiration of this period (point 2). Besides, it is usual to include trading windows or previously fixed periods (closed periods) to determine the exercise of stock-options or restrict it. The third point in time is the date of the actual exercise of the option by the beneficiary. On this day the subscription rights cease to exist with the exercise of the options and the beneficiary either receives his concerted stocks (real stock-options) or a cash settlement (stock appreciation rights). If the beneficiary does not exercise his option during the exercising period, they are forfeited.[202]

Furthermore, a holding period is foreseen in the stock-option programs. During this period it is interdicted for the beneficiary to sell the shares he received. Despite this interdiction the beneficiary is entitled to payment of dividends and to exercise his voting rights until he sells the shares.[203] The last two points in time, five and six, represent the first possible and the actual day of sale of shares.

Lock-up period

The lock-up period is the period from the grant of the stock-options to the time of the first possible exercise of the stock-option. During this lock-up period the option cannot be exercised. When stock-option programs are financed through conditional increase of capital or the retirement of stock, a lock-up period of at least two years has to be determined (sec. 193 subs. 2 n°. 4 in connection with sec. 71 subs. 1 n°. 8). The lock-up period should not be longer than five years, because the members of the executive board are appointed for only five years (sec. 84 subs. 1 sentence 1 AktG) and the executive board should have the possibility to exercise the option within the runtime of the sock option programs. The use of staggered lock-up periods is also possible. For example, from the point in time of the first grant on, the participant in the option is allowed to exercise:

- 30 percent of the option after two years,
- 40 percent of the option after three years ,
- 30 percent of the option after four years.[204]

The following Fig. displays the different possibilities to use lock-up periods.

201 Comp. Engelsing, 2001, p. 60

202 Comp. Friedrichsen, 2000, p. 25

203 Comp. Engelsing, 2001, p. 61

204 Comp. Bösl, 2004, p. 108

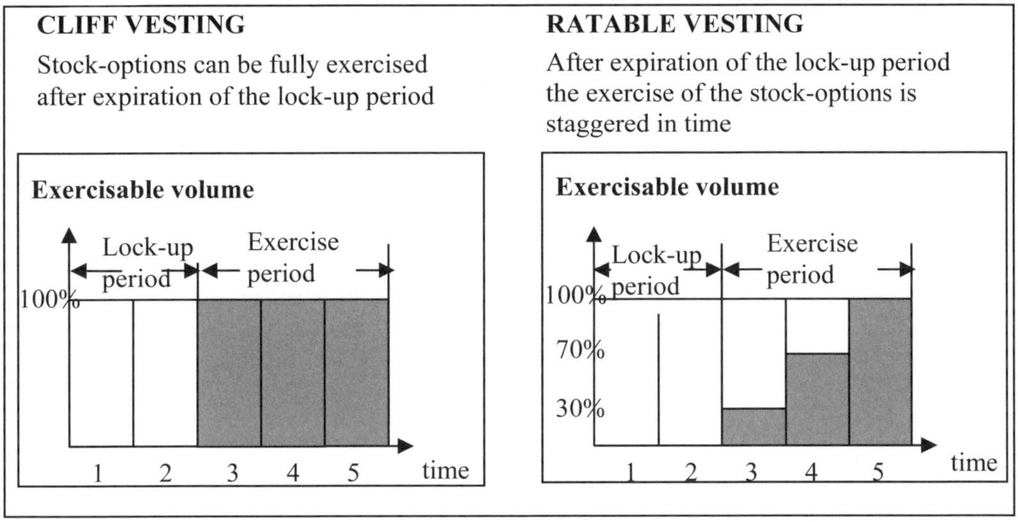

Fig. 11.6: Types of lock-up periods[205]

In Germany, the exercise of the stock-option is generally quite close to the termination of the lock-up period. A timely staggered execution could motivate managers to a longer holding period of the stock-options and bind the employees longer to the company.[206]

Exercise periods and holding periods

In order to avoid any insider suspicions against the participants, it is sensible to permit the execution of the option only after the expiration of a certain lock-up period and within certain periods. This limitation of exercise has two variations: The first is the possibility of the use of trading windows and the second the use of blocking periods.

In the variation of trading windows, the exercise of stock-options is limited to certain periods in a year. The concept emanates from the assumption that there is an informational balance between all market participants in this period of time.

Trading windows consist mostly of five to 21 trading days after critical reporting days as publications of quarterly reports or annual balance sheets.[207] For example, the Metro AG uses for its stock-option program a trading window. The trading window determined by the company limits the exercise of the stock-options by the beneficiary to eight weeks after the day of the general assembly.

---

[205] Following Kramarsch, 2004, p. 180

[206] Comp. ibid, p. 180

[207] Comp. Kramarsch, 2004, p. 88

In the second option, the blocking periods, it is assumed that, in contrast to the concept of trading windows, there is an informational disequilibrium in certain periods between all market participants. The managers have a lead in terms of information over other market participants because of their position in the company. The blocking periods consist in Germany mostly of one to fours weeks before the publication of information e.g. quarterly reports. In these periods of time, the participants of the stock-option programs may not exercise their stock-options.[208]

Furthermore a holding period can be agreed in the stock-option program. The holding period indicates the period of time during which the shares which a manager after the exercise of the options has received may not be sold. The shares are then kept e.g. in blocked deposits. However, the shareholder is entitled to receive dividends and to vote during this time.[209]

11.5 Personal investment and cap

A well designed stock-option program should in any case require a personal investment by the participant. The design of the personal investment can be done in different variations. The BASF AG e.g. has determined as personal investment that the employees have to make an investment in shares of their own company. The participants in the option program have to invest in one share in order to receive four options afterwards.

Furthermore, there is the possibility to stagger the personal investment at the different levels of management or directors. With diminishing responsibility at the hierarchic level, generally the amount of personal investment by the beneficiary is reduced.[210] Through the investment in shares and the holding period connected with it, the beneficiaries are affected by rising as well as sinking share prices. The employees and managers in question are thereby also engaged in the managerial risks.

A further criterion which may not miss in stock-option programs is the use of caps. The Corporate Governance Codex recommends for extraordinary, unforeseeable developments the fixation of caps.[211] Hereby the amount of attribution of stock-options is limited to a certain share price maximum of the stock-options.

---

[208] Comp. ibid, p. 88 et seq.

[209] Comp. Engelsing, 2001, p. 61

[210] Comp. Kramarsch, 2004, p. 234

[211] Comp. DCGK, section 4.2.3., 2008

11.6 Dividend policy

Dividend payments can have a negative influence on the stock quotation. An example for this is the dividend policy of the former DAX-30 company ALTANA. By the distribution of a dividend of EUR 1.30, an extraordinary dividend of EUR 33.00 and a bonus dividend of EUR 0.50 per individual share certificate entitled to dividends[212], a considerable stock market loss was caused.

The closing price of a share after one day of dividend payment was only at EUR 16.80, while the final price on the previous day had been at EUR 46.56. The high dividend payments were due to the fact that the company had sold its pharmaceutical sector for EUR 4.6 billion and had decided to fully distribute the sale proceeds.[213]

The value of stock-option increases when the stock quotation of the company goes up. As exemplified above, the distribution of dividends can have negative effects on the stock quotation of the company. Besides, there is generally no dividend protection for stock-options.[214] Furthermore, the owners of stock-options do not have any right to claim for the payment of dividends. These reasons imply that the executive board has no incentive to take a decision in favor of the high dividends. The interest in avoiding a dividend payment on the other hand is big.

A study by Lambert, Lanen and Larcker proved that the introduction of a stock-option program can lead to the reduction of dividend payment. The basis for the examination was the expected future dividend distributions. These were estimated with the help of a prognosis model and put into relation to the factual dividend payments.[215]

In a study by Jolls it was discovered that the executive board could decide – because of the ownership of stock-options – to use the freely disposable cashflows for a buyback of shares instead of using them for dividend payments to the shareholders. Jolls examined the dividend policy and undertaken buybacks of shares of 324 US companies in the year 1993. The author came to the conclusion that the probability of a repurchase of stock is growing when the manager is in the possession of stock-options, because the share prices rise this way.[216]

However, other studies come to the conclusion that dividend distributions rise when the management has own stock-options.[217] This means that the results of the individual studies are not definite. The use of free cashflows for the buyback of shares of the company does not

---

[212] Comp. Mußler, 2007

[213] Comp. ibid

[214] Comp. Lazar, 2007, p. 51

[215] Comp. Lambert/Lanen/Larcker, 1989, p. 411 et seq.

[216] Comp. Lazar, 2007, p. 51

[217] Comp. DeFusco/Zorn/Johnson, 1991, p. 36 et seq.

have to be assessed as negative for the shareholders if the achieved return of the company is higher than the return of an alternative investment.[218]

In order to influence the executive board in such a way that dividend distributions are not limited, stock-option plans can be designed so that the exercise price is adapted after the distribution of dividends.[219]

11.7 Dilution effect

Shareholders participate in the company's success in double respect. The shareholder can participate in the increase of the company value on the one hand through their shares and at the other through the profit distributed to them. A dilution effect for the existing shareholder is created when a capital increase is undertaken. New shares are created via capital increase and cannot by purchased by the existing shareholders because of the exclusion of the buying option. The company profit has therefore to be distributed to several new shareholders. This causes that the share of the company profit for the existing shareholders is reduced. Furthermore, the dilution effect implies also a dilution of the voting rights, which is generally rather small.[220] The dilution effect provides therefore important information on the costs of the remuneration for the shareholders.[221]

12 Possibilities of financing stock-option programs

The introduction of stock-option programs is connected to legal requirements which are set down in the stock corporation act. If a company uses a "real" stock-option program it has to provide the required shares of the own company for the participants of the stock-option program. The company has several possibilities to do so, as will be explained in more detail.

---

[218] Comp. Lazar, 2007, p. 51

[219] Comp. Pellens/Crassselt/Rockholtz, 1998, p. 16

[220] Comp. Kramarsch, 2004, p. 115

[221] Comp. ibid, p. 120

12.1 Overview

The company has the possibility to finance the stock-option program trough an increase in capital. For this it has three possibilities: the ordinary capital increase (sec. 182 et seq. AktG), the authorized capital increase (sec. 202 et seq. AktG) as well as the conditional capital increase (sec. 192 et seq. AktG). Besides the different possibilities to create new shares with the help of an increase in capital the company has further possibilities to underlay the stock-option program with shares. This includes the purchase of own shares and the cooperation with third parties. The following overview gives a first insight into the different variations.

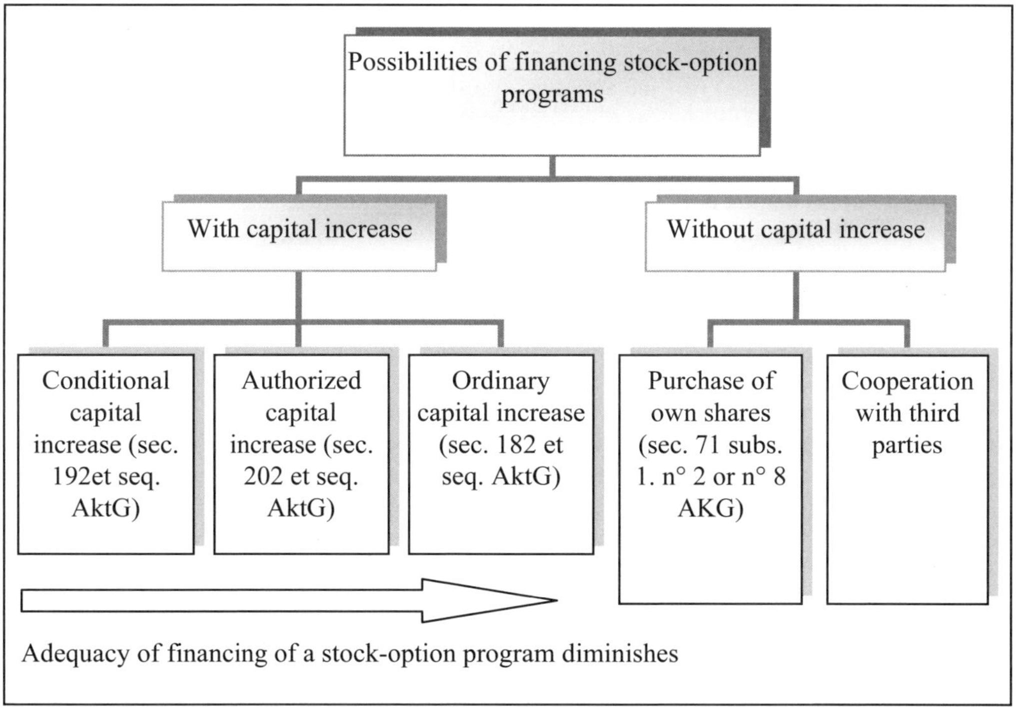

Fig. 12.1: Possibilities of financing a stock-option program[222]

[222] Following Dietz, 2004, p. 58

12.2 Conditional increase in capital

The conditional increase in capital is carried out via resolution of the General Assembly and is ring-fenced. Sec. 192 Subs. 2 AktG restricts the use of capital as follows:

- Guarantee of purchase rights for the owners of convertible bonds,
- Preparation of the merger of several companies,[223]
- Guarantee of purchase rights to employees and members of the management board (naked warrants).

The limitation of purchase rights to employees and members of the management board excludes the supervisory board from the circle of beneficiaries, because members of the supervisory board are neither employees nor management members. The legislator justifies the exclusion with the fact that the supervisory board can not determine the conditions for options for itself.[224]

With a conditional increase in capital the authorized capital is only increased to the extent in which the owners of stock-options use their right of share purchase and conversion (sec. 192 subs. 1 AktG).[225] The authorized capital increases directly with the distribution of shares.[226] The conditional capital increase is considered as authorized if at least three quarter of the authorized capital represented at the moment of resolution consents. This is only valid under the condition that the articles of association do not provide for a higher capital majority (sec.193 subs. 1 AktG).

Furthermore, the executive board and the chairman of the supervisory board have to register the resolution on the conditional capital increase in the commercial registry (sec. 195 subs. 1 AktG). Once the registration of the resolution in the commercial registry is carried out, the executive board can grant stock-options to the employees. If the beneficiary is the executive board, the options are granted by the supervisory board. The subsequently necessary entry of the distributed shares in the commercial registry has only declaratory meaning.[227]

The advantage of the conditional capital increase is that there is no timely limitation in contrast to the authorized capital increase which is limited to five years (sections 202, 221 AktG).[228] This is why this form of capital increase is especially appropriate for stock-option programs with long runtimes. Furthermore, there is no possibility to reverse the resolution

---

[223] sec. 192 subs. 2 n° 2 AktG does not apply to stock-option programs

[224] Comp. Bösl, 2004, p. 103

[225] Comp. Engelsing, 2001, p. 71

[226] Comp. Schmeisser/Hahn/Schindler, 2004, p. 73

[227] Comp. Einem/Pajunk, 2002, p. 93 et seq.

[228] Comp. Schmeisser/Hahn/Schindler, 2004, p. 74

for the conditional capital increase at a later point in time or to change it (sec. 192 subs. 4 AktG).

In the case of conditional capital increase the existing shareholders have no right to purchase new shares. By issuing new shares and the exclusion of existing shareholders from the purchase right, the number of shares in circulation is increased. The result is that the share of the voting rights of the existing shareholders diminishes. Furthermore, the property of the existing shareholders diminishes, because the participants of the stock-option program can subscribe the shares at a lower price at the current stock quotation (dilution effect). Because of these reasons the existing shareholders will only consent in the general assembly when essential information on the stock-option program, as e.g. the number of granted purchase rights is disclosed.

The legislator distinguishes different forms of the conditional capital increase in accordance with the purpose. There are different regulations for issuing convertible bonds or naked warrants.[229]

For convertible bonds and warrants in accordance with sec. 193 subs. 2 n° 1–3 AktG essential corner points of the stock-option program have to be determined. This information is:

- Purpose of the conditional increase in capital,
- Circle of beneficiaries,
- Information on the amount issued and the foundation on which this amount is calculated

Furthermore, the exclusion of existing shareholders from share purchase rights in accordance with sec. 3 sentence 4 AktG has to be agreed with three quarter majority of the represented authorized capital. In this case, it has to be made sure that the exclusion of the purchase rights is in the company's interest. The exclusion of purchase rights has to be appropriate and necessary and be in an appropriate relation to the pursued target.[230] An exclusion of the purchase right is especially appropriate when the capital increase does not surpass 10 percent of the authorized capital. Furthermore, it shall not be essentially lower than the base price of the shares (sec. 186 subs. 3 sentence 4 AktG).

Moreover, the executive board has to present a written report on the reason of the purchase right exclusion in accordance with sec. 186 subs. 4 sentence 2 AktG.[231] Besides, one has to pay attention that the amount of the conditional capital increase shall not surpass fifty percent of the authorized capital represented at the moment of resolution in accordance with sec. 192 subs. 3 n°. 1 AktG. If the company decides to introduce its stock-option program in connection with naked warrants, further details have to be fixed in the resolution of the general assembly in addition to the described corner points under sec. 193 subs. 2 sentence 1-

[229] Comp. Dietz, 2004, p. 59 et seq.

[230] Comp. Engelsing, 2001, p. 73

[231] Comp. Winter, 2000, p.192

3 AktG. In accordance with sec. 193 Abs. 2 S. 4 AktG the resolution of the general assembly has to contain:

- the repartition of the purchase rights which are given to employees or members of the management,
- performance targets,
- purchase periods,
- exercise periods and
- lock-up periods.

The German Stock Companies Act leaves a wide scope for the design of company performance targets. The use of absolute and relative performance targets as well as the use of accountancy key figures is possible.

Furthermore, the purchase and exercise periods have to be determined in the resolution, in order to counter the insider problematic sufficiently. Moreover, the legislator requires that the resolution contains the waiting period for the first time exercise. In accordance with stock company law the waiting period is at least two years long. A further bill requires a longer term.

Moreover it has to be respected that the amount of the conditional capital increase may not be higher than ten percent of the authorized capital present when the resolution is passed in accordance with sec. 192 subs. 3 n° 1 AktG. In this case, there is no need for a special exclusion of purchase rights for existing shareholders in contrast to the procurement of shares through convertible bonds or warrants. For this reason, there are no reporting duties of the executive board to the general assembly in accordance with sec. 186 subs. 4 sentence 2 AktG. In the interest of transparency and publicity one should not abandon a comprehensive reporting policy.[232]

This form of capital increase is often chosen because own shares do not have to be bought back at the market, which is positive for the company's liquidity.

The above mentioned details on the conditional capital increase in connection with naked warrants and convertible bonds have shown that it represents an appropriate measure for financing long-term stock-option programs. Furthermore, the introduction of naked warrants has to be preferred to issuing convertible bonds, because no exclusion of subscription rights is necessary which makes the procedure less complicated.

12.3 Authorized capital increase

The authorized capital increase under sec. 202 et seq. AktG is a further possibility to procure new shares in the context of a stock-option programs. The executive board is empowered by

232 Comp. Schmeisser/Hahn/Schindler, 2004, p. 75 et seq.

the articles of association, or a resolution of the general assembly with three-quarter-majority, to increase the authorized capital up to a certain amount through the issue of new shares against contributions. The empowerment requires a change of the articles of association. The change of the articles of association requires the entry into the commercial registry in accordance with sec. 181 AktG. The nominal amount, by which the authorized capital is increased, shall not exceed fifty percent of the authorized capital in accordance with sec. 202 subs. 3 sentence 1 AktG.

When the authorized capital increase is carried out, the executive board decides over the content and the conditions of the capital increase, as long as no requirements by law or the articles of association are opposed to the empowerment. The supervisory board needs to consent to the decisions of the executive board in accordance with sec. 204 subs. 1 sentence 2 AktG. This shall encounter the danger of a possible insider deal by the executive board.[233]

The subscription right of the existing shareholders is not excluded per se. An exclusion of the right of share purchase of existing shareholders is, however, necessary to introduce stock-option programs in the company. In accordance with sec. 203 subs. 2 AktG the executive board can be empowered to exclude the existing shareholders from subscription rights, which means that the provisions of sec. 186 subs. 3 and sec. 4 AktG apply analogously.

The timely limitation of the empowerment of the executive board to a maximum of five years is disadvantageous and problematic. The increase in capital can only be carried out in this period of time.

This means that the lock-up period and exercising period of the option cannot be longer than five years. However, stock-option programs should have a runtime of more than five years, in order to have a long-term incentive effect.[234]

Furthermore, it is necessary for stock-option programs with a longer runtime than five years, to have a resolution of the general assembly to decide on a new authorized capital increase before the resolution of empowerment expires. The general assembly is not bound by former resolutions. This means that the managers do not have absolute certainty with respect to the procurement of the necessary amount of shares when a stock-option program is introduced.[235]

Furthermore, the authorized capital increase is hardly appropriate for stock-option programs, because only with the entry of the increase of the authorized capital in the commercial register, the capital increase becomes effective (in accordance with sec. 189 AktG). This implies that the new shares may only be given to the managers after the entry in the commercial register (in accordance with sec. 191 sentence 1 AktG). All new shares which were given out before the registration are void (in accordance with sec. 191 sentence 2 AktG). For the participants of the stock-option program, there is therefore the danger, that

233 Comp. Engelsing, 2001, p. 70

234 Comp. Engelsing, 2001, p. 71

235 Comp. ibid, p. 71

they might not receive new shares when exercising their options, if the increase of capital has not been entered into the commercial register.[236]

The presented requirements for the execution of an authorized capital increase have shown that this form of capital increase is basically appropriate for the introduction of stock-option programs. The time and practical restrictions, however, make the authorized capital increase less appropriate for financing stock-option programs.

12.4 Ordinary capital increase

A possibility to use new shares for the options issued is an ordinary capital increase against contribution in accordance with sec. 182 et seq. AktG.

The general assembly has to decide an ordinary capital increase with three-quarter majority of the represented authorized capital. The decision and the execution of the increase of basic capital have to be entered into the commercial register by the executive board and the supervisory board of the company. Moreover, the exclusion of subscription of existing shareholders in accordance with sec. 186 AktG as well as a report in writing by the executive board on the reason for the exclusion by the general assembly is necessary. The decision of the general assembly on the exclusion of the prescription right does not contain special regulations which would make the distribution to participants of the stock-option program easier. Thereby, higher formal and material requirements are made for the decisions.[237]

Furthermore, it has to be seen as disadvantage that the exact amount of the capital increase has to be fixed when the resolution is taken by the general assembly.[238] However, when the stock-option program is introduced, it is still uncertain which number of stock-options the participant in the stock-option program will exercise at all. This means that is not clear how many shares are needed to cover the stock-options, when the stock-option program is introduced.

A possible solution of this problem is the fixation of a minimum and a maximum amount or only of a maximum amount in the general assembly resolution. The resolution on the authorized capital increase has to additionally contain a subscription period of six months in this case.[239] Via the interdiction of self-subscription in accordance with sec. 56 AktG the takeover of new shares by the company is not permitted until a need due to the stock-option program comes up..[240] The new shares can be taken over by a fiduciary, but this is seen as

---

[236] Comp. Kau/Leverenz, 1998, p. 2273

[237] Comp. Kau/Leverenz, 1998, p. 2273

[238] Comp. Engelsing, 2001, p. 69

[239] Comp. Kau/Leverenz, 1998, p. 2273

[240] Comp. Engelsing, 2001, p. 69 et seq.

inappropriate in practice, because a fiduciary must take over the complete contribution in accordance with sec. 56 subs. 3 sentence 2 AktG.[241]

On the ground of the above mentioned deliberations, we can say that the ordinary capital increase is very inflexible and therefore very rarely applied in practice.[242]

12.5 Purchase of own shares

A further possibility to cover the stock-option program is the distribution of already existing shares which the company would have to acquire for this purpose. The purchase of own shares, however, is interdicted by the following regulations:[243]

- Interdiction of self-subscription in accordance with sec. 56 subs. 1 AktG
- Interdiction of reuse of reserves in accordance with sec. 57 subs. 1AktG
- Interdiction of repurchase of own shares in accordance with sec. 71 subs. 1 AktG.

The general regulations on the interdiction of repurchases of own shares are based on the emergency decree of 19 September 1931 which was passed on the grounds of the experiences of the world economic crisis from 1929 to 1931. At that time the incorporated companies had tried to avoid the dropdown of prices of company shares via the repurchase of own shares. This lead in the best case to liquidity problems of the companies and in the worst case it meant their bankruptcy.[244]

However, the legislator has passed exceptional regulations (sec. 71 subs. 1–8 AktG), which make the purchase of own shares by the company possible. The repurchase of own shares by the company is permitted for the purpose of distribution to employees in accordance with sec. 71 subs. 1 n° 2 AktG or to managers in accordance with sec.71 subs. 1 n° 8 AktG.

12.5.1 Repurchase in accordance with sec. 71 subs. 1 n° 2 AktG

The repurchase of own shares by the company is allowed if the beneficiaries have a working contract with the company or an associate company and the shares shall be offered for purchase (sec. 71 subs. 1 n° 2 AktG). In the context of the amendment of stock company law

---

241 Comp. Klahold, 1999, p. 134

242 Comp. Klahold, 1999, p. 135

243 Comp. Engelsing, 2001, p. 87 et seq.

244 Comp. ibid, p. 86

of 1959 this regulation was introduced in order to make the issue of staff shares for the employees easier.[245]

In accordance with sec. 71 subs. 3 sentence 2 AktG the company has to hand down the repurchased shares to the beneficiary within one year after purchase. The amount of this measure of coverage shall not be higher than ten percent of the share capital (sec. 71 subs. 2 S. 1 AktG). One has to keep in mind that these 10 percent comprehend the purchased shares as well as the already purchased shares which are still in the company. For the repurchase of own shares and the design of the conditions the executive board is the only responsible. It can decide the repurchase of shares in accordance with the performing competence under sec. 76 subs. 1 AktG without consent of the general assembly.[246]

The repurchase of own shares is appropriate for the coverage of stock-option programs, when the stock-option program is only introduced for employees of subordinated management levels. This method of coverage is not appropriate for members of the executive board, because they can not be considered as employees of the company.[247] Employees are only those persons who are subject to instructions. The members of the executive board, however, are not subject to instructions. It is rather their responsibility to direct the company. Therefore, they do not have the characteristic of an employee and do not have the right to receive shares under sec. 71 subs. 1 n° 2 AktG.[248]

The purchase of own shares in accordance with sec. 71 subs. 1 n° AktG is nor appropriate for stock-option programs which entitle the executive board to purchase shares, because the legislator has narrowed the circle of participants to employees only. Also the time limitation to hand the purchased shares down to the beneficiaries within one year has to be regarded as disadvantage. Because of the short period of one year this method of coverage is not appropriate for long-term stock-option programs.

With the introduction of the KonTraG (Corporate Sector Supervision and Transparency Act) in the year 1998 the new regulation sec. 71 subs. 1 n° 8 AktG was introduced, because it makes it possible for members of the executive board and managers to take part in stock-option programs which are financed by the buyback of own shares. This regulation is explained in more detail in the following section.

12.5.2 Repurchase in accordance with sec. 7 subs. 1 n° 8 AktG

With the introduction of the sec. 7 subs. 1 n° 8 AktG the purchase of own shares by the company has been legally simplified. This regulation has proved to be very flexible, because there is no need to indicate a purpose. Under sec. 7 subs. 1 n°8 sentence 2 AktG the trade

---

245 Comp. ibid, p. 87

246 Comp. Klahold, 1999, p. 184

247 Comp. Engelsing, 2001, p. 88

248 Comp. ibid, p. 88

with own shares is excluded as only purpose.[249] If the purchase of own shares is undertaken in the context of a stock-option program, sec. 71 subs. 1 n°8 sentence 5 AktG points to sec. 193 subs. 2 AktG. The requirements for a conditional increase in capital mentioned in sec. 193 subs. 2 are also valid for own shares.

The permission to the executive board to buy back own shares has to be given in a resolution of the general assembly. This resolution is limited to 18 months in accordance with sec. 7 subs. 1 n° 8 AktG. This term of 18 months is only valid for the purchase of shares, but not for holding them. This makes it possible for the company to hold its own purchased shares for an unlimited period of time. In the authorization resolution, the minimum and maximum price for which the own shares are bought back have to be specified. These two values can be either determined in absolute or in relative numbers.[250] Furthermore, the resolution has to contain the volume for the purchase of own shares which must not be higher than ten percent of the share capital (sec. 71 subs. 1 n°1 AktG). It has to be reminded that the upper limit of ten percent comprehends the purchased shares as well as the already purchased shares which are still kept in the company (sec. 71 subs. 2 sentence 1 AktG).

In accordance with sec. 71b AktG the company is neither entitled to voting rights nor to dividend payments from own shares. Furthermore, evasive transactions are forbidden in accordance with sec. 71a AktG. If the company purchased own shares illegally the shares have to be sold within one year after their purchase.[251]

Because of the insider problematic, the Federal Authority for the Supervision of Services (Bundesanstalt für Dienstleistungsaufsicht) has to be informed on the empowerment of the executive board in accordance with sec. 71 subs.3 sentence 3 AktG. In the next general assembly the executive board has to inform the shareholders over the following points in accordance with sec. 71 subs. 3 sentence 1 AktG).

- reasons for and purpose of the purchase of own shares
- number of acquired shares and
- the amount of share capital used on them,
- their share in share capital
- their counter value in shares

When the executive board is empowered, it may buy back own shares up to a maximum amount of ten percent (sec. 71 subs. 1 n° 8 s. 6 AktG). In practice, we can distinguish between different procedures and moments for the repurchase of own shares. The following points in time are possible for the repurchase of own shares:

---

[249] Comp. Zitzewitz, 2003, p. 89

[250] Comp. Engelsing, 2001, p. 89

[251] Comp. ibid, p. 90

- when the stock-option is granted,
- during the runtime of stock-option programs
- during the execution of stock-options.

The advantage of repurchase of own shares at the moment of granting the stock-options is that the price risk is avoided if the management chooses to exercise the option at a later point in time.[252] The reason for this is that the participants in the stock-option program purchase the shares at a base price at the moment of exercising the option which corresponds to the purchase price of the shares by the company. If the stock-options are not exercised, the company bears the loss of the price dropdown in the case of necessity to sell the shares. A fact which has to be considered as disadvantage is that the shares have to be purchased at a very early point in time. This implies that liquidity is bound in the company which is not available for investments and payment of interests for existing credit. The positive side is that possible dividend payments are saved during the runtime of the stock-options (sec. 71b AktG).[253]

Altogether, it can be assumed that shares are bought back when stock-options are exercised.[254] The company has a lower risk for the case that stock-options are not exercised and the liquidity is not affected during the complete runtime of the option.[255]

Furthermore, there are different possibilities how the share can be purchased. In practice the following methods are distinguished:[256]

1. purchase of the shares on the stock market ("open market purchase")
2. purchase offer by the executive boards to the shareholders ("self tender offer")
3. purchase of shares by a majority shareholder ("negotiated repurchase").

In the case of repurchase of own shares on the stock market, the company buys its own shares anonymously at the stock market rate. In this case the company acts as enquirer at the stock market.[257]

In the case of purchase of own shares via a "self tender offer" the company suggests an offer to the shareholders to buy back own shares in a certain period. In contrast to the buyback of shares at the stock market the repurchase is not undertaken over a longer period of time, but is determined in time. The offer to the shareholders can have two different forms: in case of a "fixed price tender offer" the offer of buyback is made at a fixed price. Hereby, the buyback price and the period of time during which the repurchase shall be made are determined by the

---

252 Comp. Kallmeyer, 1999, p. 101

253 Comp. Engelsing, 2001, p. 93

254 Comp. Schmeisser/Hahn/Schindler, 2004, p. 78

255 Comp. Engelsing, 2001, p. 93

256 Comp. Schmeisser/Hahn/Schindler, 2004, p. 78

257 Comp. Engelsing, 2001, p. 91

company. The buyback price is composed of the stock quotation and a sales bonus. This motivates the shareholder to sell the shares within a determined period of time.[258]

In contrast to the "fixed price tender offer" the "Dutch auction tender offer" only determines a price margin for the repurchase. The shareholders make their offers for sales within this margin. The company then determines the lowest price ("market clearing price"), at which it wants to buy the desired number of shares back. If the number of sales offers is higher than the desired number of shares that shall be bought back, the individual sales orders are carried out proportionately.[259]

The third option consists in the possibility to purchase options from a majority shareholder. This type of repurchase is generally legally prohibited in Germany, because it offends the principle of equality in accordance with sec. 53a AktG.[260] In this case only the major shareholder receive an offer for the repurchase of own shares, while the other shareholders do not get this offer. This means that a small circle of shareholders is favored. The unequal treatment of the other shareholders, however, can not be justified.[261]

The deliberations of this section have shown that the purchase of own shares by the company is appropriate to cover shares for a stock-option program Furthermore, this type of financing does not lead to dilution of capital[262] nor of the voting rights of the existing shareholders.[263] This means that the repurchase of own shares to cover the stock-option program is an adequate method of coverage – depending on the situation of the company.

12.6 Cooperation with third parties

The financing methods described in the previous sections all have in common that the company acquires all the shares for the coverage on its own account at the capital market and makes them available afterwards. In the case of cooperation with third parties the company can cooperate with a credit institute or with a financing company belonging to the group.[264]

The company contractually agrees the grant of stock-options with a credit institute against one time payment of a fee. The company then hands the stock-options down to the

---

258 Comp. Rudolph, 2006, p. 463

259 Comp. Engelsing, 2001, p. 92

260 Comp. Rudolph, 2006, p. 463

261 Comp. Engelsing, 2001, p. 92

262 Friedrichsen is sceptical with regard to this practice, 2000, p. 232.

263 Comp. Kallmeyer, 1999, p. 101

264 Comp. Dietz, 2004, p. 69

participants of the stock-option program.[265] In the case of exercise of the stock-option by the participant of the program the credit institute has to make the required number of shares available that it had previously acquired at the stock market on its own account and in its own name.[266] There is no need of a resolution of the general assembly for the cooperation with third parties. The reason is that the structure of the company capital is not changed.[267]

The disadvantage of cooperation with third parties is that the company has to pay a fee to the credit institute. The amount of the fee depends on the type of procurement of the stock-option program.

Furthermore, it depends on the risks that the third party takes over by purchasing the shares.[268] The payment of the fee means that the company's liquidity is reduced. The fee must also be paid if the stock-option is not exercised because the determined conditions in the option contract were not fulfilled by the participant of the program.[269] The positive aspect is that the transfer of options does not imply any dilution of capital and voting rights of the present shareholders.[270]

12.7 Stock Appreciation Rights

If the participant of a stock-option program receives virtual options and the program's conditions are fulfilled, the manager can exercise the virtual option and receives a cash settlement. In contrast to the previously described financing method, no shares have to be procured in this case for covering the stock-option program.

The cash payment to be made to the participant of the stock-option program reduces directly and in full extent the operating results of the company. Indirectly this liquidity reduction means lower dividend payments for the shareholders and a smaller growth of the company. The virtual options can be basically freely designed. In the design of the SAR is has to be avoided that the company profit is exhausted. In order to avoid this, there is the possibility to put a cap for the maximum profit of exercise.[271]

---

265 Comp. Engelsing, 2001, p. 95

266 Comp. Pellens/Crassselt, 1998, p. 150 et seq.

267 Comp. Zitzewitz, 2003, p. 139

268 Comp. Engelsing, 2001, p. 96

269 Comp. Kallmeyer, 1999, p. 102 et seq.

270 Comp. Schmeisser/Hahn/Schindler, 2004, p. 82

271 Comp. Engelsing, 2001, p. 98 et seq.

13 Execution of a stock-option program in the company

In the previous chapter appropriate possibilities to cover stock-option programs were presented. Now, the question is, how and by whom the introduction of the stock-option program is carried out. It has to be clarified who has which competence in the stock-listed company and especially the responsibilities of executive board and supervisory board have to be examined. These questions are now looked into in more detail.

13.1 Decision on the introduction of a stock-option program

The first phase of the implementation of stock-option program for the management of a company commences with the decision to introduce it at all. In accordance with sec. 76 AktG the executive board manages the company in own responsibility. The introduction of a stock-option program as instrument of the implementation of the shareholder value concept means amongst other things to align the complete company strategy with this concept. This means the introduction of a stock-option program is at first a management decisions and has to be attributed to the area of responsibility of the executive board.[272] The question is whether the executive board can also decide on the introduction of the program when it is the beneficiary itself.

The members of the executive board are appointed by the supervisory board for five years in according to sec. 84 subs. 1 sentence 1 AktG. On the basis of the staff competence of the supervisory board it negotiates with the members of the executive board regarding their salary and decides on the corresponding agreements in the articles of employment. This means that only the supervisory board is responsible for the introduction of a stock-option program as remuneration for the executive board members. If the regulation of sec. 84 subs.1 sentence 5 AktG, 112 AktG did not exist, the executive board could determine the conditions of the option itself as representative of the company. The criticism of self service by the executive board, however, would be justified. For that reason it is not possible for the members of the executive board to have influence on the stock-option program and its detailed design.[273]

The general assembly does generally not take part in the decision regarding the introduction of a stock-option program. However it has a right of co-determination and decision

[272] Comp. Zitzewitz, 2003, p. 95

[273] Comp. Engelsing, 2001, p.39 et seq.

competences when the stock-option program is meant to be covered through a conditional or authorized capital increase or the repurchase of own shares.[274]

13.2 Preparation of the decisions of the general assembly

The second phase includes the preparation of the resolutions of the general assembly which are necessary for the introduction of the stock-option program.[275] Depending on the type of financial coverage of the stock-option program, different types of general assembly resolutions are necessary. In case of stock appreciation rights and when financing the program in cooperation with third parties no resolutions of the general assembly are needed for the introduction of the stock-option program into the company. The decision of the executive board is sufficient for implementation.[276]

In accordance with sec. 121 subs. 2 sentence 1 AktG the executive board is responsible for the convocation of the general assembly. Specialized literature considers this as part of the business management in the sense of sec. 77 AktG.[277] The executive board is also responsible for the preparation of the resolutions of the general assembly. Furthermore, its area of responsibility includes the preparation and circularization of the agenda (sec. 124 subs. 1 sentence 1 AktG). However, these tasks are handed down to the supervisory board and the general assembly. The general assembly has only certain rights to say in the determination of the conditions of the stock-option program in accordance with sec. 118 et seq. AktG.[278]

For the preparation of general assembly decisions by the supervisory board, the following resolutions are necessary for issuing a convertible bond or warrant:

[274] Comp. Holland, 2000, p. 41

[275] Comp. Klahold, 1999, p. 202

[276] Comp. Schmeisser/Hahn/Schindler, 2004, p. 33

[277] Comp. Zitzewitz, 2003, p. 95

[278] Comp. Schmeisser/Hahn/Schindler, 2004, p. 31 et seq.

- Decision on the execution of a conditional capital increase (sec. 192 subs. 2 n°1 AktG)
- resolution of approval or authorization of the issue of the bond (sec. 221 AktG),[279] which has to be presented to the general assembly (sec. 186 subs. 4 sentence 2 AktG).
- exclusion of prescription rights for the existing shareholders (sec. 221 subs. 4 in connection with sec. 186 subs. 3 and 4 AktG) as well as
- a written report on the reasons for the exclusion of prescription rights which has to be presented to the general assembly (sec. 186 subs. 4 sentence 2 AktG).

The issue of naked warrants via a conditional capital increase also makes a resolution of approval or authorization necessary in accordance with sec. 192 subs. 3 AktG.

In the case of sourcing the stock-option program with the purchase of own options in accordance with sec. 71 subs. 1 n° AktG the following resolutions have to be prepared:

- a resolution of authorization in accordance with sec. 71. subs. 1 n°. 8 AktG
- a corresponding exclusion of prescription rights in accordance with sec. 186 subs. 3, 4, 71 subs. 1 n°. 8 sentence 5 AktG as well as
- a report on the reason for exclusion of the prescription right (sec. 186 subs. 4 sentence 2 AktG).

Afterwards at least a three-quarter majority of the share capital when the resolution is made is necessary to put the resolution into action. The supervisory board and the executive board then announce the resolution for conditional increase in capital for entry in the commercial register (sec. 195 subs. 1 AktG). When convertible bonds or warrants are issued, the decree of the supervisory board and the executive board has to be entered into the commercial register (sec. 22 subs. 2 sentence 2 AktG). In case of a resolution for purchase of own shares the Federal Institute for the Supervision of Financial Services has to be informed (sec. 71 subs. 3 sentence 3 AktG).[280]

13.3 Grant of stock-options

After the necessary resolution has been procured, the next step is to grant the convertible options or warrants. One has to be pay attention whether the members of the executive board are also participants in the program. If employees receive convertible options or warrants, the issue is part of the executive board's responsibility. If the members of the executive board are beneficiaries of the program, the supervisory board is responsible to issue the convertible bonds (sec. 84 subs. 1 sentence 5 in connection with sec. 112 AktG).[281] In practice the

---

279 The authorization resolution gives the supervisory board the necessary freedom for the introduction and execution of a stock-option program when stock-options are granted to the executive board. In the case of a resolution of approval the executive board has the duty of execution. Comp. Kelle, 2002, p. 36

280 Comp. Zitzewitz, 2003, p. 95 et seq.

281 Comp. Schmeisser/Hahn/Schindler, 2004, p. 33

company would partially use a credit institute. The credit institute takes the convertible bonds over and offers them to the beneficiary for purchase.[282]

If pure naked warrants are given to the executive board, the supervisory board is also responsible. The grant of stock-options via conditional capital increase (sec. 192 subs. 2 n° 3 AktG) or the purchase of own shares (sec. 71 subs. N° 8 AktG) makes it necessary to conclude a new employment contract with the executive board or to change the old employment contract with the executive board.[283]

13.4 Assignment of shares

If the conditions for exercise as e.g. lock-up periods or performance targets for stock-options or bonds are fulfilled, they can be exercised by the participants of stock-option programs. In the case of exercise of real stock-options, the participants are provided with shares. For shares which are distributed via conditional capital increase, a declaration of subscription between the supervisory board and the participants in the stock-option program is made in accordance with sec. 198 AktG. The issue of shares has to be entered into the commercial register by the supervisory board within one month after the end of the business year (sec. 201 subs. 1 AktG).[284]

If shares are issued which are funded by the purchase of own shares in accordance with sec. 71 subs. 1 n° 8 AktG, the supervisory board has to give detailed information in the next general assembly. The information to the general assembly has to include the following corner points in accordance with sec. 71 subs. 3 AktG:

- reason and purpose of the purchase of shares,
- number of shares acquired
- the amount of basic capital used for the purchase of shares
- the percentage of the shares acquired in basic capital
- the corresponding value of shares.

If the participants of the stock-option programs receive virtual options, they do not receive shares, but an immediate cash settlement.

---

282 Comp. Zitzewitz, 2003, p. 96

283 Comp. ibid, p. 96, 99

284 Comp. Friedrichsen, 2000, p. 225

14 Accountancy of stock-option programs

There has been much controversy for a long time whether and how to balance stock-option programs. In 2002 there were prominent advocates, as the investment guru Warren Buffett and the ex-head of US central bank Alan Greenspran, who were in favor of including stock-option programs as expenses in the balance sheet. Especially technology companies in the US had vehemently refused to include their programs in the balance sheets. However, these were exactly the companies that had introduced especially expensive stock-option programs. The inclusion of stock-option programs as expense would have lead to an important reduction of the company profits in their cases.[285]

In the following section relevant accountancy rules for German companies for share-based remuneration are shortly described. In Germany capital market-oriented companies with their head office in the European Union have to design their consolidated account in accordance with IFRS.[286]

14.1 Accounting under IFRS

The popularity of stock-option programs was formerly based on the fact that stock-option programs were not recorded as personnel expenditure. This was possible for companies that prepared their balance sheet in accordance with IFRS, because the regulation IAS 19 "payments to employees" does not impose principles on recognition and valuation.[287]
With the introduction of IFRS 2 (share-based payment) in the year 2004 the balancing requirements for stock-option programs became clearer. IFRS came into effect on 1st January 2005. For stock-option based remunerations which were or are paid after the 7 November 2002 and for which the lock-up period has not yet expired, the new regulations are valid, too.[288]

14.1.1 Field of application

IFRS 2 regulates the accounting treatment of remuneration of employees and managers by means of shares and stock-options. The field of application of the standards, however, is not only limited to the remuneration of employee remuneration. The standard regulates all transactions in which company proportions (e.g. shares) or options as remuneration for

285 Comp. Schubert, 2002

286 Comp. Kramarsch, 2004, p. 94 et seq.

287 Comp. Pellens, 2008, p. 502

288 Comp. IFRS 2.53

goods, work or other services are granted. Another example would be the one of a business consultant who receives stock-options of the company instead of monetary payment.[289]

Moreover, IFRS has to be applied in the case of share-based remuneration, when the equity instruments are not created by the company itself, but by the parent company in case of affiliated companies. It has to be taken into consideration that the standard may not be used for transaction for which the accounting is expressly regulated in other regulations of the IFRS rules.[290]

The standard distinguishes between the following types of share-based remunerations:

- equity-settled share-based payment transactions, like shares or stock-options
- cash-settled share-based payment transactions, e.g. stock appreciation rights
- share based remunerations with a choice between equity settlement or cash settlement (combination model).[291]

It can be understood from the above deliberations that the standard distinguishes between real and virtual equity settlement. This differentiation is necessary in order to recognize whether the beneficiaries have an equity or foreign capital claim against the company.

The following table gives an overview over the different forms of design of share-based remuneration.

	Character of share	**Character of option**
Real equity settlement	Restricted Stocks	Stock-options
Virtual equity settlement	Phantom Shares	Stock Appreciation Rights

Table 14.1: share-based remunerations[292]

The following deliberations focus the accounting of share based remuneration of employees and managers of the company.

289 Comp. Pellens, 2008, p. 502

290 Comp. IFRS 2.5–2.6

291 Comp. IFRS 2.2

292 Following: Pellens, 2008, p. 503

14.1.2 Real equity settlement

Approach

The assessment of a share-based remuneration which is settled with equity instruments (*equity-settled share-based payment transactions*) is undertaken at the fair value which is determined at the moment of grant.[293] This type of accounting is valid for shares as well as stock-options which are sourced via conditional capital increase as well as via repurchase of own shares.[294]

The problem with accountancy of employee remuneration with real equity instruments is that the grant of shares or stock-options is a type of work remuneration. Work – independent of whether it is already performed or not – may not be included in the balance sheet. The problem with that impossibility to include work into the balance sheet is solved via an expense entry. If the employee remuneration is solved via stock and if the employees have to carry out an additional own investment for the shares, the following entry has to be made:[295]

personnel expense debited	*capital stock credited*
cash debited	*capital reserve credited*

The recognition of stock-options in the balance sheet on the other hand is more complex because the moment of granting the stock-option as well as the moment of exercising it have to be recognized. The corresponding journal entry in the case of granting a stock-option is as follows:

Personnel expense debited	*capital reserves credited*

If the stock-options are exercised after the lock-up period through the employees, the one-time payments by the beneficiaries for the stock-option have to be debited on cash.

Cash debited	*capital stock credited*
	capital reserve credited

The exercising profit which the beneficiary makes by exercising his stock-option does not lead to a retroactive adaption of the capital expense.[296]

The previously stated thoughts include the assumption that the work has already been performed. This is based on the assumption that the stock and stock-options are freely disposable and exercisable and do not depend on an employment relationship in the

293 Comp. IFRS 2.10, status 24.02.2005

294 Comp. IFRS 2 BC331

295 Comp. Pellens, 2008, p. 508

296 Comp. Pellens, 2008, p. 509

company.[297] This is mostly not the case in stock-option programs. In practice, stock-options are only granted, if there is an employment relationship. It may therefore be assumed that stock-options which are given out under this precondition, correspond to future work performances. The expenses must therefore be evenly distributed in the personnel expenses with effect on the net income over the period of service. In accordance with IFRS 2.15 the end of the performance period corresponds to the moment from which on the stock-options are freely disposable and exercisable. In the case of stock-option programs this period corresponds usually to the lock-up period.[298] If the stock-option program is based on performance targets, the period of performance is estimated on the day of grant.[299] If capital market related performance targets (*market conditions*) are set, the estimated period of performance may not be corrected at a later point in time. If non-capital market related performance targets (*performance conditions*) are set, the performance period and the personnel expenses to be entered have to be adapted to the actual duration up to the fulfillment of the targets, if necessary in each period.[300]

Assessment

Real equity capital instruments used as remuneration have to be recognized with the fair value of the services received by the company. The fair value of the named equity instruments (stocks and stock-options) are generally calculated from the current market price.[301]

The use of current market prices is unproblematic for companies that give out employee stock. In contrast, in the case of stock-options the use of market prices is usually not possible. This is due to the specific design of options, like the longer run-time and the possibility of an early exercise of the stock-option before the end of the run-time. The stock-options distributed to the managers are therefore very different to the stock-options traded on the stock market. For this reason a generally recognized assessment method can be used for the calculation of stock-options:[302]

In the assessment of stock-options it has to be differentiated between the intrinsic value and the fair value . The intrinsic value is determined by the value by which the current stock quotation of the share surpasses the subscription price. Therefore, it demonstrates the possible profit for the case of immediate exercise. The fair value includes besides the inner value also the time value. The time value takes the positive as well as the negative price developments into account. This means that the time value is positive at any moment before

---

297 Comp. IFRS 2.14

298 Comp. Pellens, 2008, p. 509

299 Comp. IFRS 2.15b

300 Comp. Coenenberg, p. 340

301 Comp. IFRS 2.16

302 Comp. Buschhüter/Striegel, 2009, p. 260 et seq.

the end of the run-time. This means, the stock-option has a positive value even if the intrinsic value is zero.[303]

IFRS name no specific option price model for the determination of the fair value. The reason is that possible developments of option price models should be considered in the option price assessment. Some models are admissible which take the following factors into consideration:[304]

- exercising price of the stock-option,
- run time of the option,
- current market price of underlying stock,
- the expected volatility of the stock,
- the expected return on dividends,
- and the risk free interest factor for the run-time of the stock-option.

IFRS 2 name the Black-Scholes model as example for the assessment of stock-options.[305] The advantage of the Black-Scholes model is its simple application. The disadvantage is that it can only be applied on stock-options which are not containing complex conditions (e.g. lock-up periods, exercising windows). An example is that the Black-Scholes model can not display stock-options which are exercised before the end of the run-time. Furthermore, the expected volatility is determined via a fixed value and takes not into consideration that the volatility can change in the course of the run-time of the stock-option. But this model can still deliver a sound value, if the stock-option program contains short contractual periods or short exercising periods.[306]

A further evaluation method to be taken into consideration for the assessment of stock-options is the binomial options pricing model by Cox, Ross and Rubinstein. The model contains the same calculation factors as the Black-Scholes model, but is more flexible and can be applied on almost all types of share and stock-option programs.[307]

To the contrary of the Black-Scholes model the binomial options pricing model can deal with changes of the assessment parameters during the run-time of the stock-option. The binomial model assumes that the stock quotation can have two different variations "in money" or "out of money" within a certain period of time.

By dividing the run-time of the stock-option in the shortest possible periods, the real distribution of the price development of the stock can be displayed via the binomial model.

---

[303] Comp. Pellens, 2008, p. 512

[304] Comp. IFRS 2.B6

[305] Comp. IFRS 2.B5

[306] Comp. Buschhüter/Striegel, 2009, p. 262

[307] Comp. Kramarsch, 2004, p. 211

The disadvantage is that the complexity of the model increases the shorter the examined periods become.[308]

Furthermore, simulation procedures as the Monte-Carlo simulation can be used for the evaluation of the option price. In this procedure a multitude of random paths is generated which represent the conduct of the relevant factors of influence. On these random paths the discounted value of the payment is calculated. The average of the discounted values equals the value of the stock-option as well as the random error. This random error can be diminished with a high number of simulations.[309]

Furthermore the IFRS 2 contains regulations on the record of performance targets. While in Germany under sec. 193 AktG success targets are obligatory for stock-option plans which are financed via a conditional increase in capital, the IFRS 2 divides them in market conditions and performance conditions.[310] Market conditions are stock market oriented targets , as the absolute increase of the share price or outperformance of a peergroup. Performance conditions on the other hand are key figure based targets as the increase of profit per share by 10 percent etc.

Only the market conditions are used for the determination of the fair value in the evaluation of the option. The expense remains the same in case of not achieving the aim. Performance conditions are not taken into consideration in the option assessment, but are considered in the determination of the number of shares which are likely to be exercised. In the case of not reaching the performance conditions, no personnel expense results and the accruals can be dissolved in cash in case of an application.[311] This type of assessment applies also to a combination of market and performance conditions.[312]

In exceptional cases it is possible to deviate from the fair value assessment. The stock-options are then assessed at their intrinsic value. One may recur to this type of evaluation, if it is not possible for the company to determine the fair value in a reliable way. The assessment of stock-option at their intrinsic value at the day of grant of the stock-option is not final. Each change of intrinsic value has to be considered in each reporting period up to the end of the run-time of the stock-option. It is not possible to change from the assessment at the intrinsic value to the assessment at the fair value.[313]

---

[308] Comp. Buschhüter/Striegel, 2009, p. 262

[309] Comp. Kramarsch, 2004, p. 211

[310] Comp. IFRS 2.15b

[311] Comp. IFRS 2.19

[312] Comp. Kramarsch, 2004, p. 102

[313] Comp. Pellens, 2008, p. 513

14.1.3 Virtual own capital instruments

Approach
The accounting of virtual equity instruments deviates in very few points from real equity instruments. The accounting of virtual stock-options, however, is less time consuming, because the equity is not affected in this case.[314]

Virtual equity instruments, as e.g. stock appreciation rights are recorded just like real equity instruments in the personnel expenses. In contrast to the real equity instruments no capital accrual is made as counter-entry, but a liability in form of an accrual is made.[315]

At the moment when the work of the employee is fully done, the company has to record the personnel expense and the accrual. Some value increasing rights are e.g. exercisable at once, with the effect that at this moment the services by the employee are fully rendered. Therefore, the stock appreciation rights have to be recorded at the moment of grant to full extent.[316]

If the services of the employees have not been completed, the personnel expense and the accruals have to be recorded otherwise. The services by the employee are generally not completed when certain exercising conditions, as e.g. lock-up periods were agreed on in the stock-option program. In this case, the personnel expense has to be assessed and recorded anew at any accounting date during the lock-up period. This type of assessment is carried out up to the moment when the services of the employee are completed.[317]

Assessment
Share-based remuneration instruments, which are paid in cash, as e.g. stock appreciation rights, are also assessed at the fair value. Like in the case of real stock-options the assessment of the virtual stock-options has to be carried out with the help of appropriate assessment models. In contrast to the real stock-options the first time assessment is not carried out at the moment of grant. The evaluation has to be updated at each accounting date during the run-time.[318] This means that personnel expenses and accruals have to be adapted with effect to net income to the changes in value.

---

[314] Comp. ibid, p. 513

[315] Comp. IFRS 2.7

[316] Comp. IFRS 2.32

[317] Comp. IFRS 2.32, 2.33

[318] Comp. Pellens, 2008, p. 518 et seq.

The corresponding accounting records at the end of the corresponding reporting period are formulated as follows:[319]

Personnel expense debited *accruals credited*

Through the regular adaptation of the assessment of virtual options, the exercising profit of the beneficiary affects net income in contrast to real stock-option programs. The exercising profit can be higher or lower depending on the situation of the stock markets. This induces a high volatility of the personnel expenses.[320]

14.1.4 Combination models

In practice often share-based remuneration models can be found in which real and virtual equity instruments are combined. In the case of share-based remunerations, in which the company or the beneficiaries have to chose between cash settlement and the exercise of the equity instruments, there is a difference depending on who has the choice of settlement.[321]

If the company has the right to choose the settlement type, it has to be examined whether there is a payment duty. A payment duty arises when corporate limitations restrict the distribution of equity capital instruments (e.g. there is a legal prohibition for the company to give out shares) or if the company often chose the cash settlement in the past.[322] Because of the named reason one can presume that the company will also chose the settlement type of cash payment in future.

Depending on underlying payment duties, the combination models have to be recorded as follows:[323]

If there is a payment duty, the balancing has to be carried out in accordance with the regulations for virtual equity instruments. If, later, equity instruments are distributed anyway, the accrual has to be assessed anew at the moment of exercise. The value has to be recorded in the equity without effect on net income.

If no payment duty is to be expected, the expense has to be recorded under the balancing regulations for real equity instruments.[324] One can distinguish between the following recording possibilities:[325]

---

[319] Comp. ibid, p. 520

[320] Comp. Pellens, 2008, p. 520

[321] Comp. ibid, p. 522

[322] Comp. ibid, p. 522

[323] Comp. ibid, p. 522

1. The company chooses cash settlement: The cash settlement has to be recorded like the buyback of real equity instruments in this case; it has to be deducted from equity without effect on net income.
2. The company chooses settlement via equity instruments. No further record in equity is necessary.

Furthermore, there are combination models in which the beneficiary (employee) has the right to choose the settlement type. This is a so called compound financial instrument, composed off an equity and a foreign capital component.[326] Because its existence is rather seldom in practice it will not be examined in more detail.

14.1.5 Disclosures

IFRS 2 contains extensive regulations on the disclosure of share-based remuneration. With the requirements in IFRS 2, it shall be made possible for those interested in the annual accountancy to comprehend information on kind and extent of share-based remuneration and to assess it. Furthermore, information on the assessment and the effects of share-based remuneration on the financial circumstances and revenue situation have to be published.[327]

These three principles contain minimum standards which have to be fulfilled by the companies. With regard to the information on type and extent of relevant transactions the following information has to be given under IFRS 2.445:

1. A presentation of all share-based remuneration systems used in the recording period. All important parameters of each program have to be displayed, as e.g. exercising conditions, lock-up period and run-time.
2. The number and the weighted average exercise price for all real stock-options:
   - stock-options outstanding at the start of the recording period,
   - stock-options granted in the reporting period,
   - stock-options forfeited in the recording period, because of unfulfilled exercising conditions,
   - stock-options exercised in the reporting period,
   - stock-options forfeited in the recording period,
   - stock-options outstanding at the end of the reporting period as well as
   - exercisable stock-options at the end of the reporting period.

324 Comp. IFRS 2.41 bis 2.43

325 Comp. IFRS 2.43

326 Comp. IFRS 2.35

327 Comp. Pellens, 2008, p. 522

3. Furthermore, information has to be given on all stock-options which have been exercised in the period, the average stock quotation at the day of exercise and the weighed average price of the recording period in the case of regularly exercised options.
4. The fluctuation margin in exercise prices and the contract run-time for the end of outstanding stock-options have to be equally published.

With regard to the assessment of share-based remuneration, the following information has to be given under IFRS 2.47 to 2.49.

1. The weighted average fair value for the stock-options granted in the recording period on the record date as well as information on the option price model including the following influencing factors:
   - the average stock quotation, exercise price, expected volatility, run-time of the option, expected dividends, risk-free interest rate, as well as further assumptions included in the option model;
   - a description on how the expected volatility was determined;
   - information on whether and how other factors (e.g. market conditions) were included in the calculation of the fair value.
2. For other share-based remuneration instruments (other than stock-options) the number and the average fair value at the record date have to be published as well as information on how the fair value was calculated.
3. In the case of changes of the conditions in the granted share-based remuneration instruments, information has to be given why these changes were made and information on the changed fair value and its determination has to be given.

With relation to the third principle the following information has to be published:

4. A display of the effects of the distribution of share-based remuneration on the income statement as well as on the capital and finance situation of the company.[328] In this context the following information has to be given:[329]
   - the total expenses recorded which have arisen in the recording period because of share-based remuneration
   - the total expenses which were recorded as accruals at the end of the reporting period because of share-based remuneration
   - in case of accruals which have arisen from share-based remuneration, the intrinsic value of the accrual at the end of the period.

The companies may give further details in the annual balance sheet when the three named principles are not fulfilled with the minimum requirements.

---

[328] Comp. IFRS 2.50

[329] Comp. IFRS 2.51

14.2 Balancing in accordance with US-GAAP

Since 1948 there has been detailed regulations on balancing of stock-options in the US. The basis for balancing stock-options under US-GAAP was provided by the regulations of the APB 25[330] up to the year 2004 and by the newer regulations of the SFAS 123[331]. For a long time, there had been the right to choose between both regulations. This choice included the possibility to record stock-options either under APB 25 or under SFAS 123. It was possible to change from APB 25 to SFAS 123 at any time. If, however, the stock-options had been recorded under SFAS 123, it was not possible anymore to return to APB 25.

With the upcoming of different company crises (e.g. Enron) the assessment method under APB 25 was more and more criticized,[332] with the result that in December 2004 a renewed version of the SFAS 123 Share-Based Payment was published by the FASB. This new version was referred to as SFAS 123(R). One of the most important new regulations of SFAS 123(R) was that the choice for all stock-options plans which were introduced after the 15 June 2005 was abolished. Companies which had started to record under APB 25 at this moment were not affected by this regulation. Now the regulations of the SFAS 123(R) are examined in more details:[333]

14.3 Regulations under SFAS 123(R)

SFAS 123 "Share-Based Payment" regulates the balancing of stock-options and similar rights. With the renewal of SFAS 123 into SFAS(R) the possibility to record new stock-option programs under APB 25 was abolished. The only method for assessment of stock-options and similar values is now based on the fair value.[334] The FASB gives the following reasons for this:[335]

1. Continuous criticism of the choice between balancing in accordance with APB 25 and SFAS 123.
2. By the elimination of choices the comparison of annual balance sheets is simplified.
3. The US-GAAP regulations are simplified by the abolition of parallel regulations.

---

330 Comp. Accounting Principles Board, Accounting Principles Board Opinion No. 25: Accounting for Stock Issued to Employees, 1972.

331 Comp. Statement of Financial Accounting Standards No. 123, published by FASB, 1995.

332 It was critizised that the APB determined only the intrinsic value of the option. This had generally the result, that the base price was set at the current stock quotation which meant that the intrinsic value often had the unrealistic value zero.

333 Comp. Coenenberg, 2005, p. 346

334 Comp. KPMG, 2007, p. 283

335 Comp. Coenenberg, 2005, p. 349

4. Many analogies between SFAS 123(R) and IFRS 2 and comparability between company balance sheets under IFRS and US-GAAP.

The SFAS 123(R) was developed in close collaboration with the IASB. The IASB published the standard IFRS (which was described above) almost at the same moment.[336]

Area of application

SFAS 123(R) regulates the accountancy for all variants of share-based remunerations which are spent by a company for received goods and services. This means that also non-employed parties as suppliers are included in the area of application. In practice, however, share-based remuneration paid to suppliers plays only a minor role.[337] For this reason the standards focuse share-based remuneration granted to employees.

Approach

SFAS 123 (R) distinguishes between two variants of share-based remuneration: equity classified awards and liability classified awards. The separation of share-based remuneration in two different types is crucial because there are important differences in the recognition of both forms in the balance sheet.[338]

Share-based remuneration which provide for a cash settlement, as e.g. stock appreciation rights are classified as liability classified awards. Share-based remuneration which do not provide for cash settlement as shares or real stock-options are classified as equity classified awards.[339]

If the company has the right to choose a form of settlement, i.e. if the exercise of the share-based remuneration is settled either in cash or in company shares, the intent of the company is decisive. If the company had already chosen the cash settlement in the past, one may assume that this type of settlement will be chosen in future. The result is that the company has to classify the chosen share-based remuneration as liability classified award.[340]

Assessment

Under SFAS 123(R) all transactions, in which the company has granted share-based remuneration as e.g. stock-options and similar rights for received goods and services to the employees, have to the assessed at fair value.[341] The Fair Value of share-based remuneration

---

336 Comp. KPMG, 2007, p. 283

337 Comp. ibid, p. 283

338 Comp. ibid, p. 284

339 Comp. ibid, p. 285 et seq.

340 Comp. KPMG, 2007, p. 285

341 Comp. SFAS 123(R). par. 7

has to be calculated with an accepted assessment model like for example the Black-Scholes-model or the binomial model.

The value of the share-based remuneration is to be assessed at the moment when the employees can exercise their stock-options or similar rights for the first time. This moment, however, lies in the future, with the result that the fair value has to be estimated on the grant day.[342]

The following parameters are integrated into the assessment model:[343]

- subscription price of the shares,
- expected volatility of the share,
- expected dividend of the share as well as
- risk free interest rate for the expected run-time of the stock-options.

Further changes of the parameters, which lead to a change of fair value, are not taken into consideration in equity classified awards.[344] The distribution of the fair value has to be linear over the lock-up period and the fair value has to be recorded as personnel expense and capital accrual. The following accounting record applies:[345]

Personnel expenses debited *capital accruals credited*

In case of liability classified awards, the assessment of the fair value, as in the case of equity classified awards, is made at the moment of grant. In contrast, the fair value has to be determined anew on each accounting date and to be recorded in the personnel expenses. The following accounting record applies in this case:[346]

Personnel expenses debited *accruals credited*

The recorded accrual at the moment of exercise of the share-based remuneration has to correspond to the amount which the company has to pay to the employee in form of cash settlement. In contrast to the equity classified awards, this form of assessment leads to higher income volatility.[347]

342 Comp. SFAS 123(R). par. 16

343 Comp. Dietz, 2004, p. 29

344 Comp. KPMG, 2007, p. 287

345 Comp. SFAS 123(R). par. 39

346 Comp. KPMG, 2007, p. 292

347 Comp. KPMG, 2007, p. 292 et seq.

Disclosures

For companies that balance their share-based remuneration in accordance with US-GAAP, the disclosure regulations are set down in SFAS 123(R) par. A240 and A241. These new regulations replace the regulations under SFAS 123, par. 45-48.

In the annex a description of the share-based remuneration and the most important conditions as lock-up period, performance targets, run-time etc. has to be made.[348]

Furthermore, the following detailed information on share-based remuneration has to be provided in accordance with 123(R). par. A240:

1. For each business year information on the number and the average exercise price of options has to be provided:
   - stock-options outstanding at the beginning of the business year,
   - stock-options outstanding at the end of the business year,
   - stock-options exercisable at the end of the business year
   - stock-options granted, exercised, forfeited of expired during the business year.
2. The average fair value at the moment of grant of the equity instruments has to be provided separately for each of the following groups:
   - not invested equity instruments at the beginning and the end of the year as well as
   - equity instruments granted, acquired or forfeited during the business year.
3. The average fair value at the moment of grant for all real stock-options granted in this year and other equity instruments.
4. A description of the assessment method for the evaluation of the fair value with the following parameters:
   - risk-free interest rate,
   - expected volatility of the share,
   - run-time of the stock-options as well as
   - expected payment of dividends.
5. The total expenses for remuneration of the business year in question for the recorded personnel expenses have to be displayed in the income statement.
6. Important modifications of existing option plans.

If a company has introduced several stock-option plans, the information has to be given in the group or annual consolidated financial statement or annual balance sheet separately for each stock-option program.

[348] Comp. SFAS 123(R). par. A240

Literature

Achleitner, A.-K. / **Wichels**, D.: Stock-Option-Pläne als Vergütungsbestandteil wertorientierter Entlohnungssysteme – Eine Einführung. in: Achleitner, A.-K. / Wollmert, P. (ed.): Stock-options: Finanzwirtschaft, Gesellschaftsrecht, Bilanzierung, Steuerrecht, Arbeitsrecht und Unternehmensbewertung. 2nd rev. and compl. edition, Stuttgart: Schäffer-Poeschel, 2002, p. 1–24.

Baums, Th.: Aktienoptionen für Vorstandsmitglieder, in: Martens, K.-P. / Westermann, H. P. / Zöllner, W. (ed.), Festschrift für Carsten Peter Claussen zum 70. Geburtstag, Cologne, 1997, p. 3–48.

Bieker, Chr. / **Haslauer**, A.: Aktienoptionen: Millionengagen ohne Risiko. in: Focus Online dated 05 October 2005, 2005, URL:
http://www.focus.de/finanzen/boerse/aktien/aktienoptionen-millionengagen-ohne-risiko_aid_255807.html
status: 28 June 2009.

Boecken W. / **Düwell**, F. J. / **Hümmerich**, K. [Arbeitsrecht]: Anwalt Kommentar: Arbeitsrecht. Band 1, 1st edition, Bonn: Deutscher Anwaltverlag, 2007.

Bösl, K. [Börsengangs]: Praxis des Börsengangs – Ein Leitfaden für mittelständische Unternehmen. 1st edition, Wiesbaden: Gabler, 2004.

Brühl, R.: Controlling – Grundlagen des Erfolgscontrollings. Munich: Oldenbourg Wissenschaftsverlag, 2004.

Buschhüter, M. / **Striegel**, A. (ed.): Internationale Rechnungslegung – IFRS Praxis. 1st edition, Wiesbaden: Gabler, 2009.

Carl, P. St.: Corporate Governance. in: Spahlinger, A. / Wegen, G. (ed.): Internationales Gesellschaftsrecht in der Praxis Kollisions- und Sachrecht wesentlicher Fälle mit Auslandsberührung, Europäisches Unternehmensrecht Wahl der Gesellschaftsform, Corporate Governance, Wichtige ausländische Rechtsformen, Munich: Beck, 2005, mn. 1623-1649.

Coenenberg, A. G.: Jahresabschluss- und Jahresabschlussanalyse: Betriebswirtschaftliche, handelsrechtliche, steuerrechtliche und internationale Grundsätze – HGB, IFRS und US-GAAP. 20th rev. and compl. edition, Stuttgart: Schäffer-Poeschel, 2005.

Christie, R. / **Homan**, T. R.: Geithner Says Government Shouldn't Set Caps on Pay. in: Bloomberg Online dated 18 May 2009, 2009, URL:
http://www.bloomberg.com/apps/news?pid=20601087&sid=aku QmELeU6hk
status: 01 June 2009.

DeFusco, R. A. / **Zorn**, Th. p. / **Johnson**, R. R.: The association between executive stock-option plan changes and managerial decision making, in: Financial Management, Vol. 20, 1991, p. 36–43.

Dietz, St.: Aktienoptionsprogramme und Corporate Governance: Ausgestaltung vergütungshalber gewährter Stock-options und die Berichterstattung in der externen Rechnungslegung, 1st edition, Wiesbaden: Gabler, 2004.

DCGK – Deutscher Corporate Governance Kodex: Deutscher Corporate Governance Kodex. In: corporate-governance-code dated 06 June 2008, 2008, URL:

http://www.corporate-governance-code.de/ger/kodex/index.html
status: 14 June 2009.

DCGK – Deutscher Corporate Governance Kodex [, 2009]: Deutscher Corporate Governance Kodex. In: corporate-governance-code dated 19 June 2009, 2009, URL: http://www.corporate-governance-code.de/ger/news/index.html
status: 30 June 2009.

Eglau, H.-O.: Manager und Märkte Sahnehäubchen aufs Gehalt: Mit Aktienoptionen können Manager ihr Einkommen erheblich verbessern. in: Zeit Online dated 02/1997, 1997, URL:
http://www.zeit.de/1997/02/Manager_und_Maerkte_Sahnehaeubchen_aufs_Gehalt_Mit_Aktienoptionen?page=2
status: 28 June 2009.

Einem, Chr. v. / **Panjunk**, A.: Zivil- und gesellschaftsrechtliche Anforderungen an die Ausgestaltung von Stock-options nach deutschem Recht und deren Ausstrahlungswirkung auf die steuerliche und bilanzielle Behandlung. in: Achleitner, A.-K. / Wollmert, P. (ed.): Stock-options: Finanzwirtschaft, Gesellschaftsrecht, Bilanzierung, Steuerrecht, Arbeitsrecht und Unternehmensbewertung, 2. rev. and compl. edition, Stuttgart: Schäffer-Poeschel, 2002, p. 85–117.

Engelsing, L.: Aktienoptionen als Vergütungsbestandteil der Führungskräfte börsennotierter Unternehmen: Analyse der handels- und steuerrechtlichen Implikationen. Band 198, Cologne: Josef Eul Verlag, 2001.

Filbert, D. / **Kramarsch**, M. H. [DRS 17, 2008]: Perspectives On Management Pay: DRS 17 – Neue Vorgaben zum Ausweis der Vergütung von Organmitgliedern. in: Towers Perrin dated June 2008, 2008, URL:
http://www.towersperrin.com/tp/getwebcachedoc?webc=DEU/2008/200808/DSR_final.pdf
status: 17.062009.

Financial Accounting statusards Board (FASB) US-GAAP: Statement of Financial Accounting statusards (SFAS) No. 123 Revised: Accounting for Stock-Based Compensation, 2004. In: FASB Online dated December 2004, 2004, URL:
http://www.fasb.org/jsp/FASB/Pronouncement_C/StatusPage &cid=900000009743
status: 15 June 2009.

Fröndhoff, B.: Radikalkur oder Shareholder Value, in: Handelsblatt Online dated 28 April 2009, 2009, URL:
http://www.handelsblatt.com/politik/chroniken/ radikalkur-und-shareholder-value; 2 232907
status: 28 June 2009.

Friedrichsen, S.: Aktienoptionsprogramme für Führungskräfte – Gesellschaftsrecht – Kapitalmarktrecht – Steuerrecht – Bilanzrecht – (zugl.: Hamburg, Univ., Diss., 1999). in: Hueck, G. et al. (ed.): Abhandlungen zum deutschen und europäischen Handels- und Wirtschaftsrecht. Band 120, Cologne: Heymanns, 2000.

Geschäftsberichte der DAX-30 Unternehmen aus 2005, 2006, 2007 und 2008.

Grattenthaler, T.: Die Vergütung von Vorstandsmitgliedern in Aktiengesellschaften: Eine grundlegende Betrachtung mit ausgesuchten Schwerpunkten aus rechtswissenschaftlicher Sicht (zugl. Göttingen, Univ., Diss., 2006). in: Costede, J. / Spindler, G. (ed.):

Schriften zum Wirtschafts- und Medienrecht, Steuerrecht und Zivilprozessrecht, Volume 25, Frankfurt on Main et al.: Lang, 2007.

Heise, S. (Flowtex): Firmenpleiten: Von Peanuts, Medienmogulen und Zockern. in: Focus Online dated 25.03.2009, 2009, URL: http://www.focus.de/finanzen/news/tid-13685/firmenpleiten-von-peanuts-medienmogulen-und-zockern_aid_381348.html status: 30 June 2009.

Herrling, E. / **Schmidt**, M. / **Zahner**, D.: Bankwirtschaft – edition Baden-Württemberg. 2nd edition, Troisdorf: Bildungsverlag Eins, 2008.

Heitzer, B. / **Klose**, F. / **Steiner**, M.: Wertorientierte Gestaltung von Aktienoptionsprogrammen für das Management. in: Finanz Betrieb, 2nd ed., volume 6, 2000, p. 345–355.

Holland, S.: Aktienoptionspläne – Eine finanzwirtschaftliche Analyse der Auswirkungen auf Manager und Kapitalgeber (zugl.: Ulm, Univ., Diss., 1999). in: Europäische Hochschulschriften, series 5, Volks- und Betriebswirtschaft, Volume 2608, Frankfurt on Main et al.: Lang, 2000.

IASB: International Financial Reporting statusards (IFRS) dated 24.02.2005, 2005.

Kau, W. / **Leverenz**, N.: Mitarbeiterbeteiligung und leistungsgerechte Vergütung durch Aktien-Options-Pläne. in: BB, Vol. 1998, p. 2269–2276.

Kallmeyer, H.: Aktienoptionspläne für Führungskräfte im Konzern. in: AG, 44. Vol., 1999, p. 97–103.

Kelle, H.: Die Bilanzierung von Stock-options (zugl.: Regensburg, Univ., Diss., 2001). Düsseldorf: IDW Verlag, 2002.

Kessler, M. H. / **Sauter**, Th.: Handbuch Stock-options: rechtliche, steuerliche und bilanzielle Darstellung von Stock-Option-Plänen. 1st edition, Munich: Beck, 2003.

Klahold, Chr.: Aktienoptionen als Vergütungselement (zugl. Münster (Westfalen), Univ., Diss., 1999). in: Europäische Hochschulschriften: series 2, Rechtswissenschaft, Volume 2705, Frankfurt on Main et al.: Lang, 1999.

Klemund, M.: Stock-options Plans – Ein Anreizsystem zur langfristigen Steigerung des Shareholder Value. in Heidorn, Th. V. / Cremers, H. / Moormann, J. (ed.): Diskussionsbeiträge zur Bankbetriebslehre, Volume 12. Frankfurt on Main: Bankakademie-Verlag, 1999.

Köster, K.: Weniger Boni bei Daimler – Zetsche verdient die Hälfte. In: Stuttgarter Zeitung Online dated 27.02.2009, 2009, URL: https://www.stuttgarter-zeitung.de/stz/page/1959915_0_9223_-weniger-boni-bei-daimler-zetsche-verdient-die-haelfte.html status: 28 June 2009.

KPMG (ed.): US-GAAP – Rechungslegung nach US-amerikanischen Grundsätzen, Grundlagen der US-GAAP und SEC-Vorschriften. 4th edition, Düsseldorf: IDW Verlag, 2007.

Kramarsch, M. H.: Aktienbasierte Managementvergütung, 2nd edition, Stuttgart: Schäffer-Poeschel, 2004.

Kramarsch, M. H. / **Schmelter**, St.: Perspectives On Management Pay: Restricted Stock – Fluch oder Segen. In: Towers Perrin dated July 2006, 2006, URL:

http://www.towersperrin.com/tp/getwebcachedoc?webc=HRS/DEU/2006/200607/PerspJuly06.pdf
status: 18 May 2009.

Kramarsch, M. H. / **Becker**, F. G.: Leistungs- und erfolgsorientierte Vergütung für Führungskräfte, Volume 11, Göttingen: Hogrefe, 2006.

Lambert, R. A. / **Lanen**, W.N. / **Larcker**, D. F.: Executive stock-option plans and corporate dividend policy. In: Journal of Financial and Quantitative Analysis, Vol. 24, 1989, p. 409–425.

Laux, H.: Unternehmensrechnung, Anreiz und Kontrolle – Die Messung, Zurechnung und Steuerung des Erfolges als Grundprobleme der Betriebswirtschaftslehre. 2., rev. and compl. edition, Berlin et al.: Springer, 1999.

Lazar, Chr.: Managementvergütung, Corporate Governance und Unternehmensperformance: Eine modelltheoretische und empirische Untersuchung (zugl.: Leipzig, HHL, Diss., 2006). 1st edition, Wiesbaden: Gabler, 2007.

Meiffert, M. T.: Strategische Personalentwicklung – Ein Programm in acht Etappen. 1st edition, Berlin: Springer, 2008.

Mußler, H.: Altana-Optionen: Dreistelliger Millionenverlust über Nacht. In: FAZ dated 01 June 2007, 2007, URL:
http://www.faz.net/s/Rub3D6040857A734996B355094EA04C76A0/Doc~ECDB8C34127834073896C325BBC549ED4~ATpl~Ecommon~Scontent.html
status: 30 June 2009.

Mußler, H.: Sonderausschüttung: Bittere Pille für Altana-Altaktionäre. in: FAZ dated 09 January 2007, 2007, URL:
http://www.faz.net/s/RubF3F7C1F630AE4F8D8326AC2A80BDBBDE/Doc~E42B017C7E88F44BD806B940C2848140E~ATpl~Ecommon~Scontent.html
status: 30 June 2009.

Mußler, H. / **Schäfers**, M.: Deutsche Bank will kein Staatsgeld – Ackermann verzichtet auf den Bonus. in: FAZFINANCE.NET dated 17 October 2008, 2008, URL:
http://www.fazfinance.net/Aktuell/Ackerman n-verzichtet-auf-den-Bonus-3626.faz
status: 30 June 2009.

N.N.: Teurer Mitarbeiter Bonus: SAP will Beteiligungen verkaufen. In: Spiegel Online dated 05 May 2000, 2000, URL:
http://www.spiegel.de/wirtschaft/0,1518,75037,00.html
status: 28 June 2009.

N.N.: Große Parallelen zum Fall Flowtex. in: manager-magazin online dated 15 April 2002, 2002, URL:
http://www.manager-magazin.de/geld/artikel/0,2828,191697,00.html
status: 30 June 2009.

N.N.: Finanzkrise: Commerzbank bekommt Milliardenspritze vom Staat. in: Tagesspiegel dated 03 November 2008, 2008, URL:
http://www.tagesspiegel.de/wirtschaft/Unternehmen-Finanzkrise-Commerzbank;art129,2652012
status: 30 June 2009.

N.N.: Große Koalition: Miese Manager sollen mit Gehalt haften. in: Der Westen Online dated 23 April 2009, 2009, URL:

http://www.derwesten.de/nachrichten/nachrichten/wirtschaft-und-finanzen/2009/4/23/news-117796511/detail.html
status: 30 June 2009.

N.N.: Managergehälter: Schlechte Leistung wird bestraft. in: Zeit Online dated 12.03.2009, 2009, URL:
http://www.zeit.de/online/2009/11/managergehaelter?page=1
status: 15 July 2009.

N.N.: Say-on-Pay: Großinvestoren fordern Mitsprache bei Vorstandsvergütung. In: Rat für Nachhaltige Entwicklung Online dated 05.02.2009, 2009, URL:
http://www.Nachhaltigkeitsrat.de/news-termine/news/2009/2009-02-05/say-on-pay-grossinvestoren-fordern-mitsprache-bei-vorstandsverguetung/
status: 01 June 2009.

N.N.: Gericht weist Wenger-Klage ab. in: welt-online dated 13 August1998, 1998, URL:
http://www.welt.de/print-welt/article624923/Gericht_weist_Wenger_Klage_ab.html
status: 29.6.2009.

Pape, T.: Vergütungs- und Abfindungszahlungen an Vorstandsmitglieder deutscher Aktiengesellschaften im Fall feindlicher Unternehmensübernahmen (zugl. Jena, Univ., Diss., 2004). Volume 80, Berlin: Tenea Verlag, 2004.

Pellens, B. / **Crasselt**, N. / **Rockholtz**, C.: Wertorientierte Entlohnungssysteme für Führungskräfte – Anforderungen und empirische Evidenz. In: Pellens, B. (ed.): Unternehmenswertorientierte Entlohnungssysteme. Stuttgart: Schäffer-Poeschel, 1998, p. 1–28.

Pellens, B. / **Crasselt**, N.: Aktienkursorientierte Entlohnungsinstrumente im Jahresabschluß. in: Pellens, B. (ed.): Unternehmenswertorientierte Entlohnungssysteme. Stuttgart: Schäffer-Poeschel, 1998, p. 125–160.

Pellens, B. et al.: Internationale Rechnungslegung, IFRS 1 bis 8, IAS 1 bis 41, IFRIC-Interpretationen, statusardentwürfe. 7. rev. and compl. edition, Stuttgart: Schäffer-Poeschel, 2008.

Rappaport, A.: Shareholder Value: Ein Handbuch für Manager und Investoren. 2nd edition, Stuttgart: Schaeffer-Poeschel, 1999.

Rickens, Chr.: Die nächste Krise wird kommen. in: manager-magazin.de dated 05.02.2009, 2009, URL:
http://www.manager-magazin.de/magazin/artikel/0,2828,591846,00.html
status: 03 July 2009.

Roels, H.: Shareholder Value – Verantwortung gegenüber den Aktionären. in: RWE Pressecenter vom 28.03.2006, 2006, URL:
http://www.rwe.com/app/Pressecenter/Download. aspx?pmid=4001043&datei=1
status: 09 June 2009.

Rudolph, B.: Unternehmensfinanzierung und Kapitalmarkt. 1st edition, Tübingen: Mohr Siebeck, 2006.

RWE AG: Der Long Term Incentive Plan 2005. in: RWE Nachrichten dated 10 May 2005, 2005, URL:
http://www.rwe.com/web/cms/de/110504/rwe/investor-relations/nachrichten/news-ad-hoc-mitteilungen/?pmid=4000575
status: 17 June 2009.

Schmeisser, W. / **Hahn**, M / **Schindler**, F.: Aktienoptionsprogramme als Vergütungskomponente – und deren Rechnungslegung nach nationalen und internationalen Recht. Mering: Rainer Hampp Verlag, 2004.

Schmeisser, W. / **Dittmann**, M.: Shareholder Value-Ansatz als Grundphilosophie eines anreizorientierten Entgeltssystems. in: Schmeisser, W. /Brinkötter, H.-O. / Krimphove, D. (ed.): Internationales Entgeltmanagement. Munich/Mering: Rainer Hampp Verlag, 2004, p. 1–44.

Schmeisser, W.: Finanzorientierte Personalwirtschaft. Munich: Oldenburg, 2008.

Schubert, S.: Steigender Druck auf High-Tech-Firmen und ihre Buchführungspraktiken: Dunkel um Aktienoptionen soll gelichtet werden. in: Handelsblatt Online dated 29 July 2002, 2002, URL:
http://www.handelsblatt.com/archiv/dunkel-um-aktienoptionen-soll-gelichtet-wirden;551095
status: 30 June 2009.

Smith, A.: Natur und Ursachen des Volkswohlstandes. Volume 2, Berlin: Verlag Von Elwin Staude, 1879.

Towers P.: Aktienoptionen verlieren weltweit weiter an Bedeutung. in: Towers Perrin dated 18 May 2009, 2009, URL:
http://www.towersperrin.com/tp/showdctmdoc.jsp?url=Master_Brand_2/DEU/Press_Releases/2009/20090618/2009_06_18.htm&country=deu
status: 08 July 2009.

Union Investment: Union Investment Studie: Viele DAX-Gesellschaften noch mit ungenügenden Optionsprogrammen. in: Union Investment Online dated 26 September 2005, 2005, URL:
http://unternehmen.union-investment.de/unioninvestment/presse/pressestelle/9787986c3cbed771e8c58879dfb02f56.0.0/Pressemitteilung05_AOP1.html
status: 05 July 2009.

Union Investment: Union Investment – Aktienoptionspläne 2006 – Pressemitteilung. In: Union Investment Online dated 24 October 2006, 2006, URL:
http://privatkunden.union-investment.de/-snm-0184233381-1234720503-055ad00000-0000000128-1234726848-enm-cgi-bin/htsearch.cgi?config=privatkunden&format=short&words=aktienoptionsprogramme
status: 18 June 2009.

Weber, M.: Formen und Ausgestaltungsmöglichkeiten von Stock-options in internationaler Praxis. in: Achleitner, A.-K. / Wollmert, P. (ed.): Stock-options. 2nd edition, Stuttgart: Schaefer-Poeschel Verlag, 2002, p. 25–44.

Weiß, D. M.: Aktienoptionsprogramme nach dem KonTraG. in: WM, 53. Vol., 1999, p. 353–363.

Werder, A. v.: Shareholder-Value-Ansatz als (einzige) richtige Wertschnur des Wertschnurhandelns. In: ZGR, 27. Vol., 1998, p. 69–91.

Winter, St.: Prinzipien der Gestaltung von Managementanreizsystemen (zugl.: Berlin, Humboldt-Univ., Diss., 1996). in: Neue betriebswirtschaftliche Forschung, Volume 178, Wiesbaden: Gabler 1996.

Winter, St.: Optionspläne als Instrument wertorientierter Managementvergütung (zugl.: Berlin, Humboldt-Univ., Habil.-Schr., 1999). in: Europäische Hochschulschriften,

Reihe 5, Volks- und Betriebswirtschaft, Volume 2554, Frankfurt on Main et al.: Lang, 2000.

Winter, St.: Aktienoptionspläne und Motivationseffekte. In: Eckardstein / Dudo v. (ed.), Handbuch variable Vergütung für Führungskräfte, Munich: Vahlen, 2001, p. 85–105.

Zitzewitz, J. v.: SOP. 2003.

IV Delegations and their consequences for labor, tax and social security law aspects

15 Aspects concerning labor contracts

15.1 Contents of labor contracts

A labor contract is a mutual exchange contract under private law and the law of obligations[349] between the employer and the employee that creates mutual rights and obligations. The employee is obliged to perform his/her work; the employer on the other hand is obliged to pay the salary. The following table shows the principal and subsidiary obligations of employer and employee.

	employer	employee
Principal obligations	Duty to pay salary Duty of equal treatment Prohibition of victimizations	Duty to perform work
Subsidiary obligations	Duty to provide work Duty to grant holidays Fiduciary duties Duty to protect the employee's health and life Duty to protect the employee's personal rights Duty to protect the employee's property Duty to protect the employee against sexual harassment at work	Obligation of confidentiality Omission of notices and information damaging reputation or creditworthiness of employer Restraint on competition Duty to inform, to render accounts and make restitution Duty to give notice of impending damage Fiduciary duty

Table 15.1: Main and ancillary duties from employment relationship[350]

---

[349] Comp. Schaub, 2004, p. 119

[350] Following Schmeisser, 2008, p. 256

The labor contract is not subject to particular formal requirements. However, under the Law of Proof of Substantial Conditions Applicable to the Employment Relationship (Nachweisgesetz), the employer is obliged to put the essential terms of the contract in writing and to deliver the document to the employee within four weeks after the start of the labor relationship.[351]

The documents have to comprise the following minimum contents

- name and address of the parties of the contract,
- starting point of the labor contract,
- in the case of a temporary work contract the estimated duration of the labor relationship,
- place of work or notice that the employees shall work in different places,
- characterization/description of work,
- composition and amount of remuneration and due-date,
- contractual hours of work,
- annual vacation,
- notice period,
- notice of existing collective agreements, company and operating agreements.

If these points are not put down in writing, a valid labor relationship results anyway[352] and both parties are obliged to fulfill the underlying duties. Should the document not contain individual regulations, the legal regulations come into effect (e.g. vacation).

The employee has an authority to give instructions resulting from the labor contract, i.e. he may determine the working hours and the place of work for the employee.[353]

However, this authority is not sufficient to send the employee for a limited or unlimited period abroad.[354] In this case, the employee's interests have to be taken into consideration and no unreasonable burdens shall be caused by the change of job location.

Because the authority to give directives is a unilateral right of performance determination, it has to correspond to reasonable discretion in accordance with s. 106 GeWO (German Trade, Commerce and Industry Regulation Act). However, the employer has the possibility to expressly agree on a possible delegation abroad in the labor contract already. Also a very short period abroad, in the sense of a business travel, does not cause problems in the sense of the unilateral authority to give directives of the employer.

---

[351] Comp. Schaub, 2004, p. 120

[352] Comp. ibid, p. 121

[353] Comp. Schmeisser, 2008, p. 257

[354] Comp. Schmeisser, 2008, p. 257

15.2 Possibilities to design contracts in the case of delegations abroad

If an employee is sent abroad, the existing labor contract with the employer in the home country has to be adapted to the new situation. Basically, there are two possibilities to achieve this: Firstly, a supplementary contract can be concluded in addition to the existing contract. Secondly, a local labor contract can be concluded with the host company abroad in connection with a dislocation agreement with the home company.[355] Both possibilities will be examined in more details in the following paragraphs.

15.2.1 Supplementary contract as addition to the labor contract

If an employee in the context of a labor relation at home is sent abroad and if the focus lies on the employer in the delegating company, mostly a supplementary contract in addition to the existing labor contract is concluded. This supplementary contract contains all essential regulations concerning the job abroad. Thus, the labor contract is maintained without changes and the home company remains the only contract partner of the employee. Commonly, the authority to give directives remains with the company at home. These circumstances mean that the contractual main and supplementary duties of the labor contract remain unchanged and, therefore, the employee continues to be obliged to work for the home company and the employer has to remunerate this work.

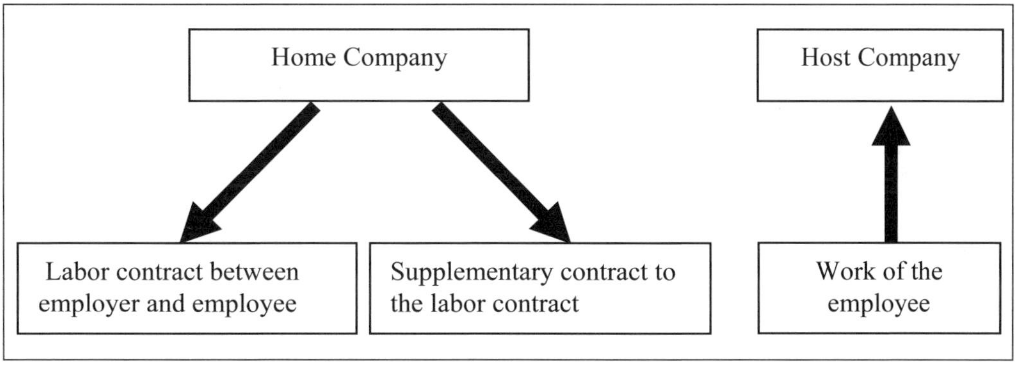

Fig. 15.1: Relations in the case of a supplementary contract in addition to the labor contract

It is thinkable to organize delegations abroad this way, if the control of the foreign company units or market research for managerial decisions is aimed at. Also for delegations abroad for

---

[355] Comp. Schmeisser, p. 257

the purpose of human resource development, especially for the promotion of young staff and for project related delegations, this type of contract is common.[356]

15.2.2 Dislocation agreement and local labor contract

As far as the employee does most of the work for the host company abroad in the context of the delegation, it is common to conclude a labor contract with the hosting company and a dislocation agreement with the delegating company. Therefore, it is also important, whether the employee fills a job in the host company which belongs as permanent job to its business organization.[357]

Both contracts regulate the correspondent relation to the home and the host company, as will be shown in more details:

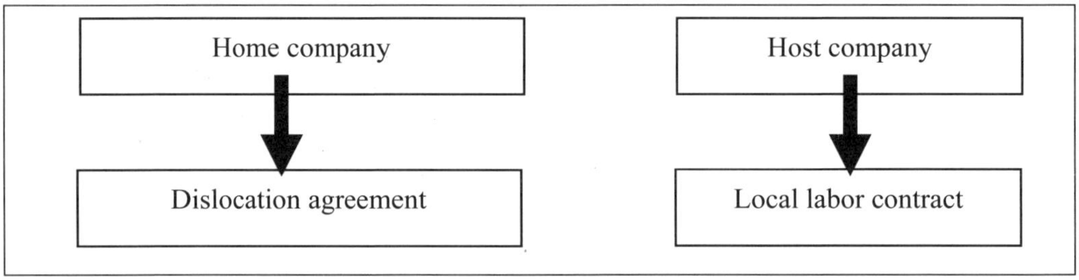

Fig. 15.2: Relations in the case of dislocation agreement and local labor contract

Relation to the host company
The employee concludes a local labor contract with the host company abroad, which is the basis for an active work relationship. All main and supplementary duties of a work relationship for the employee and employer result from this. In this local labor contract all essential topics of an employment relationship are regulated. Among those are the duration of the relationship, the type of work and working hours. Normally the contract conditions take the local circumstances into consideration.[358]

Relation to the home company
The relation to the home company is defined by the conclusion of a dislocation agreement with the delegating employer in addition to the existing labor contract and an additional local

356 Comp. Schmeisser, 2008, p. 257

357 Comp. ibid, p. 258

358 Comp. Schmeisser, 2008, p. 258

labor contract with the host company. This dislocation agreement regulates the inactive work as well as further additional payments and bonuses, if applicable.[359]

This way the employee can be sure that he can re-enter into his original labor contract after returning from his delegation abroad or at least to be re-employed under the former conditions. Under certain circumstances also the continuance of German social security is possible. This, however, will be described later in more details. Furthermore, the employment affiliation with the delegating society is not interrupted because of the design of the contract. This contract design bears also advantages for the delegating company: In case of need it can get a qualified employee back into the head office. Usually, a call-back right is therefore regulated, which permits the employer to call the employee back at an early point in time. Furthermore, the head office can represent certain interests in the branch offices abroad via the integration of the employee.[360]

Agreement on inactive work

Through the agreement on inactive work the existing labor contract with the delegating company is transformed in a so called passive contract.[361] This is due to the fact that the contractual main duties – for the employee the duty to deliver the work and for the employer the duty to pay the remuneration for the work – do not apply for the duration of the delegation. The supplementary duties, however, continue to exist, as for example fiduciary duties. In this form of contractual design the employee remains in a legal connection with the home company.

Return Guarantee

The return guarantee or resettlement commitment is the employer's promise towards the delegated employee to re-employ him in the home company after the termination of the delegation. The conditions under which the labor contract is carried out can be negotiated in variable ways. Essentially, there are three possibilities: The employer can assure an employment under comparable conditions as before the delegation for the employee's return. A further possibility is to consent an employment under consideration of the experience gained abroad without concretizing the exact job after return. Finally, the employer can promise a concrete position to the employee. However, this last type of return guarantee is rather seldom used.[362]

Additional remuneration and provisions

The delegated shall be integrated as far as possible into the salary structure of the host company during his time abroad in order to avoid a negative impact on the atmosphere between the employees based on the often big differences between the salaries and to evade

---

359 Comp. ibid, p. 258

360 Comp. ibid, p. 258

361 Comp. ibid, p. 258

362 Comp. Schmeisser, 2008, p. 259

tensions. It is common for the delegating company to pay additional salary and make additional provisions to the delegated employee, despite the fact that there is no proper payment duty on its side.[363]

16 Tax law aspects

Besides labor law questions also tax law consequences are caused by the delegation concerning the employer as well as the employee. It has to be clarified in which state under which conditions and to which extent taxes have to be paid and whether tax duties arise in both states and which consequences a possibly existing double tax agreement has.

16.1 Income tax liability

The extent of tax liability depends on the place of residence. A person is said to have the residency in the Federal Republic of Germany, if he is domiciled or has his habitual residence in this country. In accordance with the German General Fiscal Code domicile and habitual residence are defined as follows: "The domicile is where someone has a habitation under circumstances that point to the fact that the habitation is to be maintained and used."[364]

"The residence which is always and from the beginning timely continuous during more than six months has to be considered as habitual residence."[365]

If an employee is resident in Germany he underlies the unlimited tax duty of this country. In the other countries he has only the duty to pay taxes for the revenues that he obtains from the other country's sources.

The following table shows under which conditions the employee is subject to which tax duty.

---

[363] Comp. ibid, p. 259

[364] sec. 8 AO

[365] sec. 9 AO

Unlimited tax duty	Limited tax duty
if domicile/habitual residence in the country → all domestic and foreign revenues subject to unlimited tax	No residence/habitual residence in the country →in accordance with sec. 1 IV EStG in connection with sec. 49 EStG limited tax duty → only duty to pay domestic taxes

Table 16.1: Unlimited and limited tax duty

According to this, the employee has to pay taxes on all revenues in Germany, if he has his domicile or habitual residence in Germany and has unlimited duty to pay taxes.[366]

If, however, the employee does not have his domicile or habitual residence in Germany, he has limited duty to pay taxes and has only the duty to pay taxes on domestic revenues.[367]

16.2 International tax law

In order to identify a possible tax duty of a person, it is common in many countries to examine the residency of the employee (residence principle) as well as the performance and utilization of the labor on its territory (source or origin principle).

If a residency results from this examination, it leads to unlimited duty to pay taxes on all domestic and foreign revenues. If, however, the source or origin principle is used, the tax duty is limited. Then, the employee has only the duty to pay taxes on domestic revenues.

With regard to the residency of an employee who is sent abroad the following situations can arise:

1. resident in Germany,
2. resident in the host country,
3. resident in Germany and in the host country (so called dual residency).

In the last case a double taxation can be the result, i.e. the same taxpayer has to pay the same or similar taxes in at least two countries.[368]

---

366 Comp. Schmeisser, 2008, p. 261

367 Comp. ibid, p. 262

368 Comp. Schmeisser, 2008, p. 262 et seq.

16.3 Double tax agreement

16.3.1 General explanations

Double tax agreements are agreements under international law which are concluded to avoid multiple taxation with similar taxes for the same taxpayer. The following possibilities exist: on the one hand, the country where the revenues accrue (country of source) can take the taxation back or limit it. On the other hand the country of source can offset the revenues which have accrued and been charged with taxes against the foreign tax.[369] Double taxation agreements are normally bilateral, i.e. they are concluded between two countries. Only in exceptional cases multilateral treaties are concluded.[370] The Organization for Economic Cooperation and Development (OECD) has created a sample agreement, which serves as orientation for the agreements concluded between the Federal Republic of Germany and other countries.

There are the following principles with regard to double taxation:

- In accordance with the residency country principle, a person is subject to taxation in the country where he has his domicile or habitual abode.
- In accordance with the source country principle, a person is subject to taxation in the country where the revenues are originated.
- Following the world revenue principle, the total world income of the taxpayer is subject to taxation.
- Under the territoriality principle, all revenues are subject to taxation that have been generated on the territory of the country in question.[371]

16.3.2 General allocation criterion: residency

Under article 4 OECD-MA (OECD model agreement) a person is considered as having its residency in the contracting country, as far as he is subject to taxation due to his residence, habitual abode, place of management or a similar taxation characteristic. In the case of dual residency, the person is deemed to be resident in the country where he has his permanent residency. If the person has his permanent residency in both countries, the residency is determined on the basis of personal and economic relationships to the countries. The person is deemed to be resident in the country to which he has closer connections – where the centre of his vital interests lies. If the person is married, this is where he has his family. If the centre of vital interests cannot be determined, the habitual abode is taken as starting point. Only if this cannot be determined, then the nationality of the person is decisive for his residency.

---

369 Bundesfinanzministerium, status: 03 May 2009, 21:12h

370 Schmeisser, 2008, p. 263

371 Bundesfinanzministerium, status: 03 May 2009, 21:26h

Finally, the model agreement provides that the contracting states shall communicate between each other if the residency cannot be determined on the basis of the named criteria.[372]

The following fig. represents a short examination schedule of the considered characteristics:

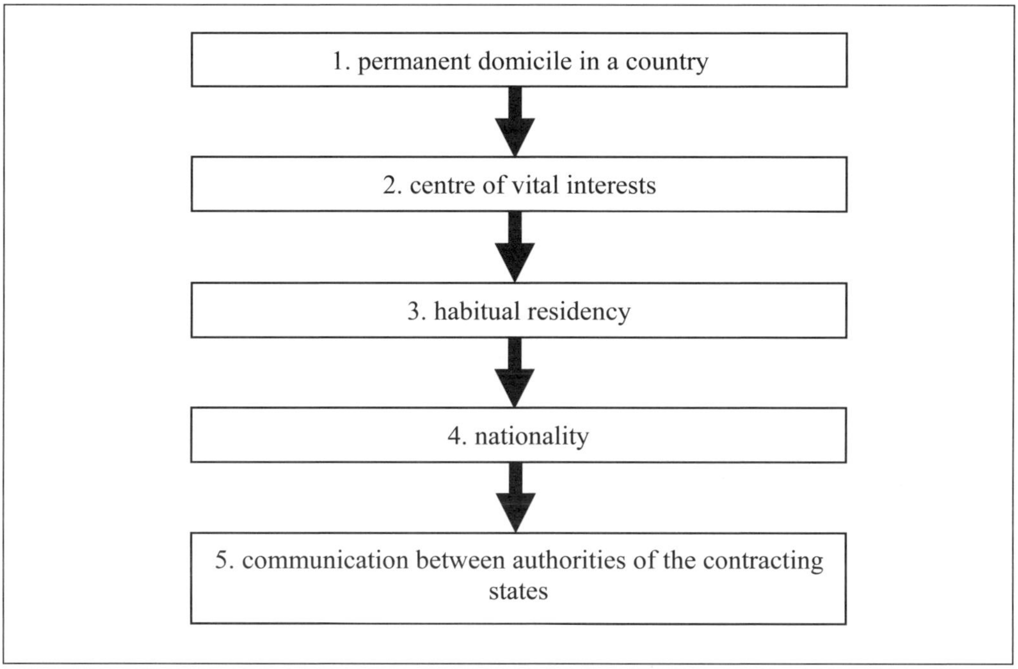

Fig. 16.1: examination scheme for residency[373]

If on the basis of the named examination scheme in the case of dual residency one of the contracting states is determined as country of residency, the taxpayer is taxed under the double taxation agreement as if he would reside only in this country. In principle, the unlimited duty to pay taxes continues to exist in both states.[374]

---

372 Comp. Schmeisser, 2008, p. 263

373 Following Schmeisser, 2008, p. 264

374 Comp. ibid, p. 263

16.3.3 Domestic tax law

Unlimited duty to pay taxes
Article 15 of the model agreement of the OECD determines after the clarification of the residency, which country has the right to charge taxes for income on employed work. The system described in this article for the allocation of the right to charge taxes is shortly displayed in the following fig.

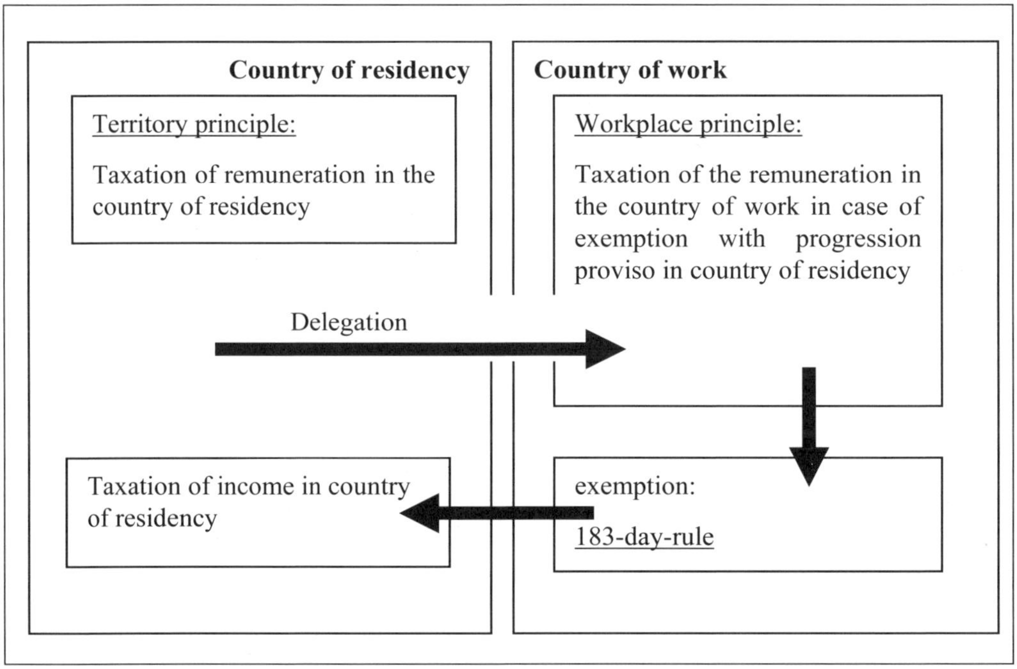

Fig. 16.2: System of Art. 15 OECD-MA[375]

According to this, the workplace principle under Art. 15 subs. 1 OECD-MA is considered as basic rule under which all remunerations as wages and salaries etc. are charged with taxes in the country in which the work is performed and in which the employee is physically located to perform this work.[376] This means for an employee who has been sent abroad that his income in Germany has to be exempt under consideration of the progression proviso. This means that all other domestic revenues, as e.g. revenues from capital assets are charged with

[375] Following Schmeisser, 2008, p. 264

[376] Comp. ibid, p. 265

the tax rate, which was calculated taking the foreign revenues into consideration. Therefore, the employee has to file a tax return.

The 183-day-rule sets a limit to the workplace principle. The right to charge taxes remains in Germany if the following conditions are fulfilled:

1. Residency in the country of work not longer than 183 days in twelve months,
2. Remuneration is paid by or on behalf of employer who is not resident in the country of work
3. Remuneration is not paid by the branch office the employer holds in the workplace country.

If these three conditions apply cumulatively, the income is subject to German taxation; otherwise the workplace principle is still applicable.[377]

Limited duty to pay taxes
If an employee has only limited duty to pay taxes in Germany, he will be subject to taxation in the country of work from the moment of moving into that country.[378] The 183-day-rule may not be applied because the model agreement assumes in article 15 section 2 that the employee resides in Germany. Besides, the income from employed work may not be subject to tax.[379]

17 Social security aspects

As up to now, aspects of labor law and tax law relating to the delegation of an employee were examined, the focus lies now on social security aspects. Before an employee takes his work in another country up, it has to be examined whether the social protection remains existent in Germany or whether it is expanded to the work abroad. Highest priority should have the aim to remain in the German social security system, in order to make sure that the delegated party can return into the German security system without problems. Without the maintenance of social security protections, gaps in old age insurance and loss of entitlement to benefits for unemployment would be the result.[380] The social protection of employees sent abroad by German companies is covered either through cross-national treaties or the so

---

[377] Comp. Schmeisser, 2008, p. 267

[378] Comp. Art. 15 Abs. 1 OECD-MA

[379] Comp. Schmeisser, 2008, p. 267

[380] Comp. ibid, p. 267

called transmission. Through this, the delegated employee has social security protection under domestic regulations.[381]

17.1 Employment abroad

17.1.1 Territory principle

The regulations on the compulsory insurance in health insurance, nursing insurance, old-age insurance and unemployment insurance are valid for all employees who work in the field of application of the Social Code (=Federal Republic of Germany) (sec. 3 SGB IV). With the exception of special cases, this territory principle applies also to foreigners occupied in Germany.[382] The duty to have social insurance originates from an employed labor relationship in Germany. This means that any person who is employed and receives salary, is a person with dependent employment in Germany under sec. 7 (1) SGB IV.[383]

17.1.2 Existence of a so-called "Ausstrahlung" (transmission)

In order to avoid that the continuity of the social protection is interrupted in the case of delegation of a German employee abroad, sec. 4 SGB IV determines that the regulations of social security apply also to the case of an assignment abroad if the delegation is limited in time due to the characteristic of the employment or previously by contract (so-called Ausstrahlung). Furthermore, the social security of German citizens abroad is guaranteed by transnational agreements and/ or socials security agreements under certain circumstances.[384]

The German social security law defines the term Ausstrahlung as the effect of a domestic employment relationship on another country.[385]

This exemption from the territory principle is valid for any place of work in the world. Finally, sec. 4 BGB IV is only valid in Germany and not binding for other countries. Because next to the territory principle also the regulations of other states may apply, the Ausstrahlung (transmission) can lead to dual insurance.[386]

---

381 Comp. Haufe Personal Office Online, Haufe Index 1564333, status: 26 April 2009, 14:38h

382 Comp. Haufe Personal Office Online, Haufe Index 1564333, status: 26 April 2009, 14:53h

383 Comp. Schmeisser, 2008, p. 268

384 Comp. Haufe Personal Office Online, Haufe Index 1564333, status: 26 April 2009, 15:08h

385 Comp. BfA., n° 24/2002, p. 4

386 Comp. Schmeisser, 2008, p. 268

Sec. 4 SGB IV: Ausstrahlung

(1) As far as the regulations on the obligation of social security and the right to social security require an occupation, they are also valid for persons, who are assigned to an area outside this application area in the context of an existing labor relationship within the application area of this code, if the delegation is limited due to the characteristic of the occupation or previously by contract.

(2) Sec. I applies correspondingly to persons who execute independent work.[387]

Therefore, the following conditions are required for the Ausstrahlung under sec. 4 SGB IV:

- The delegation has to be made into an area outside the area of application of the German Social Security Code (outside of Germany).
- The delegation is made in the context of an existing labor relationship in Germany.
- The delegation has to be timely limited in advance either by contract or by the characteristic of the occupation (e.g. for a certain project).
- It is possible for the employer to get the employee back from the other country at any moment. By this, it is guaranteed that the employee is integrated in the German organizational structure for the complete duration of the assignment abroad.
- The employer is entitled to give directives to the employees.
- The remuneration for work is paid by the employer in Germany, i.e. it is not handed down to the foreign company.[388]

If one of these conditions is not fulfilled, there is no duty to have social security protection in Germany in accordance with sec. 4 SGB IV. This result is achieved by the following examination scheme:[389]

[387] Juris, status: 26 April 2009, 15:35h

[388] Wikipedia, status: 26 April 2009, 16:04h

[389] Comp. Haufe Personal Office Online, Haufe Index 1572178, status: 26 April 2009, 16:16h

Is the employee sent abroad in the context of a domestic labor relationship?	→	no	→	no transmission
↓				
yes				
↓				
Is there a timely limitation (by contract or characteristic of the work)?	→	No	→	No transmission
↓				
yes				
↓				
Does the employee remain integrated in the domestic company structure?	→	No	→	No transmission
↓				
yes				↓
↓				
Transmission applies				Transmission does not apply
↓				↓
The German law continues to apply; consideration of agreement regulations if necessary.				German law does not apply.

Fig. 17.1: examination scheme for obligatory social security

For a better understanding the described conditions are explained in more detail in the following paragraphs.

17.2 Domestic labor relation

The labor relationship must be with an employer based in Germany. If the employer is based in another country, the Ausstrahlung does not apply.[390] This means that the condition is that there had been a previous legal relation to the employer in the Federal Republic of Germany and the centre of vital interests had been Germany before the delegation. These limitations are necessary, because the regulations on the transmission have the effect that the employee does not lose his social protection in Germany and has no gaps in the course of insurance.

<u>Example: delegation also in case of lack of duty to pay income taxes:</u>

An employer based in Germany delegated two employees (A to Egypt, B to Paraguay) and their salary is paid by the German company (disclosure in the pay role). For employer A no

[390] Comp. Haufe Personal Office Online, Haufe Index 1564346, status: 26 April 2009, 16:46h

income tax is deducted under the German-Egyptian Agreement for Double Taxation Avoidance. This difference when paying remuneration changes nothing for the delegation in the sense of Ausstrahlung. For this, all other conditions have to be fulfilled.[391]

Also the occupation in other countries in representations is seen as Ausstrahlung, because those are considered as independent branches of the domestic company and the employees remain to full extent employees of the domestic employer.[392]

17.3 Limitation in time of delegations

An employer, who is working abroad for a domestic employee is deemed to be subject to social security in the sense of transmission under the following circumstances: The occupation in the foreign country (=delegation) has to be temporarily limited by its characteristics or by contract.[393] A delegation is considered to be limited in time when from the beginning it is determined that there is a significant limitation.[394] There are no fixed time limitations and there is no danger if the delegation is restricted to several years.[395] What counts is whether there is any limitation at all. In 1994 the Federal Social Court of Germany has decided that the maximum duration of the limitation cannot be derived from legal texts, because such presumption would run couther the economic interests in a flexible, temporary assignment abroad of employees.[396] If an employee reaches a certain age limit, it is not considered as a timely limitation in this sense. Also if in advance, a contractual agreement for a further limited period is made, this prolongation can be regarded as contractually timely limited in advance.[397]

Example: timely limitation in advance

An employee is sent to Brazil for two years and the contract foresees that a prolongation of the delegation for further two years is possible. This is a case of a delegation limited in time in advance.[398]

It is important to consider: if the limitation in time is only arising in the course of the delegation and not under the named circumstances, then the transmission in the sense of sec. 4 SGB IV does not apply.

---

391 Comp. ibid, Haufe Index 1564347, status: 26 April 2009, 17:21h

392 Comp. ibid, Haufe Index 1564349, status: 26 April 2009, 17:36h

393 Comp. ibid, Haufe Index 1564351, status: 26 April 2009, 18:02h

394 Comp. ibid, Haufe Index 1564352, status: 26 April 2009, 18:24h

395 Comp. ibid, Haufe Index 1564352, status: 26 April 2009, 18:26h

396 Comp. BSG, decision of 4 May 1994, 11 file 55/93

397 Comp. Haufe Personal Office Online, Haufe Index 1564351, status: 26 April 2009, 18:13h

398 Haufe Personal Office Online, Haufe Index 1564354, status: 26 April 2009, 19:10h

Example: contractually agreed limitation in time of the delegation

A domestic employer delegates A and B in the context of their labor relationship to Pakistan. The duration of the delegation of employee A is restricted to three years from the beginning. The employer B is at first delegated for an unknown period of time. After one year however, it comes out – contrary to expectations – that the delegation will come to an end in the next year.

The delegation of the employee A is timely limited in advance; therefore, it is a case of a delegation in the sense of transmission. Because the timely limitation of the delegation of employee B had not existed in advance, but only arisen in the course of the delegation, it is not a case of delegation in the sense of transmission.[399]

Limitations on the basis of the characteristics of the delegation have to be presumed for certain work relationships which in accordance with general experience are no long-term relationships. This is the case in project work, installation work and briefings as well as in the case of construction of buildings and industrial installations.[400]

Example: timely limitation due to job characteristics

A domestic company has the obligation to build an embankment dam in India in a period of no longer than five years and delegates employees in the context of their existing labor relationship to this company. There is no written agreement in the labor contract on the duration of the delegation.

These are delegations which are timely limited by their characteristics. Therefore the regulation on transmission has to be applied.[401]

17.4 Trans- and international agreements with Germany

Besides the above mentioned conditions for the existence of a transmission in the sense of sec. 4 SGB IV (domestic labor relation, time limitation of the delegation), there are so calls trans- and international agreements of individual countries with the Federal Republic of Germany.

These agreements prevail over the Social Security Code and have influence on all or individual areas of social security.[402] The argument for the conclusion of such agreements was that the delegated person may have social insurance in Germany on the basis of the

---

399 Comp. ibid, Haufe Index 1564352, status: 26 April 2009, 18:48h

400 Comp. ibid, Haufe Index 1564353, status: 26 April 2009, 19:02h

401 Comp. ibid, Haufe Index 1564353, status: 26 April 2009, 19:05h

402 Comp. Haufe Personal Office Online, Haufe Index 1564376, status: 27 April 2009, 08:27h

territory principle as well as at the place of occupation. This can lead to double payment of contributions.[403]

The EEC agreements count among these agreements as transnational law and bilateral agreement count among them as international law. This shall prevent that the delegated person has to change the insurance system for the time of his occupation abroad and has to pay double contributions.[404]

EEC regulations

For countries belonging to the European Economic area it is assured through regulations that in the case of delegations the social insurance is to be carried out in accordance with domestic law.[405] Among these countries are Belgium, Bulgaria, Denmark, Germany, Estonia, Finland, France, Greece, Great Britain and Northern Ireland, Iceland, Italy, Latvia, Liechtenstein, Latonia, Luxembourg, Malta, Netherland, Norway, Austria, Poland, Portugal, Rumania, Sweden, Slovakia, Slovenia, Czech Republic, Hungary and Cyprus.[406] For this, the delegated person has to work in a member state of the European Economic Area. The core of the regulation – EEC regulation n° 1408/71 applies here and is valid for all member states of the European Union. The area of application was expanded to Iceland, Norway and Liechtenstein and is also applied to Swiss citizens.[407] Under Art. 2 sec. 1 EEC-Regulation N°1408/71 the personal scope of application includes employees, independent workers, students and refugees. For this reason, an employee who is citizen of a third country does not fall under the area of application of the EEC regulations.[408] In this case the regulations on transmission apply, as long as no regulations on competences from agreements on social security between the EU members state exist.[409] The factual area of application is regulated in Art. 4 sect. 1 of the EEC Regulation 1408/71 and includes insurance branches that offer the following benefits:

- Benefits in case of illness and maternity,
- Benefits in case of disability, including benefits determined for maintenance or improvement of the employment capacity,
- benefits in old-age,
- benefits to surviving dependants,
- benefits in case of working accidents and professional diseases,
- burial allowances,

---

403 Schmeisser, 2008, p. 271

404 Ibid, p. 271 et seq.

405 Comp. Haufe Personal Office Online, Haufe Index 1564341, status: 27 April 2009, 09:03h

406 Comp. ibid, Haufe Index 1564341, status: 27 April 2009, 09:45h

407 Schmeisser, 2008, p. 272

408 Ibid, p. 272

409 Haufe Personal Office Online, Haufe Index 1564341, status: 27 April 2009, 09:51h

- benefits for unemployment and
- family benefits.[410]

Via the enumeration of the above mentioned benefits it becomes clear that the German Social Security (health, old-age, accident and unemployment insurance) is subject of the factual area of application of the agreement. It includes also the nursing insurance, whose benefits have to be counted among the health care benefits in the context of the EEC regulations.[411]

A duration of twelve months is considered as limit in accordance with Art. 14 Sect.1a of the EEC Regulation 1408/71 if the employee does not replace another employee for whom the delegation time has expired. If the work is longer than twelve years, the German legal provisions still apply until the end of the work. At the longest, however, for further twelve months, as far as the responsible authority in an EU state or the designated authority has consented to it before expiration of the time.[412] For this, the German employer has to apply for the approval declaration before the end of the 12-months deadline.

17.5 Bilateral social security agreements

Because the EEC regulations have joint regulations for the member states, there are additional agreements between Germany and the other country in question.[413] Through these agreements with countries outside the European Union it is guaranteed that the German Social Security is still covering the delegated person. The duration can vary in the individual agreements.[414] In the following table[415] the individual contracting states with the contained insurance branches for each country are listed:

410 Schmeisser, 2008, p. 273

411 Ibid, p. 273

412 Haufe Personal Office Online, Haufe Index 1564341, status: 27 April 2009, 09:58h

413 Schmeisser, 2008, p. 275

414 Comp. Haufe Personal Office Online, Haufe Index 1564341, status: 27 April 2009, 21:00h

415 Ibid, Haufe Index 1564341, status: 27 April 2009, 21:06h

Country	Insurance branch
Australia	only pension and unemployment insurance
Bosnia-Herzegovina	without nursing insurance
Chile	only pension and unemployment insurance
China	only pension and unemployment insurance
Israel	without nursing and unemployment insurance
Japan	only pension and unemployment insurance
Canada excluding Quebec	only pension and unemployment insurance
Korea	only pension and unemployment insurance
Croatia	only health and pension insurance
Morocco	without nursing insurance
Macedonia	without nursing insurance
Quebec	only pension and unemployment insurance
Republic Montenegro and Republic Serbia	without nursing insurance
Turkey	without nursing insurance
Tunisia	without nursing and unemployment insurance
USA	only pension insurance

Table 17.1: Contracting states and the included insurance branches

The individual agreements can be separated into closed and open agreements. Closed agreements (as e.g. Tunisia and Morocco) are restricted to the citizens of the contracting states as well as refugees and stateless persons. In open agreements all persons are integrated which have a connection to the social security law of one of the two contracting states.[416]

416 Comp. Schmeisser, 2008, p. 276

The factual application area is very different for all agreements, because hardly any contracting country includes all insurance branches. Many agreements with contracting countries offer only limited possibilities. It is striking that all social security agreements include the legal pension insurance.[417]

17.6 Payment of contributions during the assignment abroad

There are no effects on the contributions to the social security carriers, if the German social security duty continues to exist during the delegation. The health, nursing, pension and unemployment contributions are deducted to the social security carrier by the employer.[418] The difficulty is that it may pose problems to calculate the amount of the salary paid in order to be able to calculate the contributions. In order to avoid disadvantages for the employee with regard to the later claims for pension benefits, there is a peculiarity for the calculation of contributions to the pension insurance. Between the real salary and a fictive remuneration a so called favorability comparison is made. If the fictive salary is higher than the real remuneration, the benefits have to be calculated on the basis of the fictive salary (sec. 166 subs. 1 n° 1 SGB IV). The aim of this procedure is to do justice to the changing income situation.[419] It has to be considered that only for the pension insurance contributions shall be calculated on the basis of a fictive salary. This calculation cannot be applied for health, nursing and pension insurance.

For the determination of the fictive salary an amount is taken as basis which can be calculated by multiplying the contribution assessment ceiling of the current calendar year with a ratio.[420]

The sum of the income of the employee during the last three calendar months fully subject to obligatory contributions before the start of the occupation abroad is divided by the sum of the amounts of the contribution assessment ceilings for this period.[421] The ratio obtained from this procedure has to be published with four decimal digits. The monthly assessment remuneration is calculated by multiplication of the contribution assessment ceiling relevant for the month of contribution with the ratio whereby at least a ratio of 0.6667 has to be used.[422]

---

[417] Comp. ibid, p. 276

[418] Comp. Haufe Personal Office Online, Haufe Index 1564379, status: 26 April 2009, 20:47h

[419] Comp. ibid, Haufe Index 1564380, status: 26 April 2009, 21:00h

[420] Haufe Personal Office Online, Haufe Index 1564381, status: 26 April 2009, 21:12h

[421] Comp. ibid, Haufe Index 1564381, status: 26 April 2009, 21:17h

[422] Ibid, Haufe Index 1564381, status: 26 April 2009, 21:17h

The ratio determined once at the beginning of the assignment abroad remains valid for the whole duration of the delegation for the determination of the fictive work remuneration.[423]

Example: Calculation of the fictive remuneration

Labor remuneration for the last three calendar months fully subject to the obligatory contributions.

Month	**Remuneration**	**Contribution assessment ceiling**
July 2009	EUR 3,800	EUR 5,400
August 2009	EUR 3,800	EUR 5,400
September 2009	EUR 3,800	EUR 5,400
	EUR 11,400	EUR 16,200

Table 17.2: Work remuneration including obligatory contributions

From the sum of the remunerations (EUR 11,400) and the sum of the contribution assessment ceilings (EUR 16,200) the ratio of (EUR 11,400/EUR 16,200 =) 0.7037 can be obtained. This ratio is higher than the required minimum ratio of 0,6667 and is therefore valid. The fictive remuneration for the year 2009 amounts to (5,400 x 0.7037 =) EUR 3,700.98. [424]
Because the calculation is based on the actually paid remuneration for the last three calendar years fully subject to social security contribution before taking up the occupation abroad, one time revenues (as e.g. Christmas allowances) can be fully taken into account.[425]

Example 2: calculation of the fictive salary in the case of exceeding the contribution assessment ceiling

Last three months fully subject to social security contributions

[423] Ibid, Haufe Index 1564383, status: 26 April 2009, 21:47h

[424] Ibid, Haufe Index 1564381, status: 26 April 2009, 21:27h

[425] Comp. ibid, Haufe Index 1564382, status: 26 April 2009, 21:32h

month	**remuneration**	**contribution assessment ceiling**
November 2008	EUR 2,600	EUR 5,300
December 2008	EUR 5,400	EUR 5,300
January 2009	EUR 2,700	EUR 5,400
	EUR 10,700 EUR	EUR 16,000
ratio (EUR 10,700 :EUR 16,000 =) 0,6688		

Table 17.3: work remuneration with contribution assessment margin

For the calendar year 2009 a monthly assessment remuneration of EUR 5,400 x 0.668 = EUR 3,611.52 can be obtained.[426]

If only two calendar months are subject with two fully obligatory contribution duties, the ratio has be calculated from the sum of the remuneration and the sum of the contribution assessment ceiling of these two months.[427]

17.7 Provisions in the case of lack of insurance obligation

However, it can happen that the delegated person is neither subject to section 4 SGB IV nor to transnational or international law and therefore cannot remain in the German social security system. Under certain circumstances the delegated person can still retain his protection. For this, national rules can be important, especially for persons working in countries without agreements.[428]

17.7.1 Pension insurance

The continuity of the pension insurance is an important aspect for the delegated person, because otherwise it comes to insurance gaps. In accordance with sec. 4 subs. 1 n° 2 SGB IV the delegated person can obtain a pension insurance obligation upon application. The condition is that the employee works only for a limited time abroad and the insurance duty is applied for by an institution based in Germany.[429]

---

[426] Haufe Personal Office Online, Haufe Index 1564382, status: 26 April 2009, 21:37h

[427] Ibid, Haufe Index 1564383, status: 26 April 2009, 21:37h

[428] Comp. Schmeisser, 2008, p. 278

[429] Comp. Schmeisser, 2008, p. 278

17.7.2 Health insurance

In accordance with sec. 9 subs. 1 n° SGB V an employee can get insurance on his own free will, if he dropped out of the insurance obligation and has been either insured by more than 24 months during the last five years before dropping out or has been continuously insured for at least 12 months before dropping out. In accordance with sec. 17 SBG V the contributions will be compensated by the employer. The employer has to compensate the healthcare contributions up to the amount which would have arisen in Germany. The potentially resulting difference has to be paid by the employee. In most cases, however, a private foreign health insurance is contracted, because the employer wants to avoid the difference of cost compensation and the extensive burocracy connected with it.[430]

If an employee had been voluntarily insured in the legal health insurance, there is the possibility of a health insurance with an on-hold position under section 240 sect. 4a SGB V. In order to be able to use the health insurance with on-hold position, the delegated person may not have co-insured family members who stay in Germany, and has to work abroad for professional reasons. The employee should avoid the termination of the voluntary legal health insurance in order to be able to return into this insurance after coming back to Germany.[431]

17.7.3 Nursing insurance

It is also possible to continue the nursing insurance on a voluntary basis. During the delegation the employee cannot claim for benefits, but the time of insurance is counted. In accordance with section 26 subs. 2 sentence 2 SBG XI the application has to be made one month after the end of the insurance duty.

17.7.4 Accident insurance

In the case of the legal accident insurance the delegated person has no possibility to remain insured on a voluntary basis. On application of the employee he can contract a foreign accident insurance with the responsible mutual insurance association of employees (in accordance with section 140 subs. 2 and 3 SGB VII).

17.7.5 Unemployment insurance

The delegated person cannot remain insured on a voluntary basis in the legal unemployment insurance. The result is that he can loose his right to unemployment benefits after his return. Generally, it has to be considered, that unemployment benefits can be obtained, if the

---

[430] Comp. ibid, p. 278

[431] Comp. ibid, p. 279

employee in period of three years had had a work relationship under social security law for at least 12 months. Therefore, the delegated person is entitled to receive unemployment benefits, if the assignment abroad was no longer than 24 months.[432]

17.8 Termination of transmission

The Ausstrahlung is terminated under the three conditions below:

1. **The foreign place of employment remains the same, but the employer at home is exchanged.** Not always when the employer at home is replaced, the delegation is terminated. If the change of employer is characterized by the fact that the former company is taken over by another national company, the change has not to be considered and the delegation is continued.[433]

   <u>Example: Delegation and change of employee</u>

   The employees A and B work due to delegation in the sense of an Ausstrahlung under sec. 4 SGB IV in Saudi-Arabia. The employer of employee A changes in the context of this employment, because the company is taken over by another national company. The employer B accepts a position in another German company in Saudi-Arabia, because the employment relation with the first German company is terminated.

   For employee A there are no changes with regard to his delegation in the sense of the Ausstrahlung in the context of the replacement of his employer. For employee B, a new work relation is taken up in a foreign country which is not based on the delegation, with the result, that he is not subject to the German insurance obligation in the sense of Ausstrahlung and therefore to German social security.[434]

2. **The workplace is temporarily dislocated from the foreign country to the home country.**
   If however the delegated person is temporarily in his home country during the delegation period (e.g. for holidays or because of part-time employment), the delegation itself is not interrupted. In sec. 8 subs. 1 n° 2 SGB IV it is determined that a duration of two month in the course of a calendar year is acceptable. If, however, it is contractually agreed that the period named in sec. 8 subs. 1 n° 2 SGB IV shall be exceeded, a new delegation would commence when the occupation abroad is taken up again.[435]

---

[432] Comp. Schmeisser, p. 280

[433] Comp. Haufe Personal Office Online, Haufe Index 1564357, status: 27 April 2009, 21:54h

[434] Ibid, Haufe Index 1564357, status: 27 April 2009, 22:09h

[435] Comp. Haufe Personal Office Online, Haufe Index 1564358, status: 27 April 2009, 22:12h

Example: Delegation with temporary return to the home country
Employee D has been sent to Norway by his national employer, employee has been sent to Panama. The delegation of employee D has no timely limitation, but the employer returns from time to time for holidays or reporting purposes to his home country for a short period of time. This means that there is no timely limitation of the activity abroad. As this is a case of unlimited delegation, there is no obligation of insurance by means of Ausstrahlung in the sense of sec. 4 SGB IV.

Employee E has the contractual obligation to return every three years to Germany for reporting purposes and the introduction into new techniques for a temporary national occupation during a period of time which exceed the duration of two months in the course of a calendar year under sec. 8 subs. 1 n° 2 SGB IV. The national activities discontinue the foreign occupation with the effect that it is a case of limited delegation in the sense of the Ausstrahlung. This means each occupation abroad is subject to the insurance duty under the German Social Security.[436]

3. A limited delegation is turned into an unlimited delegation.

18 Conclusion

The individual design of the labor contract has essential influence on labor, tax and social security aspects. It is important to know the contractual connections with the national employer, the timely limitation or the duration of the delegation as well as the employee's organizational integration, in order to make a competent judgment over the individual case of the delegated person. In some cases, and when the employer does not dispose over the necessary know-how, advice should be obtained from competent legal experts, consultants or authorities, because delegation issues are highly complex.[437] Especially for small and medium-sized companies a delegation is always a financial and personnel burden. Therefore, alternatives should be taken into consideration, like e.g. cooperations with other companies, company pool memberships in the chambers of foreign trades etc.[438] However, even in times of worldwide economic crisis the following trend can be observed: even if the delegation of employees is connected with high expenses, also in future many employees will be sent abroad by their companies.[439] 40 percent of interviewed companies assumed to send more employees to positions in foreign offices in future. 51 internationally active companies took

[436] Ibid, Haufe Index 1564358, status: 27 April 2009, 22:19h

[437] Comp. Schmeisser, 2008, p. 280

[438] IHK-Nordwestfalen, status: 26 April 2009, 22:47h

[439] Haufe , status: 27 April 2009, 22:32h

part in the study. 25 percent had between 100 and 500 German employees in foreign offices, 10 percent disposed over more than 500 expatriates.[440]

Therefore, it will be unavoidable for the companies in question to deal with the issues of delegation with regard to labor, social and tax law aspects.

Literature

Bundesfinanzministerium: URL: http://www.bundesfinanzministerium.de
http://www.bundesfinanzministerium.de/nn_39818/DE/BMF__Startseite/Service/Glossar/D/004__Doppelbesteuerungsabkommen.html
status: 03 May 2009, 21:12h.
http://www.bundesfinanzministerium.de/nn_39818/DE/BMF__Startseite/Service/Glossar/D/004__Doppelbesteuerungsabkommen.html
status: 03 May 2009, 21:26h.

Haufe Personal Office Online: Haufe Index. Version 13.4.22.0, 2009.

Haufe: URL: https://www.haufe.de
https://www.haufe.de/SID61.Xq1OL0n2FSo/personal/newsDetails?newsID=1240419184.52
status: 27 April 2009, 22:32h.
https://www.haufe.de/SID61.Xq1OL0n2FSo/personal/newsDetails?newsID=1240419184.52
status: 27 April 2009, 22:41h.

IHK-Nordwestfalen: URL: http://www.ihk-nordwestfalen.de
http://www.ihk-nordwestfalen.de/marktkontakt/Auslandsentsendung.cfm
status: 26 April 2009, 22:47h.

Juris: URL: http://bundesrecht.juris.de
http://bundesrecht.juris.de/sgb_4/__4.html
status: 26 April 2009, 15:35h.

Küfner-Schmitt, I.: Arbeitsrecht. 5th edition, Planegg, 2007.

Schaub, G.: Arbeitsrecht von A-Z. 17th edition, dtv, Munich, 2004.

Schmeisser, W.: Finanzorientierte Personalwirtschaft. Oldenbourg Wissenschaftsverlag, Munich, 2008.

Wikipedia: URL: http://www.wikipedia.de
http://de.wikipedia.org/wiki/Entsendung
status: 26 April 2009, 16:04h.

---

[440] Ibid, status: 27 April 2009, 22:41h

V International remuneration systems for companies listed on stock exchange in flux

The German industry and banks are currently up in arms against the German legislative initiative regarding the regulation and adequacy of manager remuneration, which is connected with the multiple regulatory attempts on EU-level.

Questions on how and whether manager remuneration should be limited and designed in a more performance related and transparent way on a long term basis, whether stock option programs because of their short term, capital market-oriented deliberations are the wrong incentives, how the staffing of boards of directors as supervisors of manager remuneration systems should be undertaken etc. are only some aspects which should be re-discussed under the Corporate Governance and Compliance standard, with the individual parts of a variable remuneration system.

The reason for these legislative initiatives is the dimensions at which managers had received golden handshakes, even though they were losers. The following questions are now open for discussion:

1. Is there a managerial and **human resources** related basic logic, which can suffice the demands of listed companies? The finance-oriented personnel management logic can be discussed as possible process in this context. Among the aspects to be discussed is e.g. the question regarding who should be responsible for the remuneration of senior managers and managing directors and whether this would be comprehensible at the stock-market and for the investors.
2. **Substance:** This term means the adequacy of variable parts of the remuneration of the manager performance, so that some sort of positive "sustainability" of manager conduct for the company can be instrumentally guaranteed, without providing wrong incentives with e.g. stock option programs. A negative form sanction has to exist, e.g. via human capital assessment or shareholding in the company, if managers fail. There is the question which other and complementary instruments apart from stock option programs could be used in the context of a "four-pillar-remuneration-management" to "punish" management failure. Could this be the Berlin Human Capital Assessment Model?
3. **Transparency**: There must be an (individual) disclosure on the performance assessment of managers, risk management, remuneration systems for senior management and

(executive) directors from quarterly reports, business reports, information from financial public relations for the stock market and other stakeholders.

The finance-oriented personnel management provides approaches for all three fields of discussion and shows with its basic logic the process for listed companies regarding how for branches, business units, managers and directors of a company the performance measurement can be logically deducted and how variable parts for the remuneration management system in the context of the Berlin Balanced Scorecard Approach and the Berlin Human Capital Assessment Model can be obtained.

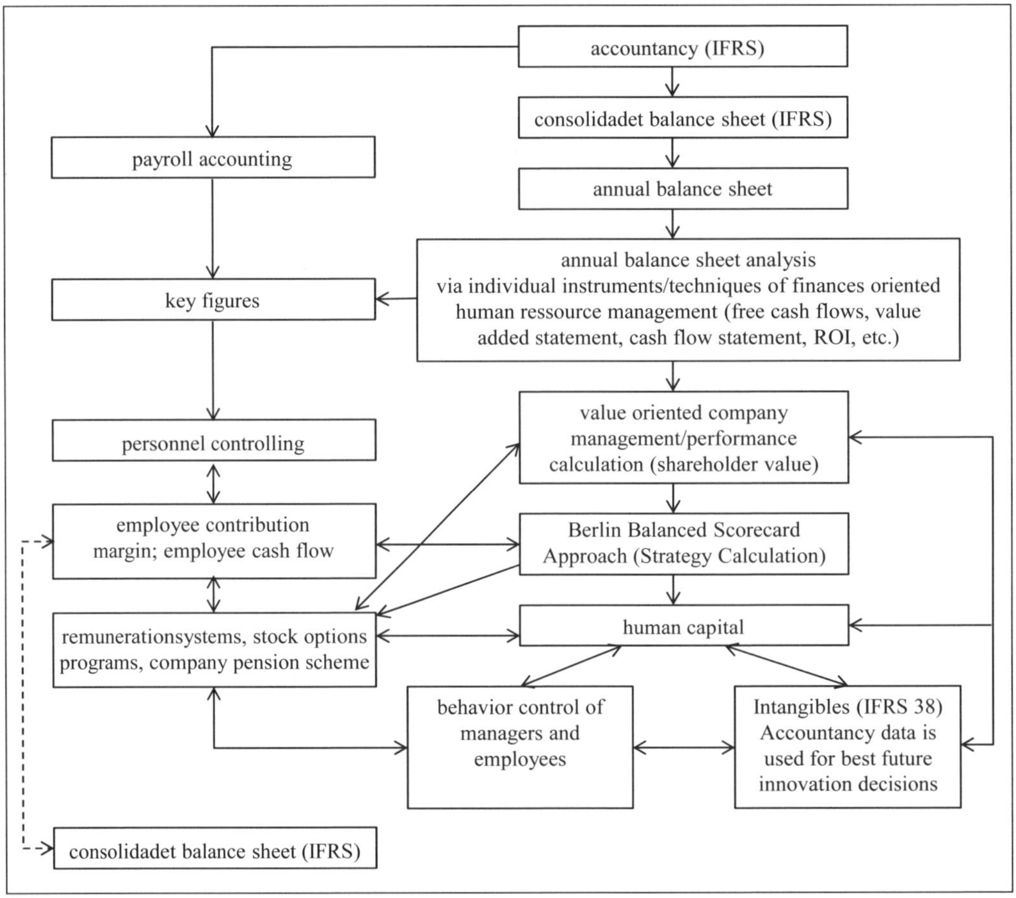

Fig. V.1: Logic of finance oriented human resource management[441]

[441] Schmeisser, 2008 and Schmeisser/Clausen, 2009.

The logic of finance-oriented personnel management is based on the trend towards internationalization of accountancy (International Financial Report standards), because internal and external accountancy are ideally integrated in such a way, as is expected and demanded by the stock markets in quarterly statements.

Thus, the accountancy, but also the finance-oriented personnel management, has to satisfy two concrete informational requests of stock markets, investors and stakeholders:

- To give account for economic actions, in this case for actions concerning personnel management.
- To make sure information is available which makes it possible to assess the economic development of the company – especially personnel management values, with the help of selected controlling techniques and key figures.

There are two types of corporate information problems for accountancy in this context:

1. For the recognition or the formulation of decision problems (regarding personnel management) accounting information in accordance with IFRS is needed, e.g. social plan, decisions with respect to company pension schemes, stock option programs, Berlin Human Capital Assessment Model etc.
2. Furthermore, information for the personnel management assessment regarding the consequences resulting from the decision alternatives is necessary, e.g. strategy calculation via innovation success calculation and the Berlin Balanced Scorecard Approach, Human capital assessment and its consequences for the research and development team (e.g. patent assessment), staff assignment, cash flow calculation, value added statements and their consequences for pursuing the strategy etc.

Groups and listed capital companies, which use IFRS, tend to use a finance-oriented personnel management without knowing the name of the concept. With the help of the following financial "indicators", the personnel management instruments, activities and key figures of such companies can be attributed to different types:

- They have a personnel controlling that deduces key figures from the IFRS consolidated accounts or IFRS annual balance sheets, the payroll etc. and use the information again as figures for quarterly reports to the stock market, rating and annual balances.
- Up to now they have a type of stock option program for managers, directors, executive managers and out-of-tariff staff etc. which is based on the result of the ordinary activities as orientation.
- They have a company pension scheme and/or employee involvement models, which are permanently scrutinized under financial aspects.
- Finance oriented personnel management looks into financial burdens caused by staff which can then be found in the profit and loss statement and the annual balance sheet.

Companies that employ at least implicitly an advanced finance-oriented personnel management, recognize that

1. Group, company and business unit strategies can be better handled and controlled with the help of financed-oriented personnel strategies. This can be achieved via the Berlin Balanced Scorecard Approach, the Berlin Human Capital Assessment Model or intangible value drivers (patents, licenses, human capital etc.).
2. Also a merger, the closure of a plant, a partial relocation of a plant, outplacements etc. have to be assessed and calculated under finance-oriented personnel management aspects.

Thus, finance oriented personnel management supports the fulfillment of the duty to inform the investors via market reports:

- As a help to decide over investments, in order to assess a current research program with relevant and reliable information with regard to its success prospects. In order to take a decision over an investment, the investor e.g. of the pharmaceutical industry has to be able to assess the development of medicine and the human capital potential of researchers with key figures via their patents and the resulting cash flows and EBITs.

- A further aim is to control the manager behavior in the sense of a shareholder-value-philosophy via the instruments of the Berlin Human Capital Assessment Model. Owners/Shareholders give relatively far reaching decision making rights and demand in return to be regularly informed about the development of the business and reserve the right to withdraw the managers from their position at the next board meeting or general assembly. In order to make sure that managers behave in the interest of the shareholders, they have the possibility to contractually guarantee a certain part of the shareholder value to the managers in the context of a management remuneration system with variable components. More precisely, these are accretions that can be deducted from a human capital assessment calculation and are reflected in the employee participation e.g. in stock option programs.

Because of the close link between finance-oriented personnel management and accountancy, this approach is especially adequate for practical use. In a four pillar remuneration system a new combination of management remuneration is suggested:

1st pillar	2nd pillar	3rd pillar	4th pillar
Composition of the total remuneration (remuneration menu)	**Timely orientation**	**Instruments for execution**	**Weighting of remuneration components**
Result-based strategy bonus	Long term (3 to 10 years)	SOP as well as other employee participation programs	**Risk salary** Up to 20 percent SOP, up to 30 percent human capital and establishment assessment
Personal, operative Berlin Human Capital Assessment, performance bonus, efficiency pay	Short term or middle term (1 to 2 years)	BBSC, Risk-Management, Compliance	
Company social security contribution	Short term and partially long term	Company pension scheme and social security contribution	**Fixed salary** Weighting with 50 to 80 percent of annual salary depending on position and country
Guaranteed management bonuses fixed special payments	Short term (several times per year)	Management by Objectives (MbO), Compliance	
Fixed basic salary	Short term (continuousl y, mostly monthly)	Agreement on the basis of wage spectrums depending on country of location	
SOP = Stock option program BBSC = Berliner Balanced Scorecard			

Table 0. : Four pillars of international remuneration management[442]

[442] Adapted from: Schmeisser, 2008.

Literature

Schmeisser, W.: Finanzorientierte Personalwirtschaft. Oldenbourg Wissenschaftsverlag. Munich 2008.

Schmeisser, W. / **Clausen**, L.: Controlling und Berliner Balanced Scorecard Ansatz. Oldenbourg Wissenschaftsverlag. ; Munich 2009.

Schmeisser, W. / **Mohnkopf**, H. / **Hartmann**, M. / **Metze**, G (ed.): Innovationserfolgs-rechnung. Springer Verlag, Berlin 2008.

VI Application of the Berlin Human Capital Assessment Model in an international bank for the control of its branches

Now it will be attempted to undertake a human capital assessment of branches via benefit analysis for an international bank on the basis of existing data from controlling and with the use of planning, control and deviation analyses. The Berlin Human Capital Assessment Model[443] has the advantage of a flexible usage of instruments of internal and external accountancy and should therefore not be understood as inflexible construct. This means the model can be individually adapted to the company, in order to do justice to the changing demands of assessment. The starting point is always the existing data within the company, which is prepared in a way that guarantees the measurement of the employee performance depending on the company performance.

19 Bank internal controlling data as basis

The bank is structured in divisions and can be separated in two big areas: "branch office" and "central office". The strategic planning and controlling is undertaken in the central office; there, the company aims are defined, planned and controlled. The operative execution of the strategic aims is done in the individual branches, which can be seen as almost autonomously acting part banks in other countries within the "whole bank".

They are defined by the responsibility for success of the branch director and their operative freedom for actions, which strongly supports a close to market company management. As they have their own responsibility with regard to costs and revenues, they are basically profit centers within the overall group.

---

443 Comp. Schmeisser/Clausen, 2009.

The following four characteristics allow considering the branch offices as profit centers:

1. market performance as goal,
2. personal responsibility,
3. individual formulation of goals
4. contribution margin calculation (sometimes also cash flow or EBIT) as center performance calculation.

The central office of the group assesses the market performance of the branches via profit center calculations, which are taken as data basis for the human capital assessment. The resulting key figures are measurable criteria for the assessment of human capital in the branches. These data are completed with figures regarding staff structure and costs resulting from the reports of the personnel development area. Under 37 branches 15 branches are taken into consideration for human capital assessment. The lack of assessment of the other branches is due to the fact that they are responsible for a transregional market area, which would distort the results of the rest of the branches.

Additionally, it must be underlined, that the data used in this model has to be seen fictively, because a uniform period of data collection does not exist and no conclusion can be made for an existing bank.

20 Selected key figures and their weighting

In order to undertake an adequate assessment of the branches, informative key figures have to be determined in advance. Special attention is paid to the fact whether the key figures permit to measure what is supposed to be measured, i.e. the performance of the branch taking the strategic goals into consideration. Furthermore, they should be little prone to assessment mistakes. In order to make the origin and the influence of key figures clear, they are bundled and divided in three perspectives on the basis of the Berlin Balanced Scorecard.

That makes it possible to discover cause-effect-relations in the target perspective and simplifies the deduction of actions. By different weighting of criteria their significance is influenced and the goals on the bank's side underlined. Below selected key figures and their weighting are presented, which are shown in table 20.1.

	Criteria	Weighting
Profit oriented perspective	Active business volume per branch	15 percent
	Deposit business volume per branch	5 percent
	Existing customers	2 percent
	New customers	4 percent
	Lending business volume per customer	2 percent
	Deposit business volume per customer	2 percent
Profit center calculations	Income/cost ratio	20 percent
	Risk profitability	20 percent
	Contribution margin 1	10 percent
HR standards	Lending business volume per employee capacity	6 percent
	Deposit business volume per employee capacity	2 percent
	Personnel costs per employee capacity per month	3 percent
	Labor turnover rate	3 percent
	Rate of further training	2 percent
	Absence rate	4 percent
	Percentage	**100** percent

Table 20.1: Weighting of target criteria

The achieved business volume per branch is chosen to clarify that a profit oriented perspective is used. For this, the division into lending and deposit business volume is made. The lending business is understood as collective term for bank business having an effect on the asset side of the bank balance sheet, as demands against customers, companies, banks as e.g. loans, time loans and overdraft facilities. I.e. it consists mainly of granting credits. The deposit business is synonymous for bank business, which is displayed at the liabilities side of the balance sheet of the bank, i.e. liabilities against customers and banks, e.g. demand deposits, time deposits and saving deposits or loans for refinancing credits. To make profit, the asset side as well as the liabilities side is important, because in order to grant credits in the lending business, foreign capital on the liabilities side is needed. The core business of the bank is still the lending business, so that it has a higher significance in the model. The achieved lending business volume per branch has a weight of 15 percent in the assessment and the deposit business per branch has a weight of 5 percent.

In order to generate lending and deposit business volume customers are needed. Therefore, the capability to identify, to acquire and to bind the right customers on a long term basis belongs to the central profitability factors. To do so, new customers are distinguished from existing customers. Nowadays, the efforts of the banks are rather concentrated on winning new customers instead of reinforcing existing customer relationships. But also a large pool of existing customers is an important leverage for revenues and at the same time an indicator for customer satisfaction and the bank's potential to bind customers on a long term. Because winning new customers demands more sales efforts and the accretion of new customers can be regarded as key figure for the growth potential of the bank, winning new customers is with 4 percent higher weighted than holding existing customers with 2 percent.

The presentation of the business volume per customer shows the size and economic strength of the customer and provides information on the quality of the customer portfolio. Also in this case, there is a differentiation between lending and deposit part of customer business volume. Both figures are valued with 2 percent each.

For the controlling of the branches the determination of objective success criteria is necessary. For this, the bank has defined two central control criteria, which are seen as essential key figures for the performance measurement of individual branches and which are subject to regular analysis in the monthly profit center calculation. The first control criterion is the *income/cost ratio*[444]. It informs about the ratio of administration expenses and operative revenues and is a key figure for efficiency measurement of a company or a branch:

$$\text{income/cost ratio} = \frac{\text{gross revenues}}{\text{costs}}$$

The gross revenue of a branch is determined by adding the conditional interest rate contributions[445] and the commissions/fees[446]. The costs subsume due staff and material costs. The conditional interest contributions, as result of the multiplication of the average volume and the interest margin, occur on the asset as well as on the liability side of the bank. A minimum income/cost ratio is defined which has to be fulfilled by each branch depending on its staff and material costs. As aggregation of all branches the benefits resulting from the income/costs ratios lead to a minimum benefit for the overall bank.

444 The cost/income ratio is the inverse value of the income-to-cost ratio.

445 Conditional interest rate contributions result for example from interest surpluses of granted loans, overdraft credits, acceptances, demand deposits or overnight and time deposits.

446 The commissions/fees include commissions for guarantees, bonds, credits and fees as costs from value assessment experts or agency commissions to third parties.

The second central controlling key figure is the *risk profitability*. Through this risk adjusted performance key figure it shall be made sure that the risk involving business transaction creates the expected revenues and only those business transactions are concluded whose benefit expectancies justify taking the risks over. It has the advantage that especially for banks an important link between profitability and risk within the financial perspective can be covered. The risk profitability is determined as ratio of the expected revenues and the risk of the branch in question.[447]

$$\text{Risk profitability} = \frac{\text{gross revenues (per year)}}{\text{weighed risk assets}}$$

When banks are granting loans they have the general risk that the liquidity of the lender gets worse in the course of time and the interest payments cannot be effected anymore. This risk, which is only involved in the lending business, has to be taken into consideration for each loan as additional cost component. The volume of the loan sums is called risk assets. The risk weighting of the existing assets is realized in accordance with the credit risk-standard approach.[448] Depending on the liquidity of the lenders, different weightings are undertaken to classify the risks, which is why finally they are called weighted risk assets.[449]

As the income/cost ratio as well as the risk profitability of the bank is seen as central controlling key figure, the Human Capital Assessment Model attaches special meaning to them, so that both key figures are integrated into the assessment with a weight of 20 percent.

A further target criterion of the profit center calculation is the Contribution Margin I, which is calculated by subtracting all costs which can be directly attributed to the branch from the gross revenues.

The Contribution Margin I sheds light on the question to which extent the variable costs are covered by the revenues. The goal should be to achieve a high contribution margin; declining contribution margins can be caused by lower revenues and/or higher costs. They indicate a deterioration of the company situation. Therefore, the Assessment Model attributes a weight of 10 percent to the Contribution Margin I.

In the third perspective, which is here called Human Resources standard, the employee specific key figures are bundled. The key figure *business volume per employee capacity* gives information on how much business volume an employee has achieved in the corresponding branch office.

---

447 For the calculation of the risk profitability in the forecasting period the gross revenue has to be calculated for one year (e.g. as forecasted quarterly value*4).

448 In this approach the credit risk evaluation is supported by external rating agencies, in accordance with the requirements of Basel II.

449 Weighted risk assets = risk weighted balance sheet assets + risk weighted „traditional“ off-balance sheet transactions + risk weighted „innovative off-balance sheet transactions.

Furthermore, it is used for the assessment of the company labor productivity, which shows the capacity to perform in a branch office. The number of employees is calculated in the unit "employee capacity" (1.0 EC), in order to get a more exact image of the labor performance in relation to the working time. A full employee capacity (1.0 EC) corresponds to an employee with a weekly working time of 39 hours. A part-time employee with a regular weekly working time of 20 hours corresponds to 0.5 EC. The business volume per employee capacity is calculated by dividing the business volume by the number of employee capacities. Also in this case, a separation into lending and deposit business volume is undertaken in accordance with its origin. Because the lending business is regarded as more attendance demanding, it is integrated with a weight of 6 percent, while the attendance demanding deposit business volume is assessed with 2 percent.

Staff costs are a further criterion for human capital assessment. Staff costs are any costs, which a bank incurs by the deployment of employees in a branch. These are largely fix costs, which regularly account for a large proportion of the costs of operation. They consist of salaries and wages, social contributions and expenses for pensions and support. Also in this case, the number of employees corresponds to the number of available employee capacities. The key figure Staff Costs per EC has a weight of 3 percent and is calculated as ratio of staff costs by number of employee capacities. If the Staff Costs per EC surpass the average, the cause could be an excessive remuneration of the employees as well as surplus.

The sum of employee entries and departs is, also in the branch offices of a bank, an indicator for employee satisfaction and engagement. Especially in times of acute lack of experts on the labor market, a high turnover rate means rising staff costs, because newly negotiated contracts are generally more cost intensive than the former contracts. The goal of the bank is therefore to keep the turnover rate low with the help of specific employee engagement measures. For the right interpretation, it is important to keep in mind that the ratio is also subject to fluctuations of the economic and labor situation. Thus, a low turnover rate can indicate staff expansions in an economically strong period.

The *turnover rate* shown in the model is calculated with the help of the Schlüter-formula and is taken into consideration with a weight of 3 percent.

It is important for the bank to show professional perspectives to the employees and to support their further advancement in the company in a target-oriented way, not least in order to intensify the employee engagement. The training measures are carried out by a management school that takes care of the employees with a network of selected experts and cooperation partners for the internal and external further training. In order to examine the intensity of further training of the employees, the key figure *Rate of Further Training is calculated* and given a weight of 2 percent. This key figure is calculated from the ratio total number of days of further training divided by total number of employees and shows the average time of further education per employee in the selected inquiry period (of one year – chosen for this case).

Only healthy, motivated and qualified employees are able to achieve extraordinary performance. In order to gain information on the health structure of the staff and the effectiveness of the initiated health supporting measures, the key figure *Absence Rate* has been integrated into the model. The key figure takes exclusively the absence hours due to illness and the due labor time in hours per branch into consideration. Because absence time due to illness causes a large amount of costs this key figure is integrated with a weight of 4 percent into the model.

21 On the calculability of the bank human capital assessment model

When the specific target criteria for the human capital assessment have been chosen and prioritized with the help of their weighting, in the further course, the assessment of the branch offices via benefit analyses is undertaken, in order to allow an attribution of the special contributions to the company benefits. The benefit analysis is a non-monetary assessment method from the area of cost accountancy and contains the previously defined, performance related target criteria, which are of subjective nature in their weighting and specification categories (see table 21.1). The identification is undertaken with the help of objective criteria for each branch office separately. Via the benefit analysis, the branch offices shall be measured in relation to the fulfillment of the target criteria, in order to give information on their benefit. In this context, the achievement of the employees is measured in relation to revenues gained. In accordance with this approach, the employee is useful for the branch office by raising the performance capacity of human capital. The higher the total performance of the branch office is, the higher is the generated benefit and the more human capital is deployed.

Three specification categories (SC) are defined for the individual criteria, to which each branch office, depending on its specification, is attributed. The specification criteria are scaled with point values which reflect the degree of fulfillment of the target criterion. The specifications in the SC 2 result from the weighted average value of all branch offices and represent the degree of target achievement. The criteria income/cost ratio and risk profitability represent a peculiarity with respect to the degree of target achievement. The pursued income/cost ratio is currently at ≥ 5 for any branch office and the aspired risk profitability per branch office has been fixed at 2.2. percent.

	Criteria	Weighting	EC 1	EC 2	EC 3
revenue related perspective	asset-side business volume of the branch office	15 percent	lower	EUR 1.502m	higher
	liability-side business volume of the branch office	5 percent	lower	EUR 536 m	higher
	existing customers	2 percent	lower	2,934	higher
	new customers	4 percent	lower	110	higher
	asset-side business volume per customer	2 percent	lower	EUR 524	higher
	liability-side business volume per customer	2 percent	lower	EUR 189k	higher
profit center calculations	income/cost ratio	20 percent	lower	5.0	higher
	risk profitability	20 percent	lower	2.2 percent	higher
	contribution margin 1	10 percent	lower	EUR 2.605k	higher
HR-standard	asset-side business volume per EC	6 percent	lower	EUR 50 m	higher
	liability-side business volume per EC	2 percent	lower	EUR 18m	higher
	staff costs per EC per month	3 percent	higher	EUR 6.0k	lower
	turnover rate	3 percent	higher	6 percent	lower
	further training rate	2 percent	lower	2.0	higher
	absence rate	4 Percent	higher	2.43 percent	lower
	point value	**100 percent**	**0–2 points**	**3 points**	**4–6 points**

Table 21.1: Specification categories of the criteria in the benefit analysis

If the result of a branch office is worse than demanded in the SC 2, it is attributed to SC 1 or SC 3 depending on the specification. In these two specification categories the point values are even further categorized in order to be able to make a more differentiated assessment of the over-fulfillment of the target value. The opposite case occurs in SC 1, with a result which lies far below the target fulfillment where no point is given. By the multiplication of the weighting of criteria with the achieved point value of the specification category, a partial benefit value in the target criterion can be calculated. The sum of the assessed part benefit values corresponds to the total benefit value of the branch.

22 Results of the bank human capital assessment

For each of the 15 branch offices of the bank via benefit analysis an individual benefit was calculated. The results are illustrated in Fig. 1 with a Box-and-Whisker-Plot[450], which allows gaining an overview over the frequency of the calculated benefit values per branch.

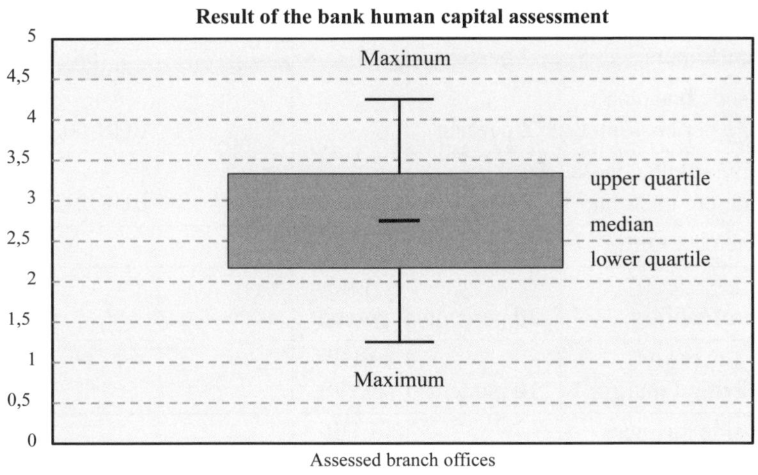

Fig. 22.1: Result comparison between the branches

The illustrated maximum in the diagram shows the branch with the highest point value of 4.3 points, whereas the minimum represents the branch with the lowest points (1.42). The so called box is divided in two by the median, which divides the totality of the branch offices at a value of 2.73 in two parts. This means that 50 percent of all branches, which are located between the median and the maximum, have a higher benefit value and 50 percent of all branches which are located between the minimum and the median have a lower benefit value. Between the minimum and the lower quartile and the maximum and the upper quartile

---

[450] The Box-and-Whisker-Plot is a graphic method of explorative data analysis in order to illustrate the frequency distribution of the calculated date.

are 25 percent of the branches at each side, which have reached the non-fulfillment or over fulfillment of the previously defined target values.

In order to exemplify the existing results, three branches are introduced which have achieved a benefit value over, under and corresponding to the average. Table 22.1. gives an example for a branch with a benefit value under average.

	Criteria	Weighting	SC 1	SC 2	SC 3	Pt.	Bene fit
revenue related perspective	asset-side business volume of the branch office	15 percent		EUR 1,502 m		3	0.45
	liability-side business volume of the branch office	5 percent		EUR 536 m		3	0.15
	existing customers	2 percent	2.495			2	0.04
	new customers	4 percent			168	5	0.2
	asset-side business volume per customer	2 percent			EUR 602k	5	0.1
	liability-side business volume per customer	2 percent			EUR 215k	4	0.08
profit center calculations	income/cost ratio	20 percent	3.10			0	0
	risk profitability	20 percent	1.43 percent			0	0
	contribution margin 1	10 percent	EUR 1,798			1	0.1
HR-standard	asset-side business volume per EC	6 percent	EUR 37 m			0	0
	liability-side business volume per EC	2 percent	EUR 13 m			0	0
	staff costs per EC per month	3 percent			5,9k EUR	4	0-12
	turnover rate	3 percent		6 percent		3	0.09
	further training rate	2 percent	1.62			0	0
	absence rate	4 percent			1.67 Percent	5	0.2
	point value	**100 percent**	**0–2 points**	**3 points**	**4–6 points**	**35**	**1.53**

Table 22.1: Example of a branch with a benefit value below average

In the assessment of the benefit related perspective, the branch achieves the target values with the asset-side as well as with the liability-side business volume and achieves three points for each in the SC 2. In the customer development, it has a below average number of existing customers, so that only 2 points in the SC 1 are given.

It can be observed that the branch attempts to get a wider customer pool by higher sales activities for new customers. Therefore, the acquisition of new customers is much higher than the requested target value and receives five points. The same can be said about the asset-side and liability-side business volume per customer. The branch disposes over economically strong customers and is classified in SC 3 with both values. The number in the profit center calculation is below average. Especially the weakness regarding the benefit/cost ratio and the high risk in the customer portfolio with an unsatisfactory contribution margin are reasons for the bad result. This is made worse by weak employee productivity in the lending as well as in the deposit business. A further considerable flaw of the branch is the low rate of further training, because the employee does not receive the requested two days of further training per year. A positive value is the staff costs per employee capacity which are slightly below average. The reason can be the low absence rate, which accounts for healthy employees and an agreeable working atmosphere, which keeps the turnover value at a constant value of 6 percent. However, it is also possible that there is not enough satisfactorily qualified staff available and that therefore the staff costs are below average.

The branch office quantified in table 22.2 has with a value of 2.73 an average benefit value. The over-fulfillment of the targets in the revenue oriented perspective as well as the fulfillment of the target of the risk profitability is sign of a customer pool with strong liquidity. However, neither the targets for the income/cost ratio nor for the required contribution margin are met, which can be explained by a disproportional growth of the costs, which is not satisfactorily covered with net revenues. The staff costs can be reason for this: with EUR 6,300 per employee per month they are above the average staff costs of all branch offices. A positive factor is the performance of the branch with regard of the generated business volume per employee capacity. The turnover rate and absence rate is average and has little influence on the labor productivity. The completed days of further training per employee completely correspond to the requirements.

	Criteria	Weighting	SC 1	SC 2	SC 3	Pt.	Bene fit
revenue related perspective	asset-side business volume of the branch office	15 percent			EUR 1,662 m	4	0.6
	liability-side business volume of the branch office	5 percent			EUR 670 m	4	0.2
	existing customers	2 percent			3,109	5	0.1
	new customers	4 percent			132	4	0.16
	asset-side business volume per customer	2 percent		EUR 535 T		3	0.06
	liability-side business volume per customer	2 percent			EUR 216 T	4	0.08
profit center calculations	income/cost ratio	20 percent	4.13			1	0.2
	risk profitability	20 percent		2.13 percent		3	0.6
	contribution margin 1	10 percent	EUR 2,417 T			2	0.2
HR-standard	asset-side business volume per EC	6 percent		EUR 50 m		3	0.18
	liability-side business volume per EC	2 percent			EUR 20 m	4	0.08
	staff costs per EC per month	3 percent	EUR 6.3 T			0	0
	turnover rate	3 percent		5.56 percent		3	0.09
	further training rate	2 percent		2.11		3	0.06
	absence rate	4 percent		2.49 percent		3	0.12
	point value	**100 percent**	**0–2 points**	**3 points**	**4–6 points**	**46**	**2.73**

Table 22.2: Example of a branch office with average benefit

The last branch offices illustrated as example in table 22.3 has a benefit value above average. It is especially characterized by the quality of the customer portfolio, which is shown by the risk profitability. With regard to the income/cost ratio and the achieved contribution margin this branch achieves values above average, too. This is also proved by the low staff costs per employee capacity with high labor capacity in the lending and deposit business activities. This high earning power of the branch, however, is at the expense of employees, which can be discovered with the help of the absence rate. The high turnover rate can be explained with the strategic orientation of the branch, which wants to achieve a better customer attendance in new target markets through specialists. In order to do justice to the risings demands on the target markets, the customers are trained at an above average rate. This means that the human capital is increased in a quantitative way by the employment of new staff and qualitatively by further training measures.

	Criteria	Weighting	SC 1	SC 2	SC 3	Pt.	Benefit value
revenue related perspective	asset-side business volume of the branch office	15 percent		EUR 1.557 m		3	0.45
	liability-side business volume of the branch office	5 percent			EUR 834 m	5	0.25
	existing customers	2 percent			3.166	5	0.1
	new customers	4 percent			207	5	0.2
	asset-side business volume per customer	2 percent	EUR 492T			2	0.04
	liability-side business volume per customer	2 percent			EUR 263 T	5	0.1
profit center calculations	income/cost ratio	20 percent			5.54	5	1
	risk profitability	20 percent			2.59 percent	5	1
	contribution margin 1	10 percent			EUR 2,791 T	5	0.5
HR-standard	asset-side business volume per EC	6 percent			EUR 55 m	4	0.24
	liability-side business volume per EC	2 percent			EUR 29 m	6	0.12
	staff costs per EC per month	3 percent			EUR 5.6 T	6	0.18
	turnover rate	3 percent	13.51 percent			0	0
	further training rate	2 percent			2.97	6	0.12
	absence rate	4 percent	5.06 percent			0	0
	point value	**100 percent**	**0–2 points**	**3 points**	**4–6 points**	**62**	**4.3**

Table 22.3: Example of a branch with above average benefit

23 Bank-Human Capital Assessment Model in the context of remuneration management

With the help of the presented results it could be proved that the developed bank human capital analysis with the help of the benefit analysis is appropriate to assess the performance potential of a branch or of a manager in dependence of the available human capital. Furthermore, the weaknesses of the branches were discovered. These flaws should be a central component for further future directed measures of control in order to increase the benefit in branches with low benefit values. The human capital assessment includes the strategic goals to be achieved on branch office level. The lower the assessment level is, the more concrete the targets are formulated, until they consist of actions on employee level which can be applied in daily business. The successful execution of strategies and aims is essentially determined by the behavior and the motivation of the branch manager. The current dynamic competitive situation requires creativity and eagerness for innovation by the employees.

By granting performance related remuneration components, it becomes easier to achieve targets for the individual branches. A bonus depending on the earnings of a branch can be of more relevance for the behavior than a bonus that depends from a variable that can neither be influenced by the branch manager nor by his employees at 100 percent.

Therefore, the remuneration system should be geared by the value creation processes or the results which can be influenced. The value contribution of an employee is measured by the bank human capital assessment. The connection to remuneration creates the necessary incentives. This way, the interests of the manager with regard to the increase of company value can be connected to the achievement of strategic goals, when the bonuses for the branch managers depend for example to 50 percent on the achievement of the requested goals and on the results of human capital assessment. This type of remuneration management corresponds especially in the expansion of the traditional, primarily capital oriented view of the factors “humans” and “customers” to an integrated, value driven control of the overall bank. It is even more important to quantify the performance of the employees to underline the relevance of human capital as significant value driver of a company and, thus, also of the shareholder value. As can be seen from the example of a bank, many promising approaches on the assessment of human capital are inclosed the presented Berlin Bank Human Capital Assessment Model.

As studies show remuneration systems and target achievement are rather badly linked. In accordance with the empiric research of Gilles at the level of upper and middle management, only half of the companies made this connection; at the lower management and on operative level the percentage was with 20 percent even lower.[451] New forms of remuneration, however, can increase the cost efficiency and performance of the company. The introduction of market-adjusted, performance-oriented remuneration systems increases the employee motivation and the personnel costs can be more efficiently directed towards the market. Through a comprehensive human capital assessment on the level of employees and on branch level, the additive, variable payments can not only be connected to the individual performances, but also to the overall performance of the company and to the business development. There are already suggestions for a new monetary incentive system in finance oriented personnel management, which contributes to the behavior control of managers in the sense of shareholder value. In a four-pillar-remuneration system a new organization of the remuneration composition is suggested, which fulfills the requirements of owners and shareholders of a variable, long-term and value-oriented manager remuneration.[452]

---

[451] Gilles, 2002, p. 198

[452] Comp. Schmeisser, 2008, p. 75

VII Development of a human resource information system of international personnel controlling

Internationally active companies need in the context of their merger and acquisition activities as well as for their internationalization strategies international personnel controlling. This is necessary in order to adjust to permanently changing customer needs and to meet the shareholder requirements in the sense of shareholder value on global capital markets.

Consequently, since more than a decade the globalization brings the requirement along to instrumentally lead such companies in accordance with the shareholder-value-approach.

This requirement and its acceptance in Germany are reinforced by the Law on Control and Transparency in Enterprises (KonTraG), which permits the introduction of an international early warning system and of stock option programs and thereby provides a material incentive and control system for managers and employees in the sense of the shareholder value approach to (German) international companies.

There are, however, still many misunderstandings and problems with regard to the concrete procedure for a value-oriented controlling. The question on the kind and specialty of international personal controlling causes the same diffuse problems.[453] Whoever hopes for an answer to the complex "international personnel controlling" from specialist literature is mostly deceived, because this topic is mostly fully avoided or only marginally touched, for example in the are of international missing hours development, the development of the labor efficiency, the international remuneration management system or the evaluation of the international personnel assignment with local cultural problems in the countries of assignment.

---

[453] Comp. Schmeisser/Clermont/Kriener, 1997, p. 529 et seq.; Schmeisser/Clermont, 1999; Schmeisser/Clermont/Protz, 1999 and DGFP, 2001, p. 188

The demand for practicable concepts for international personnel controlling exists regardless of whether this situation is connected with a substantial realization deficit of international personnel management or not.[454] Therefore, the question has to be discussed how the managerial contents of a human resource information system should be structured to do justice to the possible needs in personnel controlling.

24 About the description and definition of international personnel controlling

The condition for the recognition of international personnel controlling as own management discipline is the identification of an individual problem that can be transformed into an own object of knowledge and practical application field. Correspondingly, it is necessary to base the international personnel controlling concept on a clearly defined international and personnel management problem for which there cannot exist any alternative solution. With the problem directed to personnel management and international issues, the following diverse company targets can be pursued.

In this context, an international personnel controlling concept is understood as a theoretical approach towards a target oriented solution of a specific problem of international personnel controlling.

International personnel controlling includes the totality of international personnel management tasks serving for a target-oriented coordination of management decisions with international effects on a company group guaranteeing trough their execution of coordination concepts an international orientation of the group as well as the provision of information for the management on international level.

In this context the controlling concept of Management Control is suitable, because it largely conforms to the intentions of the KonTraG, which will be described further below. The Management Control is a partial field of Management Accounting[455], whose specific problematic includes amongst other things the coordination of international management decisions as well as target oriented influence on the behavior/actions of international managers on subordinated company group levels of the worldwide management organization.

The solution of this specific problem of International Controlling is partially limited to the provision of international company management with information. Management Control, however, is also connected with the execution and the utilization of international

[454] Comp. DGFP, 2001, p. 188 et seq.

[455] Comp. Atkinson, 1997

coordination concepts. Management Control also provides information for international reporting to company external interested persons (e.g. shareholders, shareholder value concept) and international internal auditing. This means that the KonTraG offers a legal and managerial base for International Personnel Controlling.

25 Targets of international personnel controlling

The approach for the solution of the problem of an International Controlling Concept with international decision coordination as specific controlling problematic is seen in the implementation of international concepts of decision coordination and the organization in the company group as well as in the certainty of provision of information to international management.

Tasks of international personnel controlling

Personnel management tasks can be attributed to two different areas depending on the reference of an international personnel decision:

- System design
- Process support

The system designing, international personnel tasks do not refer to a concrete operative or strategic decision, but to decisions which will be regularly made in foreign branches in future. For these international personnel decisions, coordination systems are created in order to guarantee the application of appropriate coordination concepts and of their effects from the point of view of the parent company in the company group.

An international coordination system in the human resource area is the totality of personnel management methods, models, technical support (hardware, personnel information software), international personnel processes as well as the application of international coordination concepts.

In the design of international coordination systems, target oriented company group regulations on the distribution and execution of international personnel management tasks in the application of coordination concepts are formulated and put down in writing. Personnel planning and controlling systems in worldwide groups, budgeting as well as control and transfer price systems in the group of companies as well as international incentives and flexible remuneration management systems are some examples for international coordination systems.

The design of such an international coordination system should therefore always be connected to the installation and development of an international personnel information system, which facilitates the personnel information regularly needed in the application of this coordination concept.

Personnel information systems include the processes of information gathering, interpretation and transfer, personnel carriers of these processes as well as information and communication equipment, instruments, methods and models for the routine facilitation of exactly defined personnel information for international management in the group.

This means that among the system designing tasks of international personnel controlling are:

- system designing and controlling of coordination systems,
- personnel information systems as well as
- international project management with special focus on international team building.

Among the process supporting tasks of personnel controlling are coordinating and information facilitating personnel tasks in concrete decision situations. Among those process-supporting controlling tasks are

- management of coordination processes in the group as well as
- problem specific facilitation of information.

26 Use of personnel information systems

Personnel information systems are socio-technical systems, consisting of human and mechanical components. A totally automated, group-wide (personnel) information system cannot be realized yet, because not all information processes in an internationally acting company are programmable and therefore automatable yet. The background of this problematic is on the one hand the content of personnel information which shall be controlling-wise recorded and, on the other hand, the language information of possibly 50 or more (foreign) languages, which cause translation, comparability and cost problems.

It is therefore necessary to introduce assumptions for further deliberations on international personnel controlling and to approach the concrete managerial contents with the help of well selected examples which are fit to contribute to an effective and efficient international personnel controlling.

1. For reasons of simplification we make the assumption of a German, stock-listed group that is internationally active, understands direct investments abroad not only as portfolio management, uses an own international personnel management in order to actively intervene as “global player” in the implementation of its strategies in the subsidiaries abroad.

2. In the context of international personnel controlling also foreign activities are examined, which have a connection with regard to their content to the activities of the head office. It will be assumed that persons, products, money, know-how and information between head office and foreign subsidiaries are transferred. Only if activities are carried out by the foreign company units which have a connection to the business of the head office or the whole group of companies, a support via corresponding personnel economic measures is important. If the foreign direct investments, however, are only portfolio investments, made only for reasons of profit-making, they have generally no consequences on the personnel functions of the company.[456]
3. Among other things, these foreign activities of the mother company can be explained with the fact that the purchase, sale, foundation and installation of subsidiaries abroad pursues the target to dislocate the group activities of the value added chain, which in accordance with Porter need supporting measures in the area of Human Resource Management.
4. The aim of the parent company is to control the management via strategies, organizational structures, managers and a (personnel) controlling system.
5. International personnel controlling should use application systems as SAP[457] or PeopleSoft[458] via the use of a working language, i.e. English or German, in order to detect, determine, process and interpret the effectiveness, efficiency and risk factors of foreign activities with the help of external and internal (international) accountancy.
6. International Personnel Controlling helps to construe an accepted, appropriate and cost-efficient international remuneration system for middle and top management worldwide and permanently controls the individual parts.[459]
7. Agreeing with the practice-based recommendations of the DGFP working group that the International Personnel Controlling is a special form of "National Personnel Controlling" [460] means that individual functions can be pursued via benchmarking as well as cost effectiveness and profitability considerations. These are the personnel functions:

   - Personnel recruitment and selection
   - Personnel development with culture and language training
   - Compensation and incentive design in the context of a remuneration management system
   - Management and (company) culture as well as
   - International labor relations.

456 Weber/Festing/Dowling/Schuler, 1998, p. 3

457 Comp. Mülder, 2001, p. 457 et seq.

458 Comp. Krawinkel, 2001, p. 447 et seq.

459 Comp. Brinkkötter, 1997, p.413 et seq. and 429 et seq.; Beynio/Krieger, 1998, p. 353 et seq.; Rosen, 1998, p. 341 et seq. and Brinkkötter, 2000, p. 663 et seq.

460 Comp. DGFP, 2001, p. 108 et seq.

Only through this persecution of personnel functions the assignments abroad can be supported and recognition of potential and personnel development of (top) potential managers in the group is possible on the long run.

27 Obtaining personnel information

Constructing personnel information systems means to imitate information processes of international group activities, especially of international personnel activities, and to organize them (in a computer processing suitable way). Personnel information processes serve to transform raw data into meaningful, decision-oriented, company political information or in other words: they serve to build the bridge between an (original) information offer in the foreign company units of the group world-wide and need of information of the head branch or the investors of the officially stock-listed group, i.e. the requirements of (international) accountancy.

Especially the Act for Control and Transparency in the Corporate Sector (KonTraG)[461] can provide a formal context and first framework for international personnel controlling.

The target of the law is to improve control and transparency in the area of officially stock-listed corporations. The scope of examination in the legal annual statement is completely renewed in sec. 317 HGB which was reformulated by the KonTraG including the individual company closing as well as the audit of consolidated financial statements. The essential reforms are:

- Focus of the examination on incorrectness and violations against legal regulations and completing regulations or the bylaws or the articles of association (sec. 317 subs. 1 sentence 3 HGB),
- Extension of the examination duties in connection with the position report (sec. 317 subs. 2 HGB) as well as
- Examination duty with regard to the supervisory system to be installed in accordance with sec. 91 subs. 2 AktG for officially stock-listed corporations (sec. 317 subs. 4 HGB).

1. In accordance with sec. 317 subs. 1 sentence 3 HGB the examination has to be performed in such a way that incorrectness or violations against legal provisions, bylaws or articles of association can be recognized. This includes only those violations which essentially affect the image of the situation of assets, finances and revenues of the company under sec. 264 subs. 2 HGB.

---

461 Comp. Schmeisser, 2001, p. 811 et seq.

Violations are deliberate deviations from accounting principles in the sense of sec. 317 subs. 1 sentence 2 HGB applicable for the preparation of the annual statement or the position report.

The following violation have an essential influence on the (consolidated) annual statement: violations of managers of regulations of the foreign company unit which question the business activities in the host country, but also violations of regulations which violate permits and licenses, violations of rules concerning form and content of the annual statement and violations of environment protection regulations which implicate a duty of removal and therefore a duty of accruals under commercial law.

Further typical examples are balance sheet adulterations, irregularities or legal violations by managers, also in foreign company branches, in the area of stocks: statement of stocks in the balance sheet even if they are not or not any more owned by the company, statement of non-existing stocks, empty containers, addition of water to oil-containers and transfer of costs which have arisen in already concluded orders to new orders.

In the case of signs of mistakes, frauds, property impairments and other violations of the law, the annual auditor has to assess which circumstances have lead to this and how the annual statement is affected, but also the international personnel controller has to decide which report he has to present to the management of the national company unit or the group top management on personnel wrongdoings.

2. In accordance with sec. 317 subs. 2 HGB the examination of the position report has to include the following reports:

    - Consistency of the position report with the annual statement and the findings of the examination
    - Generally applicable presentation of the position of the company,
    - Correct presentation of risks of future developments.[462]

The reform – in connection with the examination of the position report – which underlies the highest requirements is the examination whether risks of the future development of the group, influenced by the foreign revenues which can often account for between 40 and 90 percent of the total revenues, are correctly presented. Thereby the legislator clarifies that the prognosis of the company development in the position report does not only imply opportunities but also risks. This requirement makes clear that companies had not totally complied with the requirements of reporting and displaying the actual situation of the company from the legislator's point of view.

Thereby, the legislator increases the requirements for the position report with regard to the provision of decision relevant information, which also has to be contributed by the (international) personnel controlling, if it is endangering growth. The position report of the group is therefore more complex than the annual statement because future oriented facts

[462] Comp. Schindler/Rabenhorst, 1998, p. 1986 et seq.

which can be connected to foreign activities as well as human resources have to be taken into consideration. An example to remember is the Greencard discussion in the information and communication technology area.

For the going concern assumption, a period of 12 months has to be taken as basis, starting from the reporting date of annual statement. For all other essential risks, like the expected development of essential foreign activities, a period of prognosis of two years is appropriate. It can be necessary to determine a correspondingly longer period of prognosis for foreign business which causes risks for the company over a longer period of time.[463]

The legislator does not substantiate the risk term neither in the legal wording nor in the explanatory memorandum. In the business management literature the risk term is not uniformly used either.[464] Risk in the sense of sec. 289 HGB means the possibility of unfavorable, future developments which can be expected with a considerable likelihood.

Risks are understood as the possibility of danger of losses in the context of business activities. This implies a negative deviation of the company development from the planned variables. This results from the general uncertainty of future events which are connected with an incomplete information status.[465]

In the broadest sense the term risk can be described as insecurity. Insecurities, i.e. possible deviations from expected values exist on all levels of business activities as well as in international personnel management and the related foreign activities. Among the possible uncertainties are e.g. changes of political or economic factors in individual countries where the company is active, as e.g. exchange rates, inflation rates and taxes, changes of the structure of the sector by new competitors, new technologies and unexpected legal dispute, complaints and claims which have to be tackled by employees or managers or have been caused by them.

---

[463] Comp. Arbeitskreis Schmalenbach-Gesellschaft für Betriebswirtschaft e.V., 2000, p. 7

[464] Comp. Baetge/Schulze, 1998, p. 937 et seq.

[465] Comp. Bitz, 2000, p. 13

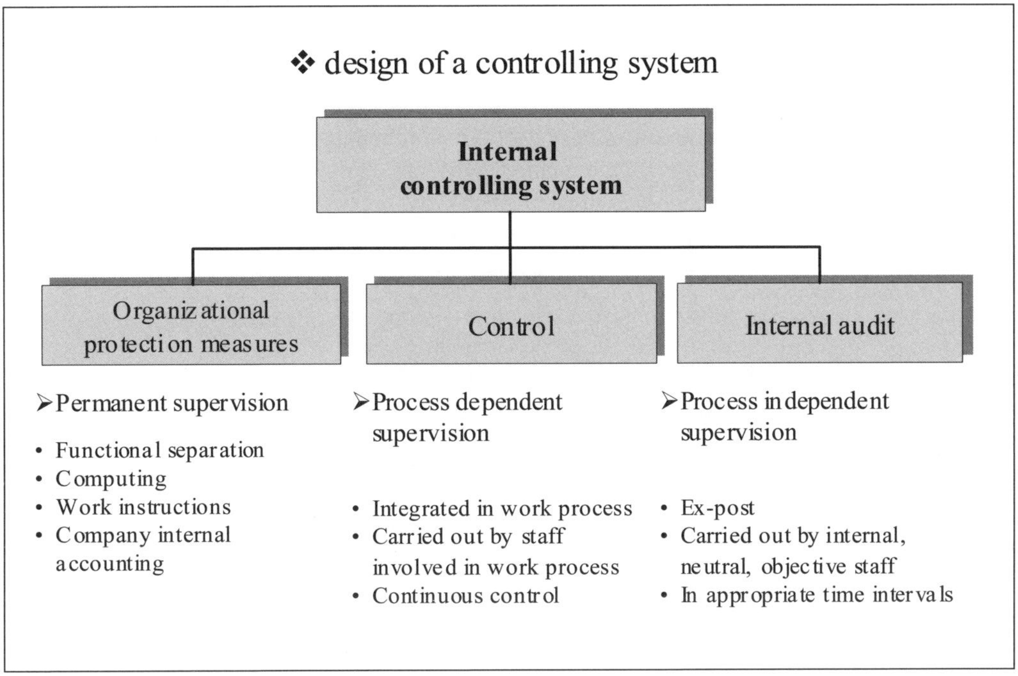

Fig. 27.1: Design of a controlling system[466]

In accordance with sec. 91 subs. 2 AktG the executive board has to install a supervisory system. The executive board has to take care of an appropriate risk management and an appropriate internal audit. The target of risk management and supervisory system is the early detection of risks. The responsible company institutions shall be enabled in due time to introduce measure to keep the company safe.

[466] Following Lück (a), 1998, p. 9

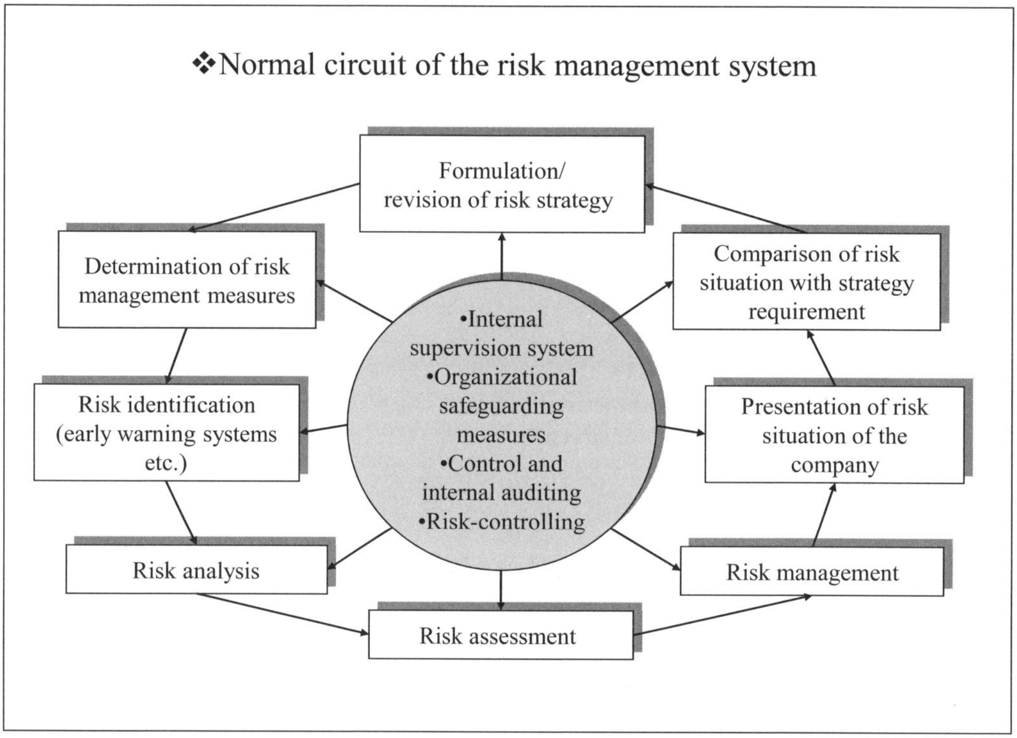

Fig. 27.2: General circuit of a risk management system[467]

In accordance with sec. 91 subs. 2 AktG the executive board has to take appropriate measures and to install a supervisory system in order to allow the detection of developments which might endanger the existence of the corporation. The application of this regulation is part of the examination of the business auditor in accordance with sec. 317 subs. 4 HGB.[468]

Because in accordance with sec. 91 subs. 2 AktG the risk detection system, that has to be installed and continued by the executive board, has to be directed to the early detection of existing endangering developments, only this aspect of the total risk management system is part of the audit. This aspect is defined by determination by the company of risk areas which can imply existence endangering developments.

As internationally active companies gain an essential part of their revenues abroad, the international personnel work and management is a risk area per se, which entails the need to take a closer look on the international personnel controlling as risk field in its aggregated and disaggregated form.

---

467 Following Lück, 1998, p. 1926

468 Comp. Scharf, 1999, p. 180

The risk management system has to make sure that existing risk (risky foreign activities in connection with staff activities) are determined, analyzed and assessed and that the risk related information are forwarded to the decision makers. Part of the risk management is a supervisory system (e.g. international personnel controlling) which supervises the compliance with the requirements. The risk detection system in the sense of sec. 91 subs. 2 AktG has to make sure that those risks which could endanger the further existence of the company are recognized at an early point in time. The auditor has to state whether the executive board fulfills its requirements.

The auditor examines the existence of measures taken by the executive board with the help of related documentations, organizational instructions and own deliberation in the context of the examination of the internal controlling system.[469] The documentation is crucial. Without these documents the auditor can only determine whether the legal regulations ware complied with by means of an organizational examination. However, such an examination is complex. The documentation should display general procedure, measure for risk identification, risk measurement, risk controlling, determined responsibilities as well as training measures for the employees. The examination of the documentation is carried out by inspecting the documents, via analysis of the reports as well as by interviewing leading staff members.

The documentation fulfills the following tasks:

- The subordinated leaders can prove a dutiful behavior in the case of appearance of a company crisis (accountability function).
- The documentation is necessary to ensure the compliance of risk management with the (legal) requirements in the course of time (maintenance function).
- The documentation is basis and condition for the examination of the supervisory system (examination function). [470]

The auditor has to tell whether the taken measures are fit for their purpose by assessing whether they are appropriate in relation to the corresponding risk situation of the company as well as the size and complexity of company structure. In sum, the measures have to be appropriate to recognize the risks of further developments at an early point in time and to internally discuss them in order to give the executive board the possibility to react. The suitability is checked by examining whether the supervisory system does justice to the company goals and strategy. Furthermore, it is important how the reporting of risk potential, e.g. frequency, assessment of risks through sensitivity analyses of worst case considerations is carried out. The suitability of the system, its aptness and appropriateness are determined by the size, structure and environment of the company.

---

[469] Comp. Adler/Düring/Schmaltz, 2000, Vol. 7, p. 123 et seq.

[470] Comp. Kromschröder/Lück, 1998, p. 1573 et seq.

❖ Surveillance of the risk management process by the internal supervision system

Continuous analysis of the risk management process

Prevention function

- Part of all segments of the internal supervisory system
- Avoidance of potential risks
- Minimization of the possible amount of damage of the existing risks

Correction function

- Is above all part of the internal audit
- Development and installation phase: Examination of execution and effectiveness of the system
- After implementation: examination of functionality

Fig. 27.3: Surveillance of the risk management process by the internal supervisory system

If the requirements of the KonTraG are transferred to the risk area "international personnel controlling", there still many creative ideas, e.g. "what does an internal supervisory system "international personnel controlling" look like? How are the risks identified and determined? What are early indicators in this risk area? Which typical key figures are appropriate to make direct and indirect measurement appropriate? What does a regular report to the management look like? How far can the aspects of international personnel controlling be adjusted to internal and external accountancy in order to comply with the KonTraG?

28 Design of personnel information systems

For Scherm/Pietsch/Süß[471] there are three different conceptions of controlling, which are the accountancy oriented, information oriented and the coordination oriented conception from which the targets and tasks and instruments of an international personnel controlling can be deducted. Unfortunately their approach was so general that no concrete key figures could be deduced.

28.1 Targets of an international personnel information system

On the management level, in the context of the functions planning, organization and provision of staff and personnel management, the selection task is perceived. The controlling function on the other hand contains comprehensive reflection tasks.

Controlling as management function can be characterized as reflection on decisions which in the context of other management functions are taken against the background of the complete target system of a company.[472]

Management, in a functional sense, means to take decisions, to execute them and to question them, as well as to take responsibility for the decisions taken. These decisions do not have to be made on the basis of rational calculations but can also result from intuition. To fulfill the management tasks, information is needed either way. The gathering, preparation and provision of this information can be regarded as leader or management support (in the sense of a controlling function.).[473] The authors do not write which information they refer to and they do not further concretize them with the help of qualitative or quantitative key figures etc.

However, some considerations are interesting for the context of international personnel controlling, i.e. that besides the still important perspective of the individual – national – company unit a not less important **international perspective on the whole company group** has to exist. It is true that in international companies, there is still a competition on the individual, national markets, but specific competitive advantages have to be created and used on an international level, in order to stand up against national and international competition on these markets on the long run. It is therefore not possible, even in the case of intense

[471] Comp. Scherm/Pietsch/Süß, 2000, p. 396 et seq.

[472] Comp. Scherm/Süß, 2001, p. 360

[473] Comp. ibid., p. 358

decentralization and differentiation to completely abstain from a central controlling of the company activities, because only in this way economies of scale and economies of scope and national differences can be used, which allows to generate competitive advantages. It is, however, possible that – partially considerable – differences between the targets related to the company as a whole (the head office) and the country market related targets of the company units exist. If these differences lead to different assessments of situation, alternatives concerning possible measures and results of measures, there are bound to be conflicts, at the same time these differences in perspective underline the need of reflection in these decisions. In any case, it has to be made sure that in these decisions related to the company as a whole, the relevant information from the relevant decisions of the foreign branches is available.[474]

If one deducts from the above deliberations the contents of international personnel controlling for delegated leading positions more concretely, the following targets result:

1. Examination of leaders on all levels of the group at home as well as abroad with regard of whether they can be taken into consideration for further management tasks on the next level, especially in the case of management needs in the parent company.
2. Proof that the manager can balance functional interests as marketing, research and development, production and country interests and international interests of the parent company in the context of operative and strategic competition strategies and value adding processes.
3. The primary reasons for failures in multinational operations can be attributed to deficits in understanding essential differences on all levels of personnel management in a foreign environment.[475] It has to be questioned how many individual functions of International Personnel Management have not been executed and how they could help to develop the manager.
4. In contrast to Scherm's understanding the international personnel controlling has to be integrated into an accounting, world wide controlling conception, e.g. in accordance with US-GAAP for stock listed companies, because otherwise the KonTraG with risk detection – as criterion for the parent company – would be questioned.
5. The international personnel (information) controlling is only possible in connection with individual national personnel information controlling systems, i.e. that these have to superpose each other and should not be managed and considered as two separated systems. The best way to demonstrate this is by means of international management remuneration systems, because the remuneration justice shall be reached nationally and internationally/worldwide in the group.
6. The international personnel controlling shall examine to which extent PeopleSoftware, SAP-Software etc. as personnel information systems can help on a world-wide basis and to which extent own company solutions are necessary.

In other words: There is much creative work to be done by companies and economic science.

---

[474] Comp. Scherm/Süß, 2001, p. 362

[475] Comp. Destanick/Bennett, 1978; comp. Weber/Festing et al., p. 11

28.2 Functional design of an international personnel controlling system

This means that international personnel controlling is a network of (national) personnel controlling in the narrower sense and of other functional and project oriented controlling types as R&D controlling, production controlling, logistics controlling and marketing controlling worldwide, which provides risks and opportunities in the coordination concept of company management. Only this way international competition strategies and competitive advantages of the group are not compromised. The target in research and development controlling with a national/international personnel controlling has to be to obtain complete transparency in national/international technology management regarding the question whether the company disposes over enough high professionals with corresponding abilities to secure future profit potentials for the company. This is necessary with regard to project oriented R&D activities, employees, regarding (remuneration/personnel) costs, degree of target achievement measured e.g. in patents and in order to control deviations if necessary.

Net work structures of internationally active companies are new challenges for international personnel management, as Perlitz[476] shows by the examples of General Motors with the corresponding production and logistic network. When components of a car are fabricated and assembled on a worldwide basis, this means that technology-oriented companies are especially in need of differentiated production controlling in connection with personnel controlling. One should only think about the target variable (personnel) costs and their compositions, efficiency, which is influenced by illness, absence hours, turn-over rates, use of technology with different qualifications of the staff, quantity of production with different labor time models and limitations with regard to culture, society, politics and legal regulations as well as quality assurance with different total quality management models and instruments in connection to approaches of personnel training or development.

The logistics performance is closely connected with international network structures. It can be measured by the achievement of the following targets, which at the same time underline the personnel service character of the logistics function. These aims are service time, service reliability and readiness for delivery in the planned service time all around the world.

The following question can be asked about (international) marketing controlling in connection with the (international) personnel controlling: to which extent can the manager systematically put the world wide company management concept into practice in the individual countries in order to be successful as global player for the group in the automobile sector, telecommunication sector or aviation sector? Quantitative targets of marketing controlling and target adaptation in the context of performance assessment of personnel controlling can be company growth, market shares or contribution margins. Qualitative key figures include high brand awareness, positive image and high levels of repurchases which

---

[476] Comp. Perlitz, 1997, p. 231 et seq.

the delegated managers can only achieve with the help of knowledge and understanding of culture.

About the national/international personnel controlling it can be asked, among other things, how a personnel development strategy for non-tariff employees and top managers should be executed nationally and internationally in the group or how a target adjustment concept for upper and top managers should be standardized nationally and on a world-wide basis in the group and how this would affect their careers. In this respect, the personnel development strategy concept can certainly be well connected with the strategy concept of personnel assessment. From this point departs the strategy concept of personnel remuneration management in the group. All concepts can be connected to the functional concepts in the research and development area, the marketing area, of national subsidiaries etc.

28.3 Process supporting tasks of international personnel controlling

Targets and tasks of process supporting personnel controlling refer to all innovative cases of functional international personnel controlling, i.e. when they are project-wise introduced in the worldwide group for the first time or reformed after several years.

In 1996, the BASF has developed a remuneration system for international delegations of upper managers. It created the following criteria or requirements for a corresponding material system of international delegations:

- Acceptance by the delegated managers who can understand all calculation steps in their remuneration statement because of its transparency and plausibility;
- Adequacy which is ensured by the flexibility of remuneration systems which make an appropriate adjustment to individually different cases possible.
- Economic efficiency, which is achieved by the fact that on the basis of its "easy handling", quick, immediately understandable results allow to take actions in individual and group-related matters on an international as well as international basis. In this context e.g. components of the remuneration management system are permanently gathered, updated, examined and adjusted, in order to guarantee the highest possible justice of upper managers in the group and in comparison to worldwide labor markets.[477]

[477] Comp. Brinkkötter, 1997, p. 443

Literature

Arbeitskreis „Externe und Interne Überwachung der Unternehmung“ der Schmalenbach-Gesellschaft für Betriebswirtschaft e.V.: Auswirkungen des KonTraG auf die Unternehmensüberwachung. in: Der Betrieb, volume 37, annex n°.11/2000, 2000.

Abts, D. / **Mülder**, W.: Grundkurs Wirtschaftsinformatik. 3rd edition Braunschweig/Wiesbaden, 2001.

Atkinson, A. A.: Management Accounting. 2nd edition, Upper Saddle River (N.J.), 1997.

Baetge, J. / **Schulze**, D.: Möglichkeiten der Objektivierung der Lageberichterstattung über „Risiken der künftigen Entwicklung“. in: Der Betrieb, volume 19, 1998, p. 937–948.

Beynio, W. / **Krieger**, W.: Stock Option Incentives von Oberen Führungskräften in der Henkel-Gruppe. in: Clermont, A. / Schmeisser, W. (ed.) 1998, p. 353–362.

Bitz, H.: Risikomanagement nach KonTraG. Stuttgart, 2000.

Brinkkötter, H.-O.: Grundfragen aus der Entsendungspraxis der BASF mit besonderer Berücksichtigung von Oberen Führungskräften. in: Clermont, A. / Schmeisser, W. (ed.): l.c, 1997, p. 413–428.

Brinkkötter, H.-O.: Das Gehaltssystem der BASF für internationale Entsendungen von Oberen Führungskräften. in: Clermont, A. / Schmeisser, W. (ed.): l.c., 1997, p. 429–444.

Brinkkötter, H.-O.: Aktienprogramme der BASF Aktiengesellschaft. „BOP“ für Obere Führungskräfte – „plus“ für die Belegschaft. in: Clermont, A. / Schmeisser, W. / Krimphove, D. (ed.). l.c., 2000, p. 663–691.

Clermont, A. / **Schmeisser**, W. (ed.): Internationales Personalmanagement. Munich, 1997.

Clermont, A./ **Schmeisser**, W. / **Krimphove**, D. (ed.): Personalführung und Organisation. Munich, 2000.

DGFP – Deutsche Gesellschaft für Personalführung e.V. (ed.): Personalcontrolling in der Praxis. Stuttgart, 2001.

Gilles, S.: BSC als Konzept. 2002.

Krawinkel, U.: PeopleSoft Collaborative Applications – Ein Überblick über globale Einsatzmöglichkeiten im Personalbereich. in: Clermont, A. / Schmeisser, W. / Krimphove, D. (ed.): l.c., 2001, p. 447–456.

Kromschröder, B./ **Lück**, W.: Grundsätze risikoorientierter Unternehmensüberwachung. in: Der Betrieb, Volume 32, 1998, p. 1573–1576.

Lück, W.: Der Umgang mit unternehmerischen Risiken durch ein Risikomanagementsystem und durch ein Überwachungssystem. in: Der Betrieb, 51. Vol., 1998, p. 1925–1930.

Lück, W. (a): Elemente eines Risiko-Managementsystems. Die Notwendigkeit eines Risiko-Managementsystems durch den Entwurf eines Gesetzes zur Kontrolle und Transparenz im Unternehmensbereich (KonTraG). in: Der Betrieb, 51. Vol., 1998, p. 8–14.

Mülder, W.: Entwicklungstendenzen beim Personalinformationssystem SAP R/3 HR. in: Clermont, A. / Schmeisser, W. / Krimphove, D. (ed.): l.c., 2001, p. 457–476.

Rosen, R. v.: Aktienoptionen für Führungskräfte. in: Clermont, A. / Schmeisser, W. (ed.): l.c., 1998, p. 341–351.

Perlitz. M.: Aspekte, Strategien und Probleme bei der Entwicklung eines Handlungssystems für das internationale Personalmanagement. in: Clermont, A. / Schmeisser, W. (ed.): l.c., 1997, p. 223–237.

Scharf, P.: Risikomanagement- und Überwachungssystem im Finanzbereich. in: Reform des Aktienrechts, der Rechnungslegung und Prüfung. Stuttgart, 1999, p. 177–201.

Scherm, E. / **Pietsch**, G. / **Süß**, S.: Internationales Personalcontrolling zwischen statusardisierung und Differenzierung. in: Personal, edition 52, volume 9, 2000, p. 470–476.

Scherm, E. / **Süß**, S.: Internationales Management. Munich, 2001.

Schindler, J./ **Rabenhorst**, D.: Auswirkungen des KonTraG auf die Abschlussprüfung (part I). in: Betriebsberater, volume 37, 1998, p. 1886-1893.

Schmeisser, W.: Finanzorientierte Personalwirtschaft. Munich 2008.

Schmeisser, W.: Flexibles Vergütungsmanagement im Rahmen des Shareholder Value-Ansatzes. in: Clermont, A. / Schmeisser, W. / Krimphove, D. (ed.): l.c., 2001, p. 811–819.

Schmeisser, W. / **Clermont**, A./ **Merle**, A.: Internationales Personalcontrolling. in: Clermont, A. / Schmeisser, W. (ed.): l.c., 1997, p. 529–544.

Schmeisser, W./ **Clermont**, A.: Personalmanagement. Praxis der Lohn- und Gehaltsabrechnung, Personalcontrolling und Arbeitsrecht. Herne/Berlin, 1999.

Schmeisser, W. / **Clermont**, A./ **Protz**, A. (ed.): Personalinformationssysteme und Personalcontrolling. Auf dem Weg zum Personalkosten-Management. Luchterhand, 1999.

VIII Design, functionality and application fields of information systems in HR[478]

29 International personnel information systems

29.1 Definition of information systems

In the case of an information system in the HR context this is a person-computer-system. The system consists of human and technical components and describes a systematically ordered network of informational relations which is developed between human and information processing machines, data and methods. This means that the personnel information systems serve to satisfy the purposeful informational needs of the participants of the worldwide company organization.

An informational system serves as instrument for companies in order to be able to act in a quicker, better and more precise way in HR work.[479] Furthermore, it helps to secure and expand strategic competitive advantages which are created through human capital.

A personnel information system is part of the organization system, in which information and communication relations between individual positions count among the central organizational appearances. By connecting them with other parts of the system, an electronic

---

[478] This chapter, including illustrations, is largely based on the book: Strohmeier, St.: Informationssysteme im Personalmanagement: Architektur -Funktionalität – Anwendung. Vieweg und Teubner Verlag, Wiesbaden 2008. Strohmeier is nearly the only author dealing with this topic in Germany.

[479] Comp. Jung, 2005

mutual data transfer is made possible. This should involve a unification of the data base and thus make adjustment actions between the individual system components superfluous. The advantage of the connection is the higher informative value through combination of different areas of the system.

A personnel information system is open to its environment, because it freely interacts with it. Besides, it is dynamic because parts of the information system might change due to this interaction.[480]

29.2 Tasks

An information system contributes to the best possible provision with information to employees and managers. Its tasks include the processing, analysis and transfer of data and information which are different from system to system. Therefore, each system has its own tasks which are shown in more detail chapter 31.4.

29.3 Reasons for introduction of personnel information systems

Two essential reasons which make the introduction of a personnel information system necessary are (1) the standardization of comparable information, e.g. in the remuneration system problematic and in the international assignment of personnel as well as (2) the context of globalization and of company internal decentralization, which makes it easier to operate an information system for international companies.

The company guideline is thereby conveyed in a unified way. However, the regional individuality and culture still have to be accepted by all groups.

Furthermore, processes have to be unified. Via the personnel information system a process becomes more structured, which saves time. The personnel information system makes it possible to save costs and to simplify processes. Via these tasks information are provided in much shorter time and are more up-to-date and more comprehensive, e.g. to answer the question whether the introduction of a personnel information system is useful, the application of a cost-benefit-analysis is suitable.

The costs can be considered in two different ways: on the one hand in accordance with Weimair who structures the cost elements in accordance with activities and development phases. He distinguishes the planning phase, implementation phase and operation phase. Domsch on the other hand differentiates one-time and running costs.

---

[480] Wirtschaftslexikon24, status: 2009

The benefit is segmented along the individual application fields of personnel work. Among the application areas are the *personnel administration* where the question is asked how to avoid the frequency of errors in the calculation of salaries, the *personnel planning* which makes it possible to create an enlarged and improved information flow for planning purposes on the basis of higher data consistency, and the *personnel controlling* which renders a quicker and more exact deviation control in relation to performance and personnel flow possible via the Berlin Balanced Scorecard Approach.[481]

29.4 Personnel information systems as condition for process optimization

An information system in the personnel department is condition for the total process optimization.

Via *Business Process Reengineering* processes are reformed to be able to work in a more economic and customer-oriented way. With the help of computer aided personnel information systems there is the possibility to fundamentally reform out-of-date, cost-intensive personnel processes.

Human resource processes have to be analyzed for this purpose. In the case of a business process analysis, mistakes and weaknesses are found out and documented. Examples for these weaknesses are wrong data entries, double work or long waiting times and downtimes.

The target of process optimization is the improvement, optimization and unification of future business processes. By this, it shall be achieved to reduce the time and costs needed and improve the quality at the same time.

Via the utilization of *IT systems* in human resource area, rationalization and automation measures of routine work can be carried out. One example of an application field is the remuneration calculation. Besides, it serves to support administrative tasks, e.g. the administration of applicants and seminars and the preparation of statistics.

The personnel recruiting process is optimized via *e-recruiting*. Hereby measures as personnel marketing, search for applicants and qualification examination are winded up via the internet. The publication of free positions or the maintenance of the candidate pool, i.e. of the platforms where applicants can register for jobs and internships, are winded up via internet.

With the help of *e-administration,* administrative HR processes are optimized. There is a shift of the data collection and maintenance from the human resources resort to the employees and managers. The maintenance of and the access to information is realized via the browser. Data misuse can be avoided through cryptographic techniques and security can be achieved via error checking and authorizations by supervisors. By the entry of routine

481 Comp. Jung, 2005

processes directly per web-based form, for example changes of address and bank details, they can be quickly registered.

In the *e-relationship management* customer satisfaction and the increase of customer benefits are focused. This means that more awareness of the human resource customer processes shall be created. Internal customers including employees and managers but also external customers as e.g. applicants and public institutions require from the human resource department a variation of services. Examples are the creation of personnel statistics or the calculation of company pension. Via automation, routine work is transferred from the human resource department to the customer. The remuneration certification for example can be printed out by the employee himself.

With regard to opportunities of personnel information systems one can distinguish qualitative and quantitative benefits.

The qualitative benefits include the acceleration and simplification of human resource processes. Through quantitative benefits more statistics are possible. This can mean that the service quality in the human resource department can be improved, because via web the "virtual" human resource department is available around the clock. In case of standard questions, the human resource department can be relieved and rationalization effects can be obtained. Furthermore, cost savings can be the result of process optimization.

Among other things, there are two basic problems: processes can only be estimated at the beginning and cost-benefit-assessment can often not be carried out after implementation.[482]

30 Basics

30.1 Architectonic basics

30.1.1 Architecture

The architecture has to be construed for hardware, software and processes. These components have to be connected on a joint data basis to the architecture as cutting point.

There is also a connection to the backup systems usually common in business to ensure data security.

[482] Comp. Schmeisser/Grothe/Hummel, 2003

30.1.2 System architecture

The system architecture describes the structure of software systems. It consists of individual software components and their relations. These are basic resources which can be accessed via the operating system, as well as basic characteristics and the organizational structure of these resources. Thereby, the internal structure of an individual system, which is also referred to as micro architecture, and the structure of a portfolio of different individual systems, which is also called macro architecture, are described.

A system is referred to as individual, if the software unit is independently loadable and if it can be individually purchased from other companies on the market. In case of portfolios of several individual systems the components depend on each other.

30.1.3 Layer model

The layer model serves as support for structuring the description of real system architectures. The individual systems are in correspondence between each other with the layers built up on each other. The basis is the data management layer. The upper layer uses services and results of the corresponding underlying layer.

Data maintenance systems are responsible for saving, administration and facilitation of persistent data of the system. These consist of encoded data, e.g. letters, numbers, special characters and not encoded data as e.g. illustrations, audio and video documents.

Application systems realize all functions of specialist functionalities of a system, e.g. the processing logic, the application core and the business logic. In this context the systems use services of the data maintenance systems. Thereby, at first a data recall from the data maintenance layer is undertaken, then data processing and finally result saving in the data maintenance layer is carried out.

The *presentation systems* serve as user interface. This is the place where an interaction between users and applications takes place. The user receives all necessary functionalities to work with the system.

Integration systems render services of communication between otherwise isolated individual systems or components of the application layer as for example function calls in the system. Furthermore, integration systems take over the direction of data transport. They also adequately describe heterogeneous macro architectures and form the facultative layer between the application and presentation systems.[483]

---

[483] Comp. Strohmeier, 2008

30.1.4 Implementation

The implementation can be achieved in an individual system. The separation in layers makes it possible to distribute individual services to different systems.

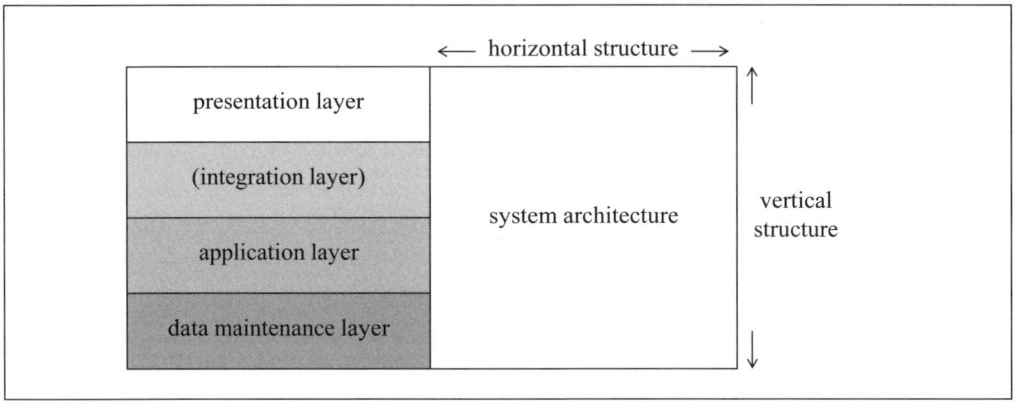

Fig. 30.1: Logic layer and physical structures of system architecture[484]

Personnel information systems can be structured and arranged from two basic points of view, the horizontal and the vertical point of view. The vertical structure shows how one or several systems are ordered over several architectonic layers. They are subdivided in layer overlapping systems and layer specific systems. The *layer overlapping systems* offer services from several architectonic layers. Services of data maintenance, application and presentation are regularly offered from such systems via corresponding components.

The layer specific system offers exclusively services from one single architectonic layer. Here, outsourcing data maintenance services to independent data base systems is very common. The data base system works in this architecture as layer specific data maintenance system while services of application and presentation remain in the corresponding personnel system.

There is, however, also the possibility of combining *layer overlapping systems and layer specific systems*. The attribution of a system to an architectonic layer can also depend from the form of application.

The horizontal structure displays the organization of one or more systems by relations to the architectonic layer in question. Services can be rendered via one or several systems. If in the data maintenance of the application layer several individual systems are used in parallel, these can be integrated. If there is no integration, an additional integration layer is necessary or useful.[485]

[484] Source: Strohmeier, 2008, p. 5

[485] Comp. Strohmeier, 2008

30.2 Organizational basics

Information systems in personnel management have to be planned, made available, implemented and applied in order to lead to the desired results. These functions are understood as processes which can be executed in different ways. A typical process can be recognized in the following figure. It starts with system planning. This includes the theoretical picturing of future information systems. The system appropriation includes the necessary activities of development and/or purchase of the planned systems. In the context of system implementation the mutual adaptation of organization and information systems takes place. After this, the activities for the implementation and maintenance of applicability of the system are carried out in the context of system operation. The last step is the system application in which diverse users take over system aided functions of personnel management.

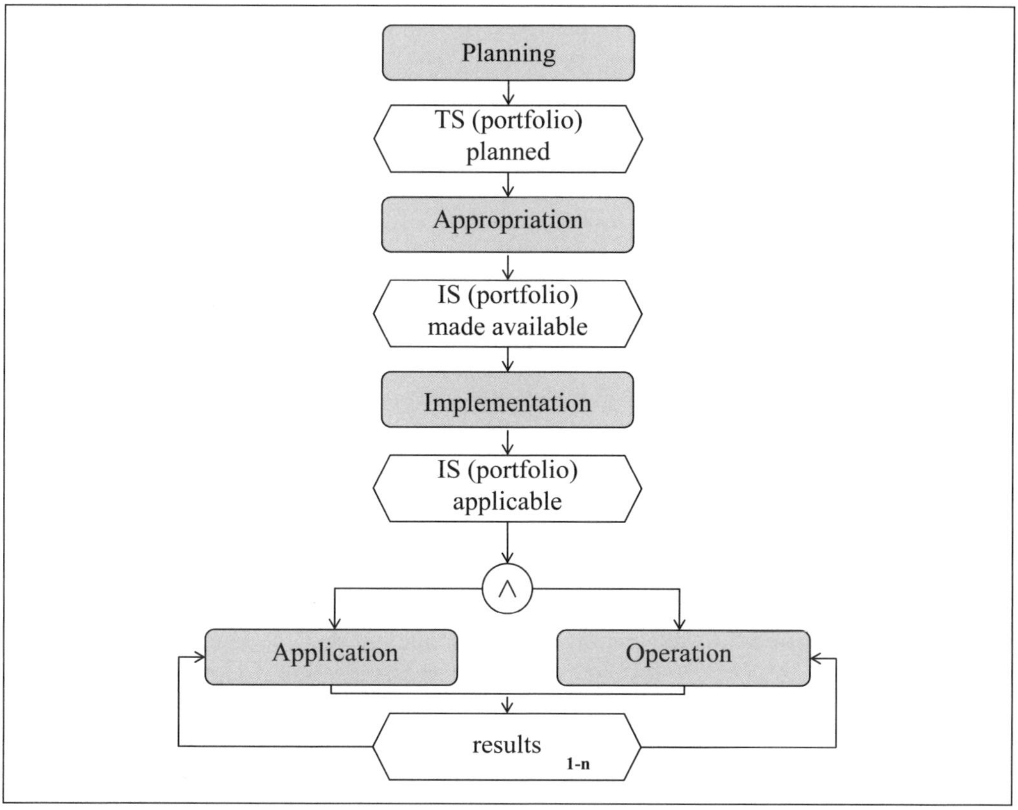

Fig. 30.2: information system related activities[486]

[486] Source: Strohmeier, 2008, p. 15

30.3 Legal basics

The following paragraphs describe the governing legal regulations on the application of information systems in personnel management.

30.3.1 Data protection

In the context of data protection, information systems in personnel management are targeted on the collection, saving and processing of data concerning areas especially worthy of protection from potential for misuse. The central data protection law is the German Data Protection Act (BDSG). The goal is to "protect the rights of individuals being affected by the use of their personal data" (sec. 1 subs. 1 BDSG). This way, personal rights of the affected individuals shall be protected and misuse shall be avoided.

Because of the sensitivity of personnel data there are high demands regarding the security for accessing, saving and transferring these data. The measures concern especially the confidentiality of personnel data for protection against unauthorized reading, the integrity of data for protection against fraud, authenticity of the communication partner for protection against masquerade, proof of data entry for non-repudiation and the regular saving of data for protection against loss of data. In order to fulfill these requirements, a preventive control by the assignment of user rights has to take place and all activities have to be recorded.[487]

30.3.2 Co-determination

The co-determination is regulated by the Industrial Constitution Law (BetrVG). These regulations are relevant for the implementation and application of information systems. Among other things the regulations are targeted on the participation of the works council. The German Act on the Speakers Committee (SprAuG) on the other hand governs the participation of Speakers Committees.

30.3.3 Equal treatment

Through equal treatment of all employees discrimination shall be avoided. The General Equal Treatment Act (AGG) distinguishes eight discrimination factors which shall be protected against discrimination and harassment. Those factors are race, ethnic origin, disability, sexual identity, religion, ideology and sex.

---

[487] Comp. Strohmeier, 2008

30.3.4 Data transfer

The ordinance on the collection and transmission of data for social insurers (DEÜV) obliges the employee to make concretely defined reports to the social insurance institutions. Hereby data have to be electronically transferred and measures for data security and data protection have to be taken.

30.3.5 Remuneration accounting

For remuneration accounting the principles of Generally Accepted Data Processing Aided Accounting Systems and the principles for Data Access and Verifiability of Digital Data apply. The system categories involved are the personnel accounting systems and document management systems, in case the wage accounting receipts are archived.

31 Structure, functionality and application area

31.1 Structure

A personnel information system is composed of the components human resource database, work place database, methods and model database, electronic data processing systems and system users. These components can be understood as follows:

The *human resource database* administrates the human resource master data. It is needed for the payroll. Examples of basic data of the human resource database are personal data, employment data as well as remuneration master data. These basic data are completed with dynamic data, e.g. data from time tickets, commission accounts and travel expense reports.

The *workplace database* is targeted towards saving workplace related data, e.g. the tasks of the position, the requirements of training and education and the workplace number.

The *method and model database* is responsible for procedures and algorithms for further processing of the saved data. It can be understood as computer programs to solve the problems in the human resource area, e.g. remuneration accounting systems or calculations of absence periods and turnover rates.

The core of a personnel information system is the *electronic data processing system.* They are the central processing unit. Further peripheral devices are attached to allow data input, saving and output. The totality of the technical equipment is called hardware.

Among the system users active users can be distinguished from passive users. *Active users* are employees who directly correspond with the data and change and connect them via use of methods. *Passive users* are employees who only take notice of the available information and may use them as basis for decisions.[488]

31.2 Functionality

The personnel information system can be organized in very different ways depending on the area of responsibility and development status. It is generally especially customized to the company needs. This way, it can be subsequently updated at any time. The staff specific tasks and problems which are fulfilled with an information system can be separated in the main areas of administrative and dispositive tasks.

With *the administrative tasks* rationalization effects shall be achieved which arise because of the relief provided in the area of routine work and mass procedures. These include especially repeated procedures like salary accounting.

The *dispositive tasks* support the decision and planning process in the human resource field via the improvement of the information basis. By processing data material, a better information basis and decision-making support is provided. It serves for future problem areas of human resource planning and controlling, e.g. for the selection of applicants.

The personnel information system fulfils the functions of personnel planning, human resource accounting, of master data administration, time determination, administration and human resource reporting. The position plan, personal resource availability plan and the plans for further training are understood as personnel planning.

Human resource accounting includes the calculation and management of remuneration costs, income tax and social security contribution accounting. *Master data administration* connects change services, labor contract data and address data. *Time determination* includes absence control and administration and assessment as well as holiday accounting. Administration includes administrative efforts for employment, dismissals and *promotions*. *Personnel reporting* connects statistical analyses, creation of internal staff statistics, e.g. the age structure or turnover rate, as well as statistics for external addressees.

Widely used data processing systems in personnel management are for example SAP or the PAISY-system. The *SAP system* offers the possibility to standardize core processes in human resource management. This means that the necessary data are available to the authorized employees at once. The *PAISY-system* is a dialogue-oriented, modularly structured data administration, calculation and information system for the DP-based human resource management. It can be quickly and cost-efficiently used and is also applied in remuneration accounting etc.

---

[488] Comp. Strohmeier, 2008 and Jung, 2005

31.3 Area of application

31.3.1 Users

Users can be divided into the groups of active users and potential addressees. Active users can be split into participants with active-indirect and active-direct contact. The *active-indirect users* can access information without entering into direct contact with the personnel information system. The *active-direct contact* involves the access with direct contact.

The information desires of potential addressees are fulfilled with the help of the information system. Within a company, the data are needed by the human resource department, the supervisors and employees. Outside the company, information from the personnel information system is accessed by public institutions and the general public, as e.g. shareholders, associations and media.

This means that information systems are used by complete workforce, managers and executive board members.

31.3.2 Examples for application areas

There is a variety of application areas where information systems are helpful. Examples for such application areas are online-queries, reports, applications, personnel development and the termination of labor contracts.

In the case of an *online-query*, a request with regard to selected employee data is possible at all times because of the direct and permanent connection of the terminals of the human resource department to the central unit. Ideally, it is already clear when installing the system how employee data are structured under certain topics, who has the authorization to access the data and how the highest possible transparency in the presentation is achieved.

Reports on the other hands are issued by the information system in order to support the human resource department in the execution of its supervisory functions. So called event connected reports, periodic reports and aperiodic reports can be distinguished. *Event-connected reports* are connected with the occurrence of certain facts, e.g. a job entry, resignation or change of address of employees. *Period reports* are regularly issued in previously defined intervals. *Aperiodic reports* on the other hand are individually issued if needed. They are mostly prepared for organizational and administrative purposes, for example address lists, telephone directories or parking space lists.

In the *application field* the applicants' data are saved in accordance with a certain scheme. With the help of the personnel data collected in the application and by further processing with the help of the information system, a variety of interpretations is possible. An example is the preparation of a list of all vacancies, an overview over the ads published in the media, supervisory functions for running applications and the recognition of recurring applications.

In the *personnel development area* the most important elements of the career of an employee are mostly saved in form of a list. Among these are the duration of the membership in the company, area of responsibility, possible further qualification measures and promotions.

The *termination of labor relations* also creates a high number of administrative tasks which are systemized and accelerated with the help of information systems. Important tasks in the termination of a labor relation are the preparation of letters of notice, the issue of employment certificates with an exit interview, changes of the personnel file or personnel data as well as archiving the personnel records, in order to be able to give information in case of further enquiries of succeeding companies or insurance-related questions and to make additional statistical evaluations.[489]

31.4 Structure, functionality and application of the model with four layers

31.4.1 Data management layer

The data management layer is divided in three areas: the database systems, the data warehouse systems and the document management systems.

A *Database System* is a tool for storing, managing and furnishing large amounts of data.

A *Data Warehouse System* serves to combine and keep personnel management data from several other sources for the purposes of information furnishing and decision support in personnel management.

A *Document Management System* is an application for issuing, digitalizing, archiving, retrieving, forwarding and destroying human resource documents of all kinds.

Now, a closer look is taken on a database system in relation to its structure, functionality and application.

The composition of a database system is to be distinguished in the specified user interface, the database component, the data management system and the programming language or programming tools. The function of a database is data storage. The database management system, on the other hand, has the functionality of definition and manipulation of the database and of scanning. The database system is used as layer specific data keeping system or as trans-layer application system. In this case data management services are rendered for systems of higher layers. The efficient storage and facilitation of very large amounts of data is guaranteed in this system category. This system is crucial for personnel management. Database systems available in office packages offer multiple application scenarios in

[489] Comp. Strohmeier, 2008 and Jung, 2005

personnel managing. They deal with general personnel data management. Many commercially offered personnel information systems are really only simple database-based applications. They relief the users from data modeling and designing the corresponding forms or user interfaces. However, they offer the possibilities to make ad-hoc queries and reports in the core functionalities of the database management system. Database systems are widely used for data management in personnel management application systems.

31.4.2 Application layer

The following system types belong to the application layer: The *personnel requirements planning system* is an application to determine future qualitative, quantitative, temporal and local manpower needs. The *staff assignment planning system* is an application for short and middle term adjustment of quantitative and qualitative personnel requirements with the available workforce or the concrete assignment of personnel. This system is quasi a "prime example" for decision-supporting systems in personnel management. The *personnel development planning system* supports the systematic development of employees according to their qualifications and interests and the requirements of the company. The *personnel cost planning system* is an application for the support of personnel management in the prognosis and analysis of personnel costs. The design, communication and administration of holistic remuneration concepts are carried out with the help of the *remuneration management system*. The *online analytical processing system* carries decision-oriented, multi-dimensional and (dis-)aggregated data analyses out. The *data mining system* is used for the identification of valid, still unknown, potentially useful and comprehensible patterns in data. The *working time management system* electronically evaluates, plans and controls individual-related times of presence and absence. *The personal remuneration system* is designed for an automated, individual-related and work-related calculation of salaries and wages and further payment-relevant factors e.g. travel expenses. The *access management system* is an application for planning, managing, recording and controlling access by staff and third persons to individual parts of the company. In order to execute and assess one or several standardized procedures for the evaluation of the values of different performance and personality characteristics, there is a *test system*. *The supply management system* helps with the facilitation of personnel. The *scenario system* is an application for the purpose of self-diagnosis and training which simulates a complex problem which is assigned to a test person and is meant to be solved with regard to certain targets in the best possible way. The *computer based training system* provides know-how through the facilitation of learning contents and the learning process related support by learning users. In order to achieve a comprehensive, administrative and didactical support of learning processes in the company, the *learning management system* is used. Via the *performance management system* individual work performance of the employees can be systematically managed. The enterprise resource planning system is a data and function-integrated company-wide application for the support of central company functions. It is used for procurement, production, financing, sales and personnel management.

Between all these systems, the *personnel assignment planning system* is now analyzed with regard to its structure, functionality and application areas in more details. The typical structure is a strongly heterogeneous, sector-oriented group of personnel assignment planning systems. The utilized data work with data on employees, requirements, shifts, shift sequences and operation schedules. Within a personnel assignment planning system there can be components for requirement planning, shift planning, employee assignment and analysis and dispositions. The following Fig. illustrates the typical architecture of a personnel assignment system. The *requirements planning component* is responsible for the determination of the temporal, local, qualitative and quantitative staff requirements. It is necessary for any assignment plan as input value. The *shift planning component* plans individual shifts that cover the corresponding requirements. Once the requirements are covered, *the shift sequence planning* is undertaken whereby the individual shifts in the course of time are composed. The interim results are saved in corresponding files. The task of *the employee assignment component* is to assign individual employees to the shift sequences. The employees' data are kept in an employee master file. In the *assignment plan data base* the previously achieved results are finally saved. The analysis and disposition component evaluates short-term manual changes in the issued assignment plans.

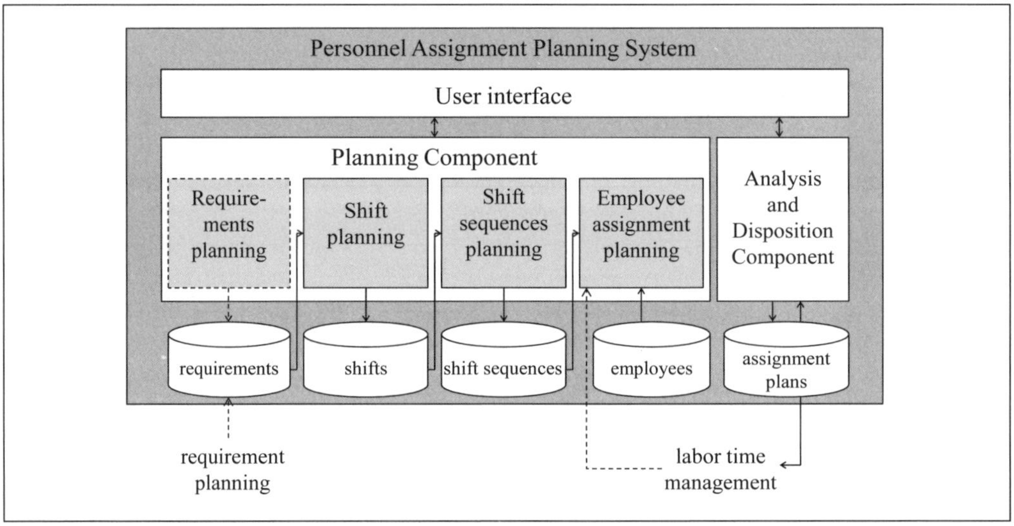

Fig. 31.1: Typical architecture of Personnel Assignment Planning Systems[490]

Application areas are sectors and company areas characterized by short-term fluctuations and requirements with need of timely fulfillment. These are for example commerce, hotels and restaurants, guard services, transport and traffic and emergency services in the public are and industrial fabrication. Around one quarter of the German companies use personnel assignment planning systems.[491]

---

[490] Source: Strohmeier, 2008, p. 107

[491] Comp. Strohmeier, 2008

31.4.3 Presentation and Integration Systems

The *business process management system* is an application for the design, execution and management for business processes over several application systems and users. The *gateway system* is a platform where all necessary applications, information and interactions are made available to the internal and external users in a personalized and requirement-adjusted way. In order to connect the telephones as end user equipment with the application systems, the *voice response system* is used. The *service center system* systematically supports specialized organizational units for the network supported processing of administrative personnel functions. The *browser system* makes the access, display and utilization of the services and applications on internet servers possible.

A detailed description of a presentation and integration system is now carried out with the help of a service center system. The structure generally contains a data keeping component with service center agencies, a knowledge database and an access history. There is a communication and process management and an analysis component. The data keeping component receives the relevant data from the *service center agencies*. The communication component combines different telecommunication services to a "contact center" and support the communication between agents and the requesting persons. Through the *process management component* the distribution and processing of the incoming request is managed. The *analysis component* makes it possible to evaluate incoming requests and to handle them and thereby to control the service center organization.

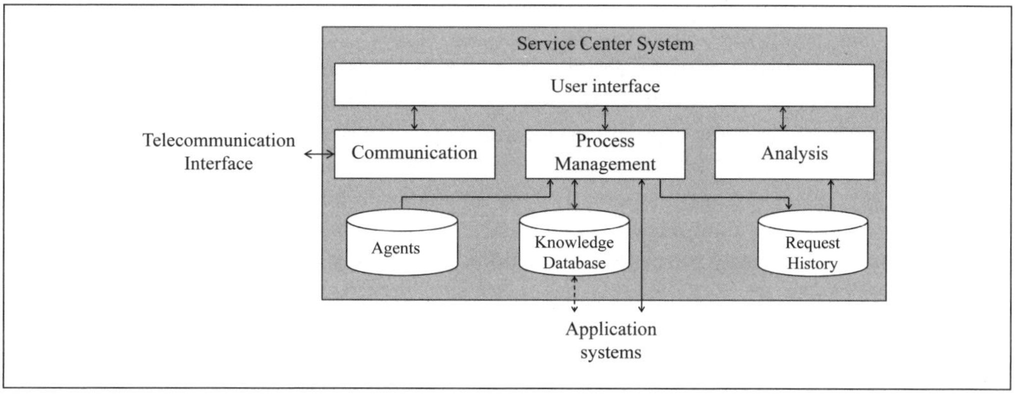

Fig. 31.2: Typical architecture of Service Center Systems[492]

Furthermore, it has to be mentioned that there is the possibility of shared service centers. There is a smooth transition to outsourcing: The services are not only offered to the own company, but also to customer companies. Service center systems are useful for handling administrative and personnel management tasks, the interaction between human resource

---

[492] Source: Strohmeier 2008, p. 339

management and employees, supervisors and applicants. They are used to deal with the proper requests, as information hotline and for reclamation management.

32 Conclusion

32.1 Opportunities

Modern IT solutions offer big opportunities to optimize the traditional, time and cost consuming personnel management processes. A profound analysis and optimization of personnel processes is indispensable for the successful and efficient utilization of information systems in human resource area. By their introduction it is possible to offer better personnel services at lower costs. Quicker administrative processes and the possibility of quicker specialized trainings belong to the opportunities, too. Furthermore, the information and communication are improved by use of internet and intranet.

32.2 Risks

However, the introduction implies certain risks as well. Different factors can be responsible for this. E.g. problems can result from tensions between different factions in the company. The employees who attribute high importance to their privacy are alienated and persons who are not open-minded with regard to reforms and not willing to learn, show skepticism and pessimism towards the introduction. The rationalization possibilities mean besides the positive aspects for the company also negative aspects for individual employees, who are afraid to lose their jobs and therefore take position against the introduction of an information system.

A profound analysis and optimization of personnel processes is indispensable for the successful and cost-efficient utilization of information systems in the human resource area. Despite of the numerous advantages, a personnel information system remains only a tool in order to be able to act in a better, quicker and more flexible way in human resource work.

Today, the introduction of a personnel information system can still offer large, strategic, competitive advantages – the introduction of an international personnel information system will be unavoidable in future.

Literature

Jung, H.: Personalwirtschaft. 6. revised edition, Oldenbourg Wissenschaftsverlag GmbH, Munich 2005.

Schmeisser, W. / **Clermont**, A.: Personalinformationssysteme und Personalcontrolling – Auf dem Weg zum Personalkosten-Management. Hermann Luchterhand Verlag, Neuwird Kriftel (Taunus) 1999.

Schmeisser, W. / **Grothe**, J. / **Hummel**, Th. R.: Internationales Personalcontrolling und internationale Personalinformationssysteme. Rainer Hampp Verlag, Munich and Mering 2003.

Strohmeier, St.: Informationssysteme im Personalmanagement: Architektur – Funktionalität – Anwendung. Vieweg und Teubner Verlag, Wiesbaden 2008.

Wirtschaftslexikon24: URL: http://www.wirtschaftslexikon24.net/d/informationssystem/informationssystem.htm status: 2009.

IX International external personnel recruitment and staff selection

33 Internet as medium of international personnel recruitment

The use of the internet as way of recruitment of personnel has strongly increased since the mid 1990s. According to a study of the University of Frankfurt on Main and the internet job exchange Monster.de 91 percent of the 1,000 biggest companies located in Germany use Electronic Recruiting by now.[493]

Similar trends were shown by a customer survey of the jobpilot AG in spring 2001 from which could be understood that already at that time the **internet** was used **as popular recruiting instrument**. 86 percent of the participating business clients (with less than 100 up to more than 10.000 employees) reported the use of online job boards, while 71 percent used print media and only 60 percent in-hose websites for the recruitment of staff (with the possibility to give multiple answers).

The consequence was that most business customers increased their planned budget for the online use from 22 to 39 percent until the year 2003 and reduced the budget in the print area from 24 percent to 8 percent.[494] See Fig. 33.1:

---

[493] Comp. Holtbrügge, 2007, p.100

[494] Comp. jobpilot AG, status: 04 June 2002

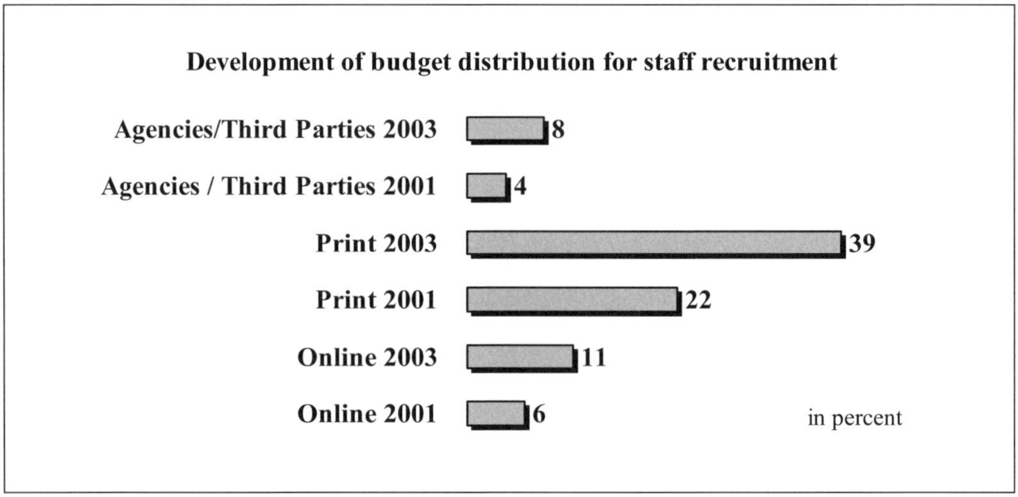

Fig. 33.1: Budget distribution in staff recruitment[495]

34 Instruments of Electronic Recruitment

Electronic Recruiting (E-Recruiting) is when the applicant is approached directly via the internet. While employment ads were formerly taken out in print media they are now digitalized. While applicants formerly sent job applications to newspapers they can now be published on the internet, and while formerly job application folders were prepared, now it is online forms, emails or applicants web pages.[496]

Among the most important activities of E-Recruiting are:

- job boards
- human resources websites

Further **possibilities** to address applicants via the internet are:

- virtual recruiting fairs,
- online-recruiting games
- newsgroups.[497]

---

[495] Beck, 2002, p. 10

[496] Comp. ibid, p. 17

[497] Comp. Brake/Zimmer, 2002, p. 50

The **advantages** of e-recruiting are obvious. Via the system based selection it is possible to save a considerable amount of time and money. It is not necessary any more to look at all application documents received. All information needed on the applicants is available in a standardized form and additional postal costs for sending application documents back do not apply.[498]

Further advantages are the speed and availability of the internet, its timeliness as well as the comfortable search and navigation function. Besides, a job offer which is published on the internet can reach special international target groups and be seen for relatively little expenses and independent from place and time by the applicants. Moreover, the internet offers the possibility for applicants willing to change jobs to leave their profile anonymously.[499]

34.1 Internet job boards

Internet job boards are generally virtual platforms where employers offer positions to be filled, but also job seekers can publish job applications. Without much effort the electronic ad can be published at once and remains over weeks and sometimes even months on the internet. A change of data is possible at any time and the ad can be removed with one click. Graphic design of the ad is not needed and it is not necessary to take the rhythm of publication of the print medium into consideration. The costs of an electronic ad are relatively low.[500] On the career website STEPSTONE the ad can be published from ca. EUR 600 for four weeks.

In comparison: the price of a print ad for example in the Frankfurter Allgemeine Zeitung amounted to up to EUR 10,000 for a one-time publication. With this comparison, opportunity costs of job boards are quickly calculated.[501]

34.1.1 Advantages and disadvantages of internet job boards

The real **advantage** of electronic job centres emerged only after more than one year after the start of the first job platform, by the matching process. Within this process it is possible to automatically preselect the job offers and applicants with predefined search criteria by comparing the criteria of the job offers with those of the job searches.[502] Furthermore, companies can place banner ads or direct links to their own websites on the job board or even

---

[498] Comp. Brökermann/Pepels, 2002, p. 86

[499] Comp. Holtbrügge, 2007, p. 102

[500] Comp. Weber, 2007, p. 59

[501] Comp. Brökermann/Pepels, 2002, p. 91

[502] Comp. Weideneder, 2001, p. 385

create an own area to publish their company profile on the website.[503] The target is to inform potential applicants and to create and maintain a company image.[504]

If a position has to be urgently filled, the ad on the internet job board can be useful, because it addresses a broad international market, because online job boards can be read supra-regionally and worldwide. It is an advantage that over the link to the company homepage the applicant can obtain essential information on the advertising company and reflect on whether the employee would be suitable. If an applicant posts a job search the company can look at it anonymously. If a potential applicant is found, he will be contacted, without need of having placed an ad.[505]

A **disadvantage**, on the other hand, is the flood of applications. Through the online-presentation, applicants can be animated to quickly apply for the job. The lowering inhibition thresholds animate also those applicants to send an application, whose requirements do not correspond to the needs of the advertising company, which can be the case for a large share of the incoming applications. This leads to high administration expenses, if no screening is installed, because all incoming applications have to be dealt with.[506]

34.1.2 Providers of electronic job boards

The number of online job markets has strongly grown in the last years. Today there are around 200 providers of electronic job boards on the market.[507] In the mid 1995 the number amounted only to five providers of job boards on the internet, e.g. the Federal Employment Agency, diverse daily and weekly papers as well some internet job boards which were mainly specialized on certain professional categories.[508]

The most popular online job markets can be found under the following internet addresses:

- http://www.monster.de;
- http://www.jobpilot.de;
- http://www.stepstone.de;
- http://jobware.de;
- http://www.jobscout.de,;
- http://www.berufsstart.de;
- http://www.stellenanzeigen.de;
- http://www.jobonline.de.[509]

---

[503] Comp. Brökermann, 2007, p. 77–79

[504] Comp. Spickschen, 2005, p. 101

[505] Comp. Weber, 2007, p. 59

[506] Comp. Brökermann/Pepels, 2002, p. 91

[507] Comp. Karle, 2001, p. 18

[508] Comp. Holtbrügge, 2007, p. 100

[509] Comp. Brake/Zimmer, 2002, p. 53

This high number of providers of job markets makes it difficult for the companies to make a decision for the right electronic job board. The job boards have recognized this problem. In the past years so called **career boards** were born where more services are offered than by normal job boards and which have therefore become complete providers for personnel recruiting.[510] They offer to compose and design job offers and take the administrative part in connection with recruitment over, as well as the pre-selection of applicants if desired.[511]

34.1.3 Arrangement and structure of internet job offers

The content and the formal design of job offers on the internet are very similar to those of printed ads and in many cases the same ads are used online and offline.[512]

In the selection of the appropriate job board besides the usual criteria like the general optical impression and the navigation of the website, the following points are important:

- regional relation and extent of the offer, quality of the offer,
- user friendliness and application as well as low error rate and exactitude,
- offer of additional services.[513]

However, one should not forget that job boards are meant to publish job offers, which cannot replace personal contacts.[514]

34.2 Human-Resources-Websites

In the context of electronic staff recruitment, the company website or the HR-homepage gains importance beside the job boards. The essential reason can be seen in the fact that companies that want to recruit personnel want to avoid the job board interface knowingly or unknowingly. The attractiveness of a company is mostly subjectively assessed and can lead to a visit to the company homepage. Whether an employer is attractive depends on the company image, the popularity, the experience of other acquaintances, the products and services etc. A further possibility to attract the applicants consists in the preparation of contents and design of the website.

---

[510] Comp. Karle, 2001, p. 18

[511] Comp. Brökermann/Pepels, 2002, p. 92

[512] Comp. Spickschen, 2005, p. 99

[513] Comp. Egle/Bens, 2004, p. 324–326

[514] Comp. Brökermann/Pepels, 2002, p. 92

In the paper age the possibilities to present oneself as employer were very limited. In the internet on the other hand, there is the possibility to obtain overall information without problems, including HR-offers and activities.[515]

An optimally designed **HR website** consists of the following four **contemplation elements**:

Following Herzberg's motivation theory, the **formal requirements** to a HR website could be referred to as hygiene factors. This means that the fulfillment of all requirement does not forcibly lead to the success of the homepage, the non-fulfillment, however, leads to its failure at once. The access to the homepage of a company has generally to be intuitively possible. The homepage must be accessible by entering a simple publicly known name. An example is the Badische Anilin und Soda Fabrik. Instead of this name the internet address www.basf.com is used. In order to make sure that the homepage is internationally available it should be accessible with pure .de or .com addresses in equal shares. In this context the homepage should offer the possibility of seeing the page in different languages. Furthermore, the layout has to support the recognition value of the presenting company by the use of a corporate design. The formal requirements of a HR-homepage also include a user-friendly navigation, the possibility of direct contact by providing an email address and name of contact persons as well as the guarantee of timeliness and data protection.[516]

The **basic elements** are all functionalities, services and information which are mandatory for the process of personnel recruiting without regard to sector and company size and which the job seekers expect nowadays.

These are the address of the target group, job offers, information with regard to job offers, company related information and the possibility to send an application.[517]

A target group specific address is only possible if it is known who visits the HR-page on the internet. By direct interviewing on the page this knowledge can be gained, which makes it possible to address the target group in accordance with motivation for a job and degree of qualification. For example, the address could be under the heading: "pupils and apprentices, students, university graduates and professionals."[518]

Additional elements on the other hand contain functionalities, services and information which give indirect support to the process of staff recruiting, clarify company HR activities and increase transparency. Among those are the following elements: the possibility of unsolicited applications, the HR-info-base, the personnel development as well as personalization.[519]

---

515 Comp. Beck, 2002, p. 168

516 Comp. ibid, p. 169–174

517 Comp. Beck, 2002, p. 175

518 Comp. Brökermann/Pepels, 2002, p. 86

519 Comp. Beck, 2002, p. 192–193

Service elements of HR sites contain information which exceeds the personal recruiting. This information has to underline the attractiveness of a company, accelerate interaction and communication with the company and/or provide additional benefits[520] and awaken the user's interest for further visits. Crazy Chicken, a computer game, which was available for download on the company website of Jonny-Walker was one of the most spectacular actions in this area and a reference with regard to individuality and creativity. Self-assessment, surveys, application tips, screensavers or competitions already belong to the classic additional offers.[521]

34.3 Virtual recruiting fairs

Under http://www.jobfair24.de it is possible to visit a job fair at home without paying entry fees. The leading online job fair of the Federal Republic – as it claims itself – can be found under this internet address. On the monthly dates of fairs graduates, diplomats, career starters and prospective trainees can have a look around the virtual fair halls with their virtual Ego and gather information on possible future employers or very generally on different career options. All information is available on one mouse-click.

At the virtual stands the personnel managers are live available for discussions and questions. It is possible for several persons at once to take part in the discussions, if however the discussion should become more intense. the recruiter and the job seeker can continue the talk in confidence in a private chat. The ideal case after a discussion in an individual chat would be that the applicant hands a digital application file to the discussion partner or even receives an invitation for a personal job interview.[522]

The advantage is that companies can directly compare the applicant's qualification with their own profile of requirements for the vacancy. Furthermore, all virtual application files have an identical basic pattern, which simplifies the comparability of the applicants, but is to the disadvantage of individuality.[523]

In addition to the fair dates, topic or sector specific company chats are offered several times per month. Furthermore, expert chats take place where help for application questions is offered. The virtual recruiting fairs are situated quasi in the niche between real recruiting fairs and internet job boards. Almost all sectors are represented on such fairs, but above all automobile, IT and telecommunication, transport and logistics. But also humanists and social scientists as well as legal practitioners have good chances on these online fairs. Among the

520 Comp. ibid, p. 200

521 Comp. Brökermann/Pepels, 2002, p. 90

522 Comp. Brink et al., 2008, p. 285–286

523 Comp. Brake/Zimmer, 2002, p. 55

participating companies are e.g. Bertelsmann, Allianz, Siemens, Porsche, Daimler AG as well as Chrysler Deutschland.[524]

The multimedia effort to be present on a virtual fair is relatively high but is willingly accepted by the participants. The price of a company presentation depends on the desired communication tool. The preparation of a virtual 3D-stand with a run-time of six months costs as package with Jobfair24 around 12,000EUR.[525] The platform was presented on the CEBIT 2000 for the first time and purchased by the Klaus Resch Verlag at the end of 2003 and integrated into the job starter network (www.berufsstart.de).[526]

34.4 Online recruiting games

34.4.1 Definition

It is uncertain whether it was the realization that personnel will have higher significance in the company in future or only the possibilities which the medium internet provides to distinguish oneself from the competitors which lead to a new innovative recruiting instrument. It is still a fact that the search for new forms of personnel recruiting has developed an innovative recruiting instrument which is defined as follows:

"Online games are a web-based link between assessment center elements and their imbedding in a game context for a target-oriented recruitment, with the possibility of documentation and matching of performance results (Hard and Soft Skills) as well as the option of automated applicant screening and ranking for the support and backup of the selection decision."[527]

34.4.2 Siemens – "Challenge Unlimited"

With the online-game "**Challenge Unlimited**" in the year 2000 Siemens went a new way of e-recruiting for the first time. Students and career starter could get into contact by means of a game. On their journey to Nouvopolis, a city in the future, the players had to survive many adventures and to solve the problems of a city in the outer space. Partially on their own or in a team, the players were exposed to different dangers and had to protect the city from threatening meteorites etc.[528]

---

[524] Comp. Brink et al., 2008, p. 286

[525] Comp. Brake/Zimmer, 2002, p. 55

[526] Comp. Brink et al., 2008, p. 286

[527] Beck, 2002, p. 212

[528] Comp. ibid, p. 212–213

New ways of addressing target groups via the internet in combination with a procedure for pre-selection were meant to be tested with this product. This instrument linked a high-performance online assessment with a fascinating and attractive online-game and took around between three and four hours. In the course of the play time the participants elaborated their competence profile. This reflected the achieved results of the test against the background that the online game is a recruiting instrument for Siemens. Almost 13,000 players took part within the game time on the internet and more than 10,000 participants released their profiles for examination.[529]

34.4.3 Possible limitations to online games

Besides the **opportunities** as the increase of the degree of familiarity, the possibility to improve the company image, as well as time and cost saving in the automated applicant pre-selection, there are certain **limitation** of the utilization of online games.[530]

The **first limitation** is the medium internet itself. The use of the internet as recruitment platform is not accepted by all applicants or companies. The second limit is the data security and data protection. The users often fear that transferred data could be misused in the internet or get lost. Besides, there is the fear of viruses, which exists in many companies despite firewall solutions and antivirus protection programs. A further limitation is the appearing capacity problems. Bottlenecks can emerge when many players access the game at the same time, e.g. when the game is new on the internet. Besides, there is the problem of authenticity of the player. There is not yet a guarantee that the solved tasks were really solved by the registered person. Furthermore, the soft skills, e.g. on social competences can not be reliably assessed via this medium. However, one should not forget that the development of online games is still in its infancy and the above mentioned limitations with the exception of the last point can be overcome in further developments.[531]

34.5 Newsgroups

In comparison with 3D-fairs or online recruiting games newsgups may seem rather old-fashioned. Newsgroups are meant for the exchange of information on certain topics in open groups. The notice board, where one could write or read notices, served as basic idea for newsgroups. Because there is a big number of newsgroups, they are divided into topics. Moreover, there is the possibility to search for keywords in the newsgroups, as Deja (www.dejanews.com). Generally it is free of charge to post an ad. The disadvantage is that there is no possibility for graphic design, company logos or other optical effects. The ads are

---

[529] Comp. ibid, p. 213

[530] Comp. Beck, 2002, p. 225

[531] Comp. ibid, p. 225–227

only published in text format. A corresponding link to the company website should not be forgotten in any case. The danger that an ad is overlooked is quite high given the amount of ads looking just the same. Among the most important German newsgroups are:

- de.markt.arbeit.angebote,
- de.markt.arbeit.gesuche,
- de.markt.arbeit.de,
- at.jobs und euro.jobs.[532]

In order to access one of the above mentioned newsgroups the user has to enter the corresponding news-server with address in his browser. An alternative way of research is possible under http://www.deja.com. Here it is possible to publish a notice in the newsgroups without charge. In this context one may often find ads which are not directly linked to the topic. Whether this form of employee recruiting is still interesting depends on the one hand on the spread of job boards on the electronic applicant data bases and their price on the other hand.[533]

34.6 Risks and opportunities of e-recruiting

34.6.1 Opportunities

Via the use of the internet, considerable amounts of time and costs can be saved. "*This aspect can become the decisive factor in the fight for so called high potentials*".[534] The internal processing and forwarding of application on a digital way is made much easier. The processing processes are shortened which again improves the application process. With the online presence companies can also contribute to the improvement of their company image. Information is available at all times and by the timely transmission it can be assumed that data are up-to-date. The internet as only medium can be used for worldwide personnel recruiting without limits. Currently, there is a strong need for specialized workers in Germany and especially for this reason an internet presence is crucial for internationally active companies. Furthermore, a creative design and additional benefits for the user on the company homepage can be a sign of capacity for innovation of the company. This effect, however, could decrease with the growing establishment of the internet and the rising offer of additional benefits. The simplified possibility to reach applicants via the internet is far better as in print media. This aspect is certainly very important, because the probability to find the best possible candidate for a vacancy increases with the number of applicants.

---

532 Comp. Brake/Zimmer, 2002, p. 56

533 Comp. ibid, p. 56

534 Preuss/Knoll, 2001, p. 130

34.6.2 Risks

The personnel recruitment via internet does not only bring opportunities but also some risks. A stable company security is condition for a communication without barriers between the user and the company: "*Applicants who suffer a technical crash when accessing a company homepage or when filling in a time-intense application form will certainly not return to the company website too soon.*"[535]

The big question regarding security when transferring personal data concerns most users of the world wide web (www). A company in which an online application is possible has to provide for absolute security for data transfer and to make the access for unauthorized persons impossible. The user is flooded with information in a continuously rising amount of offers. In order to limit them and to guarantee an efficient use of job offers, a user-friendly navigation has to be available. Table 35.1 presents **opportunities and risks** of e-recruiting again in catchwords:

Opportunities	**risks**
o Potential to save costs	o Company security
o Potential to save time	o Data security
o Stabilization of image	o Information flood
o Worldwide recruiting	o Identification in the staff selection procedures
o Timeliness	
o Communication of capacity for innovation	o Pressure for quick processing
o Broad range of contacts	
o Neutrality in personnel selection	
o Unlimited timely utilization	

Table. 34.1: Risks and opportunities of e-recruiting[536]

Especially the risks have to be considered in online staff recruiting and have to be integrated in the recruitment strategy. However, with the rapidly growing development of communication and information technologies a reduction of risks can be assumed in the long run. The opportunities offered via personnel recruitment on the internet should be used because there will be enormous progresses in this area.[537]

---

[535] Dix/Witrahm, 2001, p. 444

[536] Bröckermann/Pepels, 2002, p. 94

[537] Comp. ibid, p. 93–94

35 Scouting

In specialized literature there is no fix definition of the term scouting . Scouting rather combines methods of personnel recruiting in such a way which allows "the use of the proactive character of addressing the applicants at an early stage and which is different to the traditional recruitment methods through newspaper ads and recruitment consultants."[538] In any case scouting is about the recruitment of career starters. Scouting is directed towards a special and important target group which is students and job starters and especially high potentials. The focus of scouting activities lies on graduate-oriented measures and university presence.[539]

35.1 Scouting through university presence

The target of university presence is to distinguish the company especially for career starters as attractive partner with unique profile on the job market on the long run. The chosen terminology of university presence shall comprehend all measures to implement an active service and communication policy with regard to chairs, universities and students. The following paragraphs shall describe several recommendations for the success of scouting at universities more closely.[540]

35.1.1 Concentration on selected disciplines and universities

One way is to focus selected disciplines and universities at a high specialist level and strong practical relevance. The so called key universities should be capable to form career starters who can be used for multiple areas and are prepared to adjust to new situations and who can always deal with new problems.[541]

35.1.2 Personalization and differentiation of offers

The next step is the **personalization**, i.e. the determination of certain key partners at the university. The first contact partners should be the chairs or deans, because they are very stable staff members. Also contacts with student groups and student councils are important.

---

538 Bröckermann/Pepels, 2002, p. 135, own translation.

539 Comp. ibid, p. 120

540 Comp. ibid, p. 122

541 Comp. Eisele/Horender, 1999, p. 29

Because of the higher turnover, these contacts have generally to be renewed each semester. It is always recommendable to use a continuously created and maintained data basis.

It is possible to distinguish **two groups of contact partners** at universities in the context of scouting via university presence: on the one hand professors, chairs and employees and on the other hand students themselves. The group of professors, chairs and employees serve as contact partners especially for lesson-related activities and offers for indirect scouting. Among these offers are offers of working on topics of research and development in the context of diploma and master theses, as well as dissertations but also by offering company visitations, excursions and assigning assistant lecturers and instructors who offer presentations and seminars. By such events often contacts with especially talented and interested students are construed, which can then be intensified with the help of the following measures:[542]

The group of students is the contact partners for direct scouting at university. The direct scouting can be undertaken in a variety of ways. The offer of internships to students has to be mentioned in this context. As in degree programs of some universities and in all universities of applied science an internship is obligatory and the student is given the right to choose an appropriate internship, companies can attract students and, thereby, also high potentials with corresponding offers and information on internships and with the offer of special trainee assistance. Very important is the support of student projects and initiatives. Initiatives as AIESEC are mostly supported by highly talented and active students and can be supported by sponsoring memberships and the offer of internships but also by means of fair participation and ads. The involved students themselves are often very demanded in recruitments.[543]

35.1.3 Use of multi-step programs

A further form of direct scouting is programs which are developed by students in several steps. These are referred to as "*systematically structured scouting activities*".[544] A first approach to be mentioned is a first contact with talented students in form of excursions (partially in connection with case studies) with regard to an intense collaboration with universities. This is followed by internships in the company and an intense after-care during the time when the students do not work for the company. Also information on the development of the company and on individual development possibilities within the company, e.g. the offer to work on a certain topic in the thesis is given. With the career start of the student the last phase is concluded.[545]

A further instrument of direct scouting is the participation in different cooperative programs. Companies recruit students generally themselves and then send them to university courses. In

---

[542] Comp. Bröckermann/Pepels, 2002, p. 123

[543] Comp. ibid, p. 123–124

[544] Bröckermann/Pepels, 2002, p. 124

[545] Comp. ibid, p. 124–125

parallel, the students have a practical part of their studies in the company. Because it will be more difficult in future to find enough qualified graduates, the participation in such cooperation will gain momentum.[546]

35.2 Scouting via graduate-oriented measures

In order to cover a short-term recruiting need graduate-oriented measures are used. In this case not the aspect of promotion and development is focused, but the interest in selection.[547]

The three most important instruments of graduate-oriented measures in this context are graduate fairs, graduate workshops and on-campus recruiting.[548]

35.2.1 Graduate fairs

"Graduate fairs" has become a collective term for large informational events between one and two days (frequently at university) with the target of informing graduates and students close to graduation about their job entry and career possibilities, the right application strategy, selection procedures of the companies etc.[549]

They can also be understood as market places where different companies present themselves with their own stands to graduates and students. This gives students the possibility to meet many potential employers and to gain additional information and to take part in the services offered, e.g. discussions or specialized presentations, application trainings etc.

Because of the mass character of such events it is hardly possible to have a direct communication or receive detailed information. The participation in a graduate fair often targets more the improvement of the company image than the real recruiting purposes.[550]

With regard to the individual functions or sectors one can observe a stronger specialization of today's graduate fairs. In this case the filling of certain positions is the main goal of the fair. Already in the run-up phase of the fair the organizers make a preselection of the visitors with help of the sent application documentation. In order to organize the processes during the fair in the best possible way, job interview meetings are arranged by the company beforehand. But problems can still arise, e.g. if the invited candidates do not comply with the

---

546 Comp. Schmeisser, 2006, p. 40

547 Comp. Bröckermann/Pepels, 2002, p. 126

548 Comp. Schmeisser, 2006, p. 40

549 Bachelor und Master, status: 11 July 2009

550 Comp. Sunter, 2000, p. 42 et seq.

assigned times or if the selection criteria for the participation in the fair were not exactly defined.[551]

35.2.2 Graduate Workshops

Graduate workshops are company related events with exclusive character for a custom-fit and tailor-made personnel recruiting.[552] In the case of a graduate workshop, generally only one company or only a small number of companies presents itself. The visitors are preselected in correspondence with the individual requirements of the companies. The workshops with a duration of between one and three days contain presentations, lectures as well as an active participation of the visitors e.g. in case studies, scenarios and other tasks. It is also possible to integrate additional framework programs to have a positive, long-lasting influence on the value of the experience of such a workshop. The participants receive a feedback at the end of the workshop. Organization and structure of the workshop are very similar to assessment centers as method for staff selection.[553]

The organization and execution of such workshops can be carried out by the company itself but also by an external service-provider. The Access AG is one of the most popular external service-providers and carries out between 30 and 40 recruiting workshops per year.[554] The number of labor contracts concluded is seen as success criterion of graduate workshops from the company's point of view, so that graduate workshop have established as quick recruiting procedures.[555]

35.2.3 On-Campus-Recruiting

In the case of on-campus-recruiting companies present themselves on the premises of a university with the goal to stimulate interesting applications of graduating students. In the US On-Campus-Recruiting is an often used instrument of personnel recruiting; in Germany, however, it is only rarely used. At the best US universities around 30 percent of the graduates decide to start their career with a company with which they were in contact in the context of on-campus-recruiting. Before the final examination phase the contact can also have developed in an internship.[556]

---

[551] Comp. ibid, p. 42 et seq.

[552] Comp. Bröckermann/Pepels, 2002, p. 127

[553] Comp. ibid, p. 127

[554] Comp. Bröckermann/Pepels, 2002, p. 138

[555] Comp. ibid, p. 127

[556] Comp. ibid, p. 128

A German example of on-campus recruiting is the **Job-Truck of the HypoVereinbank**. The Job-Truck was driven from university to university across Germany in the last years, in order to contact students and present the company as modern and attractive employee.[557]

35.3 Future perspectives of Scouting

The demographic development in Germany makes clear that in a few years the number of high-potentials will strongly decrease. The simplification of the distribution of Green-Cards and the long-term promotion of the inclination to study are factors which can attenuate the situation but cannot stop the process. As soon as the number of highly qualified staff decreases the competition around them increases. One can therefore assume that the company presence at universities will rise in the next years. This leads to a further differentiation of existing scouting strategies and the development of new ones, in order to reach a monopoly or at least positioning advantages. On the one hand, it is possible to offer more **service-oriented support** to students in form of provision of scientific data and information as well as literatures. On the other hand there is the possibility to show more interest in the participation in **cooperations** in university programs.[558]

In the context of university-related scouting special attention should be paid to female students who had been integrated in all marketing activities in the university area, but are rarely taken into consideration and activated with regard to their own special potential. Companies can also become more attractive via participation in alumni initiatives of the university, which does not exactly support the search for career starters but the one of experienced graduates. Nowadays the term "alumni" refers to university graduates who still have contact to their university and their former fellow students.

36 Personnel leasing

A further modern external way of staff provision is personnel leasing. . This type of staff provision is very different to the already mentioned ways of personnel recruiting. In this form, the personnel seeking company acts as client. The client is referred to as lessee and contacts the lesser. The lesser lends a workforce to the lessee against a fee for a limited period of time. The legislator refers to this type of contract as **temporary employment.** Personnel leasing is also called temp-work or freelance work. The Act on Personnel Lease

557 Comp. Schmeisser, 2006, p. 41

558 Comp. Bröckermann/Pepels, 2002, p. 132

requires that temporary work Agencies dispose over a license and the authorization by the Federal Employment Agency.[559]

Fig. 36.1 displays **the economic and legal relations** of the parties involved.

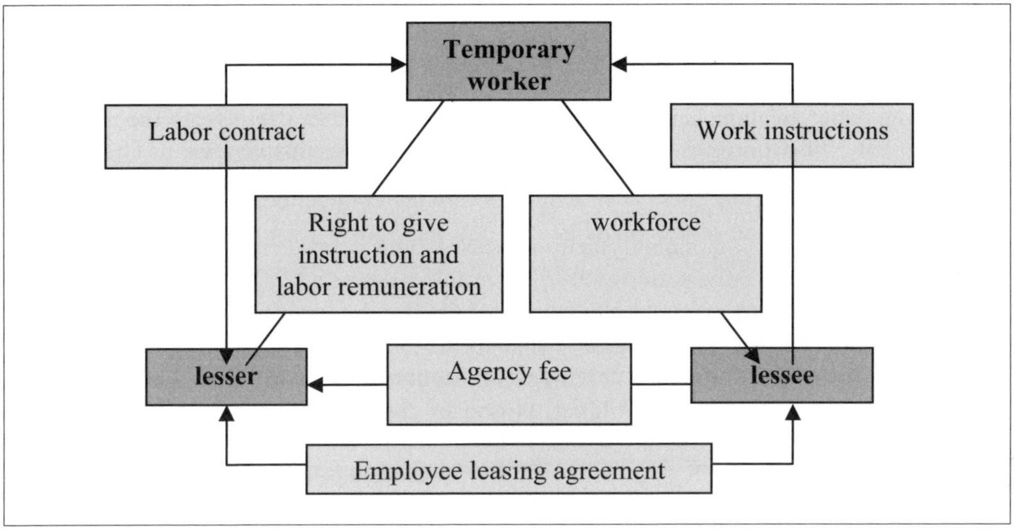

Fig. 36.1: personnel leasing[560]

There is a labor contract between the temporary worker and the lesser. There are no special regulations for this contract. All special regulations including the limitation possibilities which generally exist for all labor relations apply. Within this working relation the usual labor law, social law and tax law conditions are valid. The temporary worker receives the agreed remuneration from the lesser who deducts the taxes and social contributions due. In accordance with sections 3 and 9 of the Act on Labor Lease (Arbeitnehmerüberlassungsgesetz) the lesser has to send the permanent worker for the leasing time under the individual labor conditions to the leasing company, which applies especially to the labor remuneration of the work exercised. Because in quickly growing companies in different sectors this is partially hard to put in practice, the legislator offers a solution to the problem. If the lesser enters as member in an employer association which has concluded a collective agreement with the unions in charge with regard to personnel leasing, the conditions fixed in the collective agreement will apply.

The lesser is the one who can give instructions to the temporary worker. In the agency, the temporary worker has a right to vote and can be elected to the labor council. Additionally, the temporary worker, as soon as he comes of age and is assigned to the lessee for more than

559 Comp. Bröckermann, 2007, p. 89

560 Bröckermann, 2007, p. 89

three months, has a right to participate in the company labor council elections in accordance with sec. 7 of the Industrial Constitutional Law.

The lessee has to pay attention that the lesser fulfills its duties with regard to payment of social insurances. If it does not fulfill the payment duties, the lessee has the duties of a primary obligor. For this reason, the lesser has to inform the health insurance in charge on each personnel leasing. The company labor council of the lessee has a right of co-determination, if a temporary worker is to be assigned to a job.

There is no labor relation between the temporary worker and the lesser. The temporary worker provides his workforce to the lessee and the lessee gives instructions.

In this **triangular relation,** there is an employee leasing contract between the lesser and the lessee through which the lessee is obliged to pay the agreed leasing fee and the lesser to provide the agreed personnel.[561]

At the end of 2002, 4,261 companies were predominantly or exclusively active in the area of temporary work leasing in Germany. Almost 320,000 employees were sent to other companies on the annual average.[562] **Randstad Deutschland GmbH & Co. KG**, which belongs to the Dutch Randstad Holding, is one of the largest temporary work agencies worldwide and has been active in Germany for 40 years. With around 700,000 employees and total revenues of EUR 17.2 billion (2008) Randstad is active in more than 50 countries.[563]

The most important **advantage** of personnel leasing is that employees can be obtained at short notice. Temporary workers are particularly used for writing work in the administrative area or as craftsmen, assemblers and technicians in the corporate area. The personnel leasing offers the possibility to cover short-time work peaks and to replace own personnel in the case of illness, holiday etc.[564] Another positive aspect of personnel leasing is that no costs for personnel recruiting or for job adjustment arise. Furthermore, the lessee has no employment risk. He can quickly release the temporary worker even without stating reasons.

Considering all these advantages, the **disadvantages** shall not be forgotten. Because their activity is limited in time, the temporary workers often feel like “second class employees”. Social problem and the high time consumption to transmit company specific knowledge are the other side of the coin.[565]

---

[561] Comp. Bröckermann, 2007, p. 89–91

[562] Comp. Holtbrügge, 2005, p. 36

[563] Comp. Randstad, status: 01 July 2009

[564] Comp. Jung, 2008, p. 145

[565] Comp. Holtbrügge, 2007, p. 99–100

37 Personnel selection process

The goal of personnel selection, the **second partial phase or task of personnel recruitment,**[566] is to find the most appropriate applicant on the labor market in the company or outside the company to fill in a vacancy. Once, staff is employed, it is often very difficult to dismiss them. Therefore, it is in the company's interest to have high entrance barriers for the applicants via a strict selection.[567] The key phrase "*the right man for the right workplace*"[568] is most appropriate to describe the object of personnel selection. In order to find this employee, different, systematical processes were developed in the context of staff selection. Among this procedures are rather classical procedures as analysis of written application files, execution of tests and job interviews.[569]

The labor council has a co-determination right in the staff selection in accordance with s. 95 subs. 2 Work Council Constitution Act (BetrVG). This means that the selection criteria a company has established need the labor council's consent. If a company employs more than 1,000 employees the works council can also demand for selection criteria to be established.[570]

37.1 Selection process in form of a personnel selection chain

The comprehensive selection process can be displayed with the help of a **personnel selection chain**: the determination of the present and estimated future requirements, which depart from a very good until acceptable fulfillment of the labor tasks are the point of departure. The thereby determined requirements should be overcome with a corresponding qualification of the applicants in question. This is ensured via a **pre-selection** with the help of the underlying application. It will then be tried to determine the **qualification of the applicant** with diverse, sometimes simultaneously applied procedures. Finally, an **overall assessment** is undertaken.[571] See Fig. 37.1:

---

[566] Comp. Holtbrügge, 2007, p. 103

[567] Comp. Jung, 2008, p. 153

[568] Manke, 2008, p. 8, own translation

[569] Comp. Bröckermann/Pepels, 2002, p. 193

[570] Comp. Olfert, 2008, p. 131

[571] Comp. Berthel/Becker, 2007, p. 262–264

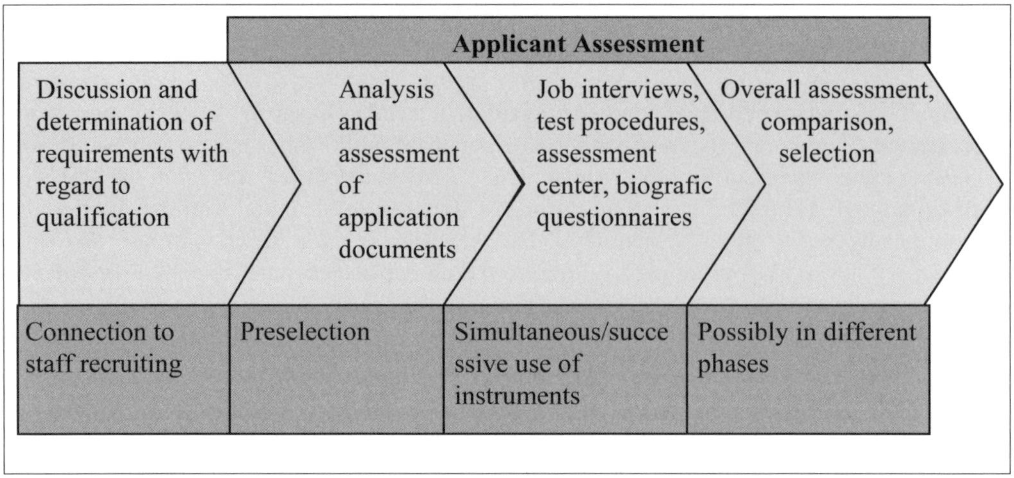

Fig. 37.1: Personnel selection chain[572]

In the best case, after the complete assessment procedure a test result from which can be deducted to which degree the applicant fulfills the job requirements is received. It has to be mentioned that without exact requirement profiles of the individual jobs no target-oriented examination of the qualification is possible.

Against this background, it is necessary in practice to overcome a multitude of **difficulties** in personnel selection, which result from the following factors:

- There are always multiple requirement categories:

1. Can criterion: aptitude (mental as well as physical abilities, knowledge, personality characteristics as e.g. concentration capability and persistence) or work experience.
2. Want criterion: desired personal goals in professional life and expectancies in connection with work.[573]

- There are **multiple sources** that show to which extent it is necessary to fulfill the requirement categories, e.g.:

1. the future labor tasks of the applicant,
2. the labor processes and tools contributing to the fulfillment of the tasks (e.g. certain IT-equipment) as well as

---

572 Ibid, p. 264

573 Comp. ibid, p. 264

3. the company internal (e.g. supervisors, colleagues) and the company external labor environment (e.g. employees of certain customer businesses or suppliers) with which the employee needs to be in contact.[574]

- It can be tried via **multiple procedures** to determine the actual qualification of the applicants:

1. analysis and assessment of application documents,
2. questioning in form of (job) interviews,
3. expertise,
4. tests,
5. observations,
6. in the case of internal job applicants: e.g. the results of performance assessment from former times as well as,
7. in the case of external job applicants: observations of behavior, test work.[575]

The **problem** is that all procedures for personnel selection lead to results concerning several types and partially categories of requirements. In order to be able to make a decision on the qualification of the applicant, the results of the examination of the qualification are attributes to the different requirement types in such a way that a conclusion can be drawn to which degree each is fulfilled. The **main problem** is that the determination of the applicant's qualification is rather an evaluation. The better the assessment is with regard to its verifiability and interpersonal comprehensibility, the closer it gets to the ideal of objectivity.[576]

37.2 Selection procedures in internal and external applications

The **personnel selection of internal applicants** turns its attention not only on the application, but especially on already available data, e.g. from personnel assessment and the performance in former or actual positions in the company. A comprehensive personnel assessment systematically evaluates the performance and personality characteristics of the employee. The data available in the human resource area are entered in the context of personnel development. The advantage is that information on the applicant is already available and the information basis is larger than in case of external applicants.

---

574 Comp. Berthel/Becker, 2007, p. 264

575 Comp. ibid, p. 256

576 Comp. ibid, p. 264–265

The **selection of external applicants** is especially based on the analysis of application documents, however these documents might have been obtained. Only afterwards, further procedures come into consideration. In practice, the continuative instruments are used alternatively or in combination. Among the procedures and instruments of external personnel selection is for example the analysis of application documents, the job interview, tests (performance, intelligence as well as personality tests), and graphological expertise. The assessment center is a combination of most of these procedures,[577] see Fig. 37.2.

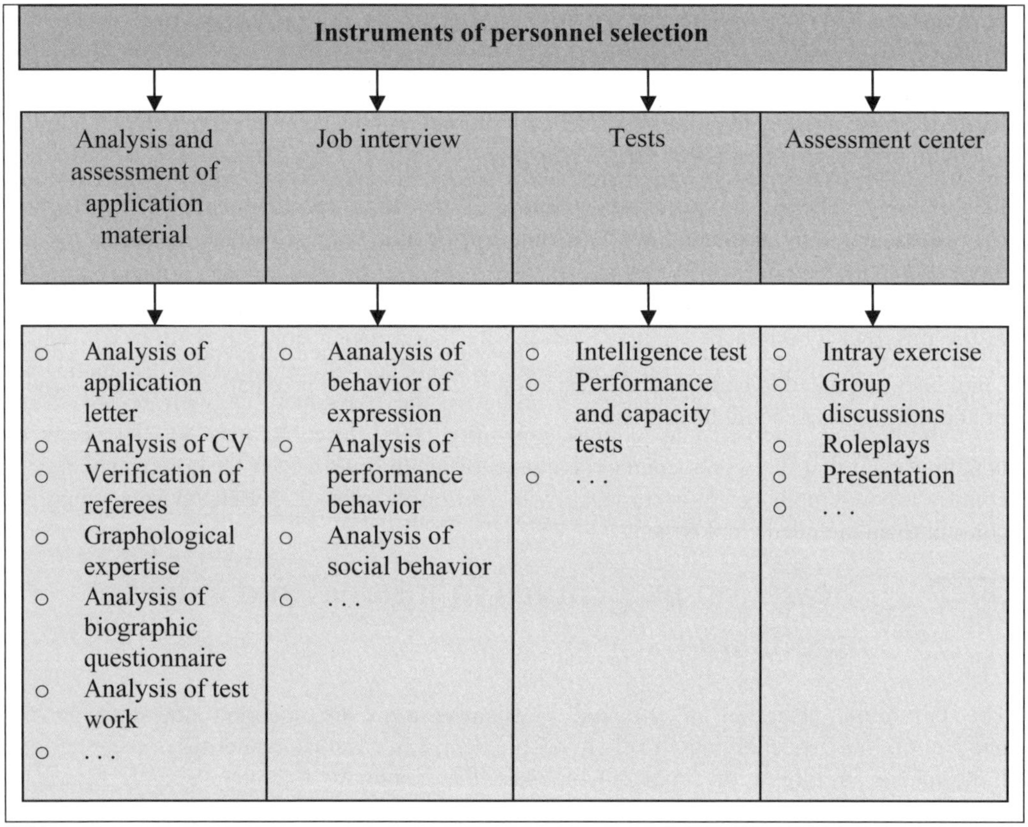

Fig. 37.2: Instruments of personnel selection[578]

Many companies still use only the analysis of application documents and job interviews. Assessment centers are often used for the selection of national and international managers.

---

[577] Comp. Bröckermann, 2007, p. 92

[578] Following Berthel/Becker, 2007, p. 266

Among the **mostly frequently used selection procedures** in Germany are:

- analysis of application material (almost 100 percent),
- job interviews (about 98 percent),
- diverse tests: IQ, behavior, personality, performance or mixed forms,
- assessment center (increasingly)
- referees and graphological expertise (max. 2 percent to 5 percent, decreasing).[579]

37.3 Process of external selection of personnel

Internal as well as external applicants come into consideration for the selection of personnel. Even if more information is available on internal applicants, they go through similar selection processes especially in large companies.

In following scheme the **necessary steps** and the **time requirements of a selection procedure** are only presented for **external applicants** because of the similarity of the procedures. See Fig. 37.3.

Through the necessary correspondence in the selection process there can be large time intervals between the individual steps. The time requirement depends on how many candidates are asked to come to the interviews. If 20 applicants take part, a time requirement of about three days should be estimated.

It is important that the works council is always officially informed if the company wants to employ a new employee. In accordance with section 99 subs. 2, 3 BetrVG it can deny its consent to an intended employment.[580]

---

579 Beska-Bewerbungsservice, status: 22 July 2009

580 Comp. Blank et al., 2009, p. 142

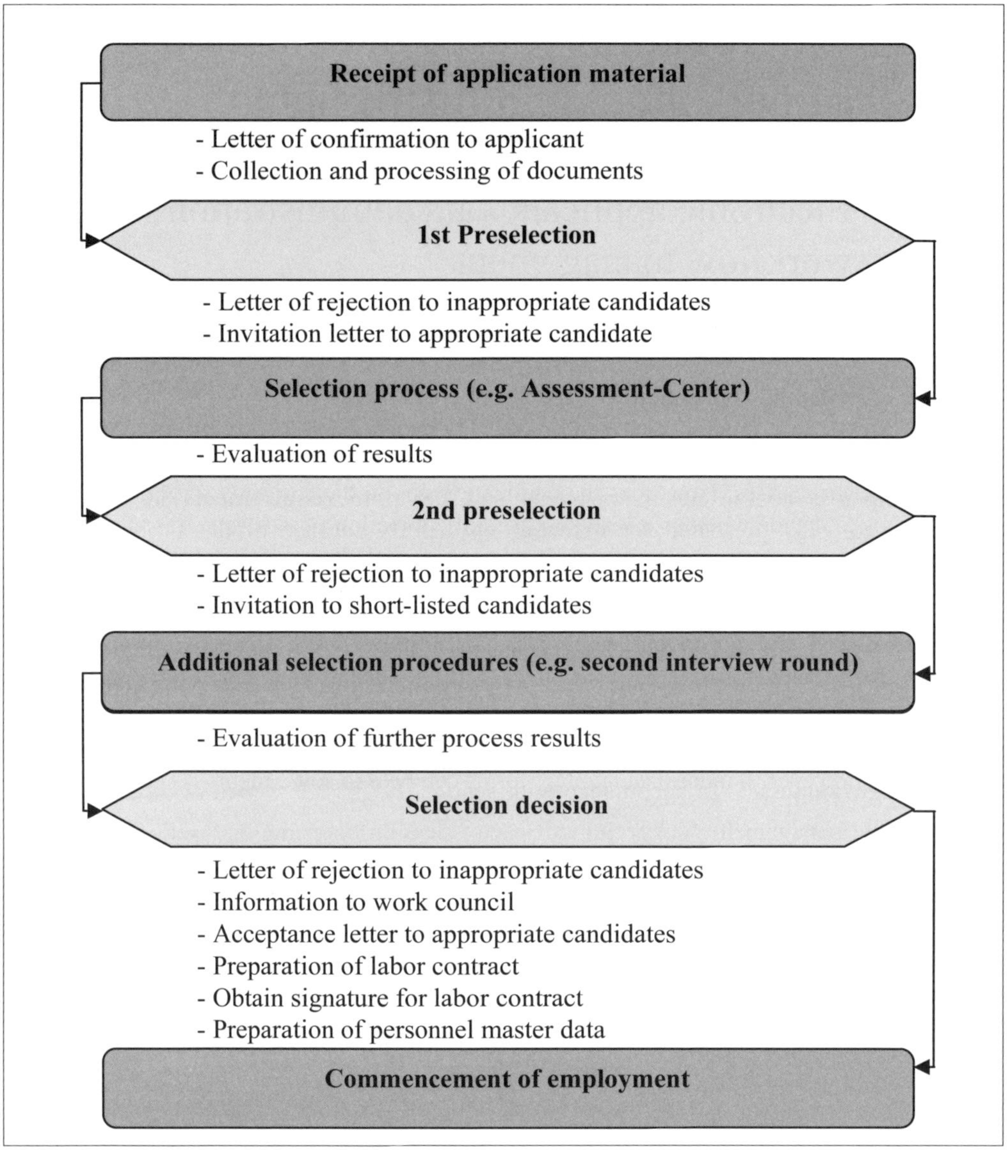

Fig. 37.3: Course of external selection of personnel[581]

[581] Blank et al., 2009, p. 142

38 External procedures of selection of personnel and instruments

38.1 Electronic applicant data administration – workflow management

The conventional applications in paper form often need a long time to circulate in the company. In bigger companies it can partially take days or even weeks until the responsible employee receives the application. If it is possible to dispose over an application in digital form, the access to it is simplified and above all accelerated. As soon as an application is deposited on the company server, it is possible to access it within few seconds. However, the following question arises: How does the application get on the company server? At first, the company has to decide whether it wants to use a combination of electronic management and paper form or pure electronic applicant data management. It has to be observed that some applicants (especially the older generation) will still apply in the traditional way in the next years. If the procedure is limited to pure online applications, a large circle of applicants is excluded. If applications sent by mail are accepted anyway, the data have to be prepared manually or per scan for the system to allow electronical management. At a first glance it seems to be easy to process digital data. But many different programs and systems often make the access to these data complicated. For example, several problems can arise in email applications; however, standardized contact forms can help to solve them.[582]

The applicant management via **special software** is less problematic. SAP/HR, Peoplesoft or BEWERBIS are popular programs. With the help of these programs correspondence, adjournments of appointments, travel expense accounting, labor contract composition and many other activities in the workflow are possible.[583] With the help of the workflow management the recruiter can distribute and forward the sighted CVs with or without comments. The workflow management is therefore an instrument which supports the individual steps in whole work process, beginning with receipt of the application up to the formulation of the labor contract.[584]

[582] Comp. Brake/Zimmer, 2002, p. 57–58

[583] Comp. ibid, p. 58

[584] Comp. Beck, 2002, p. 149

38.2 Online application

Keeping in mind to which extent internet has established itself as procurement method, it is no surprise that **the possibility of online applications** becomes more and more popular (especially among the applicants).

Against the background of a steadily increasing cost pressure and competition for well trained and qualified employees, companies are more and more prepared to accept electronic applications.

Online applications, depending on how detailed they are, receive diverse feedback from the human resource department. Anyway, human resource managers still like complete written application files.[585]

Especially in case of job offers on the internet, applicants are often asked to hand in online or internet applications. In order to be able to apply, applicants sometimes have to go through an online assessment with internet-based exercises or test procedures or have to fill in forms for the determination of their aptitude, i.e. a type of questionnaire. Personnel managers or corresponding software compare the qualifications stated by the applicant with the requirement profile of the vacancy. If there is no analogy, the possible candidate will not be allowed to apply at all.[586]

38.2.1 Possibilities of online application

First of all there is the **brief application** which is received by the company via **email**. It should contain a traditional application letter and a short presentation of the applicant's profile.[587] This form of application meets little acceptance in companies, because the care leaves often a lot to be desired and important information on professional and personal background of the applicant are missing.

Another point of critique is that often general formulations are used, which bears the danger that the applicant might send a mass circular which again could lead to the disappearance of inexact applications in the flood of information.

E-mail applications with CV are a little more popular. Thus, the personnel seeking company can obtain a first impression of the personal and professional background of the applicant. However, this form of application does still not contain all documents which are usually comprised in a written application.[588]

---

585 Comp. Bröckermann/Pepels, 2002, p. 216

586 Comp. Bröckermann, 2007, p. 94

587 Comp. Olfert, 2008, p. 129

588 Comp. Bröckermann/Pepels, 2002, p. 216–217

In an **email application with attachment** the application material is sent as attachment (e.g. text file with CV or scanned certificates). Reason for criticism is the danger to transfer viruses or problems when opening the attachments. This form of online application, however, is the most popular one in companies and has nowadays the **same significance** as written applications. Therefore, the same assessment criteria apply for both application forms.[589]

Besides, there is the possibility of an **own website** or **application website** to which the applicant can refer in an email or written job application letter. This page is mostly created by the job seeking person himself and serves to present his own person. In most cases, however, it is not designed in a position-specific way.[590]

38.2.2 Default in online applications

Several **disadvantages** in connection with online applications need to be mentioned:

- Because of technical problems attachments of an email application cannot be opened or contain viruses.
- Often online applications contain insufficient information on the applicant and are quickly excluded from the application process.
- Online applications and offered position do not match.
- The speed of the internet leads the applicant to neglect the care with which a written application is created.[591]

Because of the named problems many companies develop and use so-called online application forms. These structured and standardized forms are positioned on online job boards or the company website as a link in order to receive all important information about an applicant which is necessary for an application. This way an equal treatment of any applicant who has applied in writing is ensured and the same assessment is possible.[592] A further advantage is that the statements made are compatible with the system to 100 percent. Unfortunately, the force of expression and the individuality of the applicant are lost this way.[593]

---

589 Comp. Olfert, 2008, p. 130 and Bröckermann/Pepels, 2002, p. 217

590 Comp. Bröckermann, 2007, p. 94–95

591 Comp. Bröckermann/Pepels, 2002, p. 217

592 Comp. Schmeisser, 2006, p. 49

593 Comp. Brake/Zimmer, 2002, p. 58

38.3 Telephone interview

Telephone interviews are seen as modern and personal procedures of staff selection. It is generally used **before** a possible job interview or other instruments of staff selection to examine language knowledge. The aim of the telephone interview is to make a selection between the potential candidates for efficiency reasons and to obtain further information if necessary.[594]

The content of the conversation is similar to the one in a job interview. This kind of interview, however, is rather shorter.[595] If the ideas of the company do not match the applicant and if a realistic solution cannot be expected, the application process can be stopped at this point. This has the advantage that the following cost and time-intense selection procedures can be left out.[596]

Well aimed questions of the interviewer and his capacity to lead a professional telephone interview are decisive for the success of such calls. It would be fatal, if a strong candidate was excluded because of a badly organized telephone interview. In order to secure the quality of the phone interview, a telephone training of the appointed interviewer is a must-have.[597]

38.4 Digital interview

After the analysis of application material and the selection of the eligible applicants usually a job interview is carried out. This step can today be digitally supported; however, the degree of utilization and acceptance is still expandable.

In order to carry out a digital interview, technical conditions, i.e. a web-cam, PC-equipment and high-performance transmission capacities must be available to the recruiters and applicants. Furthermore, both sides must have the point of view that the chosen medium is an appropriate instrument.

The recruiter must be completely convinced to be able to take an employment decision with this kind of media utilization and a full renouncement to personal contact. Against this background the question arises to which extent things can be evaluated in a digital interview which are important in a personal, private conversion.

Certainly, the digital interview reduces the travel expenses of personal job interviews, but the practicality of the use of this instrument is still at least questionable. Today it is imaginable

---

594 Comp. Bröckermann/Pepels, 2002, p. 253

595 Comp. Bröckermann, 2007, p. 119

596 Comp. Schmeisser, 2006, p. 52

597 Comp. Bröckermann/Pepels, 2002, p. 253

that interviews serve as preselection for the **completion** of application material. It can be concluded that digital interviews still have rather the status of a phone interview and not the one of a selection or job interview.[598]

Literature

Achouri, C.: Recruiting und Placement, Methoden und Instrumente der Personalauswahl und -platzierung. Gabler, Wiesbaden 2007.

Anger, G. / **Christ**, H. / **Kiel**, I. / **Müller**, H.: in Blank, A. / Christ, H. / Schneider, K.-H. (ed.): Personalwirtschaft. 4th ed., Bildungsverlag Eins, Troisdorf 2009.

Arbeitsagentur: Der Arbeits- und Ausbildungsmarkt in Deutschland – Juli 2009 – Der Monatsbericht der Bundesagentur für Arbeit, URL: http://www.pub.arbeitsagentur.de/hst/services/statistik/000000/html/start/monat/aktuell.pdf
Status: 06 August2009.

Bachelor und Master: URL: http://www.bachelorundmaster.de/karriere/jobsuche/absolventenmessen.html
Status: 11 July 2009.

Beck, C.: Professionelles E-Recruiting: Strategien – Instrumente – Beispiele. Luchterhand Neuwied, Kriftel 2002.

Berthel, J. / **Becker**, F. G.: Personal – Management Grundzüge für Konzeptionen betrieblicher Personalarbeit. 8th edition, Schäffer-Poeschel, Stuttgart 2007.

Beska-Bewerbungsservice: Auswahlverfahren in Deutschland für Bewerbungen, URL: http://www.beska-bewerbungsservice.de/bewerbertipps/auswahlverfahren_d.htm
Status: 22 July 2009.

Blank, A. / **Christ**, H. / **Schneider**, K.-H. (ed.): Personalwirtschaft. 4th edition, Bildungsverlag Eins, Troisdorf 2009.

Böck, R.: Personalmanagement. Oldenbourg Wissenschaftsverlag, Munich 2002.

Brink, A. / **Ernst-Auch**, U. / **Hamm**, M. / **Faber**, M. / **Hess**, J. / **Jünger**, A.: Gabler/MLP Berufs- und Karriereplaner Wirtschaft 2008/2009. 11th edition, Wiesbaden: Gabler 2008.

Brake, J. / **Zimmer**, D. (1993): Ganzheitliche Personalauswahl. Bamberg: Bayerische Verlagsanstalt 1993.

Brake, J. / **Zimmer**, D.: Praxis der Personalauswahl: So wählen Sie den idealen Bewerber aus. Würzburg: Lexika 2002.

Brenner, D.: Neue Mitarbeiter – suchen, auswählen, einstellen. Munich; Unterschleißheim: Luchterhand 2003.

---

598 Comp. Beck, 2002, p. 151

Bröckermann, R.: Auswertung und Erkenntnisse. in: Bröckermann, R. / Pepels, W. (ed.): Handbuch Recruitment: die neuen Wege moderner Personalakquisition; Planung, Beschaffungswege, Auswahlverfahren; Beiträge aus Forschung und Praxis. 1st edition, Berlin: Cornelsen 2002.

Bröckermann, R.: Personalwirtschaft Lehr- und Übungsbuch für Human Ressource Management. 4th edition, Stuttgart: Schäffer-Poeschel 2007.

Bröckermann, R. / **Pepels**, W. (ed.): Handbuch Recruitment: die neuen Wege moderner Personalakquisition ; Planung, Beschaffungswege, Auswahlverfahren; Beiträge aus Forschung und Praxis. 1st ed., Berlin: Cornelsen 2002.

Deters, U.: Externe Personalbeschaffungswege: Klassisches Posting – Forschungsbericht. in: Bröckermann, R. / Pepels, W. (ed.): Handbuch Recruitment: die neuen Wege moderner Personalakquisition; Planung, Beschaffungswege, Auswahlverfahren; Beiträge aus Forschung und Praxis. 1st ed., Berlin: Cornelsen 2002.

Dix, M. / **Witrahm**, A.: Das Internet als Instrument des Personalrecruiting. in: Clermont, A. / Schmeisser, W. / Krimphove, D. (ed.): Strategisches Personalmanagement in Globalen Unternehmen. Munich: Vahlen 2001.

Domsch, M. E. / **Ladwig**, D. H.: Handbuch Mitarbeiterbefragung. 2nd ed., Berlin; Heidelberg: Springer 2006.

Egle, F. / **Bens**, W.: Talentmarketing: Strategien für Job-search und Selbstvermarktung und Fallmanagement. Wiesbaden: Gabler, 2nd ed., 2004.

Eisele, D. / **Horender**, U.: Auf der Suche nach den High Potentials. in: Personalwirtschaft, 12, 1999, p. 27–34.

Fiedler, R.: Organisation kompakt. Munich: Oldenbourg Wissenschaftsverlag 2007.

Hardenacke, H.: Personalauswahl: Unpersönliche Auswahlverfahren – Forschungsbericht. in: Bröckermann, R. / Pepels, W. (ed.): Handbuch Recruitment: die neuen Wege moderner Personalakquisition; Planung, Beschaffungswege, Auswahlverfahren; Beiträge aus Forschung und Praxis. 1st ed., Berlin: Cornelsen 2002.

Hartmann, G.: Personalbedarfsanalyse – Forschungsbericht. in: Bröckermann, R. / Pepels, W. (ed.): Handbuch Recruitment: die neuen Wege moderner Personalakquisition; Planung, Beschaffungswege, Auswahlverfahren; Beiträge aus Forschung und Praxis. 1st ed., Berlin: Cornelsen 2002.

Hentze, J. / **Kammel**, A.: Personalwirtschaftslehre. 1st ed., Bern; Stuttgart; Wien: Haupt 2001.

Hesse, J. / **Schrader**, H. C.: Das neue Test-Trainings-Programm. Einstellungs- und Eignungstests erfolgreich bestehen: Die wichtigsten Testaufgaben – und wie man sie löst. Frankfurt on Main: Eichborn 1991.

Hesse, J. / **Schrader**, H. C.: Assessment Center. Das härteste Personalauswahlverfahren. Frankfurt on Main: Eichborn 1994.

Hesse, J. / **Schrader**, H. C.: Assessment Center für Hochschulabsolventen, Bewältigungsstrategien für das härteste Personalausleseverfahren. Frankfurt on Main: Eichborn 1998.

Hesse, J. / **Schrader**, H. C.: Das Hesse/Schrader Bewerbungshandbuch: Alles, was Sie für ein erfolgreiches Berufsleben wissen müssen. Frankfurt on Main: Eichborn 2002.

Heyse, V. / **Erpenbeck**, J. (ed.): Kompetenzmanagement: Methoden, Vorgehen, Kode(r) und Kode(r)x im Praxistest. Waxmann 2007.

Holtbrügge, D.: Personalmanagement. 2nd ed., Berlin; Heidelberg: Springer 2005.

Holtbrügge, D.: Personalmanagement. 3nd edition, Berlin; Heidelberg: Springer 2007.

Jeserich, W.: Mitarbeiter auswählen und fördern: Assessment-Center-Verfahren. 6th ed., Munich and Wien: Hanser 1991.

jobpilot AG: URL:
http:www.jobpilot.de/content/service/pr_ir/news/grafik-rekrutierungs-budgets.html
Status: 04 June 2002.

Jung, H.: Allgemeine Betriebswirtschaft. 10th edition, Munich: Oldenbourg Wissenschaftsverlag, 2006.

Jung, H. J.: Personalwirtschaft. 8th edition, Munich: Oldenbourg Wissenschaftsverlag, 2008.

Karle, R.: The Next Generation. in Personalwirtschaft, special volume 5, 2001, p. 30–38.

Kleinmann, M.: Assessment-Center. Göttingen: Hogrefe 2003.

Klimecki, R. G. / **Gmür**, M.: Personalmanagement. 3rd edition, Stuttgart: Lucius & Lucius 2005.

Kolb, M.: Personalmanagement: Grundlagen – Anwendung – Umsetzung. Wiesbaden: Gabler 2008.

Kropp, W.: Systematische Personalauswahl. 2nd edition, Munich: Oldenbourg Wissenschaftsverlag 2001.

Krüger, K.-H.: Personalauswahl: Angebotssichtung – Forschungsbericht. in: Bröckermann, R. / Pepels, W. (ed.): Handbuch Recruitment: die neuen Wege moderner Personalakquisition ; Planung, Beschaffungswege, Auswahlverfahren; Beiträge aus Forschung und Praxis. 1st edition, Berlin: Cornelsen 2002.

Manke, T.: Personalauswahlverfahren unter der Lupe: Vom Assessment Center bis zum Zeugnis. Norderstedt: Books on Demand 2008.

Obermann, C.: Assessment Center: Entwicklung, Durchführung, Trends. Mit originalen AC-Übungen. 3rd edition, Wiesbaden: Gabler 2006.

Oechsler, W. A.: Personal und Arbeit: Einführung in die Personalwirtschaft unter Einbeziehung des Arbeitsrechts. 6th edition, Munich: Oldenbourg Wissenschaftsverlag 1997.

Olfert, K.: Personalwirtschaft. 13th edition, Ludwigshafen (Rhein): Kiehl 2008.

Olfert, K. / **Rahn**, H. J.: Lexikon der Betriebswirtschaftslehre. 4th edition, Ludwigshafen (Rhein): Kiehl 2001.

Preuss, A. / **Knoll**, T.: Computergestützte Assessmentsysteme. in: Personal, n° 3, 2001, p. 128–132.

Randstad: Unternehmensprofil, URL:
http://www.randstad.de/content/aboutrandstad/unternehmensprofil/
Status: 01 July 2009.

Rationalisierungskuratorium der Deutschen Wirtschaft: RKW – Handbuch Personalplanung. 2nd edition, Neuwied, Frankfurt on Main 1990.

Reuschenbach, B.: Personalgewinnung und Personalauswahl für die Pflege. 1st edition, Munich: Elsevier, Urban & Fischer 2004.

Rieck, W.: Externe Personalbeschaffungswege: Scouting – Forschungsbericht: high Potentials durch Scouting gewinnen. in: Bröckermann, R. / Pepels, W. (ed.): Handbuch Recruitment: die neuen Wege moderner Personalakquisition; Planung,

Beschaffungswege, Auswahlverfahren; Beiträge aus Forschung und Praxis. 1st edition, Berlin: Cornelsen 2002.

Rump, P.: Kaufmännisches Lexikon – Wirtschaft für jedermann. Munich: Lexikographisches Institut 1985.

Schmeisser, W.: Einfach lernen! Personalmanagement. www.studentensupport.de 2006.

Schmeisser, W. / **Eckstein**, P. / **Klugmann**, P.: Externe Personalbeschaffungswege: Progressives Posting – Forschungsbericht: Personalrecruiting im Internet. in: Bröckermann, R. / Pepels, W. (ed.): Handbuch Recruitment: die neuen Wege moderner Personalakquisition ; Planung, Beschaffungswege, Auswahlverfahren ; Beiträge aus Forschung und Praxis. 1st edition, Berlin: Cornelsen 2002.

Scholz, C.: Personalmanagement. 5th edition, Munich: Vahlen 2000.

Schuler, H.: Assessment Center als Auswahl- und Erfolgsinstrument – Einleitung und Überblick. Stuttgart: Eichborn 1987.

Schuler, H.: Das Einstellungsinterview. Göttingen, Bern, Toronto, Seattle: Hogrefe 2002.

Schuler, H. / **Frier**, D. / **Kaufmann**, M.: Personalauswahl im europäischen Vergleich. Göttingen: Verlag für Angewandte Psychologie 1993.

Schüller, A.: Praxisbericht: Scouting – die neuen Methoden der Bewerberrekrutierung. in: Bröckermann, R. / Pepels, W. (ed.): Handbuch Recruitment: die neuen Wege moderner Personalakquisition; Planung, Beschaffungswege, Auswahlverfahren; Beiträge aus Forschung und Praxis. 1st edition, Berlin: Cornelsen 2002.

Spickschen, E.: Internes Unternehmertum und Recruiting von High Potentials: Theoretische und empirische Untersuchung. 1st edition, Wiesbaden: Deutscher Universitäts-Verlag 2005.

Steinmann, H. / **Schreyögg**, G.: Management – Grundlagen der Unternehmensführung. Wiesbaden: Gabler 1997.

Stelzer-Rothe, T.: Personalauswahl: Persönliche Auswahlverfahren – Forschungsbericht. in: Bröckermann, R. / Pepels, W. (ed.): Handbuch Recruitment: die neuen Wege moderner Personalakquisition; Planung, Beschaffungswege, Auswahlverfahren; Beiträge aus Forschung und Praxis. 1st edition, Berlin: Cornelsen 2002.

Stelzer-Rothe, T. / **Hohmeister**: Personalwirtschaft. 1st edition, Stuttgart; Berlin; Cologne: Kohlhammer 2007.

Stock-Homburg, R.: Personalmanagement: Theorien – Konzepte – Instrumente. Wiesbaden: Gabler 2008.

Sunter, S.: Kontakte knüpfen auf Recruiting-Messen. in: Personalwirtschaft, 2000, S. 42–45.

Weber, S.: Das professionelle 1x1, Den besten Mitarbeiter finden: Bewerberflut zielsicher bewältigen. Anforderungsprofil und Stellenbeschreibung. Wege der Personalbeschaffung. Bewerbungsunterlagen und Bewerbungsgespräch. Berlin: Cornelsen 2007.

Weideneder, M.: Erfahrungsbericht: Personalvermittlung im Internet. In: Personal, Heft 7, 2001, S. 384–387.

Weuster, A.: Personalauswahl. Wiesbaden: Gabler Verlag 2004.

Wickel-Kirch, S. / **Janusch**, M. / **Knorr**, E.: Personalwirtschaft, Grundlagen der Personalarbeit im Unternehmen. Wiesbaden: Gabler 2008.

Zapp, R. M.: Systematische Personalauswahl und ihre rechtlichen Rahmenbedingungen. Martin Meidenbauer 2006.

X Company culture and country culture

The term *Company Culture* contains cultural values in form of company specific basis convictions, imagination and orientation patterns that coin the behavior of and relations between company members. In specialized literature, company culture is also referred to as organizational culture. In the context of increasing pressure of competition and in the process of globalization it becomes more and more important to look into the concept of company culture. On the one hand, company culture can be a decisive competitive advantage on national and international markets, with which companies can distinguish themselves from the competitors and improve their own image. A strong company culture can help to improve labor structures, motivate employees and contribute to efficient work within the organization, and finally it makes a positive contribution to the company growth and its success. On the other hand, an internationally active company which has subsidiaries abroad or starts a subsidiary with a multinational company automatically needs to deal with its company culture. The formation of Joint Ventures steadily increased in the context of growing globalization and today confronts companies with big challenges, because a "joint organization-culture framework" has to be found.[599] This means that the company has to decide whether to follow the concept of worldwide standardization with a uniform presentation of the company to the outside or of local differentiation.[600]

The company culture shall create "circumstances for the best possible conditions for the viability of the system company" [601] or to exist as competitive dimension on the national and international market.

Jones and Bouncken understand company culture as "system of joint values and rules, the relation of the organization members between each other, as well as interactions with external partners".[602] The company culture shall give a common understanding to the organization members or employees on how to react to certain situations. Certain norms, standards, work instructions and rituals embody thereby joint basic convictions, values and ideals in the company. The company culture reflects the moral concepts of the management

---

599 Sackmann, 2002, p. XIII, own translation

600 Comp. Welge/Holtbrügge, 2003, p. 190

601 Sackmann, 2002, p. XIII

602 Jones/Bouncken, 2008, p. 408, own translation

as well as the behavior to each other and to employees and external partners. The goal of company culture is to coordinate company targets, employee satisfaction and customer orientation and to achieve a competitive advantage over the competitors and to increase company effectiveness.[603]

In certain situations it becomes necessary for the company to question the existing culture, to analyze it and to change it, if necessary, or to set priorities. The reasons can be in the external sphere, the company environment or the internal sphere, i.e. in the company itself.[604] Now, some of these situations will be described.

39 Factors in the company environment

39.1 Economic factors

Increasing internationalization/globalization: By opening up new markets abroad, the international competitive pressure in some sectors rises. The company needs to create a strong company culture as competitive advantage which is hard to copy.

Strategic alliances/mergers: these find their reason in the global presence (propagation of the image), enlargement of the own product portfolio (creation of new market segments) and the creation of a competitive dimension (higher market share) and serve to maintain the competitive capacity. A joint culture has to be created.

Technological progress: new technologies as the internet access demand for new sales channels. The company should adjust its culture to the new developments.

39.2 Socio-cultural values

Moral values: One can distinguish moral values of customers and of employees. The moral values of customers have an influence on their needs, desires and interest in certain products. The moral values of potential employees have an influence on their expectations towards the company culture and thus, on the selection of the future employers. Through a diversity of values with high expectations, the company should create a culture which awakens the interest of customers and potential employees.

---

[603] bpb, status: 13 June 2009, 16:15h

[604] Comp. Sackmann, 2002, p. 11 et seq.

Demographic developments: The decreasing birth rate causes a low availability/number of highly qualified employees especially in the area of IT and engineering. Companies compete for these potential employees and should try to get their attention via company culture and to win persons with high know-how over.

40 Company-internal factors

Strong company growth: In case of strong company growth the former company culture and the internal company environment should be maintained and communicated to new employees. Typical growth sectors are currently the IT and communication sector.

Stagnation: Stagnation causes changes in the company strategies and –tactics as well as in the company culture.

Productivity problems: these are mostly signs of internal friction losses. Also in this case, thoughts should be given to a change of company culture.

Management changes: a change in the top management can either lead to continuance of existing strategies, aims and the company culture, a reorientation or setting of new priorities.

41 Cultural values

A company culture consists of many cultural values which influence the behavior of employees, e.g. how they perceive their environment, how they take information and process them, how they react to and behave in certain situations. Cultural values can be understood as general criteria, standards, principles and directives which make clear to the employees which behavior is desirable in certain situations. These values are mostly not formalized and not written down. Special norms, rules and standard work instructions help the employees and external partners like customers, suppliers and the society to understand and internalize the company culture.

41.1 Types of cultural values

There are two kinds of cultural values: target-oriented performance values and instrumental values (see Fig. 41.1).

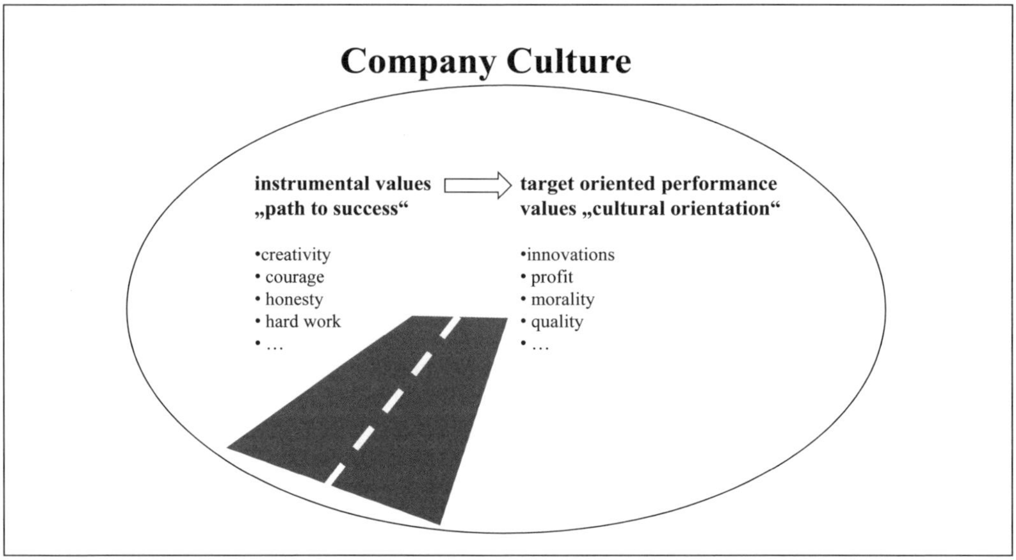

Fig. 41.1: Company Culture

In accordance with Jones and Bouncken target-oriented performance values are the aspired targets and results of a company or an organization.[605] This is also referred to as lead-culture or cultural orientation in a company. This lead culture depends on the company and the sector and is reflected in economic efficiency, reliability, responsibility for one's actions, innovations, morality etc. The *instrumental values* represent the desired behavior to reach the target-oriented performance values. Among these values is respect for authority figures, preparedness to take risks, honesty, creativity etc. To make it clear: A company cannot launch innovative products if the employees in the development department do not have creative ideas and are not willing to take certain risks, because the company knows only later – via sales revenues and profits – whether the product was really well accepted by customers and whether it was beneficial for the purchasers.

41.2 Functions of cultural values

Cultural values fulfill diverse tasks. They serve as "important intermediary for mutual adaptation of people in organizations."[606] If employees in organizations share cultural values, a joint basis is created in order to reduce tensions within the organization, overcome

---

605 Comp. Jones/Bouncken, 2008, p. 409

606 Jones/Bouncken, 2008, p. 416

misunderstandings, strengthen the collaboration and to create a harmonic company atmosphere.

Besides, team-related work within an organization can mean higher identification of the employees with the organization and its aims and have positive effects on its motivation and performance.

For partners outside the company, the organizational team seems to be a coherent unity and within the company, the individual self-esteem of the team-employees rises. Cultural values, as informal structure, simplify working within the organizational structure. The company culture points the desired behavior patterns of the employees out (via rules, instructions, directives) and show which cultural values should be followed – and therefore serves as behavioral guidelines. Furthermore, cultural values give orientation support in case of ambiguities and uncertainties within and outside the organization. If an employee is insecure in a certain situation whether the decision is wrong or right, or which action is in the company's interest, he can base the decision on the cultural values of the company.

41.3 International aspect

The global interconnections grew constantly in the past years, which challenges internationally active companies, because different countries and cultures come together and need to be aligned with each other. This way, the values and norms of different countries, so called national cultures, have influence on the organization culture. If two or more companies collaborate in form of a joint venture, the challenge it to create a cross-country, joint company culture, which respects the values and norms of each member company. However, the companies have to make compromises. The difficulties shall be pointed out with the help of the following example: if a Mexican and a US-American company merge, completely controversial work methods and behavior can be observed. For Mexican companies a moderate work pace with long lunch breaks with the family and slow decision taking in small teams of top managers are characteristic. Relations between companies are only built up after some time, because several personal contacts are necessary to create a basis for mutual trust. In the US, on the other hand, high working speed with short or no breaks prevail, and decisions are quickly taken by managers on the lower level (peripheral decision-taking).[607] Because of the controversial ways of working, a takeover-country-culture expert can help in order to avoid failure because of national cultures.

Also the delegation of managers and experts abroad, e.g. in subsidiaries, belongs to the central topics in companies today. The employees should learn in trainings, country studies or country lectures the values, norms, customs and behavioral forms in the foreign society before being delegated, in order to be able to react to certain local situations in an appropriate and target-oriented way and to avoid a failure of national cultures.

---

[607] Comp. Jones/Bouncken, 2008, p. 412

42 Communication of culture

This chapter presents, how the company culture is communicated to employees and groups of stakeholders (suppliers, customers, society). This is undertaken in form of socialization tactics, stories, rituals, the company language etc. Now different forms of communication of culture will be described

42.1 Communication of culture via socialization

Socialization is needed to learn, communicate and internalize existing company values, norms, regulations and role expectancies towards new employees. The socialization as instrument for direction of people contains the "cognitive, affective and behavior-related assimilation by employees based on conscious and unconscious psychical influence".[608] The employee is subject to unconscious influence if he observes the behavior of employees who have been working for the company for more time, analyzes their conduct and deduces for himself a situation-adjusted, appropriate behavior. A conscious influence can be exercised on the employee by the purposeful use of socialization tactics, as will now be explained.

In order to clarify the term socialization there are different model approaches in specialized literature. In this book the socialization model of Van Maanen and Schein will be examined. This model shows the socialization process and points out which measures an organization can take to influence new employees, to internalize desired values and to act accordingly. This influence on the behavior of new employees and the role orientation is taken via socialization tactics, an instrument of directed socialization.

Van Maanen and Schein distinguish twelve socialization tactics, while always two confront each other and lead to controversial role orientations.[609] The tactics named in Fig. 42.1 on the left side lead to an institutionalized role orientation of new employees, i.e. employees are required to accept certain norms and regulations, to adjust themselves and to obey. They shall use the "autochthonic" employees as guidance and react in the same way – as desired. "A disadvantage is the connected danger of mental unification of the employees, which means that their sensitivity for important changes of environmental conditions is reduced."[610] It becomes clear that in this role orientation there is no space for individual reactions and creative suggestions for problem solving. Companies run the risk to miss the entry in new markets because of lack of innovation and customer orientation. Therefore, it is not recommendable to use this style in creative, innovative sectors. However, these tactics are advantageous in order to prevent erroneous decisions, to simplify coordination, take

---

608 Welge, 2003, p. 169, own translation

609 Comp. Jones/Bouncken, 2008, p. 421

610 Welge, 2003, p. 169, own translation

decisions faster and to reduce the control efforts. In practice, the institutionalized role orientation is above all relevant in military, in the financial sector etc. where values as honesty and trustworthiness come in first place. The danger of institutionalized role orientation lies in the manipulation of the employees by the leaders. Leaders might play on the behavior – adaptation, obedience, subordination – and act against the law.

collective communication of company-typical learning experience → standardized reactions to certain circumstances	↔	**individual** learning experience is unique → individual reactions by new employees
formal learning of new tasks in isolation from „old" personnel (e.g. external trainings)	↔	**informal** experiencing new tasks in the organizational team as new team member
sequential new employee obtains overview over task process and functions	↔	**random** sequences of tasks and order of functions directed by employee needs (individual)
fix new employee knows time schedule for tasks	↔	**variable** Learning and tasks directed by the individual employee needs
directed old employees as mentors and advisors for new employees	↔	**individual** Employees individually analyzes own actions without instructions from others
separating no social support by old employees (ignoring, derision)	↔	**integrating** Social support by existing team

Fig. 42.1: Socialization tactics of Maan and Schein[611]

The tactics on the right (see above) lead to an individual role orientation. In this case individual and creative ways of working of the employees are targeted and promoted. The employees can experiment with changing norms and values within their scope of action and create innovative solutions for problems. This type of role orientation is typical of creative, innovative sectors as the IT or automobile sector.

611 Comp. Jones/Bouncken, 2008, p. 421

42.2 Communication of culture via stories, rituals and company language

With increasing age and growth of the company, stories, rituals and ceremonies emerge[612] and coin the company image. These means of communication tighten company networks up, as will be explained in the following paragraphs.

Stories, partially true, partially invented, can reflect behavior and methods of persons – which appear in the story – and show what employees estimate or condemn and therefore, have influence on the employees' behavior.[613] The company stories are very different, but always have a "true" core and morality.

Rituals on the other hand are returning ceremonial, solemn actions on a certain place at a certain time. These serve to communicate and strengthen certain company specific values and norms and to strengthen inter-personal connections. The communication between the employees can improve the company atmosphere and contribute to higher employee motivation and satisfaction. Finally, motivated and satisfied employees have a positive effect on the company success, because they are more willing to perform and to identify with the company culture.

In practice three different rituals are used: transition rituals, integration rituals and expansion/improvement rituals. Transition rituals are timely limited, from entrance, over promotion up to leaving the company and comprehend e.g. special job titles (e.g. fashionable titles) as well as badges of honors for the most effective and creative employees of the year. Integration rituals strengthen the relations between employees in form of Christmas parties, company parties, company excursions, Halloween, barbecue, eating together or traditional weekly meetings. In case of expansion/ improvement rituals certain employees are complimented for their work performance/ efforts in awarding events, newspaper publications or employee promotions, which motivates them to continue to commit themselves to their work.

The company culture is also reflected in **company language** in form of company typical abbreviations and verbiage. The meaning behind the term company language is a uniform language in a company or an organization whose meaning can be understood by all organizational members. In specialized literature the term company parlance is also common.

"The knowledge of a specific language within a sector, a company or a certain department is of central importance for the daily work, the communication with colleagues, employees and the supervisor, i.e. for the daily 'survival' in a company"[614], because the uniform company language as communication basis simplifies the coordination of tasks, avoids communication

612 Comp. Sackmann, 2002, p. 57

613 Comp. Jones/Bouncken, 2008, p. 427

614 Sackmann, 2002, p. 28, own translation

problems and reduces communication barriers and strengthens the identification feeling of employees within the company.

Examples for sector-typical company language are:

- For example, "bottom fishing" is a verb that refers to a practice of buying up stock that is undervalued due to overtrading or a falling market.[615]
- "Chasing Returns" refers to the practice of taking great risks on the feeling that one can gain great returns.[616]

42.3 Communication of culture via other means of communication

Besides the above mentioned means of communication to communicate the company culture, there are still diverse others, e.g. company symbols, company dress, brands of company cars, manuals etc.

Company symbols are certain characteristics of the company which are consciously used by the company to recreate a recognition value and to exclude confusions with other companies, to create a corporate representation to the outside and project a good company image. Examples are a good company location (view over the city), design of the building or location of the offices on a certain symbolic floor.

Also uniform **company dresses** or a certain **brand of company cars** (e.g. Mercedes for prestige in upper segments) get company values across to the outside.

Furthermore, new employees can gain information on work directives, rules and typical conduct from **company manuals** and get to know the company culture.

43 Formation of company culture

After having explained the term company culture and the means of communication, now the formation of company culture is analysed. It is composed of four factors: the characteristics of the employees, the ethics within an organization or company, the disposition rights of the employees and the company structure. In the following chapter these factors of influence will briefly be described.

---

[615] http://www.ehow.com/about_7426498_stock-market-slang-termp.html, Status: 30 June 2011, 13:15h

[616] Ibid.

43.1 Employee characteristics

The most important influence factor for company culture is the people who constitute the company.[617] Different moral ideas of people account for the company culture, but also for their interest in the company. Each potential employee chooses a company which suits his moral ideas. On the other hand, companies search on purpose for suitable employees who identify with company values and ensure effective work. The company founders have the most important influence on the company culture, because their ideas and convictions are reflected in the company culture and provide the basis for the desired conduct of the employees.

43.2 Ethics within the organization

Also ethics of individual personalities in the companies, of professional groups and of the society coin the company culture. The term company ethics comprises moral values, convictions and rules which ensure an appropriate behavior of the employees towards each other and to other company stakeholders.[618] Ethical values are used by leaders in certain situations to support the decision-making process. If the managers have to decide whether a measure is morally right or wrong, they observe the ethical values of the company and act accordingly.

43.3 Disposition rights

Besides the characteristics of the employees and the company ethics, disposition rights of the employees coin the company culture. Demsetz defines disposition rights as rights to dispose and use company resources and responsibilities which a company gives to its employees.[619] Disposition rights also function as employee motivation. If the employees perform well, the company can concede disposition rights as award and further motivation. Furthermore, employees are bound to the company e.g. through employee shares or profit participation options. If a company wants to keep highly qualified employees, this aspect is of great relevance. Some examples of disposition rights are displayed in the following table 43.1:

---

617 Comp. Jones/Bouncken, 2008, p. 431

618 Comp. ibid, p. 434

619 Comp. Demsetz, 1967, p. 347 et seq.

	Rights of disposition over company resources	Rights of utilization of company resources
Manager	High salaries, stock options, high compensations	authorization to take decisions, to control resources
Employee	permanent position, profit participation systems, non-monetary benefits, employee shares, compensation payments, company pension, etc.	Responsibility, degree of control over one's own actions

Table 43.1: Disposition rights of managers and employees[620]

A good practical example for a company that provides certain disposition rights to its employees is the Google group. It offers numerous non-monetary benefits – two T-shirts for free per week, free all-you-can-eat buffets and snack machines, a company own fitness studio etc. Furthermore, Google employees receive free broadband internet access and mobile phones for their disposition. In addition to these non-monetary benefits, a higher salary in case of good work assessment makes a positive contribution to the motivation and company connection of the employees.[621]

43.4 Organizational structure

The fourth influence factor of the company culture is the organizational structure – a system for the coordination of company processes via the use of tasks and rights to give instructions.

Similar to the role orientation of the employees in the socialization tactics, mechanistic and organic structures are opposed to each other. Typical of mechanistic structures are a steep structure caused through many company levels, a strong centralization (e.g. central managerial powers) or the standardization of processes. Typical conducts within these mechanistic structures are carefulness, obedience, traditions and conformity of the employees. The company culture of mechanistic structures is coined by stability and predictability and is practiced military and financial institution and power plants, in order to avoid errors and to make quick decision-making possible. Organic structures on the other hand tend to be flat and decentralized (e.g. decentralized managerial powers) and are characterized by a larger freedom to deal with tasks. The employees are meant to work creatively and with courage and willingness to take risks. Because innovation and flexibility

---

620 Comp. Jones/Bouncken, 2008, p. 439

621 Google, status: 13 June 2009, 16:15h and onpulson, status: 13 June 2009, 16:16h

are characteristics of the company culture, the decentralized structure can often be found in the IT and marketing sector.

Literature

bpb: URL: http://www.bpb.de/popup/popup_lemmata.html?guid=M9O9WM
Status: 13 June 2009, 16:15h.

Demsetz, H.: Towards a Theory of Property Rights. American Economic Review, 1967, 57, p. 347–359.

E-How: http://www.ehow.com/about_7426498_stock-market-slang-terms.html, Status 30 June 2012, 13:15h.

Google: URL: http://www.google.de/intl/de/corporate/culture.html
Status: 13 June 2009, 16:15h.

Jones, G. R. / **Bouncken**, R. B.: Organisation. Theorie, Design und Wandel [book], Pearson Studium Verlag, Munich 2008.

onpulson: URL: http://www.onpulson.de/karriere/magazin/fachartikel/show--727-1.htm
Status: 13 June 2009, 16:16h.

Sackmann, S. A.: Erkennen – Entwickeln – Verändern [Buch], Hermann Luchterhand Verlag, Neuwied Kriftel (Taunus) 2002.

Welge, M. K. / **Holtbrügge**, D.: Internationales Management. Theorien, Funktionen, Fallstudien, Schäffer-Poeschel Verlag, 3rd edition, Stuttgart 2003.

XI Country research at the example of South Africa: history, population and culture

South Africa is "a world in a country". This is the slogan that the tourist board uses to advertise the outstanding landscapes and the diverse climate, numerous peoples and various cultures as well as religions and mentalities. There is hardly any other country that allows such different impressions. Two worlds, the rich and the poor – still called third world – often collide and there is still harmony. The country is intriguing, but characterized by the social contrasts, AIDS and political change.

44 Early history

"The history of South Africa is the history of a battle for land and political power."[622] It starts with the San, the so-called native people of South Africa. The Bushmen were mere gatherers and hunters and lived in the whole Southern area of Africa. Characteristic for their language are the click-consonants. Later on came the Khoikhoi, the human-humans, to the South Africa of today. Their traditional way of life combined nomadism and foraging. Both peoples were similar to each other regarding their looks and their language, which is why, nowadays, they are often called Khoisan. The hunting Khoikhoi had first and foremost conflicts with the San with regard to cattle. However, the Khoikhoi successfully drove away the San from their settlement or subdued them.

The great migration movement of the Bantu peoples began already before Christ in the Cameroon of today. The Nguni-language group with the languages Swazi, Tsonga, Zulu, Xhosa and Ndebele have spread furthest south from the different Bantu peoples. They reached the area between today's city of East London and Bloemfontein around 1500 post Christ.[623]

---

622 Hagemann, 2001, p. 7 et seq., own translation.

623 Comp. Wiese, 1999, p. 88

Given the great Bantu-expansion, the original inhabitants were squeezed out to other territories increasingly. Therefore, the only living space left for the San was the central Kalahari. The advancing expulsion and impoverishment of this population group came about as a consequence of their exploitation by the political and economic elite. To this day, the only surviving tribe of the Khoikhoi is the Nama tribe, which migrated to the south of Namibia. All others were traded as slaves and driven away from their settlements by the Boers since the 17th century. Diseases which were introduced, such as pox, and wars have further reduced the tribes. A vast part also mixed with the Caucasians and the Bantu.

45 Arrival of the Europeans and development of the partitioning of the land

It was not until 1652 after the foundation of Cape Town that more European settlers came to South Africa. Jan van Riebeeck, commissioned by the Dutch East Indian Company, built a supply station for their ships and crew at the Cape of Good Hope in order to reduce the human toll on their long sea routes. Initially, they did not think to expand the station in order to make it a settlement colony. Some farms and gardens administered by the "free citizens", former employees, were supposed to suffice in order to cater for the ships. However, the Cape colony became larger step by step and also the geographical research increased in the back country. This was because on the one hand, more settlers travelled to the country from Europe, and on the other hand, expeditions were initiated to instigate the cattle trade with the Khoikhoi and in order to search for natural resources. Missionaries were also interested in the country and brought the Christian faith and the European civilization to the population.[624]

From 1657, the Dutch brought numerous slaves from Indonesia and Madagascar into the country, which brought a new population group into life, which has contributed to today's ethnic diversity.

Also, politically undesirable individuals from East India were left at the Cape who brought with them the Islamic religion and still constitute as "Cape Malay" a population group with an own culture. The white settlers led may wars with the Xhosa which lasted for years.

After the Cape had been occupied by the Dutch in the 17th and 18th centuries, the British conquered the land in 1806 and built up a British Crown Colony. Most of the white settlers spoke a modification of the Dutch language, which later developed into an own language, Afrikaans. It is only after 1820 that a new large wave of English speaking settlers came into

[624] Comp. Jürgens/Bähr, 2002, p. 42 et seq. and Hagemann, 2001, p. 26 et seq.

the country. This gave rise to two language groups with English being determined as the official language.[625]

The British parliament abolished slavery in 1834. First and foremost the "Boers"[626] which were native to the border regions of the colony considered their existence threatened thereby. Therefore, they moved as "Voortrekker" to the largely unknown inland in order to escape the British law.[627]

The Boer republics received much attention with the discovery of diamonds near Kimberly in 1867 and of gold at the Witwatersrand in 1886 which attracted thousands led by diamond fever and gold rush. Finally, in 1899 the so-called Boer War (South African War) erupted, which the militarily superior English won only with great effort. However, this impaired the relationship between Caucasians of English and Boer origin for a long time.

The South Africa of today has only been a unified state since 1910 which has formed by the unification of the four colonies Natal, Transvaal, Oranje-republic and the Cape Colony to the South African Union (see Fig. 45.1).

Fig. 45.1: Former provinces of South Africa[628]

[625] Comp. Pabst, 1997, p. 36

[626] Derived from the Dutch *boer* for *peasant.*

[627] Comp. Pabst, 1997, p. 36–37

[628] South Africa – Infoweb, 2007

46 Population groups

The division of the population into four groups which was established at the times of the apartheid is still common as of today. However, only the Whites belonged to the privileged people, followed by the Coloureds and the Indians, which, again, were superior to the Blacks.

"Statistics of South Africa" carried out a population census in 2001. As shown on Fig. 46.1 79.0 percent of the about 45 million inhabitants of South Africa belong to the Blacks. In contrast to that, the Whites constitute a small population group at 9.6 percent, similar to the Coloureds at 8.9 percent. The smallest group is that of the Indians at 2.5 percent.

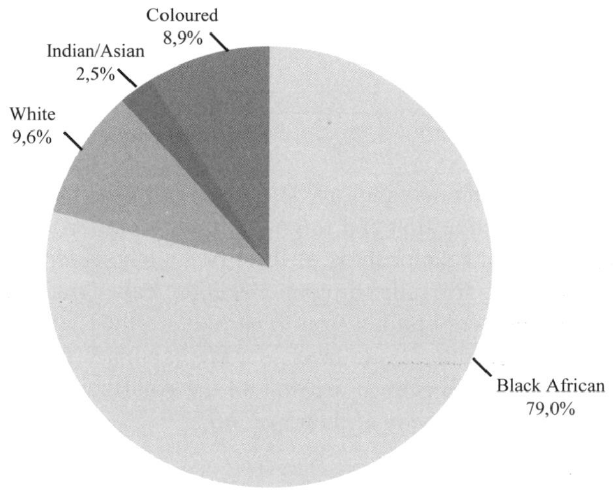

Fig. 46.1: Population groups[629]

Within the four categories there are further sub-groups with different languages, ethnic groups and religious beliefs. The skin color dominated the existence of the South African population up until the elections in April 1994. During the apartheid the skin color influenced the choice of the place of domicile, the school, the means of transport, the workplace, just to name a few examples.

In the following four chapters the population of South Africa will be presented group by group, categorized into Whites, Coloureds, Indians/Asians and Blacks.

629 Statistics South Africa, Census, 2001

46.1 Whites

The 4.5 million white South Africans are of Dutch origin, also called “Boers”, and were the first to arrive at the Cape in the mid 17th century. They were followed by settlers from England from 1820.

Both groups battled each other mercilessly in the Boer Wars. From 1857, over 5,000 German men, women and children stepped on land at East London. Cities such as Braunschweig, Hamburg or Heidelberg developed. Today, around 1 million Germans live in South Africa, along with French, Italians and 120,000 Jews.[630]

The mother tongue of these “South Africans” is the Afrikaans similar to Dutch. There is also an English speaking part of the population which is concentrated in the western and eastern Cape area and in Natal. This population group lives primarily in cities and controls the commercial and economic sectors to a large extent.

46.2 Coloureds

The children of black and white couples are still often called Coloureds. These are around 4 million South Africans that live first and foremost in the Cape Province and in Cape Town. They are the descendants of relationships of the first Dutch settlers with the Khoikhoi, the native people on the Cape. Also, the around 200,000 “Cape Malay” are added. They are descendants of slaves that were brought from East India to the Cape in the 18th century. The Coloreds have their own cultural heritage. As an example, the Coloreds which were privileged as compared to the Blacks at the time of the Apartheid define themselves by way of their profession rather than by way of their origin.[631]

46.3 Indians/Asians

Given the introduction of large-scale sugar cane farming, the expansion of mining activities and the commencing railway construction, the need for workforce in Natal increased considerably in the middle of the 19th century. This need was partly satisfied with the hiring of Indians. The Indian population of South Africa derives from the descendants of the original day laborers. The around 1.2 million people dominated a vast part of commerce in Durban. Mainly Chinese people settled in the area around Johannesburg. They came to South Africa as miners at the end of the 19th century and were also classified in this category.[632]

---

630 Comp. Nohlen, 1996, p. 631

631 Comp. Wiese, 1999, p. 85–86

632 Comp. Pabst, 1997, p. 46–47

The Indians were bound by the British in five year work contracts. After the contract terminated, many of them stayed in the country as "Free Indians". They had the freedom of establishment and were entitled to run a business. From 1870 further Indians, so-called "Passenger Indians" were attracted to the country without work contracts and oftentimes brought their wives with them. The non-native Indians did not form a homogeneous population group given also their religion, language and classification into a certain caste. They were overwhelmingly Hindus and came from rather low castes. Among the languages commonly spoken were initially Tamil, Hindi and Gujerati. Later on, English became their official language.[633]

Up until today the vast part of the Indian population lives in Natal. In the system of the Apartheid the Indians were categorized as a separate population group since 1959. They, too, were subject to similar restrictions as the other non-white population groups. Especially, they had to face compulsory relocations of inhabitants and business people. There were tensions between the Indian and the black population which turned into violence at times.[634]

46.4 Blacks

Nine different Bantu peoples with equally as many official languages belong to the black population. Around 10 million people form part of the largest South African people, the Zulu. This is almost a third of the black population. They settle first and foremost in KwaZulu-Natal and in the wider area of Johannesburg. Regarding their number, they are followed by the Xhosa (7.2 million). They live primarily between Port Elisabeth and Durban. 6 million people belong to the North and South Sotho – followed by the Tswana (3.3 million), Tsonga (1.8 million), Swazi (1 million), the Ndebele (0.6 million) East of Johannesburg and the Venda (0.9 million) in the furthermost North East on the border to Zimbabwe. The traditional cultures also contain different larger and smaller sub-groups. Most of the Blacks live in rural areas of the former ten Homelands or still in a very small space in the Townships at the border of the cities.[635]

47 Religion

Missionaries have contributed "their share" over many centuries. The largest part, around 80 percent, of all South Africans is Christians. The Dutch-Reformed Church, NG-Kerk, of the

---

633 Comp. Bilger, 1986, p. 172

634 Comp. Jürgens/Bähr, 2002, p. 57

635 Comp. Bilger, 1986, p. 215–216

"Boers" did not only convey its strictly Calvinistic way of life, but also promoted the Apartheid, which was also palpable in the NG-Kerk through the division into the four known groups.[636]

Already in 1910 the Zion Church was founded as a "Church without the White". There, its members found recognition and understanding. Most communities of faith of the Blacks combine Christian with traditional African religious beliefs. The proportion of Hindus and Muslims as part of the population is relatively low at a total of about 0.5 million people.[637]

48 Culture

Many different ethnic groups with different language, culture and origin are native to South Africa. Among these are eleven official languages. The population mainly speaks English, Afrikaans, seSotho, isiXhosa and isiZulu.

The fusion and commingling taking place in the urban areas of South Africa lead to the dilution of old, at the same time to the creation of new cultures. However, in many rural areas, the black traditions are still alive.

All of them have as common trait the belief in a male God, in the spirits of the ancestors and in supernatural powers. Polygamy is legal and usually a bride price is paid. In many cultures cattle plays an important role, both as status symbol and for sacrificial rituals.

Oftentimes the artistic heritage of the native South African people provides the only possibility to learn about a vanished culture. The petrography and cave art of the San constitute an example. Further, traditional crafts, such as the pearl crafts of the Zulu, were developed in times of need for the purpose of securing the existence of such people. The culture of the Xhosa, the so-called "red people", is likewise prominent and comes from the red scarves which are worn by most adult Xhosa. The Ndebele which are related to them live in Northern-Transvaal in self-painted homes (see Fig. 48.1).[638]

---

636 Comp. Wiese, 1999, p. 98–99

637 Comp. Wilke-Launer/Kühne, 1993, p. 421

638 Comp. Jürgens/Bähr, 2002, p. 39.

Fig. 48.1: Painted house of the Ndebele[639]

The totally independent culture of the white Africans formed in self-inflicted isolation. They tramped around Africa among their herd of cows and with a bible in their hand. To this day the conservative Dutch-Reformed Church is the center of all attention.

Even though many cultures originated in South Africa, most of them became obsolete or were destroyed during the Apartheid when the conventional and contemporary cultures were not looked after.[640]

The new South Africa arises in the streets of the Townships and cities. Retrospectives of black, contemporary and traditional artists are on display in galleries and musicians from all of Africa make an appearance at big festivals.

49 Socio-political aspects

South Africa is still far away from the ideal of a multi-cultural society. There is still a large division in the population. Blacks and Whites live largely separate from one another. There are few points of contact. The Apartheid regime made an effort to stress the differences between the ethnic groups and to ignore the common grounds. The politics of racial segregation served the purpose to secure the political and economic governance of the white minority. South Africa is still battling the consequences of these politics as of today.

---

[639] Saembassy, 2004. own translation

[640] Comp. Südafrika Guide, 2008

49.1 Concept of Apartheid

In the system of Apartheid the white minority population had developed a certain ideology. The goal was to differentiate between ethnic groups according to the origin such as language, religion, customs and traditions and to separate them physically to safeguard their respective cultural identity. Thereby, discriminatory methods used by the Whites against the Blacks in an economic and political context could disappear. The white "Boers" held the view that Whites and non-Whites would be forced to be equal. The black population had the chance to escape the superiority of the Whites by way of self-administration and political independence.[641]

Skin color became the most important distinguishing criterion for the definition of "groups".

Other phenotypical characteristics led to a legal differentiation of the population in Whites, Blacks, Coloreds and Indians/Asians in the 1950s. The country dominated by the white population pursued a specific political, social and economic discrimination against people with non-white skin color. Also, competitors could be controlled in their businesses and at their workplace in economic life. The isolation of the black population in Homelands revealed the concept of ethnicity as a mere political construction in order to achieve a demographic and also politically legitimate majority for the white minority.[642]

49.2 Reservation- and Homeland Policy

With the "Native Land Act" of 1913, the existing voting rights of the non-white population were abolished step by step and a first step was taken in the direction of Apartheid with regard to a separation in space. The "native reserves" created back then occupied 7.3 percent of the land. A sale of this land to Whites was prohibited and, on the other hand, Blacks were not allowed to purchase land in the "white" area. The area available to Blacks proved much too small from the start, and with the increasing population pressure, the problem became increasingly significant.

A first trial to consolidate the reservations was undertaken with the "Native Trust and Land Act" of 1936. The share of area should be brought to 12.4 percent via purchase of land. The reality, however, proved to be otherwise, as the area consisted of only 9.6 percent in 1945 and the split-up remained important, too. For the first time, the state obtained the right to arrange resettlements. The country which had been the property of black people since the time before 1936 and which lay in the "white" area was dissolved especially after 1948.

---

641 Comp. Pabst, 1997, p. 102–104

642 Comp. Jürgens/Bähr, 2002, p. 188

Finally, the inhabitants had to accept resettlement to the reservations. The total number of forced resettlements was 3.5 million only between 1960 and 1982/83.[643]

The introduction of the "Homeland-Concept" after 1948 was finally the continuance of the reservation policy. The "Bantu Authorities Act" of 1951 and the "Promotion of Bantu Self Government Act" at first accounted for eight, later ten Homelands, and in parallel abolished the last representation of black people in the "white" parliament. Then, efforts of consolidation and frontier displacements of the Homelands were continued. Around 16.2 million people lived in the Homelands in 1990 – among them around 1.6 million, mostly male migrant workers, who were active outside their Homelands. Around 800,000 commuters traveled, partially long distances, into the "white" region. This made clear that the economic basis of the Homeland was insufficient. The migrant workers and commuters constituted over 50 percent of the black workforce in "white" South Africa. However, the "homeland-concept" still did not make a real independence of the black population possible.[644]

Since 1994, the government has tried to reverse this massive injustice regarding the repartition of land without endangering the economic effectiveness. A land reform was planned to be carried out in three stages: the return of land, the redistribution of property and the reform of land and tenancy rights. The return of dispossessed land to the former owners, which was regulated in corresponding Act on the Return of Land of 1994, had highest priority.[645]

In order to avoid losses of productivity, cultivated farmland was protected against it at first. It will still not be easy to reach a balance between the high expectation of landless people, the interests of white farmers and the financial possibilities of the state.[646]

49.3 Democratization

The wave of democratization started in 1991 and affected the whole country. In order to point out the importance of this change, it is sometimes referred to as "second independence" or "second liberation". Especially church groups, influential legal associations and the free media, which has been flourishing since 1990, have decisively contributed to the construction of a civil society.[647]

---

[643] Comp. Wilke-Launer/Kühne, 1993, p. 431

[644] Comp. Jürgens/Bähr, 2002, p. 47

[645] Comp. Wilke-Launer/Kühne, 1993, p. 457

[646] Comp. Wiese, 1999, p. 189

[647] Comp. Ansprenger, 1999, p. 137

For the first time, a government was elected by all South Africans in 1994 which changed the political map. The independent Homelands were dissolved, but the outlines remain existent. Almost 50 percent of the population, i.e. around 19 million people, live in these areas.

Fig. 49.1 shows that today's South Africa divided into nine provinces. They do not have the same status as German Bundesländer. The provinces have each their own parliament with a Premier.

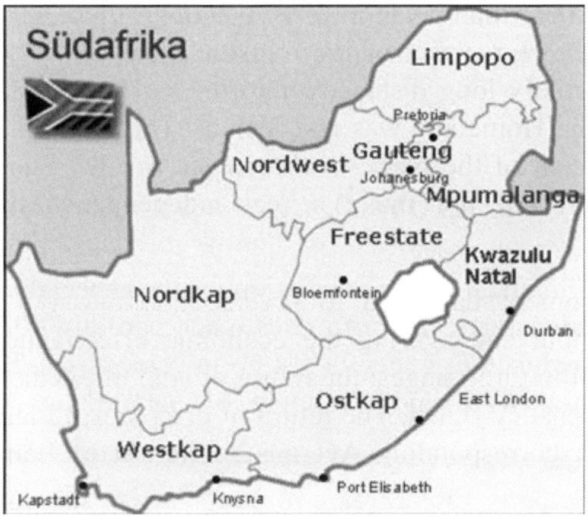

Fig. 49.1: Today's provinces in South Africa[648]

In South Africa a back-breaking preponderance of the African National Congress (ANC) as new leading force and party is recognizable. The explanation is that the voting conduct of the population is still largely driven by their clanship. But also the political opposition is more or less driven by ethnic belonging.[649]

50 Demographic and social aspects

Since the 1950s, it was possible to rise the mortality age from 40 over 55 up to 60 years in 1990. The improvement of survival chances is above all the result of exogenous impacts. Cheap medical-hygienic agents and practices were imported from industrialized countries which were partially with international support extensively distributed. This way, many infectious diseases could be partially kept under control or even eradicated. An example for

648 South Africa – Infoweb, 2007

649 Comp. Jürgens/Bähr, 2002, p. 62 et seq.

this is the worldwide anti-Malaria program which was started by the World Health Organization (WHO) with the help of several foundations in 1955. They also installed a project for the eradication of pox in 1967. Pox stopped to exist worldwide from 1977.

With the start of the 1990s the mortality rate rose all over South Africa again. Many of the returning infectious diseases started to spread again. The reasons were the agents that have become immune as well as the unhygienic living circumstances in a big part of the country, which caused a successful fight against diseases to fail. A simple solution would be the improvement of sanitary installations and the connection to a water network. However, this can still not be realized for monetary reasons in every household.[650]

50.1 Population growth

The dynamic of the development of the population results especially from the big difference between birth and mortality rate. The birth rates reach partially 40 ‰ and more, while the mortality rate has been reduced to under 13–14 ‰ in some cases. South Africa can already be attributed to a phase of distinct dropping of birth rates and strongly decreasing mortality.[651]

Other than expected by experts, the decrease of mortality has not continued since the end of the 1980s. Further progress was not achieved and the life expectation is with 55 years even below the one of the end of the 1970s. It looks as if South Africa had fallen back and showed a situation similar to the phase of high birth and death numbers. The most important factor for this negative development is the high number of AIDS cases in the population.

50.2 Development of the population under the influence of AIDS

Scientists assume that the disease originated in Central Africa and was brought in the country by migrant workers who spread it in the accommodations of mine workers. In the list of countries with the highest HIV/AIDS rates among adult population (15-49), South Africa joins the row of sad top places of south African countries with over 20 percent.[652]

A purposeful education on the risk of AIDS infections takes place only in rare cases or not at all. First steps were undertaken and a way for cheap medicine and free distribution of condoms was paved. However, there is still not enough money to put it in practice all over the country. Therefore, AIDS will still gain influence on the mortality rate.

---

650 Comp. ibid, p. 177

651 Comp. Wiese, 1999, p. 121

652 Comp. Weltbank, 2007

AIDS does not only lead to mortality, but also to a deterioration of the general state of health, the outbreak of other diseases like tuberculosis and the early death of the badly fed and cared AIDS orphans. Up to the year 2010, the life expectation will drop to 48 years. Because of the loss of human resources and the strongly rising health costs. the economic consequences of the disease are considerable. Already today, partially three persons have to be trained for each higher position, in order to take precautions for the case of deterioration of performance and death as result of AIDS. Many companies complain about warning signals as insufficient employee morale and disturbed production processes. The abatement of mass poverty and a betterment of population groups which have been disadvantaged for a long time have receded into dim distance.[653]

50.3 Birth control and decline in the rate of birth

Family planning programs started already in the 1960s in South Africa. It became apparent that not only the white population, but with some delay also Indians and Colored would go through a demographic transition similar to the one in Europe. A reduction of the fast growth of the black population on the other hand could not be established.[654]

Despite of massively supported immigration, a decrease of the share of Whites in the total population was lurking. In order to counteract this trend, the Apartheid governance started to especially support family planning, for example via the "National Family Planning Programme" of 1974. Despite the political background of the measures, the offers were mostly accepted. The context for the decrease of birth rates was rather favorable, with higher incomes and a higher share of urban population. Many black women have decided in favor of family planning because of different social and economic limits of the Apartheid politics, because pregnancy and child education would have narrowed their employment possibilities in the cities. Today the background of family planning is not political goals, but a comprehensive health program.[655]

An early pregnancy often conflicts with a successful graduation. The information on contraceptives should therefore be given as early as possible. A better formation of women has a positive influence on the reduction of birth numbers.

50.4 Age pyramid and future growth

It is difficult to predict the future growth of the population. It has not only to be taken into consideration how further reproduction develops, with a high number of children and

---

653 Comp. Wiese, 1999, p. 119 et seq.
654 Comp. Caldwell/Caldwell, 1993, p. 225–262
655 Comp. ibid, p. 225–262

adolescences coming into reproductive age, but also the influence of AIDS on the mortality and fertility has to be estimated.[656] Up to now the reduction of births has had only little effect on the age structure, as can be seen in Fig. 50.1.

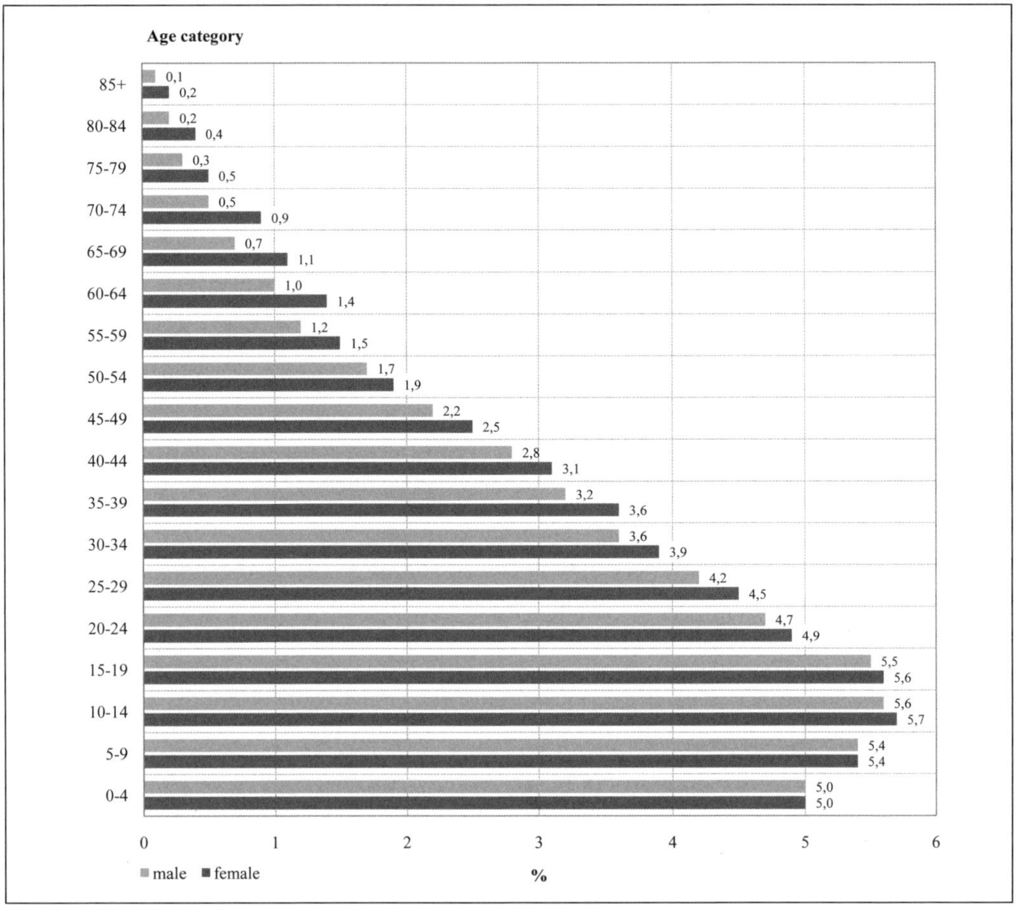

Fig. 50.1: Total population in accordance with sex and age groups[657]

50.5 Education deficit

The term "lost generation" has become a keyword for almost one third of the population. 17.9 percent of young people have no school education (comp. Fig. 50.2). Therefore, they

[656] Comp. Wiese, 1999, p. 130–133
[657] Statistics South Africa, Census, 2001

have no job and no proper home. For example in the township Soweto near Johannesburg more than 90 percent of all acts of violence are committed by young persons between 14 and 25 years. The realization by leading ANC members that the boycott of schools and university for decades during the Apartheid had harmed a whole generation is not sufficient. Training programs, school and university education will have to give new opportunities and hope to these young people.[658]

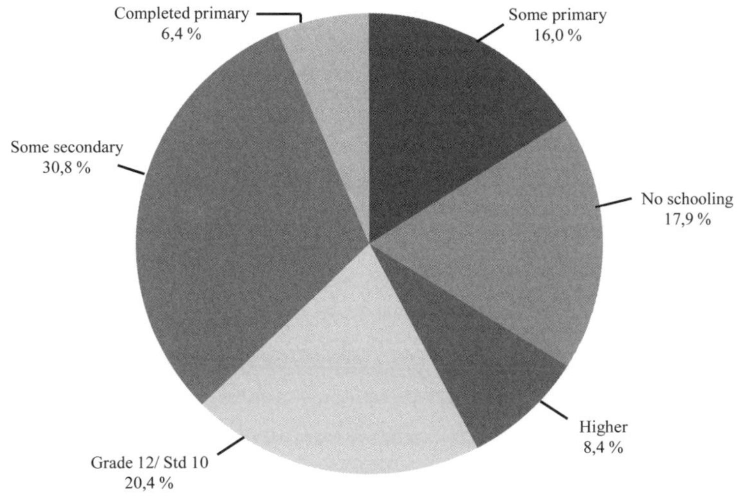

Fig. 50.2: Population aged 20 and older and highest school certificate [659]

Education in South Africa is an example for all the problems which arise in the context of restructuring the Apartheid society. Since April 1994 all children have access to state schools and general compulsory schooling was introduced. Many hundred million rand are necessary for a school developmental program, in order to guarantee the necessary human resources for further economic progress.[660]

With the Higher Education Act 1997, the government started a reform of universities. By combining universities and higher technical colleges in 2004 and 2005, the target is pursued to get rid of the former separation into "white" and "black" universities. The complex redesign is a difficult management task.[661]

---

658 Comp. Wilke-Launer/Kühne, 1993, p. 458

659 Statistics South Africa, Census, 2001

660 Comp. Wiese, 1999, p. 115–119

661 Comp. Auswärtiges Amt, 2007a

51 Economic aspects and employment policy

Among the most important economic sectors in South Africa is industry, services, mining industry and agriculture.

After the Apartheid was abolished and international sanctions were given up, the country could record a stronger economic growth. The Gross Domestic Product (GDP) increased by around 5 percent in 2006. A further growth increase from 5.1 to 5.4 percent is predicted for the next three years. The manufacturing industry and financial services account with 20 percent each for the largest part of the GDP.[662]

The economy is strongly oriented towards foreign commerce. More than half of the Gross Domestic Product is achieved with export trade. By investing in modern production technologies after the end of the Apartheid, South African products became competitive on the world market again.

Already in 1996 an export surplus of about Rand 8.7 billion was achieved. Mainly raw material, agricultural products, chemical products, machines, electronic devices and vehicles are exported. Among the most important imports are machines, synthetic material, chemical products and vehicles.

In the meantime the emerging market South Africa counts among the biggest investors in Africa and has become an economic driver on the African continent.

51.1 Unemployment

The unemployment rate of the white population is relatively low with around 2 percent. In contrast, each second Black person is unemployed or lives from occasional jobs. The main reason is the educational deficit. Furthermore, criminality and diseases – like AIDS – are further reasons.[663]

Fig. 51.1 illustrates the unemployment in South Africa over a period from 1999 until 2007. It can be perceived that since 2001 the unemployment rate has decreased by 12.8 percent. This success can be explained by the creation of new workplaces and an increasing level of education. However, there is still need for further measures.

---

[662] Comp. Auswärtiges Amt, 2007b

[663] Comp. ibid

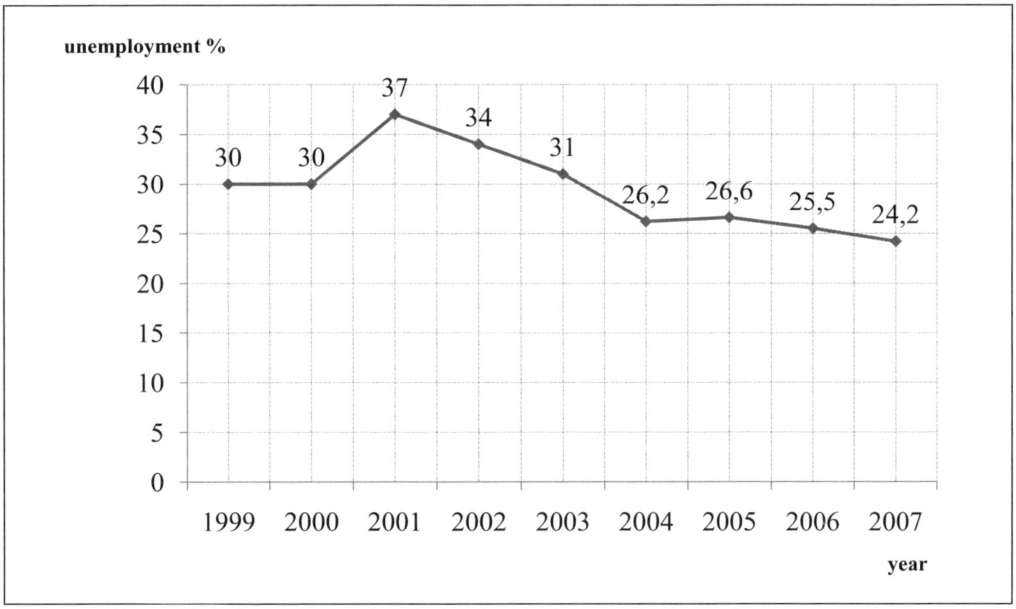

Fig. 51.1: Unemployment in South Africa[664]

A public job program shall provide training positions and work to the young generation between 6 and 19 years that accounts for one third of the black populations.[665] The "Reconstruction and Development Programme" aims at a faster and stronger industrialization of the country. This means especially an improvement of technological capabilities and thus of international capacity to compete.[666]

Economic experts like Clem Sunter – ex manager of the gold and uranium division of the mining group "Anglo American Corporation" – hold the view that small and middle-sized companies should be supported in order to avoid a reduction of workplaces and to create new jobs. This should not aim at a racial separation conform to the former line "white corporations" and "black corner shops". "This is about giving black people their place in large-scale industry as well as promoting small business."[667]

Beside the official labor market there is also an expanding "informal" sector. Examples are private taxi companies who drive their customers from the townships to work, traders with portable junk-shops or car mechanics who open a "garage" on the side of the road. If young

---

[664] CIA The World Factbook, 2008

[665] Comp. Behren/Rimscha, 1994, p. 180

[666] Comp. Wiese, 1999, p. 15

[667] Bond, 1994, p. 34, own translation

people should not achieve to get access to an official labor relation, they will also try to earn their lives with this "African" form of economy.[668]

51.2 Work relationships and Black Economic Empowerment

Even if statutory racial discrimination belongs to the past, social Apartheid will continue to exist for a long time. The reason is that the white population has collected a lot of capital during their preferential treatment over years; however, it represents only a small political power in times of black government. This leads to a tricky situation in the country. In the corridors of power in industry and economy there are mainly Whites; the political power on the other hand is in black hands.[669]

In order to escape this situation, the South African Government pursues the concept of Black Economic Empowerment (BEE), in order to increase the "black" influence with the help of statutory regulations and a preferred acceptance of tenders in an economic environment which is dominated by white people. An essential part of the BEE Act of 2004 is the Balanced Scorecard, which assesses companies in accordance with their progress in the field of empowerment of employees in four areas:[670]

- direct empowerment through ownership and control of enterprises and assets
- management at senior level
- human resource development and employment equality
- indirect empowerment through: preferential procurement, enterprise development, and corporate social investment – a residual and open-ended category.

Despite of this, many Blacks still need to resort to the knowledge and experience of white people. The Whites again cannot exist without the black potential of workforce and purchasers.

51.3 Reasons for the discrimination against women

Despite of the fact that the South African constitution explicitly prohibits discrimination on the grounds of sex, it is not ensured that the principle is put into effect. Women have a dual legal status. On the one hand they have the same codified civil rights as men and on the other hand they are subject to traditional and unwritten clan rituals. These rituals often contradict

---

668 Comp. Behrens/Rimscha, 1994, p. 181 et seq. and Wiese, 1999, p. 183 et seq.

669 Comp. Alexander, 2008

670 Comp. ibid

the state's legal system and limit the principle of equal treatment and liberal decision-making of women.[671]

The socially weak position of women also results from a lower educational standard. The reasons are early pregnancies, the compulsion to work or to take care of younger siblings. The social and geographical mobility is limited for many women. They are disadvantaged on the formal residential market, because single educating persons under 21 years or women over 21 years without children have no chance to receive state housing loans.[672]

The percentage of women in white-collar-professions (employees) is very low with 21 percent. There are better chances in the blue-collar-sectors (workers) with 88 percent. Here, women can clearly catch up with men. Because of the bad education of women from the country, their activities are concentrated on the informal sector, where little seed capital is needed. Formerly they were disadvantaged by the white colonial power and after the independence they are discriminated in a black world of men if they want to gain access to the formal labor market. The consequence is that women cover local supply deficits in unattended townships as sellers of grocery and household equipment or prostitutes.

The participation of women and the co-determination of women in political decision committees are still of little importance. The ANC supports them, and therefore has one of the highest female representation rates in parliament.

52 South Africa: automotive industry

The export of resources as gold and diamonds had been the determinate income source of South African economy for many years. Today, on the contrary, many products of automotive industry have surpassed mining products as export hit. With the new black government new production capacities are built up.

The automotive sector already counts for 29 percent of the manufacturing industry and contributes together with sub-suppliers, retailers and garages to 7.2 percent of the Gross Domestic Product (GDP). The exports amount to five billion Euros. Including sub-suppliers, retailers and garages, the sector employs around 310,000 people.

Leaders on the South African automotive market are Volkswagen, BMW and Mercedes-Benz. Companies export from there to places all around the world and account for a big part of the most important success story in South African industry. Only in 2006, the seven international car manufacturers Mercedes-Benz, BMW, Volkswagen, Toyota, Nissan, Ford and General Motors invested almost EUR 750 million in their South African factories.

---

[671] Comp. Jürgens/Bähr, 2002, p. 196

[672] Comp. Lohnert, 2000, p. 67

52.1 International automotive manufacturers in South Africa

At the beginning of the 1920s international automotive manufacturers expanded to South Africa. The reasons were cheap labor and a lucrative domestic market which made the country attractive for US company groups. Ford was first to open a production manufactory close to Port Elizabeth. General Motors followed in 1927. Around the year 1930 96 percent of registered cars were US-models.

After the Second World War further manufacturers entered the market. Mercedes-Benz, Toyota and Volkswagen had above all imported components, assembled them and therefore preferred locations close to the coast as East London, Durbain and Uitenhage next to Port Elizabeth. The state import subsidies supported the expansion of local sub-suppliers in the 1960s and 1970s. Since then the part of locally fabricated assembled components has been growing.

The province Eastern Cape had become a centre of car production where several providers concentrate up to today. The transport distances to the main purchasers in the metropolitan area of Johannesburg, however, are long and because of the geographical landform differences, it is not possible to drive to these locations with car transporters. Therefore, BMW and Nissan have switched to locations in the Pretoria area since the 1960s.

The expansion of European and Japanese companies caused that the US-market share accounted for only 35.6 percent in 1975. The following decrease in investments of American companies strengthened the dominance of Japanese and German automotive manufacturers. Today, they enter into more and more cooperations, like Mitsubishi and Daimler. At the same time, this practice led to a reduction by 50 percent of the number of manufacturers.

After the end of the Apartheid the automotive companies expanded even further, because an increasing motorization of the upcoming black middle class became apparent.

The international automotive manufacturers Daimler, General Motors, BMW, Volkswagen, Land Rover, Daewoo, Peugeot, Toyota, Subaru, Ford, Nissan and Fiat fabricate in South Africa original equipment manufacturer (OEM)-components as well as completely built up (CBU)-cars. Other companies, as Renault, supply the market exclusively with imports. From an international point of view the South African business market is still relatively small despite of favorable tendencies. Therefore, besides small numbers for the local market, production takes place especially for export business.

52.2 Motor Industry Development Program

The South African context makes the subsidiaries of international automotive groups profitable despite of the small South African market. In 1995 the "Motor Industry Development Programme" (MIDP) came into effect. MIDP reduced the import customs for imported cars drastically. At the same, time incentives were given for a stronger local share in automotive fabrication. This way, companies could export at cheap prices, and models which were only required in small numbers could be imported at low costs. Therefore, many companies decided to fabricate in South Africa.

Presently the MIDP which is effective until 2012 is examined by the government, industry and trade unions. The government, however, has already assured that it would continue its support to the automotive industry after 2012.

The export of components and the emergence of an own sub-supplier industry complete the growing automotive production and have become an important economic factor. In the surroundings of Uitenhage and Rosslyn, where several manufacturers have their business, Supplier Parks have been created. Joint Ventures between local companies and traditional suppliers fabricate the necessary components like catalytic converters, seats and airbags in these Supplier Parks.

52.3 Influence of the unions on the automotive manufacturers

The National Union of Metalworkers of South Africa (NUMSA) is largest union in South Africa. Also in the name of the government, it is committed to the following issues:

- defend rights of the members against the companies,
- process reclamations of members,
- represent members in hearings,
- fight against unfair labor conditions, limitations and all types of discrimination,
- negotiate for better wages and labor conditions with the employers and
- for death benefits for members and support of their families.

NUMSA especially wants to improve the life of worker in factory buildings. Therefore, it pursues its aims persistently:

- Close wage gape from Apartheid. With the Apartheidm a big wage difference between qualified and non-qualified workers was created. This difference existed mainly between white and black workers.
- Gain workers' interest in trainings which are recognized in other companies, in case they lose their workplace.
- Reduce training grades at the workplace to five stages of training and introduce a grading system based on capabilities. This way, anybody who attends the trainings could be awarded with higher remunerations because of his further qualification.
- Make employers aware of capabilities which workers have gained in their position in the company and encourage payment for these capabilities.

- Make sure that women receive jobs which traditionally are occupied by men, if they have the appropriate qualification. Women have to receive the same payment as men, if both sexes carry out the same work.
- Receive more control over investment plans and strategies from management. If the company plans massive workplace reductions, NUMSA wants to be informed in order to reduce the consequences for its members.

NUMSA believes in comprehensive affirmative action.

The workers need technical training and basic formation. The basic formation and training have to be provided by the employer to the employees during working hours. Furthermore, there must not be any form of discrimination against workers because of their race, sex or religion.

52.4 Future of the automotive industry

The South African car market still booms. However, one can only count on new investments in the automotive sector if the government has finally decided whether and to which extent it will further subsidize the sector after 2012. For the sub-suppliers, good business opportunities come up. The automotive manufacturers hope for a higher share of OEM-components from local fabrication. Supplier Parks like Rosslyn near Pretoria offer interesting possibilities for the customers in this context.

Because the number of cars has risen in South Africa in the last year, the demand for component parts has increased as well, e.g. for spoilers, light alloy rims, seat systems or audio equipment. These components were not offered by the manufacturers so far and the sub-suppliers can benefit from this fact. Especially among the car owners of black middle class the demand for such equipment is big. Furthermore, this group of costumers becomes more and more important in South Africa.

Although if sales have increased considerably in the last years in South Africa, the automotive manufacturers focus the fabrication of few models and export business, to achieve profitable piece numbers. BMW has started the fabrication and the export of the new BMW 3 series in its factory in Rosslyn near Pretoria. Mercedes-Benz fabricates the new C-series in its manufactory in East London and exports heavily as well.

Today, step by step other companies follow the example of German manufacturers. In 2002 BMW, Mercedes Benz and VW accounted for 97 percent of the car export business. Now, Toyota, Nissan and Ford are expanding their production and export as well.

There will be increasing pressure by cheap manufacturers from China and India that offer cars for less than 10,000 and want to take up production in South Africa.

Literature

Alexander, M.: Black Economic Empowerment, URL: http://www.southafrica.info/doing_business/trends/empowerment/bee.htm Status: 2008.

Ansprenger, F.: Politische Geschichte Afrikas im 20. Jahrhundert. 3rd edition, Munich: Beck, 1999.

Auswärtiges Amt: Bildungspolitik und Erziehungswesen, URL: http://www.auswaertiges-amt.de/diplo/de/Laenderinformationen/Suedafrika/Kultur-UndBildungspolitik.html Status: 09/2007a.

Auswärtiges Amt: Gesamtwirtschaftliche Lage, URL: http://www.auswaertiges-amt.de/diplo/de/Laenderinformationen/Suedafrika/Wirtschaft.html status: 09/2007b.

Becker, M.: Personalentwicklung: Bildung, Förderung und Organisationsentwicklung in Theorie und Praxis. 4th edition, Stuttgart: Schäffer-Poeschel, 2005.

Becker, M.: Systematische Personalentwicklung: Planung, Steuerung und Kontrolle im Funktionszyklus. Stuttgart: Schäffer-Poeschel, 2005.

Behrens, M. / **Rimscha**, R. v.: Gute Hoffnung am Kap?: Das neue Südafrika. Osnabrück: Fromm, 1994.

Bilger, H. R.: 111mal Südafrika. 3rd edition, Munich: Piper, 1986.

Bittorf, S.: Auto-Industrie: Auch in Südafrika droht Streik, URL: http://www.sueddeutsche.de/wirtschaft/artikel/356/35321/1/ Status: 15 July 2004.

Bond, P.: Es braucht keine Mauern um unsere Villen. in: Neue Züricher Zeitung Folio, n° 4, April 1994, p. 34.

Caldwell, J. C. / **Caldwell**, P.: The South African fertility decline. in: Population and Development Review, Vol. 19, No. 2, 1993, p. 225–262.

CIA The World Factbook: South Africa, last access: 22.02.2008, URL: https://www.cia.gov/library/publications/the-world-factbook/geos/sf.html status: 12.02.2008.

Daimler: Diversity: Broad Based Black Economic Empowerment, URL: http://www.daimler.com/dccom/0-5-659599-49-632764-1-0-0-661957-0-0-135-7145-0-0-0-0-0-0-0.html status: 2008.

Ehlers, C.: Branche kompakt: Kfz-Industrie und Kfz-Teile – Südafrika, URL: http://www.bfai.de/DE/Content/__SharedDocs/Links-Einzeldokumente-Datenbanken/fachdokument.html?fIdent=MKT20071022142044 status: 24 October 2007.

Hageman, A.: Kleine Geschichte Südafrikas. Munich: Beck, 2001.

Jürgens, U. / **Bähr**, J.: Das südliche Afrika: Gesellschaftliche Umbrüche zu Beginn des 21. Jahrhunderts – Zusammenwachsen einer Region im Schatten Südafrikas. Gotha: Klett-Perthes, 2002.

Lohnert, B.: Die Wohnungskrise im Post-Apartheid Südafrika: Politische Instrumente und soziale Realität: Das Beispiel Kapstadt. in: **Bähr**, J. / **Jürgens**, U. (ed.), Transformationsprozesse im Südlichen Afrika – Konsequenzen für Gesellschaft und Natur. Kiel: Geographische Schriften, 2000, p. 63–78.

Mayer, C.-H. / **Boness**, Chr. M. / **Thomas**, A.: Beruflich in Südafrika: Trainingsprogramm für Manager, Fach- und Führungskräfte. Göttingen: Vandenhoeck & Ruprecht, 2004.

Mercedes-Benz South Africa: History, last access: 10.02.2008, URL:
http://www.mercedes-benz.co.za/content/south_africa/mpc/mpc_south_africa_website/en/home_mpc/passenger_cars/home/rep_mbworld/rep_history.html
status: 2000-2007a.

Mercedes-Benz South Africa: News and Events: DaimlerChrysler South Africa renamed Mercedes-Benz South Africa, URL:
http://www.mercedesbenz.co.za/content/south_africa/mpc/mpc_south_africa_website/en/home_mpc/passenger_cars/home/rep_mbworld/rep_news_events.html#mainnav
status: 2000-2007b.

Mercedes-Benz South Africa: Overview, URL:
http://www.mercedes-benzsa.co.za/
status: 2000-2007c.

Mercedes-Benz South Africa: Manufacturing plant, URL:
http://www.mercedesbenzsa.co.za/MercedesBenzSA/portal/portalsintegra/Modules_FE/Layout1/displayFull.asp?ID=153&menu=1&submenu=1
status: 2000–2007d.

Mercedes-Benz South Africa: People, URL:
http://www.mercedesbenzsa.co.za/MercedesBenzSA/portal/portalsintegra/Modules_FE/Layout1/careers/displayFullCareers.asp?id=14
status: 2000–2007e.

Nohlen, D. (ed.): Lexikon Dritte Welt: Länder, Organisationen, Theorien, Begriffe, Personen. 9th edition, Reinbek: Rowohlt, 1996.

NUMSA: What is NUMSA?, URL:
http://www.numsa.org.za/index.php?page=whatis
status: 2004a.

NUMSA: NUMSA's Policies, URL:
http://www.numsa.org.za/index.php?page=pandp
status: 2004b.

Oberhauser, A. M.: Semiperipheral Industrialization in the Global Economy: Transition in the South African Automobile Industry. in: Geoforum 24 (2), 1993, p. 99–114.

Pabst, M.: Südafrika. Munich: Beck, 1997.

Räther, F.: Auto-Boom am Kap. in: Süd-Afrika: Magazin für Reisen, Wirtschaft und Kultur im südlichen Afrika, ed. 12, volume 3, 2007, p. 70–71.

Saembassy: Arts and Culture: Ndebele, URL:
http://www.saembassy.org/Art%20and%20Culture.htm
status: 16.12.2004.

SAQA: NQF Objectives, URL:
http://www.saqa.org.za/
status: 2008.

Schmeisser, W.: Finanzorientierte Personalwirtschaft. Munich 2008.

South Africa-Infoweb: Südafrika Karte, 2007, URL:
http://www.southafrica-infoweb.com/directory/index.php
status: 10.02.2008.

Südafrika Guide: Kultur und Geschichte. URL:
http://www.suedafrika-guide.de/kultur/index.html
status: 2008.

Rath, W. v. / **Fueter**, W. / **Cockwell**, A.: Internationale Managemententwicklung im Volkswagen-Konzern. in: Riekhof, H.-Chr. (ed.): Strategien der Personalentwicklung: Mit Praxisbeispielen von Bosch, Linde, Philips, Siemens, Volkswagen und Weka. 6th edition, Wiesbaden: Gabler, 2006, p. 263–278.

Weltbank: Weltentwicklungsbericht 2007: Entwicklung und die nächste Generation, Düsseldorf: Droste, 2007.

Wiese, B.: Südafrika mit Lesotho und Swasiland, Gotha: Klett-Perthes, 1999.

Wilke-Launer, R. / **Kühne**, W.: Südafrika. in: Nohlen, D. / Nuscheler, F. (ed.): Handbuch der Dritten Welt. Volume 5, Ostafrika und Südafrika, 3rd edition, Bonn: J.H.W. Dietz Nachf., 1993, p. 417–463.

XII The AGG and Diversity Management – a managerial and European-national legal response to the internationalization of labor markets in Europe and worldwide

Changes of environmental circumstances for companies and their employees have never been faster. Because of the increasing internationalization, the demographic changes and the fast growing technological progress, companies need to react ever faster to the changes in their environment. Especially with regard to employees, there are strong deviations from what had been known before which require new reactions to the cultural, social and experience-related characteristics of the staff. The German legislator reacted to the cultural and other variety among the employees with the introduction of the General Equal Treatment Act (AGG), which was meant to avoid discrimination in the company. At the same time, it is not only necessary to comply with the General Equal Treatment Act, but to utilize people as resources in all their varieties in order to be successful on the market. In order to use the variety of employees for own purposes, it is necessary to react via Diversity Management to the personal "characteristics" of each single employee. The following chapter will provide an overview over the most important aspects of the General Equal Treatment Act and explain the process of Diversity Management.

53 Changes in the company environment

53.1 Internationalization of labor markets

Worldwide, international economic relations are being built up like never before. The international interrelations have been intensifying since the 1990s because of the technological development, the elimination of trade, capital and employment barriers as well as because of political and economical integration trends.[673]

In Germany, the international orientation of the companies is becoming more and more important, because Germany is an export-oriented country on the one hand and on the other hand needs to import a variety of goods because of the scarcity of raw material. From Fig. 53.1 can be seen that the import and export numbers of German goods and services have more than doubled since 1997.

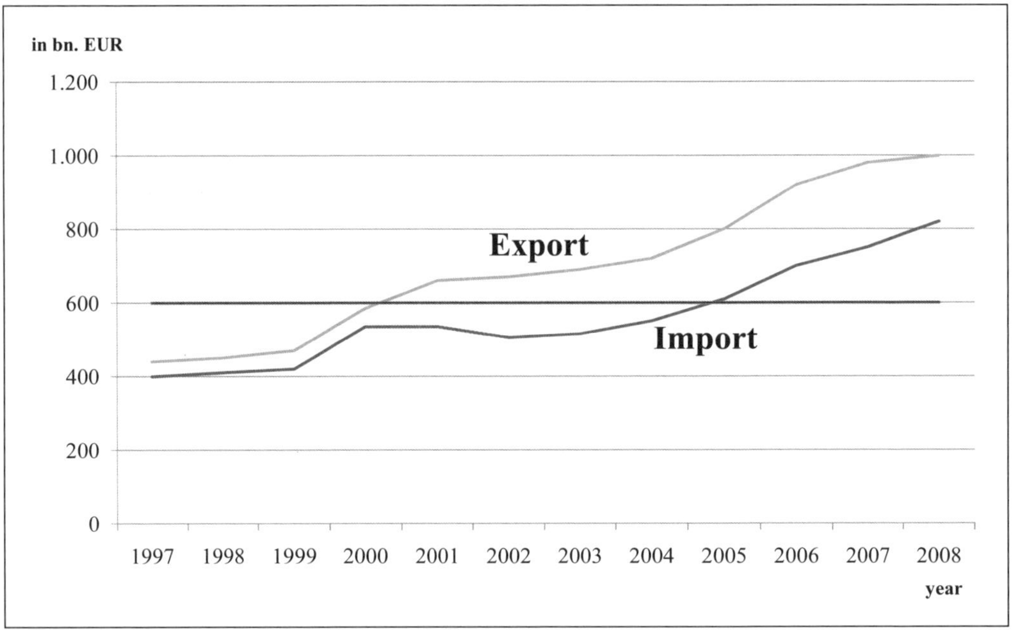

Fig. 53.1: Key figures of German foreign trade from 1997 to 2008[674]

---

[673] Comp. Oechsler, 2006, p. 98

[674] Following Statistisches Bundesamt 2009b

This ever stronger international orientation of companies is also reflected in the rising number of cross-border mergers all over the world. As can be seen in Fig. 53.2 the number of mergers and acquisitions has steadily grown since 1993, and thereby the "purchase" of foreign employees by mergers in the group.

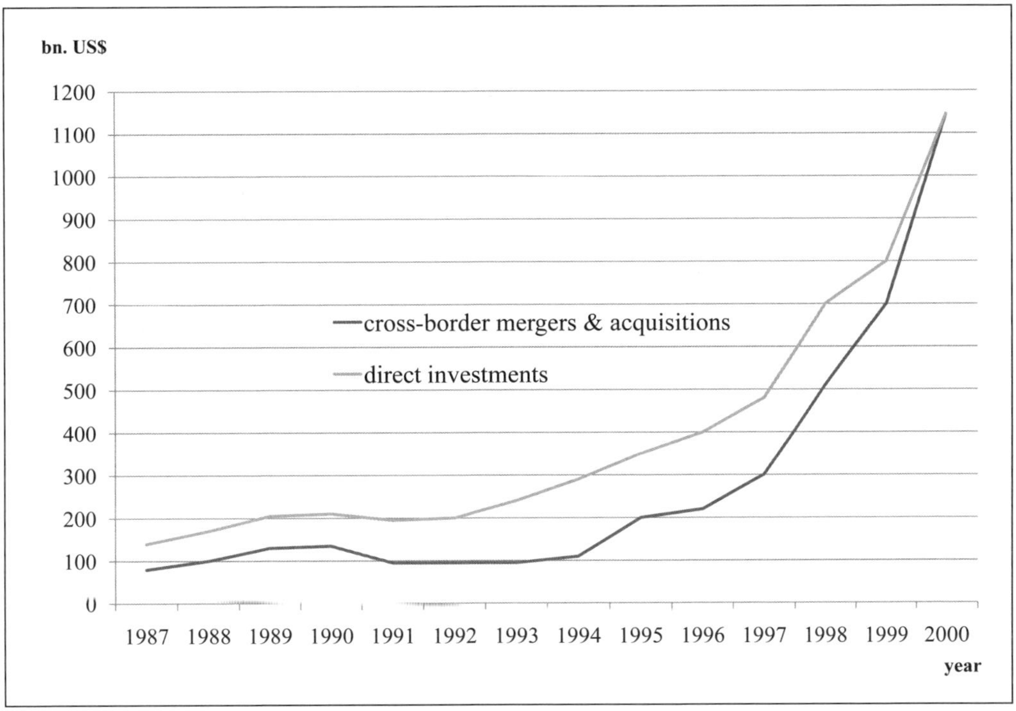

Fig. 53.2: Worldwide direct investments and cross-border mergers and acquisition[675]

Now it can be deflected that companies have to face ever more variable customer and supplier structures. It is crucial to respond to the diverse cultural differences with different employees, in order to make a long-term collaboration possible. At the same time, personnel delegations take place in the context of joint ventures and strategic alliances with foreign companies, and – what is especially problematic for developing countries and emerging nations –also an emigration of qualified employees, e.g. to Germany.

Many industrialized countries headhunt qualified specialized workforce from the health sectors in India, China or the Philippines, because there is a lack in qualified labor in the own country.[676] At the same time the emigration does not only concern developing and emerging

---

675 Andruchowitz, 2003

676 Comp. Stewart/Clark/Clark, 2007

nations. In Germany alone 500,000 people have emigrated since 1990.[677] The disadvantage is that economically weak countries or regions that pay for the formation of specialized workforce do not benefit from their formation in order to push their own economic development.

It has to be mentioned in this context that the number of foreign citizens amounted to 6.73 million in the Federal Republic of Germany at the end of 2008. This meant a small decline in comparison with the year before, but this might be caused by the high number of naturalizations of Turkish citizens.

What is noteworthy, on the other hand, is the increase by 1 percent of the persons who belong to an EU-member state.[678] It is noticeable that the formation of the European Union in 1993 has been widely contributing to the internationalization of the labor market, as could be seen in other country unions. The emergence of an unlimited labor market leads to an increasing number of foreign applicants and the possibility for companies, to start balancing the lack in specialized and managing personnel on an international level,[679] and causes a higher variety among the workforce.[680]

It can be seen that via the internationalization in general and the internationalization of the labor markets, e.g. in the European Union, an intense reflection of the companies on the values, cultural backgrounds and points of view of the applicants, employees and merger partners has become necessary to be able to continue successful work.

53.2 Technological advance

Because of the rapid technologic progress and the ever shorter product life cycles, the factor time has become a competitive factor. In order to overcome this challenge, companies need more highly qualified staff. The requirements towards the employees change in ever shorter intervals and repetitive jobs are more and more eliminated. The challenge for personnel recruitment is to win such highly qualified personnel.[681]

---

[677] Comp. Heckmann, 2006

[678] Comp. Statistisches Bundesamt, 2009a, p. 194

[679] Comp. Kirchgeorg/Günther, 2006

[680] Comp. Gloger, 2000, p. 37

[681] Comp. Oechsler, 2006, p. 100

53.3 Demographic development

Around 82.2 million people lived in Germany at the end of 2007.[682] In accordance with estimations of the German Federal Statistic Office the number will drastically decrease until 2050 as can be seen in Fig. 53.3.

It is assumed that only between 69 and 74 million people will live in Germany in 2050.[683] This fact can be attributed to the decreasing birth rates on the one hand and the rising mortality rates on the other hand.

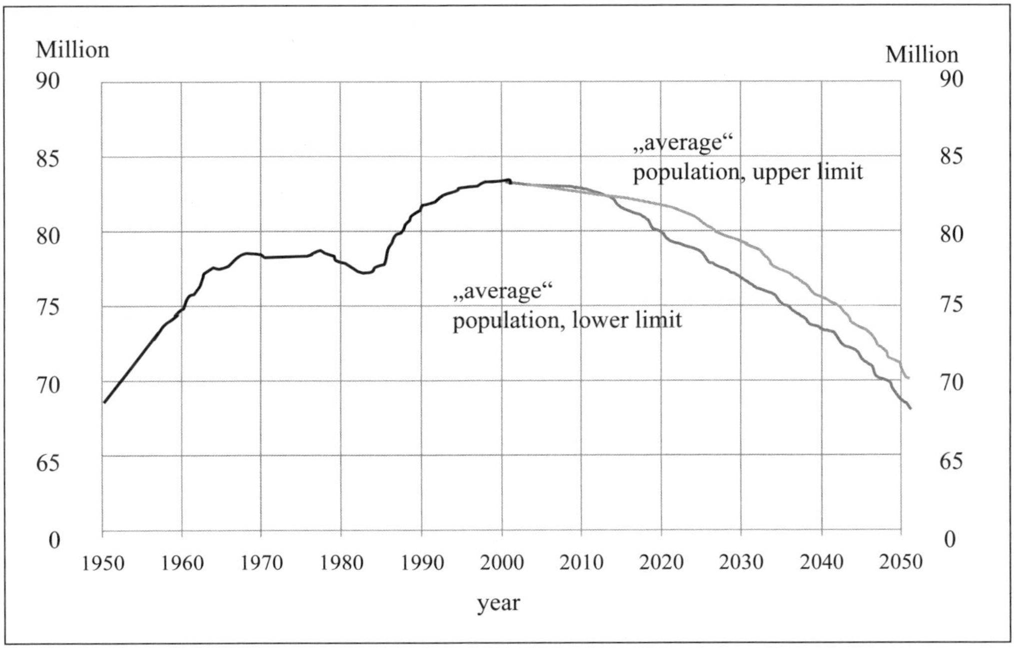

Fig. 53.3: Forecast of development of population numbers in Germany until 2050 [684]

The reason for the growing birth deficit is that with the low birth rate, the number of potential mothers decreases. It is foreseeable that the number of people in school and training age will quickly decrease and that in 2050 twice as many old as young people will live in Germany.

[682] Comp. Statistisches Bundesamt, 2008

[683] Comp. Statistisches Bundesamt, 2006, p. 15

[684] Ibid, p. 15

The mortality rate increases despite increasing life expectations, because the age groups with high numbers are getting into old age and therefore many deaths occur at once per time unit. Today the mortality rate is already higher than the birth rate.

Also the migration balance of 100,000 to 200,000 people per year cannot really encounter this tendency. Therefore it has to be expected that the population in working age will get older and less numerous. Until 2014 it can be assumed that the number of working population remains at around 50 million. The number of young people diminishes at the same time and will be compensated by a higher number of older people. In the years after, however, the total number of employable people will decrease, with the result that in 2050 only 35-39 million employable people are expected to live in Germany. Discussions on e.g. how to integrate older employees in the company work organization, because young graduates become a decreasing group, show that companies will have to adjust to this tendency in any case during the next years. As a result, older employees will have to be given special consideration especially in the case of new hires and a longer duration of employment has to be granted to them.[685]

Because of the decrease of population it will become unavoidable to integrate also women with and without children and ethnic minorities in the employment, because despite of technological progress the potential of these groups will be demanded especially in company management and administration.

53.4 EU-antidiscrimination directives

Because of the European integration, the growing globalization and the accompanying internationalization of companies and demographic change, the European Community issued directives to avoid discrimination on diverse labor markets and to promote equal chances and treatment and justice for a good life. The basis is provided in Art. 13 subs. 1 of the Treaty establishing the European Community (TEC), in accordance to which the "Council may take appropriate action to combat discrimination based on sex, racial or ethnic origin, religion, disability, age or sexual orientation."[686] This is the basis of the four EU-antidiscrimination directives of the EU council of ministers:[687]

- Council Directive 2000/43/EC of 29 June 2000 implementing the principle of equal treatment between persons irrespective of racial or ethnic origin
- Council Directive 2000/78/EC of 27 November 2000 establishing a general framework for equal treatment in employment and occupation;

---

685 Comp. Statistisches Bundesamt, 2006, p. 5 et seq.

686 Europäische Gemeinschaft, 2002, p. 11, own translation

687 DGB Bildungswerk e.V., o. J.

- Directive 2002/73/EC of the European Parliament and of the Council of 23 September 2002 amending Council Directive 76/207/EEC on the implementation of the principle of equal treatment for men and women as regards access to employment, vocational training and promotion, and working conditions; and
- Council Directive 2004/113/EC of 13 December 2004 implementing the principle of equal treatment between men and women in the access to and supply of goods and services

The directives oblige the EU member countries to ensure the implementation of the directives. In Germany they were implemented with the coming into effect of the General Equal Treatment Act (AGG) on 18th August 2006[688].

54 General Equal Treatment Act

54.1 Objective and content of AGG (Sec. 1 AGG)

Already before the General Equal Treatment Act came into effect in August 2006, discrimination because of sex, parentage, race, language, homeland and origin, faith, religious and political opinions or because of disability had been prohibited in accordance with Art. 3 of German Basic Law. Reality, however, does not correspond to the regulations of the constitution no to international law.[689] The AGG can be considered as the implementation of the EU antidiscrimination directives into national law. This means that the objective of the law, as shown in sec. 1 AGG, is to prevent or eliminate adverse treatment on the grounds of race or ethnic origin, gender, religion or secular belief, a disability, age or sexual identity. The legislator uses the terminology "adverse treatment" and not "discrimination" on purpose to make clear that not any unequal treatment being connected with a disadvantage has discriminating character.[690]

The AGG is structured in seven parts, starting with Part 1 (sec. 1 to sec. 5 AGG), containing purpose, scope of application and definitions of terms. The second part comprehends sections 6 to 18 and regulates the protection of employees against unequal treatment as well as the personal scope of application in the context of employment. Part 3 regulates the protection against unequal treatment under civil law (sec. 19 to sec. 21 AGG). Regulations on legal protection can be found in Part 4 (sec. 22 to sec. 23 AGG). Part 5, consisting of sec. 24 AGG, regulates special provisions for civil service positions, which will not be analyzed in this context. Part 6, consisting of sec. 25 to sec. 28, provides explanations on the

688 Comp. Raasch, 2007

689 Comp. Busch/Kocher/Wolkoborsky, 2009, p. 17

690 Comp. Wisskirchen, 2007, p. 9

installation of antidiscrimination offices in the Federal Republic of Germany and on this basis the last three paragraphs contain final provisions.[691]

54.2 Elements of unequal treatment

The AGG names eight elements of unequal treatment whose application will now be explained.

54.2.1 Race

The term race has to be treated with special care in the interpretation of the AGG, because the legislator wants to distance itself from the existence of differences with regard to races. The utilization of such terminology does not imply acceptance, which was meant to be clarified by the formulation "on the ground of" instead of "because of".[692]

54.2.2 Ethnic origin

The wording "ethnic origin" on the other hand implies a wide definition in the sense of parentage, national origin or folklore. This means, that ethnic origin cannot be derived from citizenship, but it also implies special ethnic groups with an own cultural identity.[693]

54.2.3 Disability

With regard to the AGG the legislation took the definition of 9th book of the German Social Code (SGB IX) and the Act on the Equal Treatment of Persons with Disability (BBG) over. These Acts define disabled people as persons whose physical functions, mental capabilities or psychological health are highly likely to deviate, for more than six months, from the condition which is typical for the respective age and whose participation in social life is therefore impaired (sec. 2. subs. 1 sentence 1 SGB IX and sec. 3 BGG).[694] Disabled people for whom adverse treatment is prohibited are therefore only persons with a severely handicapped pass.[695]

---

691 Comp. Hoffjan/Bramann/Kentrup, 2008, p. 33–38

692 Comp. Schiek, 2007, sec. 1 AGG, mn. 9

693 Comp. Wisskirchen, 2007, p. 10–11

694 Comp. Schiek, 2007, sec. 1 AGG, mn. 36

695 Comp. Wisskirchen, 2007, p. 11

54.2.4 Sexual identity

A final statutory definition of the term sexual identity is still missing. There is only an interpretation of the term, which means that homosexual and bisexual men and women, as well as transsexual and intersexual persons shall be especially protected in the AGG.[696]

54.2.5 Age

Age in this context means biological age. Not only younger but especially also older people shall be protected. The basis is that each person is ageing at individual speed and the physical characteristics of growing age cannot be generalized. Additionally, disadvantages can be caused by age itself in today's society on the ground of high degrees of automation and the change towards a society based on the provision of services.[697]

54.2.6 Religion and secular belief

Religion in the sense of the AGG is a very wide term that includes minority religions. On the other hand, there are still country-specific differences regarding how certain groups like for example the Scientology Church, can be categorized. However, it shall be underlined that rarely an adverse treatment because of religion alone takes place, but normally discrimination for multiple reasons is to be expected.[698]

Secular belief has a narrow understanding as "a certainty connected with a person on certain statements on the totality of the world and on the origin and aim of human life, which is limited to inner-worldly references and explicitly excludes the general political ideology". In order to avoid the access to denied business for example for right-wing followers, secular belief is not protected against discrimination in sec. 19 AGG and therefore under Civil law.[699]

54.2.7 Sex

Furthermore, sexual discrimination is prohibited under the AGG. This includes the male and female sex, as well as transsexuals and hermaphrodites.[700]

---

[696] Comp. Schiek, 2007, sec. 1 AGG, mn. 31–32

[697] Comp. ibid, sec. 1 AGG, mn. 44–46

[698] Comp. ibid, sec. 1 AGG, mn. 19–21

[699] Comp. Bundestagsdrucksache 16/2022, p. 13

[700] Comp. Wisskirchen, 2007, p. 12

54.3 Scope of Application

In the personal and objective scope of application it is explained by whom and when the AGG can be applied.

54.3.1 Objective scope of application (sec. 2 AGG)

In sec 2 AGG it becomes clear that the focus of the AGG lies on labor law, but that large part of civil and social law are also included. This becomes obvious because in sec. 2 subs. 1 n° 1-4 labor law aspects are determined, while in sec. 2 subs. 1 n° 5–8 AGG aspects with relation to civil and social law are established.[701] Discrimination on the grounds of one of the elements in any phase of the labor relation, from selection of staff until the termination of the labor relation, is prohibited.[702]

The objective scope of application embraces:

- access to employment and to professional advancement,
- conditions of employment, work and dismissal,
- access to vocational training
- membership and participation in organizations,
- social security,
- social advantages
- education and
- access and supply of goods and services.

With regard to the termination to employment, the Act on the Protection against Unfair Dismissal applies in accordance with section 2 subs. 4 AGG. Benefits under the Social Code and company pensions are regulated by the company Pension Act.

Access to employment and professional advancement
The conditions for the access to independent or employed work, regardless of the area of activity or the professional position, as well as the professional advancement shall be free of discrimination. The selection and employment criteria obtain special attention in this context, because they offer special potential for unequal treatment. Therefore, unequal treatment is prohibited in the advertisement of a vacancy, in selection of personnel and promotion of staff as well as in the employment conditions.[703] The inclusion of access to professional advancement shall avoid that unequal treatment is not only eliminated in the decision on a

[701] Comp. Worzalla, 2006, p. 43

[702] Comp. Schütt/Wolf, 2006, p. 17

[703] Comp. Schütt/Wolf, 2006, p. 18

promotion, but already in the professional assessment, which provides the basis for the decision.[704]

Conditions of employment, work and dismissal

Sec. 2 n° 2 AGG includes any employment and work conditions including remuneration and reasons for dismissal, especially individual and collective agreements and measures in relation to the execution and termination of the labor relation e.g. in form of dissolution agreements, as well as in relation to professional advancement. Therefore, the subsequent consequences of the terminated labor relation are included as well. This means that e.g. labor contract agreements, regulations of a company agreement, tariff regulations, promotions, individual instructions for the concretization of the duties of employment or the unilateral duties of the employer are protected by the AGG.[705]

Access to vocational training

Discrimination is prohibited with regard to the access to all forms and levels of vocational training, vocational guidance, advanced vocational training and retraining, including practical work experience. In accordance with sec. 1. German Vocational Training Act (BBiG), the preparation for vocational training, vocational training, vocational advanced training and vocational retraining are comprised in the term professional training. Practical work experiences are measures that are suitable to improve a person's chances on the labor market by imparting practical professional experience.[706]

Membership and involvement in organizations

The scope of application of the AGG also includes the membership and involvement in a labor union, an employers' association or professional organizations. The prohibition of discrimination also refers to the benefits provided by these organizations.[707]

Social protection, social benefits and education

Discrimination for a reason named in sec. 1 AGG is also not admissible in reaction to social protection, including social security and health services, socials benefits and education. A big part of these factors is regulated under public law, because to a large extent, they are state benefits. But also benefits on a private law basis in form of private medical contracts or private educational services by private providers are covered by the AGG.[708]

---

[704] Comp. Schleusner/Suckow/Voigt, 2007, sec. 2 AGG, mn. 6

[705] Comp. Worzalla, 2006, p. 45

[706] Comp. ibid, p. 46

[707] Comp. Schleusner/Suckow/Voigt, 2007, sec. 2 AGG, mn. 13

[708] Comp. ibid, sec. 2 AGG, mn. 15

Access to and supply with goods and services
Sec. 2 subs. 1 n° AGG includes the access to and supply with goods and services provided to the public including living space. The term services means not only service and labor contracts, but also agency agreements, financial services, rental agreements, loan and insurance contracts as well as leasing agreements. The protection is limited to the extent that the goods and services have to be available to the public, whereby not the size of the named public is decisive, but the fact that the declaration of intent of the provider exceeds his privacy.[709]

54.3.2 Personal scope of application (sec. 6 AGG)

Sec. 6 AGG specifies the protected persons as well as the employers

Persons covered
Among the employees and therefore, the possible claimants under the AGG are:

- employees,
- persons employed for vocational training,
- persons of similar status on account of their dependent economic status,
- applicants for an employed work relation,
- persons whose employed work relation has ended

in public economy and public service.[710]

An employee is defined as a person who is obliged to exercise independent work in personal dependence on the basis of a contract under private law. Thereby, the employee is subject to the employer's authority to give instructions, which can include content, place, time and duration of work. Furthermore, the employee is integrated in the employer's labor organization.[711] Among the employees under the AGG are also persons who have a factual work relation. [712] Part-time employees who work at least 10 hours a week and people with marginal occupation are also covered by the AGG[713].

Among the people employed for vocational training are under sect. 1 subs. 1 German Vocational Training Act (BBiG) not only trainees, but also employees in preparation for their profession, in vocational advanced training and vocational retraining.[714] If the person in question has concluded a contract under private law in dependence of another person and is

---

709 Comp. Worzalla, 2006, p. 48

710 Comp. Däubler/Bertzbach, 2008, sec. 6 AGG, mn. 1

711 Comp. Schleusner/Suckow/Voigt, 2007, sec. 6 AGG, mn. 2

712 Comp. ibid, sec. 6 AGG, mn. 4

713 Comp. Rust/Falke, 2007, sec. 6 AGG, mn. 12

714 Comp. sec. 1 subs. 1 BBiG

regularly bound by the instructions of that person, this constitutes a case of employment in the sense of the AGG. Also working students,[715] student apprentices and voluntary workers are included in the personal scope of application, because remuneration is not relevant.[716]

The circle of employees includes also persons with similar status on account of their dependent economic status, as well as people engaged in home work and those equal in law to home workers,[717] as for example persons occupied in sheltered workshops and persons in rehabilitation. Basically, people with similar status are no employees in the sense of the labor law. A definition of the term can be found in sec. 12a Collective Bargaining Agreements Act: persons with similar status are employees, who are economically dependent and similar to employees in need of social protection and who have a personal obligation to fulfill their work on the basis of service or labor agreements. Additionally, they have to work predominantly for one person or to obtain on average more than half of their salary from one person.[718] There is no need for social protection if the obligor himself just like an entrepreneur employs people under his own name.[719]

The Discrimination elements named in the AGG cover equally applicants to an employee relation. It is regarded to be sufficient when a person applies with an employer for an employment, by which the same must receive the desire for contract conclusion.[720] However, claims are only possible under the AGG if the application was subjectively serious and if the applicant can objectively be taken into consideration for the vacancy,[721] which can also be the case in unsolicited applications.[722]

The legislator also included people into the scope of the AGG whose employment has come to an end to encounter the case of subsequent consequences of the labor relation, especially with regard to the company pension.[723]

Because they have no contract under private law, civil servants and trainees with civil service contracts, judges, soldiers, community servants and war resisters are not considered in this part of the AGG. However, the AGG covers them with special consideration to their special legal position under sec. 24 AGG.[724]

---

715 Comp. Bährle, 2007, p. 12

716 Comp. Schleusner/Suckow/Voigt, 2007, sec. 6 AGG, mn. 9

717 Comp. Schiek, 2007, sec. 6 AGG, mn. 5

718 Comp. Worzalla, 2006, p. 77

719 Comp. Schleusner/Suckow/Voigt, 2007, sec. 6 AGG, mn. 10

720 Comp. Worzalla, 2006, p. 78

721 Comp. Schiek, 2007, sec. 6 AGG, mn. 6

722 Comp. Worzalla, 2006, p. 78

723 Comp. Bundestagsdrucksache 16/1789, p. 34

724 Comp. Schleusner/Suckow/Voigt, 2007, sec. 6 AGG, mn. 6

Employer in the sense of the AGG
Employers are natural and legal persons as well as partnerships with legal capacity that employ one or more persons in accordance with sec. 6 subs. 1 AGG. In the case of people engaged in home work and those equal in law, the employee is replaced by the principal or foreman.[725]

A special case it the one of temporary agency workers, because in addition to the contractual employer, i.e. the leasing agency. the leaser is deemed to be an employer in the sense of the AGG.[726]

Self-employed people and board members
Self-employed persons and board members, especially managers and members of the executive board,[727] are only covered in connection with the access to employment and with professional advancement by the regulations of Part 2 of the AGG. In this context, discrimination in relation to selection criteria and employment conditions should be avoided, which also corresponds to the objective scope of application. Self-employed people and board members are not part of the persons covered by sec. 6 subs. 1 AGG, because they are basically considered as legal representatives of the employer and therefore are deemed to have the position of employers.[728]

54.4 Elements of discrimination (sec. 3 AGG)

Sec. 3 AGG contains definitions to the different elements of discrimination. Unequal treatment can be effected in form of direct and indirect discrimination, harassment or sexual harassment as well as instructions to discriminate.

54.4.1 Direct discrimination

There is a case of direct discrimination in accordance with sec. 3 subs. 1 AGG, if one person is treated less favorably than another is, has been or would be treated in a comparable situation, whereby the unequal treatment can also be an omission.[729] The prohibition of direct discrimination shall forbid express unequal treatment, whereby structural unequal treatment is not taken into account. This means that actual differences are not considered.[730]

---

[725] Comp. Worzalla, 2006, p. 82

[726] Comp. Kolmhuber/Schreiner, 2006, mn. 64

[727] Comp. Schiek, 2007, sec. 6 AGG, mn. 11

[728] Comp. Schleusner/Suckow/Voigt, 2007, sec. 6 AGG, mn. 16

[729] Comp. Wisskirchen, 2007, p. 12

[730] Comp. Busch/Kocher/Welkoborsky, 2009, p. 36

Examples for direct unequal treatment are:

- A vacancy advertisement states that "a young saleswoman" is required (this is a direct discrimination against older applicants and males).
- The labor contract of a pregnant woman is not extended, even though this was assured before the pregnancy had come to knowledge (direct discrimination against women).
- Graduated salary in accordance with age groups (direct discrimination of younger employees).

54.4.2 Indirect discrimination

Indirect discrimination shall be taken to occur where an apparently neutral provision, criterion or practice would put persons at a particular disadvantage compared with other persons, while in reality the unequal treatment has different reasons. In this case an examination is necessary whether the disadvantage can be justified by a legitimate aim and whether it is appropriate.

Examples for indirect discrimination are:

- Part-time employees may not obtain certain benefits without work time-related reasons (because part-time employees are predominantly females, this can be a case of an indirect discrimination against women).[731]
- A social plan provides that parent time has no relevance for the staff membership time and thus, for the amount of compensation (this can be a case of indirect discrimination against women).

54.4.3 Harassment

On the basis of Art. 2 subs. 3 of the anti-racism directive 2000/43/EC[732] there is deemed to be an unlawful harassment in the sense of the AGG when an unwanted conduct takes place with the effect or purpose of violating the dignity of another person.[733] The stage of violation of human dignity under Art. 1 German Basic Law does not have to be reached in this case. With regard to the degree of violation, a conduct is sufficient which meets the criteria of defamation in accordance with sec. 185ff. Criminal Code.[734] Furthermore it has to be a case where an intimidating, hostile, degrading, humiliating or offensive environment is created. The conduct can be of verbal or non-verbal nature with the result that the point of view of an uninvolved third person decides whether there is a case of harassment under the above mentioned criteria. The harassment can also take place in form of defamation, insult, hostility, threatening or physical assault and does not have to be carried out neither

---

731 Comp. Busch/Kocher/Welkoborsky, 2009, p. 36 et seq.

732 Comp. Gaier/Wendtland, 2006, sec. 2 AGG, mn. 87

733 Comp. Däubler/Bertzbach, 2008, sec. 3 AGG, mn. 67

734 Comp. Rust/Falke, 2007, sec. 3 AGG, mn. 57

intentionally nor repeatedly. Also a one time violation of a person's dignity is sufficient if a hostile environment is the consequence.[735]

Examples for harassment are:

- An employer makes a tasteless joke on gypsies without knowing that the employee is a Roma (harassment on grounds of ethnic origin),[736]
- Abusing certain groups of people.

54.4.4 Sexual harassment

Sexual harassment is part of harassment in accordance with sec. 3 subs. 3 AGG, with the difference that the harassment has sexual motives. The description of harassment is therefore also valid for sexual harassment, with the exception that for sexual harassment the hurdles are lower, because the intrusion into privacy is bigger. Therefore, no humiliating environment has to be created, because the sexual harassment itself can be taken as example for humiliation. No danger of repetition has to be existent to fulfill this criterion;[737] however, there has to be a connection to the labor relation or a connection to the company.[738]

Examples for sexual harassment are:[739]

- smacking on an employee's bottom,
- imposing kisses on the employee,
- the supervisor hugs a trainee with sexual intentions.

54.4.5 Instructions to discriminate

Also giving instructions to discriminate against a person on any on the grounds referred to under sec. 1 AGG is deemed to be an unlawful act in accordance with the AGG. Although the instruction to discriminate has to be given intentionally, it is not necessary that the executing person has to know this prohibition. Furthermore, the instruction is also deemed to be unlawful when the instructed person does not carry out the instruction.[740]

---

735 Comp. Däubler/Bertzbach, 2008, sec. 3 AGG, mn. 67–71

736 Comp. ibid, sec. 3 AGG, mn. 71

737 Comp. ibid, sec. 3 AGG, mn. 77

738 Comp. Bauer/Göpfert/Krieger, 2008, sec. 3 AGG, mn 50

739 Comp. Däubler/Bertzbach, 2008, sec. 3 AGG, mn. 77a

740 Comp. Wisskirchen, 2007, p. 14

Examples for instruction to discriminate are:

- the instructions of a house owner to a estate agent not to sell the house to colored people (instruction to discriminate on the grounds of ethnic origin),[741]
- the instruction of an employer to the Federal Employment Office not to contract women for a vacancy (instruction to discriminate on the grounds of sex).[742]

54.5 Permissible differences of treatment

The AGG provides exceptions when a different treatment on the ground referred to in sec. 1 AGG is justified. Differentiated treatment is permissible because of occupational requirements, religion or belief, age and because of positive action.

If the discrimination against a person is undertaken because of multiple grounds stated in sec. 1 AGG, the justification has to be valid for all grounds in accordance with sec. 4 AGG. This means, if a woman is discriminated because of her sex and her religion, there has to be a reason for justification for each discrimination element.[743]

54.5.1 Positive action

Sec. 5 AGG gives the employers, parties of collective agreements, business partners and parties of a contract under private law the possibility to promote formerly disadvantaged groups. The aim is to prevent or compensate for disadvantages arising on any of the grounds of sec. 1 AGG or to avoid them for the future. However, fixed rates of affirmative action are not admissible according to the European Court of Justice, because in that case persons of other groups would be subject to discrimination.[744] It has to be noticed that in case of positive action there is freedom of assessment, which has to be partially filled by an assessment of proportionality.[745]

54.5.2 Permissible differences of treatment because of occupational requirements

Unequal treatment is admissible for reasons of occupational requirements in accordance with sec. 8 AGG, if:

---

[741] Comp. Schiek, 2007, sec. 3 AGG, mn. 76

[742] Comp. Busch/Kocher/Welkoborsky, 2009, p. 53

[743] Comp. Schütt/Wolf, 2006, p. 31

[744] Comp. ibid, p. 32 et seq.

[745] Comp. Steinkühler, 2007, mn. 84

- a differentiation is made because of the nature of the occupational activities or the context in which they are carried out, if the grounds constitute a genuine and determining occupational requirement and
- the occupational objective is legitimate
- the requirement is proportionate

Bare considerations of practicality, however, do not justify unequal treatment. The requirement made has to be necessary and proportionate.[746] For this point there is no final statement possible on when special occupational requirements exist, because these can also exist by determination of an organizational concept by the employer. Therefore, it is always necessary to examine the individual case.[747]

There is no justification for harassment or sexual harassment, because no occupational purpose is imaginable for these cases.[748]

The equality of remuneration regulated in sec. 2 is now expanded to the rest of the grounds, with the result that not only unequal treatment on the ground of sex, as regulated in the Civil Code, is prohibited. Differences with regard to remuneration are also not admissible if special regulations apply for the element of discrimination in question. On the other hand, there is a right to equal remuneration for the same and equivalent work only. The same work is performed when the usual activities of the compared persons are identical or if the employees could replace each other, i.e. if responsibility, qualification, work conditions and workload are corresponding. In order to examine whether there is a case of equivalent work, factors like education, experience, capabilities as well as mental, physical and intellectual demand and the responsibility for people and objects have to be compared.[749]

54.5.3 Admissible differences of treatment on grounds of religion or belief

Religious communities or facilities founded on religious principles are privileged by the legislator to a certain extent in the so called "church clause". Also organizations which have formed to practice a religion or belief have the right to differential treatment on grounds of religion or belief in accordance with sec. 9 AGG, as far as the grounds constitute a justified occupational requirement for the form of activities by reason of their right to self-determination or by the nature of the particular activity.[750]

The scope of application of sec. 9 AGG comprises also facilities like confessional schools, hospitals or the Caritas which therefore have the possibility to make a certain religion or

[746] Comp. Wisskirchen, 2007, p. 15

[747] Comp. Schütt/Wolf, 2006, p. 39 et seq.

[748] Comp. Steinkühler, 2007, mn. 91

[749] Comp. Schütt/Wolf, 2006, p. 40 et seq.

[750] Comp. Steinkühler, 2007, mn. 94

belief an occupational requirement for the employees. It is still not clear how far the justification goes, e.g. whether auxiliary activities can be included. Subs. 2 extends the privileges of the above mentioned facilities even further and makes it possible for them to demand for a certain conduct of their employees in the sense of acting in good faith and with loyalty. Legal consequences of violations of these requirements are determined by the facilities themselves; the courts, however, may examine whether they are adequate and proportionate.[751]

54.5.4 Admissible differences in treatment on the grounds of age

An objectively and reasonably justified and legitimate aim is the condition for a justification of unequal treatment on grounds of age. In this case the means for achieving this aim have to be appropriate and necessary.[752] The aims in accordance with sec. 10 AGG can be of general nature, for example employment policy, labor market policy or vocational training or concerning only one company.[753]

The grounds of the Act show that there are complex correlations in the case of differences in treatment because of age, with the result that it is not possible for the legislator to provide general principles to avoid discrimination, so that a flexible application is necessary.[754] In order to point out at least a certain direction, the legislator provides eight examples; the enumeration of the reasons for justification, however, is not conclusive.[755]

Example 1: Setting of special conditions of employment
Example 1 declares differential treatment to be admissible if it includes the setting of special conditions for access to employment and vocational training, as well as particular employment and working conditions, including remuneration and dismissal conditions, as far as it serves the vocational integration of certain groups. The purpose of differential treatment has to be:

- vocational integration of young people,
- vocational integration of older workers,
- vocational integration of persons with caring responsibilities or
- to ensure the protection of young people, older workers or persons with caring responsibilities .

---

[751] Comp. Schütt/Wolf, 2006, p. 42

[752] Comp. Wisskirchen, 2007, p. 16

[753] Comp. Steinkühler, 2007, mn. 105

[754] Comp. Bundestagsdruckssache 16/1780, p. 36

[755] Comp. Schütt/Wolf, 2006, p. 43

The term young people can be understood in accordance with sec. 2 subs. 2 German Youth Employment Protection Act as natural persons aged between 15 and 18 years. The term of older workers, however, has not been defined by law so far. Person with caring responsibilities include especially persons who care for children, older or handicapped relatives.[756]

Example 2: Minimum requirements to age and work experience
The AGG also admits to fix minimum conditions regarding age, professional experience or seniority in service for access to employment or to certain advantages linked to employment. It has to be observed that it has to be examined beforehand whether the unequal treatment is justified by a legitimate aim.[757] This rule can for example be applied to the remuneration, because a connection to professional experience and seniority is easier to justify as a connection to age.[758]

Example 3: Fixing of a maximum age
The fixing of a maximum age for recruitment which is based on specific training requirements of the post in question or the need for a reasonable period of employment before retirement is admissible in accordance with sec. 10 sentence 3 n° 3 AGG. The background of this example is that an extensive introduction into the work shall be confronted with productive performance.[759].

Example 4: Company pension
It is not a case of discrimination on the ground of age in accordance with sec. 10 sentence 3 n° 4 AGG, if upper age limits in the company social security systems, especially for company pensions, are fixed. However, no discrimination on other grounds may result from this.[760] This part of the law, however, conflicts with sec. 2 subs. 2 AGG where the company pension is excluded from the objective scope of application, with the result that it has to be seen to which extent this clause is applied.[761]

Example 5: General age limits
Sec. 10 sentence 3 n° 5 AGG makes clear that it is possible to fix age limits in labor contracts, company agreements or collective agreements, as it was already admissible under sec. 41 SGB VI. N° 5 determines that agreements are admissible under which the termination of the labor relation is possible without dismissal at a point in time when the employee may

756 Comp. Worzalla, 2006, p. 113

757 Comp. Steinkühler, 2007, mn. 112

758 Comp. Bundestagsdrucksache 16/1780, p. 36

759 Comp. ibid, p. 36

760 Comp. ibid, p. 36

761 Comp. Schütt/Wolf, 2006, p. 45

apply for payment of old-age pension. This rule is meant to give younger persons the chance to integrate into the labor market, while the older workers have already reached financial security.[762]

Example 6: Differentiation between benefits of social plans
Furthermore, a differentiation between social benefits within the meaning of the Works Constitution Act is admissible. A graduation of the compensation regulations based on age or length of service, whereby the employee's chances on the labor market which are dependent on his age, have been taken into consideration by means of emphasizing age relatively strongly, is legitimate. Furthermore, it is admissible to exclude the employee from certain benefits of the social plan, because they are economically secure. However, both differentiations are only valid in the context of the Works Constitution Act. Collective Agreements are currently not admissible in this form.[763]

54.6 Organizational obligations of the employer

Sec. 11 and 12 AGG provide for organizational obligations of the employer, which can be divided into preventive and reactive protection, warning and information duties.[764] The organizational duties are:

- obligation for neutral advertisement of vacancies (sec. 11 AGG),
- take necessary measures for protection (sec. 12 subs. 1 AGG),
- training obligations (sec. 12 subs. 2 AGG),
- obligation to react in case of violation against employees (sec. 12 subs. 3 AGG),
- protection of employees against discrimination by third parties (sec. 12 subs. 4 AGG) and
- information (sec. 12 subs. 5 AGG).

The employees are entitled to claim against the employer on the basis of this clause and can oblige him to fulfill his duties.[765] If the employer takes no or inappropriate measures for the elimination of harassment or sexual harassment, the affected employees are entitled to use their right to refuse performance without loss of pay in accordance with sec. 14 subs. 1 AGG. Furthermore, they have the right to claim for compensation in accordance with sec. 280 subs. 1 Civil Code (BGB), if the employer violates his obligations under sec. 12 subs. 1-4 AGG.[766] In the case of a violation of the duty to inform there are no implications concerning damages,

[762] Comp. Steinkühler, 2007, mn. 122

[763] Comp. Worzalla, 2006, p. 120 et seq.

[764] Comp. Kolmhuber/Schreiner, 2006, mn. 155

[765] Schleusner/Suckow/Voigt, 2007, sec. 12 AGG, mn. 46

[766] Comp. ibid, sec. 12 AGG, mn. 47 et seq.

i.e. it constitutes a penalty-free provision.[767] If the employer, on the other hand, violates the duty of neutral formulation of advertisements for vacancies, a violation against sec. 7 AGG can be assumed,[768] which means that all affected applicants are entitled to claim for compensation.[769]

54.6.1 Preventive measures

Among the preventive measures is the duty of neutral advertisement of vacancies, taking necessary measures of protection and training measures.

In accordance with sec. 11 AGG the advertisement of the vacancy shall not violate the prohibition of unequal treatment under sect. 7 subs. 1 AGG, also in the context of vocational training and advanced training. The possible discrimination of certain groups of applicants shall be avoided from the start. This clause, however, constitutes a mere intensification of the comparable sec. 611b BGB.[770] If an external third party is assigned with the advertisement of the vacancy, as the Federal Employment Office or a human resource agent, the employer has the duty to supervise that the advertisement is conform to the AGG. If he does not fulfill this obligation, the violations will be attributed to the employer.[771]

In accordance to sec. 12 subs. 1 AGG the employer has the obligation to take measures necessary to ensure protection against discrimination on any on the grounds referred to under sec. 1 AGG. Also preventive measures are included, because they can be regarded as more efficient than a direct intervention in case of unequal treatment. It has to be determined under objective aspects when a case of discrimination is given, which is also dependent on company structures.[772]

In sec. 12 subs. 2 AGG the employer is required to draw attention to the inadmissibility of such discrimination in a suitable manner.[773] Because this duty exists before the discrimination, preventive measures in the sense of training or further training or similar measures have to be taken by the employer. The training shall be appropriate with regard to kind, content, extent and frequency to sensitize the employees in such a way that they recognize discrimination and react to it. Also in this case, the degree of necessary efforts depends on the structure of the company. As soon as the employer has provided necessary training for his employees in accordance with sec. 12 subs. 2 sentence 2 AGG he has

767 Comp. Schleusner/Suckow/Voigt, 2007, sec. 12 AGG, mn. 57–59

768 Comp. Worzalla, 2006, p. 131

769 Comp. Bährle, 2007, p. 36

770 Comp. Bundestagsdrucksache 16/1780, p. 36

771 Comp. Worzalla, 2006, p. 124

772 Comp. Bundestagsdrucksache 16/1780, p. 37

773 Comp. Schütt/ Wolf, 2006, p. 57

fulfilled his duty, which means that he cannot be claimed against for discriminations by his employees.[774]

54.6.2 Reactive measures

Reactive measures include the obligation to react to violations by employees as well as protection of the employees against discrimination by third parties.

Where employees violate the prohibition of discrimination, the employer has the obligation under sec. 12 subs. 3 AGG to take necessary and appropriate measures to put a stop to the discrimination under sec. 7 subs. 1 AGG. If there is a violation against the prohibition of discrimination, this means in accordance with sec. 7 subs. 3 AGG that the discriminating employee has violated the contractual obligations. Possible measures taken by the employee may include cautioning, moving, relocating or dismissing the employee in question.[775]

Under sec. 12 subs. 4 AGG the employer has the obligation to protect his employees while pursuing their profession. Also in this case appropriate counter measures[776] have to be taken against third persons, e.g. customers and suppliers. The reaction by the employer, however, is expected to be proportionate e.g. to avoid the loss of important customers.[777]

54.6.3 Information obligations

Furthermore, employers have the obligation under sec. 12 subs. 5 AGG to make the General Equal Treatment Act, as well as the departments competent to handle complaints pursuant to sec. 13 AGG known to all employees. However, the employees shall only get notice of these regulations, the employer has no obligation to take care that the employees are aware of them. Possible ways of communication are putting up a notice or displaying information leaflets in a suitable place or using the internet to display a text.[778]

54.7 Legal consequences of violations of the AGG

In case a discrimination or harassment in connection with a running labor relation has taken place, the AGG provides the following penalties for the discriminated persons:

---

[774] Comp. Schütt/Wolf, 2006, p. 59

[775] Comp. Worzalla, 2006, p. 140–142

[776] Comp. Bundestagsdrucksache 16/1780, p. 37

[777] Comp. Schütt/Wolf, 2006, p. 61

[778] Comp. Bährle, 2007, p. 44

- right of appeal (sec. 13 AGG)
- right to refuse performance (sec. 14 AGG)
- right to compensation and damages (sec. 15 AGG)
- prohibition of victimization (§ 16 AGG)

In the context of these penalties, the employer is granted liability alleviation in case of discrimination in the application of collective bargaining agreements in accordance with sec. 15 subs. 3 AGG. Furthermore, a two months waiting period and the period for filing a claim under sec. 61 Labor Courts Act apply.[779]

54.7.1 Right of appeal

Under sec. 13 subs. 1 AGG the employee has the right to complain about a discrimination against him, which creates an "instrument of in-company settlement of disputes". The employee is entitled to lodge an oral or written complaint with the competent department of the company, which has to be named by the employer.[780] The competent authority can also be a department installed on the grounds of the Employee Protection Act. The legislator names a superior, an equality ombudsman or a company department for complaints as examples for competent authorities.[781] In case of an informal complaint, the employee does not have to respect any terms, however, he can forfeit his right. For the right to be forfeited, the employee has have omitted to complain for some time, although it would have been possible for him, with the result that the employer could assume that no complaint would be made.[782] An examination of the content of the claims has to be undertaken and the complaining person has to be informally informed on the result and the decisive grounds.[783] If the notice is negative, the procedure of complaint is deemed to be completed with the notification to the complaining person. If the complaint is judged as being justified, the employer has the duty to take adequate measures to remedy the complaint. At least the discriminating person has to be instructed to stop his duty violating behavior.[784]

54.7.2 Right to refuse performance

Under sec.14 AGG the employee is entitled to refuse performance without loss of payment, if the employer does not take the necessary measures in order to eliminate discrimination or sexual harassment. So far, sexual harassment has been penalized under sec. 4 of the Employee Protection Act, which was widened by the extension to all elements of

---

[779] Comp. Kolmhuber/Schreiner, 2006, mn. 203–207 und Punkt 3.8.1

[780] Comp. Schleusner/Suckow/Voigt, 2007, sec. 13 AGG, mn. 2

[781] Comp. Bundestagsdrucksache 16/1780, p. 37

[782] Comp. Schleusner/Suckow/Voigt, 2007, sec. 13 AGG, mn. 21–22

[783] Comp. Schütt/Wolf, 2006, p. 62

[784] Comp. Schleusner/Suckow/Voigt, 2007, sec. 13 AGG, mn. 39–40

discrimination under sec. 1 AGG. The condition for a possible right to refuse performance is that a complaint has been previously filed with the employer and the necessity for the protection of the employee, e.g. if health problems can be expected.[785] Sec. 14 AGG can be applied if the employer does not sufficiently react to a complaint, commits the harassment himself or if he has taken notice of a harassment with danger of repetition.[786] If an employee uses his right to refuse performance, the employer is still obliged to take care of the elimination of discrimination in order to make the employee's return to his workplace possible.[787]

The reference to the general right to refuse performance under sec. 273 BGB was integrated to underline, that this remains intact for other case constellations. The background is that the purposes of both regulations are different. Sec. 273 BGB shall enforce the fulfillment of an obligation by means of pressure, while sec. 14 AGG is meant to protect employees against further harassment or sexual harassment.[788]

54.7.3 Right to compensation

The compensation regulated in sec. 15 subs. 1 AGG corresponds to the claim under sec. 280 subs. 1 BGB and therefore constitutes a contractual right which arises from the violation of sec. 7 subs. 3 AGG. In contrast to the right for damage, the compensation compensates material loss in accordance with sec. 15 subs. 1 AGG as far as the employee is responsible for the violation of duty. In accordance with the regulations on general determination of responsibility under sec. 276 BGB the employee has to have acted intentionally or negligently. If discrimination is established, the employer is obliged under sec. 22 AGG to prove that he is not responsible for it. The compensation of the damage can only be monetary, because under sec. 15 subs. 6 AGG there is no claim to establish a work relationship, vocational relationship or professional promotion.

Furthermore, the employer is liable under sec. 278 BGB for discrimination on the part of his legal representative or persons whom he uses to perform his obligation. This means persons who become active in his name under the will of the employer. The representative must be exercising the rights or fulfill the duties of the employer and have a right to give instructions to the discriminated person.[789]

The extent of the compensation is determined in accordance with sec. 249 BGB. If an employment relationship already exists, there are primarily rights to fulfillment and equal treatment, which leads only in rare cases to rights of compensation. Claims for compensation are more common in cases of failed applications. In these cases, however, only the so called

---

785 Comp. Busch/Kocher/Wolkoborsky, 2009, p. 58

786 Comp. Worzalla, 2006, p. 154 et seq.

787 Comp. Busch/Kocher/Wolkoborsky, 2009, p. 58

788 Comp. Bundestagsdrucksache 16/1780, p. 37

789 Comp. Kolmhuber/Schreiner, 2006, mn. 209–215

best qualified applicant is entitled to compensation and has to prove that damages really occurred. This would be possible for example, when he had terminated another work relationship on the basis of the awakened confidence[790] or if travel costs for the job interview had arisen.

54.7.4 Right to claim damages

In sec. 15 subs. 2 AGG the legislator opens the possibility for a claim to damages because of discrimination on a ground under sec. 1 AGG, which however constitutes only a special form of sec. 253 BGB. This means damages for intangible loss can be required,[791] with the former exclusiveness of material and intangible damages being eliminated.[792] There is already a claim for damages if the motive for discrimination has contributed to the damage.[793]

The claim for damages can arise in case of execution or termination of a labor contracts and subsequent duties, which corresponds to an expansion of sec. 611a BGB. In contrast to the right for compensation, in case of the right for damages it is not necessary that the employer is responsible.[794]

Also in case of damages, no upper limit has been set yet; however, the damages have to be appropriate. The courts have a freedom of assessment with regard to the amount of damages, as already regulated under sec. 253 BGB. Correspondingly, damages shall be paid which are in an appropriate relation to the loss, but which have at the same time a deterring effect on the employer.[795] A limitation of the amount to three monthly salaries is only given in the course of selection procedures in which the applicant was not employed due to a discriminating selection.[796]

If in the execution of a collective bargaining agreement or a company agreement, discrimination takes place, there is only a claim for damages under sec. 15 subs. 3 AGG if the employer acts with intention or gross negligence.[797] However, agreements which violate the prohibition of discrimination under sec. 7 subs. 2 AGG, are void.[798]

---

790 Comp. Schleusner/Suckow/Voigt, 2007, sec. 15 AGG, mn. 15–18

791 Comp. Bundestagsdrucksache 16/1780, p. 38

792 Comp. Schütt/Wolf, 2006, p. 69

793 Comp. Schleusner/Suckow/Voigt, 2007, sec. 15 AGG, mn. 30

794 Comp. Worzalla, 2006, p. 165

795 Comp. Bundestagsdrucksache 16/1780, p. 38

796 Comp. Worzalla, 2006, p. 165

797 Comp. Schütt/Wolf, 2006, p. 70

798 Comp. Kolmhuber/Schreiner, 2006, mn. 222

54.7.5 Prohibition of victimization

The prohibition of victimization under sec. 16 AGG determines that employees, persons who support the employees as well as witnesses shall not be disadvantaged because of utilizing their rights under the AGG. Furthermore, the employee has no right to penalize persons who refuse to carry out unlawful instructions.[799] Sec. 16 subs. 2 AGG determines that neither the tolerance nor the rejection of discriminatory conduct by an affected employee may be used as the basis for a decision affecting that employee. For example, an employee may not be discriminated against, because he formerly rejected harassment by a superior or, in the contrary case, tolerated it. Therefore, the last case cannot be judged as silent consent by the affected person in any case.[800]

54.8 Court enforcement of claims under the AGG

Employees can go to court to enforce the above mentioned claims, as far as the employer does not fulfill them on his own accord. The following chapter will shortly analyze the terms to be respected as well as the burden of proof.

54.8.1 Terms

In order to claim for damages, compensation, omission or elimination of a discrimination a two-month term has to be respected, after whose expiration the claims under sec. 15 subs. 4 AGG and sec. 21 subs. 5 AGG expire. However, it is possible to agree on deviating regulations in collective agreements. The term commences as soon as the employee takes notice of the discrimination. In case of a rejected application or the rejection of professional promotion, the legislator assumes that the affected person gains knowledge when receiving the rejection.[801] However, it is often not yet recognizable at moment that this is a case of discrimination.[802] After the expiry of the time limit the claim may only be asserted when the disadvantaged person was prevented from meeting the deadline through no fault of their own.[803]

Furthermore, there is a three month term of assertion before a court in the sense of sec. 61b Labor Courts Act, which starts with the moment of the assertion of claims in due form and

---

[799] Comp. Bundestagsdrucksache 16/1780, p. 39

[800] Comp. Schleusner/Suckow/Voigt, 2007, sec. 16 AGG, mn. 14 et seq.

[801] Comp. Schütt/Wolf, 2006, p. 71 et seq.

[802] Comp. Busch/Kocher/Welkoborsky, 2009, p. 61

[803] Comp. Worzalla, 2006, p. 219

meeting the terms. The period for claims only affects claims for damages in accordance with sec. 15 AGG.[804]

54.8.2 Statement of facts and burden of proof

The repartition of the burden of proof deviates from the usual procedures under civil law under which the plaintiff has to prove the facts on which the claim is based.[805] The burden of proof regulated under sec. 22 ABB imitates sec. 611a subs. 1 sentence 3 BGB and determines the obligation to state the facts and the burden of proof in dependence of the scope of influence of the incidences. The discriminated person has to fully prove with the help of evidence that he was discriminated against in comparison to other people and that the unequal treatment was based on one of the elements of discrimination under sec. 1 AGG.[806] The case of defective advertisement of a vacancy receives special attention in this context. In the case of harassment or sexual harassment it has to be proven that the harassment has actually taken place, which leads to a lack of evidence in the case of absence of witnesses.[807]

In the next step the burden of proof is reversed, i.e. if discrimination on one of the grounds referred to under sec. 1 AGG has been established, the discriminating person has to prove that no discrimination has taken place or that the discrimination was admissible.[808]

Literature

Bährle, R. J.: Das Allgemeine Gleichbehandlungsgesetz in der arbeitsrechtlichen Praxis. Vol. 235 of the series: Das Recht der Wirtschaft, Richard Boorberg Verlag, Stuttgart etc. 2007.

Bauer, J.-H. / **Göpfert**, B. / **Krieger**, St.: Allgemeines Gleichbehandlungsgesetz. Kommentar. 2nd edition, Verlag C. H. Beck, Munich 2008.

Busch, Sebastian; Kocher, Eva; Welkoborsky, Horst: Das Allgemeine Gleichbehandlungsgesetz. Benachteiligung verhindern – Gleichstellung fördern. 1st edition, Bund-Verlag GmbH, Frankfurt on Main 2009.

Cox, T. Jr.: Cultural Diversity in Organzations – Theory, Research & Practise, 1st ed., Berrett-Koehler Publishers Inc., San Francisco 1993.

---

804 Comp. Kolmhuber/Schreiner, 2006, mn. 275

805 Comp. Busch/Kocher/Welkoborsky, 2009, p. 61

806 Comp. Bundestagsdrucksache 16/1780, p. 47

807 Comp. Schleusner/Suckow/Voigt, 2007, sec. 22 AGG, mn. 21

808 Comp. Schütt/Wolf, 2006, p. 86

Däubler, W. / **Bertzbach**, M. (ed.): Allgemeines Gleichbehandlungsgesetz. Handkommentar, 2nd ed., Nomos Verlagsgesellschaft, Baden-Baden 2008.

Deutscher Bundestag: Bundestagsdrucksache 16/2022. Beschlussempfehlung und Bericht des Rechtsausschusses, 6th publ., 06/2006, URL:
http://dip21.bundestag.de/dip21/btd/16/020/1602022.pdf
status: 21 April 2009.

Deutscher Bundestag: Bundestagsdrucksache 16/1780. Gesetzentwurf der Bundesregierung. Entwurf eines Gesetzes zur Umsetzung europäischer Richtlinien zur Verwirklichung des Grundsatzes der Gleichbehandlung, 06/2006, URL:
http://dip21.bundestag.de/dip21/btd/16/017/1601780.pdf
status: 23 April 2009.

DGB Bildungswerk e.V. (ed.): EU-Antidiskriminierungsrichtlinien. o. J., URL:
http://gutearbeitweltweit.de/index.php?option=com_glossary&func=view&Itemid=67&catid=14&term=EU-Antidiskriminierungsrichtlinien
Status: 21 April 2009.

Europäische Gemeinschaft: Amtsblatt der Europäischen Gemeinschaften C 325/33. Konsolidierte Fassung des Vertrages zur Gründung der Europäischen Gemeinschaft, 12/2002, URL:
http://eur-lex.europa.eu/de/treaties/dat/12002E/pdf/12002E_DE.pdf
Status: 21 Feburary 2009.

Gaier, R.; **Wendtland**, H.: Allgemeines Gleichbehandlungsgesetz AGG. Eine Einführung in das Zivilrecht. Verlag C. H. Beck, Munich 2006.

Gloger, A.: Multi-Kulti in der Arbeitswelt. in: Manager Seminare, Volume 41, 03/2000, p. 34–39.

Hartweg, D. M. R.: Individualität und Diversität als Determinanten von Personalmanagementkonzepten – Diversity Management in den Institutionen der Europäischen Union. in: Vedder, G. (ed.): Diversity-orientiertes Personalmanagement. 1st edition, Rainer Hampp Verlag, 2006, Munich und Mering, p. 1–76.

Haselier, J. / **Thiel**, M.: Diversity Management – unternehmerische Stärke durch personelle Vielfalt. 1st edition, Bund Verlag GmbH, Frankfurt on Main, 2005.

Heckmann, Carsten: „Brain drain“ aus Ostdeutschland, 11/2006, URL:
http://idw-online.de/pages/de/news183050
Status: 16 April 2009.

Hoffjan, A. / **Bramann**, A. / **Kentrup**, St.: Folgekosten von Gesetzen. Beispiel AGG, Band 35 der Schriftenreihe Controlling und Management, Internationaler Verlag der Wissenschaften, Frankfurt on Main 2008.

Kolmhuber, M. / **Schreiner**, P.: Antidiskriminierung und Arbeitsrecht – Das neue Gleichbehandlungsgesetz in der Praxis. 1st edition, Verlag für Rechts- und Anwaltspraxis, Cologne 2006.

Kirchgeorg, M. / **Günther**, E.: Employer Brands zur Unternehmensprofilierung im Personalmarkt. Eine Analyse der Wahrnehmung von Unternehmensmarken auf der Grundlage einer deutschlandweiten Befragung von High Potentials, 2006, URL:
http://www.hhl.de/fileadmin/LS/marketing/Download_Publikationen/Arbeitspapier___Nr._74.pdf
Status: 16 April 2009.

Loden, M. / **Rosener**, J.: Workforce America!, Business One Irwin, Homewood, Illinois 1991.

Monzel, M. et al.: Personelle Vielfalt am Finanzplatz Luxemburg. in: Vedder, G. (ed.): Diversity Management und Interkulturalität, 2nd ed., Rainer Hampp Verlag, Munich and Mering, 2005, p. 163–231.

Oechsler, W. A.: Personal und Arbeit. Grundlagen des Human Ressource Management und der Arbeitgeber-Arbeitnehmer-Beziehungen. 8th thoroughly rev. ed., R Oldenbourg Wissenschaftsverlag, Munich 2006.

Peters, S. / **Bensel**, N. (ed.): Frauen und Männer im Management – Diversity in Diskurs und Praxis. 2nd ed., Gabler Verlag, w.p. 2002.

Raasch, S.: Das Allgemeine Gleichbehandlungsgesetz (AGG): seit August 2006 in Kraft. in: aktuelle Informationen 2007 Volume 1, 2007, URL:
http://www.djb.de/publikationen/zeitschrift/ai-2007-2/ai-2007-2f1/
Status 21 April 2009.

Rust, U. / **Falke**, J. (ed.): Allgemeines Gleichbehandlungsgesetz mit weiterführenden Vorschriften. Kommentar, Erich Schmidt Verlag, Berlin 2007.

Sandner, D : Diversity Management – Vielfalt als Ressource Betriebswirtschaftliche Begründungen, VDM Verlag Dr. Müller, Saarbrücken 2006.

Schiek, D. (ed.): Allgemeines Gleichbehandlungsgesetz (AGG). Ein Kommentar aus europäischer Perspektive, Sellier. European Law Publisher, Oldenburg 2007.

Schleusner, A. / **Suckow**, J. / **Voigt**, B.: AGG. Kommentar zum Allgemeinen Gleichbehandlungsgesetz, Luchterhand Fachverlag, Neuwied 2007.

Schütt, K. / **Wolf**, R.: Das neue Allgemeine Gleichbehandlungsgesetz. Pflichten – Risiken – Gestaltungsmöglichkeiten. 2nd edition, GDA Verlag, Berlin 2006.

Sepehri, P.: Diversity und Managing Diversity in internationalen Unternehmen – Wahrnehmungen zum Verständnis und ökonomischer Relevanz. 1st edition, Rainer Hampp Verlag, Munich und Mering 2002.

Statistisches Bundesamt (ed.): Wirtschaft und Statistik 03/2009, URL:
http://www.destatis.de/jetspeed/portal/cms/Sites/destatis/Internet/DE/Content/Publikatninen/Querschnittsveroeffentlichungen/WirtschaftStatistik/Leseprobe/WistaMaerz09,property=file.pdf
Status: 16 April 2009.

Statistisches Bundesamt: Außenhandel, 02/2009, URL:
http://www.destatis.de/jetspeed/portal/cms/Sites/destatis/Internet/DE/Content/Statistiken/Zeitreihen/LangeReihen/Aussenhandel/Content75/lrahl01a,templateId=renderPrint.psml
Status: 19 April 2009.

Statistisches Bundesamt: Bevölkerung Deutschlands bis 2050. 11. koordinierte Bevölkerungsvorausberechnung (Pressexemplar), 11/2006, URL:
http://www.destatis.de/jetspeed/portal/cms/Sites/destatis/Internet/DE/Presse/pk/2006/Bevoelkerungsentwicklung/bevoelkerungsprojektion2050,property=file.pdf
Status: 20 April 2009.

Statistisches Bundesamt: Gebiet und Bevölkerung – Fläche und Bevölkerung, 10/2008, URL:

http://www.statistik-portal.de/Statistik-Portal/de_jb01_jahrtab1.asp
Status: 20 April 2009.

Steinkühler, B.: Allgemeines Gleichbehandlungsgesetz (AGG). Die Umsetzung des AGG im Betrieb mit Handlungsempfehlungen für die Praxis, Band 38 der Schriftenreihe Arbeitsrecht in der betrieblichen Praxis, Erich Schmidt Verlag, Berlin 2007.

Stewart, J. / **Clark**, D. / **Clark**, P. F.: focus MIGRATION. Abwanderung und Anwerbung von Fachkräften im Gesundheitswesen: Ursachen, Konsequenzen und politische Reaktionen, 08/2007, URL:
http://www.focusmigration.de/typo3_upload/groups/3/focus_Migration_Publikatione n/Kurzdossiers/KD07_Gesundheit.pdf
Status: 16 April 2009.

Stuber, Michael: Diversity – Das Potential von Vielfalt nutzen – den Erfolg durch Offenheit steigern. 1st edition, Wolters Kluwer Deutschland GmbH, Munich and Unterschleißheim, 2004.

Thomas, D. / **Ely**, R. J.: Making Differences Matter: A New Paradigm for Managing Diversity. in: Harvey, C. P. / **Allard**, M. June: Understatusing and Managing Diversity – Readings, Cases and Exercises, 3rd ed., Pearson Education Inc., New Jersey, 2005,
p. 211–227.

Wagner, D. / **Sepehri**, P.: Managing Diversity – Alter Wein in neuen Schläuchen?. In: Personal, 32. series., vol. 05/1999, p. 18–21.

Wagner, D. / **Sepehri**, P.: Managing Diversity – Wahrnehmung und Verständnis im Internationalen Management. In: Personal, 52. series., Volume 09/2000, p. 456–462.

Wagner, D. / **Voigt**, B.: Diversity – Management als Leitbild von Personalpolitik DUV, 2007, p. 21ff.

Wisskirchen, G.: Das Allgemeine Gleichbehandlungsgesetz – Auswirkungen auf die Praxis. 3rd revised and completed edition, Datakontext-Fachverlag GmbH, Frechen 2007.

Worzalla, Michael: Das neue Allgemeine Gleichbehandlungsgesetz. Neue Vorschriften, Handlungsanweisungen, Muster, Rudolf Haufe Verlag, Freiburg et al. 2006.

XIII International/European Labor Law

55 On a personnel resource management approach in international, European and national labor law

The economic and personnel management approach of this book was especially supported by the subsequent practice-related considerations concerning international and European Labor Law. Labor law does not only provide company internal conditions for actions. The extent of regulations itself constitutes a reaction to economic changes on the labor market. This means that in different economic systems and time spans, only that labor law prevails which is best to solve issues with regard to employment law, because employees, labor unions or employers vote only for the law which they assume to be most appropriate to solve an employment law conflict in the above mentioned sense.[809]

The condition for such an "evolutionarily" developing labor law is that the parties are free to choose between different rules and regulations. The national labor regulations have therefore to open the access to other labor laws.[810] At least under German law, this condition is largely fulfilled, because in principle the parties are allowed to agree on other labor law regulations. This freedom is limited by the fact that they have to respect the obligatory standards of the German labor protection (Art. 30 Introductory Law to the Civil Code/EGBGB[811]). Secondly, the described choice of law is only possible, if the parties have sufficient knowledge on the other legal system with regard to labor law. Therefore, it seems to be unavoidable to concentrate not only on the horizon of German labor law, but to look out for possible choices of regulatory alternatives in other countries and international organizations.

---

809 See in detail: Krimphove, 2006, p. 10 et seq., 17 et seq. (with further references); Krimphove (a), 2001, p. 497 et seq., 509, 528

810 Krimphove, 2006, p. 10 et seq., 14 (with further references)

811 See below: Chapter III. C (with further references)

Companies with activities in several countries have been using this point of view into international labor law for years. This way, they have been able to obtain considerable cost advantages.

The following chapter shall make a contribution to this point of view and make the capacity to choose also available to the personnel management policy of small and medium-sized companies.

56 On the terminology "labor law"

At second sight it proves to be especially difficult to provide a comprehensible and plausible definition for the term "labor law". Its scope of regulations comprehends for example the Dismissal Protection Law, vacation regulations, the Work Council Constitution Law, strike regulations etc. The regulations are so varied and diverse that definitions of labor law – for example as law on employed work – cannot do justice to the phenomenon of labor law and seem to be too superficial. This problem does not only concern German labor law, but even more European and international labor law. Especially the last mentioned fields of law regulate and determine not only legal relations with regard to the relation between employer and employee but also with regard to relations in the community of states.

No examination so far had put the international/European and national labor law under a personnel expense-economic approach, despite the fact that structuring labor law under these aspects seems especially relevant in practice, because in case of conflict the attribution to international, European or German labor law is decisive for the circle of persons affected by this field of law: in international labor law, for example, neither the individual employer nor the individual employee is addressed. Addressees of this field of law are merely countries and organizations under international law.

The overall view on labor law – which also includes international regulations on labor law – is of special importance for academics as well as practitioners. The knowledge of the existing and developing international regulations on labor law make it possible to foresee labor law trends, developments and their future implementation into national/German law.

This chapter conveys a comprehensive description of the existing international/European and German labor law. The purpose is not only to provide a guideline for the description of the rights of employees and employers arising from labor relations, but also to focus the relation between international and European and national labor law and to make it possible for the first time to assess its cross-border, worldwide importance in a globalised reality of labor law.

57 International labor law

The existing regulations under international labor law are very varied. They can especially be systematized according to the fact whether and how far an individual regulation of individual labor law grants an individual and independently enforceable legal position to the individual person, employee and employer.

57.1 International labor law as public international law

International legal regulations are characterized by the fact that in principle, they bind only the legal entities of international law, i.e. countries or entities under the law of nations with legal capacity, like confederations and country unions. The international labor law only obliges states and state organizations. The individual citizen as legal entity does not appear in international labor law. This means that employers and especially employees can benefit from international labor law regulations, but only states and state organizations are bound by them and therefore, employers and employees do not have any possibility to refer to the benefits in the context of their existing labor relation. The benefits are only available under international labor law as "reflex" to individuals, but not as independent rights.

It seems to be consequent that citizens cannot take court actions to enforce their legal advantages in front of an international court.

A very limited exception is provided by Art. 31 of the European Convention on Human Rights. Here, the citizen may have a right to claim, when he wants to take actions against state limitations. However, Art. 31 European Convention on Human rights does not give the possibility to take actions against an employer in an international labor law conflict, for example. Moreover, the judgments of the European Court for Human Rights have only "political" significance, because the ECtHR has no power to oblige the member state to implement its judgments (Art. 47 subs. 1 ECHR). Therefore, they are not binding.

Among the international regulations which can immediately influence labor law are:

Object of regulation		*Universal Declaration of Human Rights (UDHR)*	*European Convention on Human Rights (ECHR)*
Prohibition of discrimination		Art. 2	Art. 14
prohibition of slavery		Art. 4	Art. 4 including prohibition of forced labor
right to			
1.	work	Art. 23 I	(-)
2.	free choice of profession	Art. 23 I	(-)
3.	remuneration and leisure time	Art. 24	(-)
4.	equal remuneration for equal work	Art. 23, II	(-)
5.	education and access to trade unions	Art. 23 III	Art. 11
right to social security		Art. 22	(-)
right to move freely between states		Art. 14	Art. 13
freedom of thought, conscience and religion		Art. 18	Art. 9
freedom of opinion		Art. 19	Art. 10
freedom of assembly		Art. 20	Art. 11

Table 57.1: relevant labor law regulation of international labor law

Among state duties are also the responsibilities of states to take care for the legal protection of their citizens and to create the corresponding institutions. Such obligations exist especially in the context of provision of a working legal protection (Art. 8,9,19 UDHR/ Art. 6, 13 ECHR).

Because the regulations under international labor law provide no claims to the individual citizens, they are able to name such generous and comprehensive legal objects as the "right to work" in Art. 23 subs. 1 UDHR.

Also the constitutions of numerous European countries[812] as well as the Länder-constitutions[813] in Germany provide the right to work – only as non-binding declarations of intent or objectives which do not entitle individual citizens.

57.2 International special regulations

Individual regulations can have direct influence on individual labor relations and give employees and employers independent legal positions. Such isolated regulations exist in special areas of occupation as the field of transportation. In this case it is especially international treaties which prescribe certain working times for steering a vehicle and are binding for the employer.

57.3 Regulations of international private law

All other conflicts of international labor law are assessed in accordance with international private law. International private law determines which national law has to be referred to in order to solve a certain conflict. This means that international private law does not provide any material law regarding labor regulations. It is rather adequate to answer the question which national labor law shall be addressed to decide about an international labor law conflict.

For German law this question is answered in accordance with Art. 27 and Art. 30 of the German Introductory Act to the Civil Code (EGBGB) However, each nation has its own International Private Law. Besides the German International Private Law, there is a French, Japanese and Syrian one. The principle of these regulations states that the national labor law of that nation is applied where the work itself takes place. Yet, the parties in international labor relations are also free to choose the application of a certain national labor law. Which international private law is applied, is principally the choice of the plaintiff when he chooses a court (lex fori). Only if he claims before a German labor court, the German International Law is applied and decides on the question which material national labor law shall solve the conflict. For the purpose of this book it seems too extensive to enumerate all possible alternatives and all national legal systems.

---

812 E.g.: Belgium: Art. 23; Finland: Art. 18; France: Preamble of 27th Oct. 1946; Greece: Art. 22; Italy: Art. 4; Luxembourg: Art. 11; Portugal: Art. 53; Spain: Art. 35

813 Bavaria: Art. 168; Berlin: Art. 12; Bremen: Art. 49; Hessen; Art. 28; North Rhine-Westphalia Art. 24 I 3

58 European Labor Law

For the last few years the international discussion on the European Labor Law has gained memento. Judgments as in the cases "Christel Schmidt"[814], "Ayse Süzen"[815], "Kalanke"[816] and "Francovich"[817] provoked an immense number of comments in legal literature.

Individual statements of the European Court of Justice in its decisions and also the numerous comments lead unfortunately to much insecurity in legal and managerial practice. The question is more than ever: which influence has the "European Labor Law" on the national labor law and the individual work relations and on the concrete formulation of obligations between employer and employee? The regulations of European Labor Law have generally a direct effect on the individual work relations, in contrast to international labor law. They do not only grant independent rights to the parties of the labor relation against the European member states and European organizations, but also provide legally enforceable rights which the employers and employees can bring forward. In contrast to international labor law, the European Labor Law constitutes a "real" labor law order.

However, this labor law order is most complex, because it does not only consist of the statutory measures, as the European treaties, the European decrees and directives, but especially also of the comprehensive and most differentiated decisions of the European Court of Justice. As the still existing principle of unanimity of the European legislator halts its work back, the jurisdiction of the European Court of Justice is currently the most important source for European Labor Law.

Besides, the access to European Labor Law seems to be so difficult in practice, because the European Labor Law did not abolish the national labor law systems in the individual member countries, but rather supersedes them. This leads to the complex question on the interaction between European and national law.

---

814 European Court of Justice 14/4/1994 (Case C-329/92) Christel Schmidt vs. Spar- und Leihkasse der früheren Ämter Bordesholm, Kiel and Cornshagen, ECR. I 1994, p. 1311

815 European Court of Justice 11/3/1997 (Case C-13/95) Ayse Süzen vs. Zehnacker-Gebäudereinigung GmbH Krankenhausser-vice

816 European Court of Justice (Case C-450/93) Eckhardt Kalanke vs. Hansestadt Bremen

817 European Court of Justice, 19/11/1991 (Case C-6/90 and C-9/90) Andrea Francovich, et al. Italian Republic, ECR 1991, p. I–5357

58.1 Application of European Labor Law in the labor law systems of the member states

Since the decisions “Costa/ENEL”, “Walt Wilhelm”, “Internationale Handelgesellschaft” und “Simmenthal”, the European Court of Justice has established it as case law that Community law has precedence over national law.[818]

In this context, the intensity of the influence of European law on national law depends on the quality of the European rules:

Law of the **European treaties** (EC; Treaty of Paris, EURATOM-Treaty) as well as the European **regulations** have direct impact on national labor law. They are binding not only for the European member states, but also for the citizens of the European Community, who are subject to direct rights and obligations on the basis of these rules.

At first European **directives** are only binding for the member states which have the duty to implement the content of the directives into national law. Only after the legislator has drafted a national statute, the parties of a labor relation can refer to it.

If a member state does not comply with its duty to implement a directive or does not comply with it in due time, a citizen who is disadvantaged by this can directly refer to this directive against the state, because the state may not take advantage from neglecting its duty.

However, the European Court of Justice has noticed in the past year that a directive which has not (yet) been implemented has never effects to the disadvantage of European citizens. An un-implemented directive has therefore no effect for citizens in a labor relation.[819]

At best the disadvantaged citizen can claim for compensation against the omitting state.[820]

---

[818] European Court of Justice 15/7/1964 (case 6/64) Flamio Costa vs. ENEL, ECR 1964, p. 1251; European Court of Justice 13/2/1969 (case 14/68) ECR 1964, p. 1251 et seq., 1269–1271; Walt Wilhelm vs. Bundeskartellamt, ECR 1969, p. 1; European Court of Justice 17/12/1970 (case 11/70) Internationale Handelsgesellschaft vs. Einfuhr – und Vorratsstelle Getreide, ECR 1970, p. 1125 et seq., 1135; European Court of Justice 9 March 1978 (case 106/77) Staatliche Finanzverwaltung vs. Simmenthal, ECR 1978, p. 629 et seq., 644, mn 17/18 ; Details: Krimphove, 1992, p. 379 et seq. (with further references)

[819] European Court of Justice 26 February 1986 (case 152/84) M. H. Marshall vs. Southampton and South-West Hampshire Area Health Authority, ECR 1986, p. 723; European Court of Justice 19 January 1982 (case 8/81) Ursula Becker vs. Finanzamt Münster-Innenstadt, ECR 1982, p. 53; European Court of Justice 14 July 1994 (case C-91/92); European Court of Justice, 19 November 1991 (case C-6/90 and C-9/90) Andrea Francovich, et al. vs. Italian Republic, ECR 1991, p. I-5357; European Court of Justice 14 July 1994 (case C-91/92) Paola Faccini Dori, ECR I 1994, p. 3225; European Court of Justice 7 March 1996 (case C-192/94), mn 19 et seq.; El Corte Inglés SA vs. Christina Blázquez Rivero printed in NJW 1996, p. 1401 et seq.; Ukrow, NJW 1994, p. 2469 et seq.; Krimphove, 2001, p. 62 et seq. (with further references)

[820] European Court of Justice, 19 November 1991 (case C-6/90 und C-9/90) Andrea Francovich, et al. vs. Italian Republic, ECR 1991, p. I-5357; European Court of Justice 14 July 1994 (case C-91/92) Paola Faccini Dori, ECR I 1994, p. 3225; European Court of Justice 7 March 1996 (case C-192/94), mn 19 et seq.; El Corte Inglés SA vs. Christina Blázquez Rivero printed in NJW 1996; European Court of Justice 8 October 1996 (case C-178/94, C-179/94, C-188/94, C-189/94–190/94) Dillenkofer et al. vs. Bundesrepublik Deutschland; European

Francovich; European Court of Justice, 19/11/1991 (Case C-6/90 and C-9/90) Andrea Francovich, et al. vs. Italian Republic, ECR. 1991, p. I–5357

The Italian state had neglected its duty to implement the Directive 80/987 ECC into national law and therefore failed to create institutions for the protection of employees whose employers have become insolvent.

The employee, Mr. Francovich, was therefore not able to enforce his rights to remuneration against his insolvent employer.

The European Court of Justice rejected the action of Mr. Francovich against his employer.

However, the court granted the corresponding claim for damages against the Italian state to Mr. Francovich.[821]

The jurisdiction of the European Court of Justice has the most important influence on labor relations:

Because the court interprets the Law of the Treaties as well as Regulations and Directives, the influence of its jurisdiction reaches across all levels of European norms. With regard to the extremely long duration of the creation of European norms (Directives, Regulations), the way of the European Court of Justice to change law by means of case law seems far more rapid and less laborious.

By means of the above described sources of law – especially the jurisdiction of the European Court of Justice – the European legislation intervenes in many central question of company labor law – for example the equal treatment of national and foreign employees, equal treatment of men and women at the workplace and technical and social work protection.

Court of Justice 5 March 1996 (case C-46/93) brasserie du pêcheur vs. Bundesrepublik Deutschland, ECR I 1996, p. 1029, mn 31

[821] Similar cases: European Court of Justice, 19 November 1991 (case C-6/90 und C-9/90) Andrea Francovich, et al. vs. Italian Republic, ECR 1991, p. I-5357; European Court of Justice 14 July 1994 (case C-91/92) Paola Faccini Dori, ECR I 1994, p. 3225; European Court of Justice 7 March 1996 (case C-192/94); El Corte Inglés SA vs. Christina Blázquez Rivero printed in NJW 1996, p. 1401 et seq.; European Court of Justice 26 March 1996 (case C-392/93 British Telecommunications, ECR I 1996, p. 1631; European Court of Justice 3 May 1996 (case C-5/94) Hedley Lomas, ECR I 1996, p. 2553; also: European Court of Justice, 10 April 1984 (case 14/83) Colson and Kamann vs. Land Nordrhein-Westfalen, ECR 1984, 1891; European Court of Justice 10 April 1984 (case 79/83) Doris Harz vs. Deutsche Tradax GmbH, ECR 1984, p. 1921; European Court of Justice 9 November 1993 (case C-132/92) Birds Eye Walls Limited vs. F. M. Roberts

58.2 Free movement of workers (Art. 39 EC) in labor relations

The historic point of departure of the development of the European Labor Law constitutes Art. 39 EC. Art. 39 EC gives "European workers" the right to choose the workplace freely and regardless of its location in each member state of the European Community, in line with the formation history of European Labor Law.

While, at first, the right of free movement of European workers was conceived against national states,[822] this entitlement under Art. 39 EC has now become a right with direct effect between the parties of a labor contract.

Especially the jurisdiction of the European Court of Justice has established the principle that European employees have to be treated just as their national colleagues by the employers.

Scholz; European Court of Justice 23/2/1994 (Case C-419/92), ECR I 1994, p. 505

The plaintiff applied for a job in the canteen of the Cagliari University. When asked about her professional experience in public service, the plaintiff could refer to her activities with the Deutsche Bundespost. The Italian employer did not want to take her professional experience into account, because Ms. Scholz had gained it in German public service and not in the Italian one.

The European court of justice considered the negation of the Italian employer to take the professional experience Ms. Scholz had gained in Germany into account to be an inadmissible discrimination of the plaintiff in accordance with the right of free access to employment in the sense of Art. 39 subs. EC.

If the employer bases employments, classifications, especially tariff classifications and professional advancement of the employee (promotion) on the circumstances of existing professional experience, professional experience gained in another European country has to be considered just as experience gained in a national context.

The case Ugliola provides a far-reaching example:

[822] The state has to refrain from anything which obstructs the employees right to move within Europe; e.g. obstructing entry procedures: European Court of Justice 27 April 1989 (case 321/87) Commission of the European Communties vs. Kingdom of Belgium, ECR 1989, p. 997; visas: European Court of Justice 14 July 1977 (case 8/77) Concetta Sagulo, Gennaro Brenca and Addelmajid Bakhouche, ECR 1977, p. 1495; European Court of Justice 3 July 1980 (case 157/79) Stanislaus Pieck, ECR 1980, p. 2171, 2184 mn 6 et seq.; employment instruction for lack of language knowledge: European Court of Justice 28 November 1989 (C-379/87) Anita Groener vs. Minister for Education and the City of Dublin Vocational Education Committee, ECR 1989, p. 3967; refusal to grant work permit: European Court of Justice 27 March 1990 (case 113/89) Société Rush Portugesa Lada vs. Office National d'Immigration (ONI), ECR I 1990, p. 1417; European Court of Justice 14 July 1977 case (8/77) Concetta Sagulo, Gennaro Brenca and Addelmajid Bakhouche, ECR 1977, p. 1495

Ugliola; European Court of Justice 15/10/1969 (Case 15/69) Württembergische Milchverwertung Südmilch-AG vs. Salvatore Ugliola, ECR 1969, p. 363

The Italian citizen, Mr. Ugliola, had worked in Germany, before he commenced his military service in Italy. After Mr. Ugliola had taken work up again with his German employer, his employer did not to take the time served in the Italian military service into consideration for the calculation of seniority in the German company.

The employer argued that the corresponding regulation of the German Workplace Protection Act (Arbeitsplatzschutzgesetz) – being valid for German citizens – on the consideration of the military service for seniority – would not be applicable to Mr. Ugliola. Firstly, Mr. Ugliola was no German citizen. Secondly, he had served his time not with the German military but with the Italian one.

The Court of Justice ruled this procedure to be a case of unequal treatment and discrimination of foreign workers, which violates European Law – especially Art. 39 EC. This means that the statutory regulation of the German Workplace Protection Act is void to the extent to which it excludes the consideration of military service times of European workers who are employed in another European country. This applies even to the case where the European worker serves his time in his home country.

Thus, the European Court of Justice approved the claim of Mr. Ugliola for consideration of his military service in Italy for the calculation of job seniority of a German employer.

The case "Ugliola" makes clear how much influence the European Labor Law and the jurisdiction of the European Court of Justice have on the content of the individual labor relation: for that reason the employer has the obligation to treat employees in the same way as their national colleagues with regard to each individual part of the labor relation – from its creation over questions regarding holidays and rights to continued remuneration to the termination of the labor contract by dismissal or cancellation agreement.

An especially controversial example of this jurisdiction is the case "Paletta":

Paletta; European Court of Justice 3/6/1992 (Case 45/90) Alberto, Vittorio, Raffaela and Carmela Paletta vs. Brennert AG [1992], n° C 166/10[823]

All four members of the Italian family Paletta were simultaneously reported sick by an Italian doctor subsequent to their holiday in their home country Italy. Thereby their annual vacation was extended.

Their joint German employer refused to continue payment of their wages in case of illness. There were serious doubts with regard to the illness-related incapacity to work of the four employees in question: Almost every year the members of the Italian family had reported sick for several years when they spent their joint family leave in Italy.

---

[823] Comp. also European Court of Justice 2 May 1996 (case C-206/94) Paletta II Brennet AG vs. Vittiorio Paletta

Firstly, the European Court of Justice determined that the payment of benefits under the "Lohnfortzahlungs-Gesetz" (German Act on Continued Payment) in case illness to an employee, forms part of the social security scheme of the employee in accordance with Art. 4 Regulation 1408/71. Regulation 1408/71 must for that reason not only be applied to public institutions of the social security system, but also to private employers.

Then, the court interprets Art. 18 subs. 1-4 of the implementing regulation (Regulation 574/72) such that the employer is bound by the declarations by a doctor of another member country with regard to the start and duration of incapacity to work.

The court stated the promotion of the right to free movement of workers as reasons for its result of interpretation, as well as the sense of Art. 18 Regulation 574/72, which is meant "to avoid difficulties of evidence for an employee whose capacity to work has been reestablished in the meantime and therefore to support the highest possible freedom to move".[824]

58.3 Equal treatment of men and women in labor relations

The European Labor Law has especially much influence on the company labor relation with regard to equal treatment of men and women.

The importance of the equal treatment of sexes under European Labor Law is shown by the fact that the obligation of equal treatment is enshrined in one of the most important sources of European law – i.e. in Art. 141 EC.

The introduction of equal treatment of men and women at work had not only "women-friendly" motives in the history of its formation. With the introduction of Art. 141 EC (former Art. 119 EEC) especially France pursued the aim to avoid emergence of competitive advantages of other member states, after having – in contrast to most of the other European member states – already introduced the principle of equal remuneration into its national labor law. France was therefore afraid of considerable disadvantages for France as business location.[825]

The European legislator tried in numerous initiatives to regulate the equal treatment in labor relations[826] – and created especially for this purpose the so called *Equal Treatment*

---

824 Similar in European Court of Justice 12 March 1987 (case 22/86) Giuseppe Ridone vs. Allgemeine Ortskrankenkasse Bad Urach-Münsingen, ECR 1987, p. 1339

825 Neri-Sperl, Traité instituant la communauté économique européenne. in: Smit Herzog: The Law of the European Economic Community, Art. 119; Jansen in, Grabitz, Kommentar zum EEC-Vertrag, Art. 119 mn 3. Jansen, in: d. Groeben et al., Kommentar zum EEC-Vertrag Art. 119, mn. 4

826 Art. 119 EC

-Council Directive of 10 February 1975 on the approximation of the laws of the Member States relating to the application of the principle of equal pay for men and women (75/117/EEC) (OJ 1975, N° L 45, p. 19)

Regulation (2000/78/EC) -, but it is preponderantly the jurisdiction of the European Court of Justice which determines rights of labor contract parties and especially duties of the employer in this field with a variety of individual decisions.

58.3.1 Equal treatment regarding formation of the labor contract

Dekker; European Court of Justice 8/11/1990 (Case C-177/88) Elisabeth Johanna Pacifica Dekker vs. Stichting Vormingscentrum voor Jong Volwassenen (VJV-Centrum) Plus, ECR. I 1990, p. 3941

Ms. Dekker applied with an educational institution operated by the VJV. Ms. Dekker affirmed the question about her pregnancy in her job interview.

The application was rejected only for reasons of the existing pregnancy.

-Council Directive 76/207/EEC of 9 February 1976 on the implementation of the principle of equal treatment for men and women as regards access to employment, vocational training and promotion, and working conditions (OJ 1976, N° L 39, p. 40)

-Council Directive 79/7/EEC of 19 December 1978 on the progressive implementation of the principle of equal treatment for men and women in matters of social security (OJ 1978, N° L 6, p. 24)

-Council Directive 86/378/EEC of 24 July 1986 on the implementation of the principle of equal treatment for men and women in occupational social security schemes (OJ 1986, N° L 225, p. 40)

-Council Directive 86/613/EEC of 11 December 1986 on the application of the principle of equal treatment between men and women engaged in an activity, including agriculture, in a self-employed capacity, and on the protection of self-employed women during pregnancy and motherhood (OJ 1986, N° L 359, p. 56)

- Commission Proposal of 27 May1989 for a Council Directive on the burden of proof in the area of equal pay and equal treatment for women and men (Com [88] 269 fin.) (OJ 1988, N° C 176, p. 3)

- Council Resolution 12 July 1982 on the promotion of equal opportunities for women (OJ 1982, N° C 186, p. 3)

- Resolution of the Council and of the representatives of the Governments of the Member States meeting within the Council of 6 December 1994 on equal participation by women in an employment-intensive economic growth strategy within the European Union (OJ 1994, N° C 368, p. 3)

- Council recommendation of 13 December 1984 on the promotion of positive action for women (OJ 1984, N° L 331, p. 34)

- Second Council Resolution of 24 July 1986 on the promotion of equal opportunities for women (OJ 1986, N° C 203, p. 2)

- Council Resolution of 29 May 1990 on the protection of the dignity of women and men at work (OJ 1990, N° C 157, p. 3)

- Resolution of the European Parliament of 22 October 1990 on the protection of the dignity of women and men at work (OJ N° C 305, p. 30)

- Commission Proposal 92/131/EEC 27 November1991 on the protection of the dignity of women and men at work (OJ 1992, N° L 49, p. 1)

- Commission Proposal 87/567/EEC of 24 November1987 on vocational training for women(OJ 1978, N° L 342, p. 35)

The European Court of Justice ruled this rejection to be a direct discrimination in the sense of Art. 3 of Regulation 76/207/EC. A woman has a sex-related disadvantage to access an employment in case of consideration of an existing pregnancy as reason for rejection

The Federal labor court interpreted the judgment of the European Court of Justices such that the question about pregnancy is inadmissible in a job interview at all:[827] The non-consideration of pregnancy for the selection leads to the general inadmissibility of the employer's question about pregnancy during the job interview. This applies irrespective of whether men and women apply simultaneously for the position. This meant that the corresponding jurisdiction of German High Courts on the admissibility of the question about pregnancy, which made a difference depending on the circle of applicants,[828] had become obsolete with this judgment.[829]

The jurisdiction of the European Court of Justice in the Case "**Dekker**" has the result for the individual employer that under German law already the question on an existing pregnancy may not be asked in a job interview.

A further important consequence for the employer is the penalty which the jurisdiction of the European Court of Justice connects to sex-related discrimination in the selection process: In order to underline the effect of Regulation 76/207/EEC, the European Court of Justice requires in its judgment in the case Dekker for any case of inadmissible sex-related discrimination in the job selection process a civil law penalty for the employer. This penalty arises even regardless of the responsibility of the employer or the justification by national statutory exculpation reasons.[830]

The amount of the penalties for the employer is still unclear. Originally, the German labor law (sec. 611 a BGB) granted a fidelity guaranty claim to the discriminated person. In the cases "**Colson**"[831] and "**Harz**"[832] the European Court of Justice modified the content of sec. 611a BGB. The German legislator had – in correspondence with the European Regulation

[827] BAG 15 October 1992, 2 AZR 227/92; Hanau, EG-Recht und deutsches Arbeitsrecht, Brennpunkte des Arbeitsrechts 1993: Thesen und Ergebnisse der 4. Arbeitsrechtlichen Jahrestagung, p. 37 et seq., p. 18

[828] In accordance with the German High Court judgment the prohibition of discrimination was only violated when men and women applied jointly for a vacancy (applicants of mixed sex). If only women or exclusively men applied for a vacancy, the prohibition of discrimination could not be violated – from the point of view of the former jurisdiction. The question about and existing pregnancy would therefore be admissible.

[829] On the effect of jurisdiction of the European Court of Justice on the courts of member states; especially on the effect of jurisdiction of the European Court of Justice on German labor courts, Schiefer, NJW 1995, p. 160 et seq., 162 et seq.

[830] European Court of Justice 8 November 1990 (case C-177/88) Elisabeth Johanna Pacifica Dekker vs. Stichting Vormingscentrum vor Jong Volwassenen (VJV-Centrum) Plus, ECR I 1990, p. 3941, mn 23–25; also European Court of Justice, 10 April 1984 (case 14/83) Colson and Kamann vs. Land Nordrhein-Westfalen, ECR 1984, 1891

[831] European Court of Justice of 10 April 1984 (case 14/83) Colson and Kamann vs. Land Nordrhein-Westfalen, ECR 1984, 1891

[832] European Court of Justice of 10 April 1984 (case 79/83) Doris Harz vs. Deutsche Tradax GmbH, ECR 1984, p. 1921

76/207/EEC – introduced sec. 611a GBG into German law. The court, however, came to the conclusion that the penalty which was originally provided in sec. 611a BGB, i.e. the compensation for a mere **breach of faith**, was not sufficient to avoid sex-related discrimination in the forefront of labor relations.

An effective implementation of the Regulation 76/207/EWG requires a "considerable penalty" to be imposed on the discriminating employer. The jurisdiction of the European Court of Justice in the case "**Birds Eye Walls Limited**"[833] makes clear when the requirement of a considerable penalty is met: The European Court of Justice ruled that the Regulation 76/207/EEC had not been effectively implemented, even if national law provided a compensation of 6.250 GBP.[834]

The European Court of Justice made the decision that the equal treatment of sexes does not always have to be applied for the purpose of supporting women:

Kalanke; European Court of Justice (Case C-450/93) Eckhardt Kalanke vs. Hansestadt Bremen

Mr. Kalanke applied simultaneously with a female applicant for a position with the Hansestadt Bremen. As both applicants had the same or equivalent qualifications, the Hansestadt Bremen took its decision – in correspondence with sec. 4 of the Landesgleichstellungsgesetz der Hansestadt Bremen (Act on Equal Treatment of the hanseatic city of Bremen) – in favor of the employment of the female applicant. Art. 4 Landesgleichstellungsgesetz regulates that – in cases of same or equivalent qualification of applicants – women should be favored.

Mr. Kalanke felt discriminated due to his sex by the choice of the female applicant – based on the Landesgleichstellungsgesetz of Bremen.

58.3.2 Equal treatment in the determination of salary

Main purpose of sex-related equal treatment is to guarantee equal remuneration of men and women for equal work.

---

833 European Court of Justice of 9 November 1993 (case C-132/92) Birds Eye Walls Limited vs. F. M. Roberts

834 The Jurisdiction of the European Court of Justice prompted the German legislator to fix compensation claims of the employer in sec.611a BGB (Civil Code) sec 61b of the Arbeitsgerichtsgesetz (Labor Act). The amount of compensation exceeds the possible amount for breach of trust by far. However, the German legislator has determined caps for the compensation claims on the grounds of sex-related discrimination in the selection of applicants. This cap amounts to 3 monthly salaries (611a subs. 2 BGB), to 6 monthly salaries (sec. 611a subs. 2 BGB, in connection with 61b subs. 2 ArbGG), as far as several employees are discriminated against or for simultaneous advertisement of several vacancies 12 monthly salaries (sec. 611 a subs. 2 BGB in connection with 61b subs. 2 ArbGG). With regard guidance value in the case "Birds Eye Walls Limited" one can have doubts on whether the cap of 3 months is a "serious" punishment in the sense of the European Labor Law. The limitation of damages with the help of caps encounters the purpose of the punishment in any case, Krimphove, 2001, p. 184 (with further references); European Court of Justice 2August1993 (case C-271/91) M. H. Marshall vs. Southhampton and, South-West Hampshire Area Health Authority. The norms of the German legislator encounter therefore the European Labor Law and are unlawful.

Jenkins; European Court of Justice 31/3/1981 (Case 96/80) J. P. Jenkins vs. Kingsgate (Clothing Production) Ltd., ECR 1981, p. 911

Ms. Jenkins worked part-time for the company Kingsgate. Mostly women have a part-time labor contract. Ms. Jenkins earned – as all part-time employees – a salary which was 10 percent lower than the one of their male full-time colleagues.

The Court considered this in its judgment as "indirect" method of inadmissible discrimination against female employees, because mostly women have part-time employments.

58.3.3 Equal treatment in the context of termination of labor relations

Also in case of termination of a labor contract, the question regarding equal treatment of sexes has to be asked.[835] In this case, this concerns especially facts which do not imply direct discrimination against women. Mostly, the discrimination against women is indirect, because they belong to a certain group of employees (here: part-time employees).

Kriesamer-Hack; European Court of Justice 30/11/1993 (Case C-189/91) P. Kriesamer-Hack vs. Nurhan Sidal

For enterprises with up to 5 employees (small companies) the Kündigungsschutzgesetz (German Dismissal Protection Act) does not apply in Germany. This means that the employees do not enjoy the same dismissal protection as this would be the case for companies with more staff in the Federal Republic of Germany.

Workers who work less than 10 hours a week are no employees in the sense of the law.

The enterprise of the respondent (dental office) employs exclusively female employees – with two full-time employees, one part-time employee (over 10 hours per week) and 4 part-time employees with less than 10 hours.

When Ms. Krisammer-Hack received her notice of dismissal – for insufficient work performance – she sued for dismissal protection before the labor court. The labor court asked the European Court of Justice the question whether the statutory regulation under which the dismissal protection for small companies is excluded constitutes a case of indirect discrimination against women.

The European Court of Justice rightfully negated this: it could be true that nearly 90 percent of all part-time workers are women. The exemption of employees due to the size of the company, however, does not only affect women **specifically**.

[835] On the special issues of social security and the different determination of age limits for pension entitlement see: Krimphove, 2001, p. 226 (with further references)

Therefore, indirect discrimination could only be assumed if it was proved that especially small companies employ considerably more women than men. This does not apply to the underlying case.

58.3.4 Equal treatment in disability law as ground for dismissal protection

Special importance under European Labor Law is not only given to youth protection, but also to disability law, despite the fact that the European legislator has not created any independent, explicit protection of disabled persons under labor law. From the variety of directives which the national legislator has to implement into national law can be understood that the protection of disabled people is given huge relevance under European Labor Law.

The jurisdiction of the European Court of Justice – at least with regard to its practical effects – even provided the basis for dismissal protection for employees who care for a severely handicapped person.

Coleman: European Court of Justice 17/7/2008 (Case C-303/06)[836]

Ms. Coleman cares for her handicapped son. Because of this care, she is regarded as little flexible by her employer and even as lazy, because she often asks for holidays. She is subject to different types of discriminating treatment by her employer. Ms. Coleman was especially accused of using her child to receive special work conditions. She brings these arguments forward in a procedure before labor court against the termination of her labor relation.

Basically, the litigation turned around the question whether the Equal Treatment Regulation 2000/78/EEC also protects people who are not handicapped themselves, but who, because of their close connection to such people, are discriminated against by their employers in comparison to their colleagues.

The European Court of Justice follows the Opinion of General Advocate Poiares Maduro.[837] Also in a situation when a non-handicapped worker is discriminated against, because he has to take care of his handicapped child, it is a case of inadmissible discrimination in the sense of Art. 2 subs. 2a of Regulation 2000/78 and even an unlawful harassment in the sense of Art. 2 subs. 3 Regulation 2000/78.

Taking a closer look on the consequences of the point of view presented in the Opinion, this means for any person taking care of another that the employer has to take the care services of the employee into consideration. Also less flexibility and availability of the caring person may not be used as criterion for discrimination or harassment by the employer – just like in

---

[836] OJ C. 223 of 30 August 2008, p.6

[837] Opinion oft he Advocate General 31 January 2008 (case C.-303/06) Coleman/Attridge Law, in: OJ C, N° 237 30September2006, p. 6 et seq.

the case of handicapped persons themselves. This view leads to a considerable valorization of care services for handicapped people. This principle, however, has certain limits: The caring person is only subject to protection against discrimination if he has an especially close connection to the handicapped person.

58.3.5 Excursion: Dismissal protection under European Labor Law

There is currently no independent dismissal protection law under the European Labor Law. This is astonishing, as it is essential from an economic point of view to quickly create a standardized European Labor Law for several reasons: the Europeanization has already covered many legal aspects, European member states are rapidly and continuously growing together to a uniform European single market and further member states are possibly integrated into the European Community. This applies especially to dismissal law. The diversity of and differences between national dismissal protection rules lead to almost insurmountable difficulties for companies with cross-border operations in connection with their personnel management. This implies high information costs for the employer.

These costs lead to a reduction of entrepreneurial commitment on the European single market and have negative effects on the international labor market. Consequently, they are contra-productive with regard to an employment-supporting unification of European and international labor conditions. Also employees who want to work abroad are deterred by the highly different national regulations on dismissal protection. From a European point of view, the existence of manifold, highly different national dismissal regulations limits the right to free movement under Art. 39 EC.[838]

The jurisdiction of the European Court of Justice has made only few contributions to the unification of the European dismissal law. It is true that the European Court of Justice has developed a detailed jurisdiction on the dismissal possibilities of female employees, especially in the cases **Wepp**[839], **Handels-og Kontorfunktionaerernes Forbund i Danmark**[840], **Larsson**[841] **Mary Brown**[842], **Seymour-Smith**[843] and in another context in the

---

838 See in detail: Krimphove, 2001, p. 110 et seq., 289 et seq. (with further references)

839 European Court of Justice 14 July 1994 (case C-32/93) Carole Louise Wepp vs. EMO Air Cargo (UK) (Ltd), ECR I 1994, p. 3567

840 European Court of Justice 8 November 1990 (case C-179/88) Handels-og Kontorfunktionaerernes Forbund i Danmark vs. Dansk Arbejdsgiverforening, ECR I 1990, p. 3979

841 European Court of Justice 29 May 1997 (case C-400/95) Handels-og Kontorfunktionaerernes Forbund i Danmark representing Helle Elisabeth Larsson vs. Dansk Handel & Service as representative of Føtex Supermarked A/S

842 European Court of Justice 30 June 1998 (case C-394/96) Mary Brown vs. Rentokil Ltd., ECR I 1998, p. 4185

843 European Court of Justice 9 February 1999 (case C-167/97) Regina vs. Secretary of State for Employment, ex parte: Nicole Seymour-Smith and Laura Perez, ECR I 1999, p. 623

case **Coleman**[844]. However, this jurisdiction is not based on general dismissal protection, but on the prohibition of discrimination of female workers in the sense of Art. 141 EC and Art. 5 of the Regulation 76/207/EEC.[845] This means that the current jurisdiction of the European Court of Justice is based on only one aspect of dismissal law.

Because of the lack of uniform, European dismissal rules one has to recur to the national dismissal law for the individual case. This has very different forms:

- For example the dismissal law in Great Britain, but also in France and Belgium is coined by the duty of employer to comply with a formal procedure for dismissal (written dismissal notice including reasons, previous hearing of the person to be dismissed, determination of a financial settlement etc.).
- Further systematic differences of national dismissal rules result from the fact – like in the case of Austria, Italy and Belgium – that some legal systems are based on the differentiation between workers and employees with regard to dismissal law, while in the European Labor Law system this differentiation has only historic significance.
- Generally the terminological and content-related categorization of the northern member countries of the European Community deviates considerably from the one of southern members. While the northern member states mostly distinguish between the terms "ordinary" and "extraordinary" (and only afterwards categorize the dismissal into groups "conduct-related", "person-related" and "business-related"), the legal systems especially in Italy and Portugal distinguish between "objective" and "subjective" dismissal. Southern European legal systems know also dismissals to discipline employees. This sounds like the extraordinary dismissal under German law, but in fact the content of this dismissal law is very different to German dismissal law.
- Especially the Spanish and Portuguese labor law knows the extraordinary dismissal only formally and for the purpose of a "disciplinary measure", while in other legal systems, for example in German law an extraordinary dismissal can also be issued for person and business-related reasons.
- Portuguese labor law provides a comprehensive dismissal protection. This is connected with the fact that even Art. 53 of the Portuguese constitution gives the employees an own "basic right for the protection of their workplace" and thereby constitutionally guarantees the dismissal protection[846] A similar constitutionally guaranteed right exists in Art. 22 subs. 1 of the Greek constitution.[847]

---

[844] OJ C. 223 of 30 August 2008, p.6

[845] Directive 76/207/EEC of the Council of 9/2/1976 on the implementation of the principle of equal treatment for men and women as regards access to employment, vocational training and promotion, and working conditions [1976], N° L39, p. 40.

[846] Art. 53 Constitution of the Portuguese Republic (2/4/1976): The right of workers to job security is safeguarded. Dismissals without just cause or for political or ideological reasons are forbidden.

[847] Art. 22 Subs. 1 Constitution of the Greek Republic (9/6/1975): Work constitutes a right and shall enjoy the protection of the State, which shall seek to create conditions of employment for all citizens and shall pursue the moral and material advancement of the rural and urban working population

- In principle, the possibility to terminate a labor relation by dismissal is not provided by Portuguese labor law. Correspondingly, the dismissal protection of employees is very strong. It consists among other things of the participation of state institutions if a dismissal is necessary. The employer can call the supervisory authority to check the dismissal. Furthermore, the limitations for dismissals to become collective redundancies are very low,[848] with the result that in practice the employer is obliged to take up consultations with the Social Ministry almost in any case of business-related redundancy (Art. 17 GD 64-A/89).
- In contrast to the German legal system, the legal systems of Portugal and Italy offer a statutory dismissal protection – e.g. existent dismissal prohibitions by a factual dismissal protection; i.e. determination of an obligation to pay a cash settlement and/or compensation in case of dismissal by the employer. This form of factual dismissal protection does not exclude the possibility of the employer to dismiss workers. However, the employer has to compare the amount of the compensation and the possible advantages of the termination of the labor relations. These compensation rules reduce the possibilities of dismissal by the employer for economic reasons.
- A highly different standard exists in the member countries with regard to protection of pregnant women and women in childbed. The corresponding regulations in France are very detailed and moreover even under penalty.
- The dismissal law of groups of persons with special need of protection (old employees, young persons and handicapped people) is very diverse, too. While the notice period under Swedish law strongly depends from the age of the employee, other legal systems do almost nothing to protect older workers from dismissal. An express dismissal protection of handicapped people exists only in the legal systems of Germany, Austria and France.

Because of the special labor law-related and economic importance of dismissal law for the employers as well as employees, the most important national dismissal conditions of the individual legal systems will subsequently be presented.

The labor law systems of the members states partially also know the termination of the labor relation by:

- mutual understanding,
- expiration of the temporary work contract,
- fulfillment of a condition,
- completion of a task,
- death of employee,
- force Majeure and
- special court termination.

---

848 2 employees for companies with 50 or less employees, 5 employees in companies with 51 or more employees (Art. 16 GD 64-A/89)

For practical reasons the following part only presents the legal conditions in the "normal case" of termination of a labor relation, i.e. of ordinary and extraordinary termination of the labor relation by an employer or employee in a unlimited labor relation.

1. Germany

Rule	Civil Code = Bürgerliches Gesetzbuch/BGB; Protection against Unfair Dismissal Act (25/8/1969) = Kündigungsschutzgesetz/KSchG; Workplace Protection Act (14/4/1980) = Arbeitsplatzschutzgesetz/ ArbPlSchG; Maternity Protection Act (17/1/1997) = Mutterschutzgesetz/ MuSchG; works constitution act (23/12/1988) = Betriebsverfassungsgesetz/BetrVG
Admissible grounds for termination	ordinary (sec. 620 BGB), extraordinary (sec. 626 BGB) extraordinary: compelling reason = taking all interests into account, further employment and compliance with the period of notice cannot be reasonably expected (sec. 626 BGB) conduct-related: culpable misconduct of the employee; [necessary: warning by the employer] (sec. 1 subs. 2 KSCHG) person-related: objective reasons outside responsibility of the employee (e.g. reduction of capability to perform, lack of eligibility, capability to adjust); permanent illness, [necessary: negative prognosis (= unforeseeable impossibility to use employee)] (sec.1 subs. 2 KSCHG) business-related: urgent economic reasons (e.g.: lack of work, lack of revenues, rationalization, limitation of production); [necessary: correct social selection by the employer] (sec. 1 subs. 3 KSCHG)
Period of notice	sec. 622 BGB (to the end of a calendar month) 2 years ⇨ 1 month 5 years ⇨ 2 months 8 years ⇨ 3 months 10 years ⇨ 4 months 12 years ⇨ 5 months 15 years ⇨ 6 months 20 years ⇨ 7 months otherwise: collective agreements possible. limited possibility of shorter periods in individual contracts (sec. 622 subs. 4, 5 BGB)
Extraordi-nary dismissal protection	members of the work council: ordinary termination inadmissible: sec. 15 KSchG; pregnant women/mothers: ordinary termination inadmissible: sec. 9 MuSchG; extraordinary (sec. 9 subs. 3 MuSchG): termination only with consent of state authority military/civil service: ordinary termination inadmissibly: sec. 2 subs. 1, 2 ArbPlSchG; sec. 2 subs. 3 ArbPlSchG severely handicapped: ordinary termination inadmissible: sec. 21 SchwbG;

	extraordinary: procedure of consent by state authority: sec. 15 SchwbG trainees: ordinary dismissal by employer inadmissible: sec. 15 subs. 2 BBiG collective redundancies: (sec. 17 KSchG = 5 employees in company with 20–60 employees; 10 percent or > 25 employees in company with > 20–60 employees; 30 employees in company with at least 300 employees): consent of Federal Labor Authority (sec. 18 KSchG)
Participation of Works Council	Information/ Advisory right consent to extraordinary termination by member of work council sec. 103 BetrVG
Procedure	Claim: Labor Court **Exclusion period:** 3 weeks (sec. 4 KSchG) in case of invalid termination unacceptable continuation; court termination of labor relation with compensation sec. 9 et seq. KSchG, termination labor relation with cash settlement sec. 9 et seq. KSchG

2. Austria

Rule	Labor Constitution Act= Arbeitsverfassungsgesetz/ArbVerfG; Maternity Protection Act = Mutterschutzgesetz/MuschG; Labor Market Support Act = Arbeitsmarktförderungsgesetz/AMFG; Vocational Formation Act = Berufsausbildungsgesetz/BArbeitgeberG
Admissible grounds for termination	written form only sec. 19 Guts-Angest.G; sec. 30 SchauspG, sec. 32 VertragsbedienstetenGBG (sec. 863 ABGB) conduct-related: culpable violations of work or fidelity duties (e.g. insult, assault against employer and/or colleagues) person-related: lack of qualification and (physical) aptitude; illness; permanent loss of employability by employer with (negative) prognosis of continuance of limitation exceeding foreseeable time. business-related: verifiable permanent business incidents which obstruct further employment (e.g.: economic difficulties, drop in orders); rightful social selection by the employer (sec. 105 subs. 3 ArbVerfG)
Notice period	Generally sec. 1159 ABGB blue-collar workers: employment < 3 months: 1 workday to end of week tenure > 3 months: 4 weeks extension possible; equal period of notice for blue-collar and white collar workers sec. 1159 c ABGB White collar-workers. 6 weeks to end of quarter 2 years ⇨ 2 months 3 years ⇨ 3 months 15 years ⇨ 4 months 25 years ⇨ 5 months

	up to 5 months for 25 (sec. 20 AngG); other agreements for employer and employees with different periods of notice possible Special regulations: [e.g.: sec. 8 InvEinstG ⇨ 4 weeks; sec. 10 HausbO ⇨ 1 months; sec. 4 JornG, 3 months; 5 years ⇨ 4 months; per further year 1 month; up to 12 months; sec. 77 GewO, sec. 17 RegieBArbeitgeber, sec. 201 ABG, sec. 15 HGehG, LandAO = 14 days contractual extension possible
special dismissal protection	inadmissible for reasons of membership in labor union (sec. 105 subs. 3 ArbVerfG); works council member up to 3 months after loss of position – without consent of labor court – inadmissible (sec. 122 et subs. ArbVerfG); dismissal for company constitution law-related activity ((sec. 130 et subs. 4 ArbVerfG) pregnant women / mothers: 4 months after birth, or 4 weeks after end of maternity leave dismissal inadmissible (sec. 10 MuschG; Adoption: sec. 15 subs. 5) military service: (sec. 6 subs. 1 ArbPlSichG) from military draft to 1 month after termination of military service inadmissible trainees: sec. 15 BArbeitgeber: dismissal only possible in first 2 months (probation period) handicapped dismissals: (employees whose earning capacity is reduced by 50 percent) dismissal only with consent of "invalidity committee" ("12 InvEinstG"); min. 4 weeks period of notice (sec. 8 InvEinstG) collective redundancies: dismissal in 30 days 5 employees in company 20–100 employees; 5 percent employees in company 100–600 employees; 30 employees in company > 600 employees; or 5 employees over 50 years (depending on Regulation sec. 45 a AMFG: VO BGBl. 1982/543 = in 4 weeks 5 percent of > 500 employees; in company > 1.000 employees = 50 employees) dismissal void without consent of Employment Office
Participation Works Council	Information and hearing consent = dismissal not rescindable objection = possibility of works council to rescind dismissal before court no statement = rescission by employee (sec. 105 ArbVerfG)
Procedure	rescission of termination before labor court within **1 week** (sec. 105 ArbVerfG)

3. Netherlands

Rule	Burgerlijk Wetboek (BW) Act on Flexibility and Security 1/1/1999 (FlexAct) Buitengewoon Besluit Arbeitsverhoudingen 1945 (BBA)
Admissible reasons for termination	Art. 7a: 1639g BW ordinary; Art. 7a: 1639o BW extraordinary ordinary (Art. 7a: 1639g BW) conduct-related: culpable violation of duties of labor relation person-related: permanent impossibility of employment of employee; illness: general prohibition of ordinary termination on grounds of illness in first 2 years; not if: illness caused by employee; no application to consent to termination (see below) before start of illness and no emergency hospitalization of employee. business-related: (FlexAct) necessary: consent to dismissal by state office, Art. 6 BBA extraordinary (Art. 7a: 1639o , 1639p; 1639q BW) compelling reasons which make further collaboration impossible in principle: consent necessary before dismissal by supervisor, regional employment office. possibility: dissolution of contract by court (Art. 7a: 1639x BW) for "compelling reasons" (extraordinary reasons; business-related/economic reasons [e.g.: necessary to save remuneration] 7a: 1639w BW, FlexAct, court dissolution shorter than termination procedure, no special dismissal protection: but compensation of employee
Period of notice	employee: 1 month, (collective) agreement for shorter period of notice possible, for employer twice as long as for employee employer: (Art. 7a: 1639g BW) < 5 years ⇨ 1 month 5–10 years ⇨ 2 months 10–15 years ⇨ 3 months > = 15 years ⇨ 4 months (collective) agreement on extension possible for weekly remuneration at least 1 week or from 50 years 3 weeks
Special dismissal protection	works council members: generally inadmissible with the exception: winding up of company area in which works council member is employed; formerly ordinary dismissal only with consent of judge; labor union member: prohibition of dismissal for execution of labor union activities (FlexAct) pregnant women /mothers: (Art. 7a: 1639 subs. 4 BW) 12 weeks after birth, dismissal inadmissible (if no objection by employee), right to compensation military service: dismissal inadmissible on grounds of military service collective redundancy: (more than 20 employees): information and

	authorization of responsible authority and works council. Information of responsible labor union.
Participa-tion of works council	information on determination of general guidelines by employer consent to collective redundancies.
Procedure	Claim before "Lower Court" termination in case of pregnancy and illness in the sense of emergency hospitalization **within period of 8 days**

4. France

Rule	Code du Travail = Labor Act 1973 = CT Act of 13/7/1973; Act of 3/1/1975; Act of 7/1/1981; Code Pénal = CP
Admissible reasons for termination	Art. L 122-14-3 CT (factual compelling reason); ordinary extraordinary (Art. L 122-9 CT = serious, culpable intentional/gross negligent, damaging action) <u>person-related</u>: Incapacity, in principle permanent disease only in case of necessity of employee replacement (Special case for work-related causes of disease or accident (see below)) <u>conduct-related</u>: culpable, serious violation of work duty (e.g. inobservance of instructions by employer; responsibility for absence, insobriety, concealment of material) <u>business-related</u>: (Art. L 321-1 et seq. CT, Act of 3/1/1975) company reorganization on grounds of economic or company-internal difficulties; For all terminations: (Art. L 122-14-2 CT; Art. R 122-2-1 CT) compliance with procedure: invitation to prior talks with employer, after min. 5 days, or 2 months after knowledge of conduct-related grounds for dismissal with statement of reason, possibility to bring assistant; requirement for validity of dismissal procedure: notification of reasons for dismissal if required by employee (within period of 10 days) (Art. L 122-14-2; R 122-3 et seq. CT); especially in case of business-related redundancy: hearing by representative of employee (Art. L 321-2/3; Art. 432-1 CT); consent to dismissal by authority (after expiration of 7 day-period dismissal is deemed to be consented; reference to duty of reemployment of employer within 4 weeks after dismissal (Art. L 321-14)
Period of notice	Art. L 122-6 CT < 6 months ⇨ provisions of collective agreement or existing local habits 6–24 months ⇨ 1 months > 2 years ⇨ 2 months collective contracts and labor contracts regulating extension possible

Special dismissal protection	members of the works council (Art. L 425-1 et seq.; L 436 et seq. CT): hearing by the works council and consent of the commercial regulatory authority (after information on decision by works council) (aftereffect 6 months after termination of employment) pregnant women/mothers: Art. L 122-25 et seq. [equiv. adoption Art. L 122-27]: exclusion of any kind of termination (also in probation period). 6 weeks before and 10 weeks after birth (equal rights for fathers of 3 children and in case of death of pregnant woman (Art. 122-25-2); exclusion ordinary dismissal. Pregnancy and 4 weeks after return to work (effect only after end of maternity leave.). Suspension of labor relation during maternity leave = 2 years (for 1 year employment in company with 100 employees) (Art. L 122-28); violation: compensation or penalty 10.000 to 20.000 EUR (Art. 131-13 CP) handicapped persons: employee-termination-period max. 3 months with possibility of collective agreements (Art. L 323-29 et seq. CT); penalty (2 years and/ or 30.000 EUR) for dismissal for grounds of handicap. (Art. 225-1/2 CP) anomalies in case of work accidents or illness due to work (Act of 7/1/1981). Exclusion of illness-related dismissal (Art. L 122-32-2 CT); in case of certification by company physician of incapacity of employee offer of new workplace. For certified impossibility of reintegration of employee dismissal possible. military and civil service: (Art. L 122-18 et seq. CT) employees born before 31/12/1978; otherwise termination of labor relation, claim for reemployment trainees: only in first 2 months possible with limitation. Afterwards unanimously or by court order.
Participa-tion of works council	Information and hearing for business-related redundancy: Art. L 321-2/3; Art. 432-1 CT collective redundancy: Art. L 321-2 CT (10 employees in 30 days) consultation works council and state labor administration. (violation: penalty per dismissal 3750 EUR Art. L 321-11 CT)
Dismissal protection procedure	Procedure before labor court (Art. L 516-1, R 516-0 CT); conciliation hearing (Art. R 516-12) **Exclusion period** of dismissal-protection-clause only in case of abuse of legal rights or 2 months in case of dispute on remuneration calculation (Art. L 122-17) unjustified and formally deficient dismissal compensation (Art. L 122-2 et seq. CT penalty in case of discriminating dismissal 2 years, 30.000 EUR (Art. 225-1/2 CP) dismissal without necessary consent by authority (per dismissal 3750 EUR (Art. L 321-11 CT)

5. Belgium

Rule	Labor Contract Act 3/7/1978, Company Constitution Act = Act of 20/9/1948 and Collective Agreement of 9/3/1972)
Admissible Reasons for termination	Ordinary conduct-related: violations of duties from labor relation person-related: incapacity, missing or lost adequacy; illness: over 6 months (start of period of notice after recovery (compensation)) business-related: economic difficulties, dissolution of company, reorganization of company, rationalization (as Force Majeure: war, natural catastrophes) without dismissal and compensation. Work relation dissoluble) extraordinary: severe culpability and impossibility of working together. (Art. 35 Labor Contract Act) notification of extraordinary reasons for termination written only dismissal of white-collar employees (Art. 37 Labor Contract Act)
Period of notice	BLUE-COLLAR WORKERS (Art. 59 Labor Contract Act): **employee-termination** ⇨ 14 days **employer-termination** ⇨ 28 days in case tenure > 20 years: **employee-termination** ⇨ 28 days **employer-termination** ⇨ 56 days employee less than 6 months in company ⇨ agreement on period of notice, not less than 7 days and notification period not shorter than half of employer-dismissal period of notice employee / WHITE-COLLAR WORKER (Art. 82–84 Labor Contract Act): employee yearly income < 22.000 EUR, (1/5/1995): < 5 years tenure ⇨ 3 months for each further 5 years tenure ⇨ further 3 months yearly income of employee > 22.000 EUR ⇨ agreement or determination by judge (generally dependent on tenure and age) for 1. and 2.: employer-dismissal not below above mentioned terms; employee-termination: income 22.000–44.000 EUR ⇨ max. 4 ½ months; > 44.000 EUR ⇨ max. 6 months Purpose: retirement (65 years): employee-termination ⇨ 3 months employer-dismissal ⇨ 6 months; shorter periods for termination for new workplace: < 22.000 EUR ⇨ 1 month; 22.000 EUR–44.000 EUR ⇨ 3 months > 44.000 EUR ⇨ determination by judge (see above)
Special dismissal protection	member of works council.: in principle dismissal inadmissible: (Act of 19/3/1991); in time: 30 days before notification of election – assumption of office by other member of works council (4-years period); dismissal for

	urgent technical and/or economic reasons (≈ business-related) possible, as far as committee with equal representation of employers/employees (otherwise: labor court) recognizes above mentioned reasons member of labor union: dismissal inadmissible because of membership of or activity in labor union (Collective Agreement of 24/5/1971) pregnant women/mothers: before notification of pregnancy to 1 months after end of maternity leave (=8 or14 weeks after birth) dismissal because of pregnancy/maternity inadmissible military service: inadmissible dismissal for military service (Art. 38 sec. 3 Labor Contract Act) employee, who via complaint or claim remind sex-related equal treatment (Art. 136 Act of 4/8/1978) collective redundancy: (Collective Agreement of 8/5/1973) business-related redundancy in 60 days of 10 employees, in company with 20–100 employees 10 percent of employees, in company with 100–300 employees 300 employees, 30 employees Notification of and advise by company council (if not previously labor union or staff) Collective Agreement N°24 of 2/10/1975; Collective Agreement N° 24 of 21/12/1993; information to director of local employment office, termination only after 30 days possible (royal decree of 24/5/1976) (punishable)
Participation works council	information and advisory right (Art. 15 et subs. Act of 20/9/1948 and Collective Agreement 9/3/1972) in case of determination of general termination directives by employer
Procedure	Act of 10/10/1967 arbitration procedure before labor court procedure Art. 734 Act of 10/10/1967 without justification (especially without reason for justification) compensation Art 39 sec. 1, 62 et seq., 54 Labor Contract Act)

6. Italy

Rule	Codice Civile = CC; Codice di procedura Civile = CPC; L 223/91; Costitutione della Repubblica Italiana = Cost.; Employee Statute = Stat. Lav Contratto Collettivo Nazionale di Lavoro per i Dipendentio di Aziende Commerciali 1/2/1983 = Collective Agreement; 1983; Contratto Collettivo Nazionale di Lavoro per i Dipendentio di Aziende Commerciali 22/12/1981 = Collective Agreement 1981; L 604/66; L 108/90; L 533/73; TV 29/4/1965
Admissible reasons for termination	Ordinary, extraordinary 2118 CC **Employee termination:** ordinary: voluntary termination (personal reasons) 2118 CC; extraordinary: compelling reasons (= employer violation of duty e.g. belated payment of remuneration or social security contributions, insult, request of work performance contra bonos mores) **Employer termination:** (Art. 2118 et seq. CC, L 604/66, Art. 18 Stat

	Lav, L 108/90: (differentiation between subjective and objective reasons for dismissal) [ordinary] dismissal "ad nutum" without reason. Possible: trainees at end of training period, managing employees, Domestic Workers, professional sportsmen, employees over 60 with entitlement to pension benefits, temporary work relations (Art. 2195 f CC Art. 10 L 604/66). [extraordinary] dismissal: (Art. 2119 CC): serious culpable violation of duties endangering or damaging goods of the employer Conduct-related "per giusta causa" Person-related- and business-related: "per giustificato motive": (Art. 3 L 604/66); before procedure for conduct-related dismissal warning, statement of employee; on request with labor union (Art. 7 Stat. Lav.) exemption: disease related termination (work accident, work-related disease) possible with considerable notice period Art. 2110 CC in connection with the company habits collective redundancy: (L 223/91) (min. 5 of 15 employees within 21 days) rightful social selection of employees by employer
Period of notice	minimum regulations under collective agreement; deviating individual contracts. Blue- and White-Collar Employees. Collective Agreement 1983 *Level of remuneration + activities: I super + I*: up to 5 years ⇨ 60 days up to 10 years ⇨ 90 days > 10 years ⇨ 120 days *Level of remuneration + activities: II + III:* up to 5 years ⇨ 30 days up to 10 years ⇨ 45 days > 10 years ⇨ 60 days *Level of remuneration + activities: IV + V:* up to 5 years ⇨ 20 days up to 10 years ⇨ 30 days > 10 years ⇨ 45 days *Level of remuneration + activities: VI + VIII:* up to 5 years ⇨ 15 days up to 10 years ⇨ 20 days > 10 years ⇨ 20 days managing personnel collective agreement 1981 up to 2 years ⇨ 6 months up to 5 years ⇨ 8 months up to 10 years ⇨ 10 months > 10 years ⇨ 12 months illness-related dismissal: employer/employee: 180 days with remuneration, prolongation 120 days without remuneration; managing

	personnel: 12 months with remuneration, extension 6 months without remuneration; work accident dismissal only after recovery
Special dismissal protection	works council member: no express special dismissal protection, Art. 22, Art. 22 Stat. Lav. prohibition "hostile" employer conduct towards works council. Prohibition discriminating dismissal because of exercise of labor union-related activity or membership: Art. 4 L 604/66, Art. 15, 18 Stat. Lav., Art. 13 L 903/77, L 108/90 Reemployment of works council or labor union representatives in dismissal protection procedure by judge (Act 300) marriage: before asking until 1 year after marriage Art. 1 L 7/63: dismissal for reasons of marriage inadmissible and void pregnant women/mothers: begin of pregnancy – first year of age of child, person-related dismissal for pregnancy, maternity inadmissible; 2 months before and 3 months after birth: ordinary dismissal needs consent of labor authority (inspetto del lavoro); exception business-related redundancy (Art. 2 L1204/71) military/civil service: (L 772/72): ordinary termination inadmissible (L 653/40 and L 370/55) collective redundancy: anal. collective agreement 5/5/1965 only with consent of labor union and labor office
Participation of works council	Usually none
Procedure	L 604/66; L 108/90; L 533/73 Collective Agreement 29/4/1965 arbitration process Art 6 L 604/66 Local Court Art. 409 et seq. CPC Rescission of dismissal **period of notice 60 days**; damages in case of non-compliance with notice period Art. 2120 CC (sec. 56 L 297 29/5/1982: cash settlement also in case or rightful dismissal ca.: ½–1 monthly salary.

7. Spain

Rule	Codigo Civil = CC; Estatuto de los Trabajadores (employee statute) 1980 = ET; Ley de Contrato de Trabajo (Employment Contract Act) = LCT; Ley de Procedimiento laboral (Labor Procedure Act) = LPL
Admissible grounds for termination	Art. 49 ET no differentiation ordinary/extraordinary termination; but differentiation (employer-dismissal): disciplinary = subjective Art. 49 subs. 11; Art. 54–56 ET culpable serious violation of work duties [work performance, disciplinary violation, breach of fiduciary duty] (≈ conduct-related); (conduct related ≈ extraordinary) objective Art. 52, 53. = economic, technical, organizational, work

	performance. Impossibility by Force Majeure Art. 51 ET (≈ business-related and person-related) grounds; person-related: lack of appropriateness of employee, also for disease-related impediment Art. 49 subs. 5 ET), lack of adjustment to necessary technical changes, absence times (beyond responsibility; but not legal: strike, activity as representative of employees, work accident, disease, leave, vacation) in 2 succeeding months 20 percent in 4 of 12 months 25 percent, for > 5 percent than average rate of absence time in company; business-related. (if not collective redundancy (see below) economic crisis procedure: written dismissal notification with reason; offer of compensation Art. 53 subs., 53 subs. 1 b ET; no compensation for disciplinary = subjective dismissal Art. 49 subs. 11; Art. 54–56 ET (≈ conduct-related) collective redundancy: (Art. 51; 49 subs. 9, 2c ET) 10 employees for companies with up to 100 employees, 10 percent of employees for 100–300 employees, 30 employees for > 300 employees, permanent production decline and/or economic difficulties, necessary termination of business, Force Majeure *Procedure I (Force Majeure):* Determination of Force Majeure and consent to termination by authority in charge (= Director provincial de trabajo / Director General de Empleo) Art 51 subs. 121 ET; Art. 6.1 BD 696/80 *Procedure II (Force Majeure)*: Notification works council and Employment Office (with employer plan for crisis-management), consultation representatives of employees (30 days); in case of agreement: approval to agreement by authority (no agreement = approval) Art. 51 subs. 5 ET; No agreement = decision of authority over termination by employer
Period of notice	disciplinary = subjective Art. 49 subs. 11; Art. 54–56 ET objective reasons (≈ business-related and person-related) ⇨ 30 days [or compensation payment]
Special dismissal protection	Member of works council Art. 68 c ET: ordinary dismissal in connection with office during activity and one year after inadmissible. For disciplinary termination hearing by and consent of works council Art. 68 a ET Labor union membership: prohibition of discriminating dismissal No explicit dismissal protection for pregnancy and maternity. Legal practice: prohibition of dismissal because pregnancy or maternity (generally 16 weeks suspension of labor contract comp. Art. 45 subs. 1d ET) Military service: suspension of labor contract; reemployment 30 days after end of military service

Participa-tion Works Council	Right to information and hearing Consent in case of disciplinary dismissal by works council members Counseling for employer (30 days) collective redundancy for economic reason
Procedure	Ley de Procedimiento Laboral (LPL) Decree 521/90 27/4/1990 Dismissal protection claim before labor court within **20 days** Art. 130 LPL

8. Portugal

Rules	Art. Constitution, Decreto-Lei 49.408 = Labor Contract Act; Decreto-Lei 64-A/89 27/2/1989 in connection with Decreto-Lei 400/91, 403/91 16/10/1991; DL 64/89 = Dismissal Protection Act
Admissible Reasons for termination	**Employee-termination** ordinary (Art. 38 DL 64-A/89); extraordinary (Art. 35 DL 64-A/89) culpable violation of labor contract duties of the employer **Employer-termination** No differentiation ordinary/extraordinary dismissal only for serious reason and compliance with dismissal process, differentiation subjective (≈ conduct-related) and objective (≈ person-related/ business-related) reasons: business-related: Art. 26 DL 64-A/89 proved reduction of sales, necessary technical reforms or necessary change of business activity person-related = dismissal for non-adjustment of employee to necessary change of workplace; (temporary illness = suspension of labor relation) generally compensation of employee 27 DG 64-A/89) extraordinary dismissal: Art. 10 Nr. 1 DG 64-A/89 conduct-related reason: impossible continuation of labor relation (e.g. counteracting work instructions; violence; opposition against employer; repetitive conflicts with colleagues, unauthorized absence causing damage (in one year 5 continuous days or 10 days in total); considerable reduction of productivity of employee, deception on reason for absence) conduct-related dismissal without warning! Procedure: written with reasons and preliminary talks with employer; notification of employer, works council and supervisory authority by employer within 5 days after intervention employee and/or works council; dismissal: date of termination, reasons for dismissal, evidence for impossibility of further employment, compensation amount and further payment conditions.
Notice period	**Employee termination** Art. 38 DL 64-A/89 < 2 years = 30 days;

	> 2 years = 60 days **Employer termination** Collective redundancy (= company related dismissal for structural, technological and economic reasons already from 2 employee for = 50 employees; 5 employees for = 51 employees (Art. 16 GD 64-A/89) after termination of "consultation procedure" 60 days
Special dismissal protection	Termination because of membership in labor union or works council inadmissible; in case of other admissible dismissal labor union members higher compensation (Art. 58 LCCT in connection with 35 DL 215-B/75; Art. 16 L 46/79) Prohibition of dismissal because of pregnancy, maternity (maternity leave 98 days of which 60 after birth) Handicapped only in case of disablement caused by work accident in company with more than 10 employees, handicap temporary or less than 50 percent Civil service: (Art. 276 subs. 7 suspension of labor contract)
Participation Works Council	Information/counseling right Initiation of procedure before supervisory authority within 3 days Art. 27, 29 GD 64-A/89 (consequence = possibly monetary penalty, not declaration invalidness of dismissal) Collective redundancy: (see above) (Art. 16 GD 64-A/89) consultation employer, works council or labor union and Social Ministry Art. 17 GD 64-A/89
Procedure	Labor court: preliminary conciliation hearing (Art. 54 et seq. Labor Court Act) **Employee termination:** ordinary (Art. 38 DL 64-A/89) application with labor court Dissolution of work relation and consent of compensation of employer **Employer termination** Employee application (within 5 days after notification of employer) before court, temporary suspension of dismissal, (company related dismissal: claim within 1 year Art. 32, 38 GD 49.408; Reemployment or compensation in case of wrongful dismissal Art. 13 GD 64-A/89 Dismissal for non-adjustment of employee to necessary change of workplace (person-related) Conduct-related dismissal: preliminary evidence procurement procedure employer/ labor union (temporary) suspension of dismissal by court aversion of temporary suspension by employer caution money (6 months-salary) Collective redundancy protection claim before labor court – 90 days period of notice (Art. 25 subs. 2 GD 64-A/89)

9. Greece

Rule	Art. 22 Constitution, Act 3198/1955; Act 21/12/1920, Royal Decree 16/18.7.1920
Admissible reasons for termination	Written (Art. 5 subs. 3 law 3198/1955); Notification of termination at local labor office and police within 8 days (no requirement for validity, but omission punishable) Art. 9 Act 3198/1955) **ordinary** termination <u>person-related:</u> incapacity, in case of illness suspension of labor contract, notice period (depending on tenure) 1 month (for 4 years) – 6 months (for > 15 years) <u>conduct-related:</u> (Art. 673 ZGB) compensation by damaging person <u>business -related</u>: serious, provable economic reasons in the sense of change of financial circumstances of employer; social selection (in acc. with jurisdiction no prerequisite for validity of dismissal); appropriate compensation of employee by employer (Art. 674 ZGB) **extraordinary** dismissal for especially serious reason for conduct-related termination (sentence for crime, considerable culpable violation of duty) or for Force Majeure (≈ company related) <u>collective redundancy</u>: company with more than 50 employees. Monthly dismissal for economic/ company-related reasons exceeding percentage determined by employment ministry (2-3 percent); max. 30 employees. Notification to employees and employment ministry (reason, number of employees, selection), attempt to negotiate with employees. If negotiations fail confirmation of dismissal by employment ministry within 10 days (law 1387/1983; 1767/1988); simultaneous dismissal of more than 30 employees inadmissible
Period of Notice	**Employer termination:** Art. 1 Act 2112/1920: 2 months – 1 year ⇨1 month 2–4 years ⇨ 2 months 4–6 years ⇨ 3 months etc. max.: 28 years ⇨ 24 months agreements on longer periods possible for <u>employees</u>: no notice period in case of compensation payment 2 months – 1 year ⇨1 monthly salary 1–2 years ⇨ 2 monthly salaries 2–3 years ⇨ 3 monthly salaries etc. ≥ 28 years ⇨ 24 months **Employee termination** ½ notice period of employer termination, max. 3 months Employee compensation (½ usual compensation (see below) for termination over 15 years activity and/or pension age (Art. 5 subs. 1 Act 435/1976)

Special dismissal protection	labor union members (anal. for members of works council): inadmissible ordinary dismissal during office and 1 year after. Extraordinary dismissal with consent of special committee (1 judge, 1 representative employee + 1 representative employer) Art. 15 Act 1264/1982 inadmissible any unlawful discriminating dismissal Art 281 ZGB; Act 1767/1988; concerns: activity in labor union, employees: demanding support by labor union/ works council pregnancy/maternity: inadmissible dismissal during pregnancy and maternity (up to 1 year after birth) military service: after 6 months tenure principally inadmissible also 1 year after resumption of work (dismissal admissible if "special committee" consents (Art. 7 Act 244/1936) war veteran and handicapped ex-service person and family members of persons killed in war : dismissal only for defamation or incapacity for certain work position and impossibility of employee to continue to employ person (Art. 7 Act 244/1936)
Participation works council	Notification by employer (as substitute for lack of labor union: information and counseling)
Procedure	claim (exclusion) period against termination= 3 months; for compensation = 6 months (Art. 5 subs. 3 Act 3198/1955) claim to compensation of the dismissed employee (Art. 2 subs. 1 Act 3198/1955) (to be made available by employer in fund) except conduct-related dismissal or unlawfully provoked compensation; Force Majeure, possible 2/3 insurance sum; ½ after bankruptcy (Art. 2, 6 Act 3198/1955) compensation different for white-collar employees (Art. 4, 5 subs. 1 law 3198/1955) and blue-collar employees Art. 5 subs. 1 Royal Decree 16/18 7. 1920

10. England

Rule	Employment Protection (Consideration) Act 1978) = EPA; Trade Union and Labor Relations (Consolidation) Act (1992) = TULRA; Employment Rights Act. (1996) = ERA; Employment Relations Act (1999) ErelA
Admissible reasons for dismissal	ordinary, extraordinary dismal (Summary Dismissal) [as long as dismissal not discriminating; no dismissal protection for employment under 2 years] person-related: sec. 57.2, 57.2 EPA: loss of capacity, incapability, disease (in case of repeated absence from work and no other employment possibility) conduct-related: (e.g.: theft, acceptance of bribe money, insobriety and drug abuse at workplace, defamation, betrayal of company secrets, unauthorized access to data bases, unauthorized absence from workplace,

	absence for unagreed vacation, inappropriate work cloths) business-related: (*Redundancy*): durable serious economic reasons, rightful social selection under documentation and consistent application of all criteria; also if further employment breach of law, if employer cannot further legally employ employee condition for all dismissal reasons = compliance with special procedure by employers "Fairness of Procedure" Hearing of employee, with assistance, equal treatment, consideration of all circumstances (constitution of employee and his former performance) in case of conduct-related dismissal renewed misconduct despite warning, alternative disciplinary possibilities exhausted, extraordinary dismissal for essential unjustified violation of contractual duties (breach of contract) with fundamental impairment of mutual trust.
Period of notice	< 1 months ⇨ – > 1 months ⇨ 1 weak 2 years ⇨ 2 weeks 3 years ⇨ 3 weeks each further year of occupation a further weak of notice period until 12 years ⇨ 12 weeks
Special dismissal protection	employee representatives: exclusion of dismissal of member of independent labor union, participation (intent) in activities of labor union sec. 152 TULRA pregnancy/ maternity: sec. 60 EPA: dismissal for pregnancy/ maternity inadmissible. But dismissal in the sense of suspension possible in case of health risks of job or night turns, and not other employment of employee possible
Participation Works Council	Right of employee to assistance by labor union which represents majority of employees in company Information and counseling duty of employer with labor union or company employee representatives in case of redundancy dismissals. collective redundancies: (dismissals of more than 20 employees in 90 days) consultation with employee representatives (generally labor union)
Procedure	Unfair Dismissal Claim (for unjustified dismissal or violation of procedure) wrongful dismissal (for non-compliance with notice period or lack of notification of employee) Redundancy Payment Claim payment of a workplace compensation sum in case of admissible redundancy (not disappearance of possibility to use employee for technical reasons or if a employee does not fulfill an admissible alternative of employment) sec. 81 et seq., 87 EPA) **claim exclusion period**: generally 3 months

	extension of period possible in case of illness of employee, wrong information of employee by labor or local court, belated notification, new evidence

11. Ireland

Rule	Unfair Dismissal Act (1977)= UDA; Minimum Notice and Terms of Employment Act (1973) = MNTEA; Maternity Protection of Employees Act (1987) = MPEA; Protection Employment Act (1977) = PEA; Parental Leave Act = PLA; Worker Protection Act = WPA
Admissible reasons for termination	ordinary, extraordinary (Summary Dismissal) sec.6.4 a-d UDA person-related: incapacity, permanent disease sec 2.4 a UDA conduct-related: culpable violation of duty to work and ancillary duties business-related: e.g.: use of capacity too low, rationalization, restructuring duty to pay compensation if employee min. 2 years continuously employed Also breach of law in case of further employment of employee if employer cannot legally continue to employ employee. Condition dismissal = compliance with special procedure by employer "Fairness of Procedure" Hearing of employee, if necessary with assistance, equal treatment, consideration of all circumstances (constitution of employee and former activity); in case of conduct-related new misconduct despite warning and alternative disciplinary measures exhausted collective redundancy: (dismissals in 30 days: 5 employees for 20–50 employees; 10 employees for 50-100 employees; 10 percent for 100–300 employees; 30 > 3000 employees) sec. 6.1 PEA: 30 days before dismissal notification to Labor Ministry (sec. 12 PEA) and information to and counseling of employee representative (generally labor union) and consultation works council
Period of notice	sec. 4.2. MNTEA 13 weeks – 2 years ⇨11 weeks 2–5 years ⇨2 weeks 5–10 years ⇨ 4 weeks 10–15 years ⇨ 6 weeks > 15 years ⇨ 8 weeks
Special dismissal protection	Dismissal for membership and activity in labor union inadmissible (sec. 14 UAD) pregnancy: sec. 6 PLA maternity leave (14 weeks); exemption of dismissal (sec. 2.2 UDA)

Participation of works council	Information and hearing in collective redundancies
Procedure	recommendation of "Rights Commissioners" (6 Weeks) appeal to Employment Appeals Tribunal sec. 8-10 UDA (or directly to Employment Appeals Tribunal); sec. 8.2 UDA **notice period**: against ordinary termination 6 months against extraordinary dismissal 12 months

12. Denmark

Rule	Work Condition Act (1976) = AML; Employees Act (1971) = FUL; Equal Treatment Act (1990) = LBHL; Bankruptcy Act = KKL
Admissible reasons for termination	ordinary, extraordinary sec. 26 MHL <u>conduct-related</u>: culpable misconduct of employee; <u>person-related</u>: lasting incapacity, [disease sec. 2 FUL, sec. 16 MHL, disease over 1 month]; <u>business-related</u>: [Force Majeure], serious economic and technical difficulties (notification of reasons for dismissal only after 9 months) extraordinary dismissal by employer: e.g.: crime of employee, causing considerable damages in violation of labor duties sec. 20 22 MHL; unauthorized absence with serious economic consequences for employer termination by employee: default to pay remuneration in case of previous warning by employee, sec. 26 MHL serious company-related beyond responsibility "Force Majeure" (e.g. destruction of company)
Period of Notice	Act (sec. 20 et seq. MHL) and collective agreement Differentiation blue-collar and white-collar worker: **blue-collar worker**: without collective/ individual agreement period of notice of white-collar workers considered as appropriate (= 3 months) employer dismissal: depending on collective agreement up to 9 months ⇨14–21 days 3–6 years ⇨49 days > 6 years ⇨70 days employees over 50 > 9 years ⇨ 90 days > 12 years ⇨120 days employee termination: depending on collective agreement from 9 months ⇨7 days 3–6 years ⇨14 days 6–9 years ⇨21 days

	> 9 years ⇨ 28 days **White-collar employees** (sec. 2.1–2.4, 2.7, 6) FUL): up to 6 months ⇨ 1 months from 6 months ⇨ 3 months 3–6 years ⇨ 4 months 6–9 years ⇨5 months > 9 years ⇨ 6 months
special dismissal protection	members of labor union and works council (depending on collective contract = exclusion of ordinary dismissal, extension of notice period) (sec. 2, 10 FUL; sec. 10.2 AML; 1–5 Dismissal Protection Act of Labor Union Members maternity inadmissible dismissal for pregnancy and maternity (sec.16, 7, 9 Adjustment LBHL; 7 FUL; sec. 19a MHL; sec. 58 EUL); compensation for violation up to 78 weekly salaries trainees exclusion dismissal sec. 60 EUL military / civil service exclusion of employer-dismissal after 9 month tenure (sec. 1, 2 VPL; sec. 6 FUL; sec. 35 MHL) collective redundancy: (min. 10 employees within 30 days in company with 20–100 employees 10 percent of employees in company with 100–300 employees 30 employees in company with min. employees (notification employment office) sec. 2.6 FUL notification of and negotiations with employee representatives and employment office
Participation Works Council	Information and counseling duty
Procedure	With participation of labor union (arbitration court procedure) negotiation over reasons of dismissal and dismissal decision: Board of Dismissal (as arbitrary court) otherwise claim before court before possibility: (compensation, cash settlement)

13. Sweden

Rule	Work Protection Act 1982
Admissible reasons for termination	**employee termination**, ordinary.: sec. 11 Work Protection Act; extraordinary employee termination serious misbehavior by employer (e.g. insult, sexual assault, default of payment of salary employee only right to refuse work) **employer dismissal**: written notice (sec. 8, 16, 19 Work Protection Act) and on demand by employees objective reasons (sec. 7 Work Protection Act): person-related: incapacity, alcohol: danger to safety and/ or image of employer; disease only permanent, essential reduction of capacity to work

	conduct-related.: repeated, gross violation of behavioral duties of employee towards employer and/or colleagues business-related.: provable durable lack of work, social selection (sec. 22 Work Protection Act), [principle: "*last in / first out*"] claim for reemployment of employee sec. 25 et seq. Work Protection Act); employees with longer tenure: preferred reemployment - general no dismissal if further employment in other position possible extraordinary dismissal (sec. 18 subs. 1 Work Protection Act) [culpable, considerable] gross violation of duty in the sense of crimes (e.g.: concealment, violence) procedure: person-related dismissal: information to employees 2 weeks before dismissal, in case of existence of trade unions information also to trade union. On demand of employee and/or trade union ⇨ explanatory hearing (sec. 30 Work Protection Act) company-related: duty to negotiate (sec.11 et seq. Work Protection Act)
Notice Period	**employee termination**: 1 month before declaration of termination (sec. 11 Work Protection Act), prolongation by collective agreement possible; employee period of notice shorter than employer period of notice **employer termination**: min. 1 month after 6 months or continuous 12 months in 2 years. Period depends on age of employee (sec. 11 Work Protection Act): from 25 years ⇨ 2 months from 30 years ⇨ 3 months from 35 years ⇨ 4 months from 40 years ⇨5 months from 45 years ⇨ 6 months
Special Dismissal Protection	Trade Union Members previous negotiation trade union – employer (sec. 8 Co-Determination Act) Pregnancy/ Maternity (sec. 25 Equal Treatment Act) Civil service: inadmissible dismissal (Work Protection Act 1939) dismissal inadmissible for ethic reasons (sec. 7 Work Protection Act, lag 1994:134) collective redundancies: (5 employees in case of dismissal for lack of work) notification of local employment office 6 months before dismissal; duty of employer to negotiate with trade union before measure is taken (sec. 11–14 Co-Determination Act)
Participation of council	Consultation with trade union (if employees organized) (sec. 10, 15 et seq. Work Protection Act)
Procedure	District Court Dismissal suspension (sec. 34 Work Protection Act); period: 2 weeks (sec. 40 Work Protection Act) in case reasons for dismissal not stated (on demand of employee) 4 weeks in case of unlawful or void dismissal compensation claim (sec. 38 Work Protection Act) until reemployment of employee (ordinary termination

	2.200 EUR; extraordinary. = 3.300 EUR employer – also invalid dismissal – right to rescind labor contract with cash settlement (sec. 39 Work Protection Act)

14. Finland

Rule	Työsopismuslaki (320/1970) Labor Contract Act Laki ammatillisesta koulutusesta (630/1998) Training Act Laki palvelukseen kutsutun asevelvollisen työ tai virkasuhteen jatkumisesta (570/1961) Workplace Protection Act Co-Determination Act
Admissible reasons for termination	ordinary Art. 38 Labor Contract Act; extraordinary Art. 43 Labor Contract Act person-related (Art. 37 subs. 2 Labor Contract Act) [disease: considerable durable deterioration of work results; jurisdiction duration of disease 1 year] conduct-related (Art. 43 Labor Contract Act) deceit when entering into labor contract, culpable endangering jeopardizing at workplace Gross violation, violation non-competition clause or company secrets; mal-performance, after corresponding warning by employer company-related (Art. 37 a, 40 et seq. Labor Contract Act) essential permanent reduction of work and impossibility of employment of employee at other workplace (employer duty consideration of dismissed person in planned new employments Art. 42 a Labor Contract Act) company restructuring: as far as compulsory to avoid bankruptcy (termination period 2 months) Extraordinary Art. 43 Labor Contract Act. If serious reasons make compliance with notice period impossible
Period of Notice	Dependent on agreement in individual contract otherwise (employer-dismissal min.) up to 1 year up to 1 year. ⇨ 1 months 1–5 years. ⇨ 2 months 6–8 years ⇨ 3 months 9–11 years ⇨ 4 months 12–14 years ⇨ 5 months > 15 years ⇨ 6 months
Special dismissal Protection	<u>Employees of works council</u>: Art. 53 subs. 2 Labor Contract Act: dismissal only with consent of majority of staff possible <u>Pregnancy/ Maternity</u> Art. 37 subs. 5 Labor Contract Act: inadmissible dismissal on grounds of pregnancy or to avoid maternity leave <u>handicapped persons:</u> Art. 37 subs. 2, 17 subs. 3 Labor Contract Act: Only special consideration in case of selection of several employees to be dismissed

	Military and civil service Art. 2 Workplace Protection Act: exclusion dismissal during service Trainees Art. 17, 18 Training Act: exemption ordinary dismissal
Participation Works Council	Information and counseling right under Co-Determination Act Collective redundancy: Art. 37a Labor Contract Act (min 10 among 20 employees concerned) Hearing and consultation of employee representatives
Procedure	Possibility of statement of employee Art. 47 b Labor Contract Act Dismissal Protection Claim (District Court) **period for claim** = 2 years Art. 47 e subs. 1 Labor Contract Act possibility for compensation payment (Art. 47 et seq. Labor Contract Act)

Table 58.1: Termination possibilities in the individual labor law systems of European member states

58.4 Consequences of "technical work protection" in Europe on labor relations

Especially comprehensive labor protection regulations imply high costs for the employer. The European legislator created numerous rules in this field of law.[849]

The European legislator abandoned its original legislative method in the field of European technical work protection.[850] This method consisted in giving detailed information with regard to the admissibility of certain tools (pressure containers, floor-borne vehicles, lifting vehicles, safety devices in construction machines) in the European source of law itself.[851] The implementation of such directives, however, made rather slow progress. Furthermore, detailed norms bear the danger to be unable to react appropriately to future innovations and deviations.[852]

The European Legislator has taken its decision in favor of the following systematic: General minimum requirements – which apply equally to any labor relation – form the basic elements of the European technical work protection. Special regulations complete these general principles.

General rules of European technical work protection are:

---

849 Details in: Krimphove, 2001, p. 231 et seq. (with further references)

850 Comp. Council Resolution of 7 May 1985, OJ C1985, N° 136, p. 1; Joerges, Festschrift für Steindorff, p. 1252 et seq. (with further references)

851 Wlotzke, NZA 1990, p. 418 (with further references)

852 Comp. BVerfG (Federal Consitutional Court), NJW 1979, p. 359

- *Council Directive 92/104/ EEC of 3 December 1992 on the minimum requirements for improving the safety and health protection of workers in surface and underground mineral-extracting industries (twelfth individual Directive within the meaning of Article 16 (1) of Directive 89/391/EEC)*[853]
- *Council Directive 89/391/EEC of 12 June 1989 on the introduction of measures to encourage improvements in the safety and health of workers at work* [854/855/856]

---

853 OJ L 404 of 31/12/1992, p. 10 et seq.

854 OJ 1989, N° L 183, p. 1

855 The following rules are based on these „General Provisions“:

-Council Directive 89/654/EEC of 30 November1989 concerning the minimum safety and health requirements for the workplace (first individual directive within the meaning of Article 16 (1) of Directive 89/391/EEC) (OJ 1989, N° L 393, p. 1)

- Council Directive 89/655/EEC of 30 November 1989 concerning the minimum safety and health requirements for the use of work equipment by workers at work (second individual Directive within the meaning of Article 16 (1) of Directive 89/391/EEC) (OJ 1989, N° L 393, p. 13)

-Council Directive 89/656/EEC of 30 November 1989 on the minimum health and safety requirements for the use by workers of personal protective equipment at the workplace (third individual directive within the meaning of Article 16 (1) of Directive 89/391/EEC) (89/656/EEC)

-Council Directive 90/679/EEC of 26 November 1990 on the protection of workers from risks related to exposure to biological agents at work (seventh individual Directive within the meaning of Article 16 (1) of Directive 89/391/EEC) (OJ 1990, N° L 374, p. 1)

-Council Directive 93/88/EEC of 12 October 1993 amending Directive 90/679/EEC on the protection of workers from risks related to exposure to biological agents at work (seventh individual Directive within the meaning of Article 16 (1) of Directive 89/391/EEC) (OJ 1993, N° L 268, p. 71)

- Council Directive 92/57/EEC of 24 June 1992 on the implementation of minimum safety and health requirements at temporary or mobile constructions sites (eighth individual Directive within the meaning of Article 16 (1) of Directive 89/391/EEC) (OJ 1992, N° L 245, p. 6)

-Council Directive 92/58/EEC of 24 June 1992 on the minimum requirements for the provision of safety and/or health signs at work (ninth individual Directive within the meaning of Article 16 (1) of Directive 89/391/EEC)

-Council Directive 92/85/EEC of 19 October 1992 on the introduction of measures to encourage improvements in the safety and health at work of pregnant workers and workers who have recently given birth or are breastfeeding (tenth individual Directive within the meaning of Article 16 (1) of Directive 89/391/EEC) (OJ 1992, N° L 348, p. 1)

-Council Directive 92/91/EEC of 3 November 1992 concerning the minimum requirements for improving the safety and health protection of workers in the mineral-extracting industries through drilling (eleventh individual Directive within the meaning of Article 16 (1) of Directive 89/391/EEC)

-Council Directive 92/104/ EEC of 3 December 1992 on the minimum requirements for improving the safety and health protection of workers in surface and underground mineral-extracting industries (twelfth individual Directive within the meaning of Article 16 (1) of Directive 89/391/EEC) (OJ 1992, N° L 404, p. 10)

-Council Directive 93/103/EC of 23 November 1993 concerning the minimum safety and health requirements for work on board fishing vessels (thirteenth individual Directive within the meaning of Article 16 (1) of Directive 89/391/EEC) (OJ 1993, N° L 307, p. 1)

856 Directive 89/391/EEC is not yet implemented into national labor law by the German legislator. The implementation by a „Work Protection Act“ is being prepared, see below, part II. Fig. XVI a

- *Council Directive 80/1107/EEC of 27 November 1980 on the protection of workers from the risks related to exposure to chemical, physical and biological agents at work*[857]

- *Council Directive of 16 December 1988 amending Directive 80/1107/EEC on the protection of workers from the risks related to exposure to chemical, physical and biological agents at work (88/642/EEC) (88/642/EEC)* [858,859,860]

Besides the European legislator had implemented numerous **special** work protection norms:

These norms refer to the dangers of a special activity (**production procedure**):

- *Council Directive 78/610/EEC of 29 June 1978 on the approximation of the laws, regulations and administrative provisions of the Member States on the protection of the health of workers exposed to vinyl chloride monomer* [861]

- *Council Directive 90/269/EEC of 29 May 1990 on the minimum health and safety requirements for the manual handling of loads where there is a risk particularly of back injury to workers (fourth individual Directive within the meaning of Article 16 (1) of Directive 89/391/EEC)*[862]

or to dangers at the **workplace**:

---

857 OJ 1980; N° L 327, p. 8; revised version: OJ 1995, N° L 1, p. 1

858 OJ 1988, N° L 356, p. 74

859 The following rules are based on these „General Provisions“:

-Council Directive of 28 July 1982 on the protection of workers from the risks related to exposure to metallic lead and its ionic compounds at work (first individual Directive within the meaning of Article 8 of Directive 80/1107/EEC) (82/605/EEC (OJ 1982, N° L 247, p. 12)

-Council Directive 83/477/EEC of 19 September 1983 on the protection of workers from the risks related to exposure to asbestos at work (second individual Directive within the meaning of Article 8 of Directive 80/1107/EEC) (OJ 1983, N° L 263, p. 25)

-Council Directive 91/382/EEC of 25 June 1991 amending Directive 83/477/EEC on the protection of workers from the risks related to exposure to asbestos at work (second individual Directive within the meaning of Article 8 of Directive 80/1107/EEC)(OJ 1991, N° L 206, p. 16)

- Council Directive 88/364/EEC of 9 June 1988 on the protection of workers by the banning of certain specified agents and/or certain work activities (Fourth individual Directive within the meaning of Article 8 of Directive 80/1107/EEC) (OJ 1988, N° L179, p. 44)

-Commission Directive 91/322/EEC of 29 May 1991 on establishing indicative limit values by implementing Council Directive 80/1107/EEC on the protection of workers from the risks related to exposure to chemical, physical and biological agents at work (OJ 1991, N° L 177, p. 22)

860 The German legislator implemented Directive 80/1107/EEC with the Gefahrstoffverordnung (Regulation on Hazardous Material) of 26/8/1986 BGBl (German Federal Gazette) I, p. 1470) and the Second Regulation on the Amendment of the Regulation on Hazardous Material of 23/4/1990 into valid German law; see above, part II. Fig. XVI a

861 OJ 1978, N° L 197, p. 12

862 OJ 1990, N° L 156, p. 9

- *Council Directive 86/188/EEC of 12 May 1986 on the protection of workers from the risks related to exposure to noise at work*[863]

- *Council Directive 92/29/EWG of 31 March 1992 on the minimum safety and health requirements for improved medical treatment on board vessels*[864]

- *Council Directive 92/57/EEC on the implementation of minimum safety and health requirements at temporary or mobile construction sites (eighth individual Directive within the meaning of Art. 16 (1) of Directive 89/391/EEC)*[865]

- *Council Directive 83/477/EEC of 19 September 1983 on the protection of workers from the risks related to exposure to asbestos at work (second individual Directive within the meaning of Article 8 of Directive 80/1107/EEC)*[866]

- *Council Directive 91/382/EEC of 25 June 1991 amending Directive 83/477/EEC on the protection of workers from the risks related to exposure to asbestos at work (second individual Directive within the meaning of Article 8 of Directive 80/1107/EEC))*[867]

or

to the existence of special dangers from the **tools**:

Council Directive 90/270/EEC of 29 May 1990 on the minimum safety and health requirements for work with display screen equipment (fifth individual Directive within the meaning of Article 16 (1) of Directive 89/391/EEC)[868]

Council Directive 89/392/EEC of 14 June 1989 on the approximation of the laws of the Member States relating to machinery[869]

Council Directive 91/368/EEC of 20 June 1991 amending Directive (89/392/EEC) on the approximation of the laws of the Member States relating to machinery[870]

Council Directive 93/44/EEC of 14 June 1993 amending Directive 89/392/EEC on the approximation of the laws of the Member States relating to machinery[871]

Council Directive 93/68/EEC of 22 June 1993 amending Directive 89/392/EEC (machinery)[872]

863 OJ 1986, N° L 137, p. 28

864 OJ 1992, N° L 113, p. 19

865 OJ 1992, N° L 245, p. 6

866 OJ 1983, N° L 263, p. 25

867 OJ 1991, N° L 206, p. 16

868 OJ 1990, N° L 156, p. 14

869 OJ 1989, N° L 183, p. 9

870 OJ 1991, N° L 198, p. 16

871 OJ 1993, N° L 175, p. 12

Council Directive 82/605/EEC of 28 July 1982 on the protection of workers from the risks related to exposure to metallic lead and its ionic compounds at work (first individual Directive within the meaning of Article 8 of Directive 80/1107/EEC)[873]

The European legislator currently pays special attention to work protection in **road traffic** and **transportation of dangerous substances in road traffic:**

- *Amending Commission Proposal (Com [93] 421 fin. – SYN 420) of 1 October 1993 for a Council Directive concerning minimum standards for safety and health protection in transport activities as well as at the workplace in vehicles*[874]

- *Commission Proposal (Com [91] 4 fin.) of 19 June 1991 for a Council Directive on the appointment of an officer for the prevention of the risks inherent in the carriage of dangerous goods in undertakings which transport such goods*[875]

Also **international treaties** determine European rules on the "technical work protection" on roads in this context:[876]

- *European Agreement concerning the Work of Crews of Vehicles Engaged in International Road Transport (AETR) of 1 July 1970*[877]

The directives of the European danger protection law are, however, only reluctantly implemented into national law by the national legislators.

58.5 Operational impacts of European "social" work protection

The European Labor Law has to overcome many different national ideas in the area of "social work protection":

Only three essential sets of rules could be issued so far:

- *Council Directive 75/129/EEC of 17 February 1975 on the approximation of the laws of the Member States relating to collective redundancies* [878]

---

872 OJ 1993, N° L 220, p. 1

873 OJ 1982, N° L 247, p. 12

874 OJ 1993, N° C 294, p. 4

875 OJ 1991, N° C 185, p. 5

876 On the position of European Labor Law in the system of public and private international law see above, Part I. Chapter D. I.

877 BGBl (German Federal Gazette) II 1985, p. 889

- *Council Directive 92/56/EEC of 24 June 1992 amending Directive 75/129/EEC on the approximation of the laws of the Member States relating to collective redundancies* [879]

- *Council Directive 77/187/EEC of 14 February 1977 on the approximation of the laws of the Member States relating to the safeguarding of employees' rights in the event of transfers of undertakings, businesses or parts of businesses*[880]

- *Council Directive 80/987/EEC of 20 October 1980 on the approximation of the laws of the Member States relating to the protection of employees in the event of the insolvency of their employer* [881]

Especially the Council Directive 77/187/EEC of 14 February 1977 on the approximation of the laws of the Member States relating to the safeguarding of employees' rights in the event of transfers of undertakings, businesses or parts of businesses[882] has lead the European Court of Justice to making various and controversial judgments:[883]

Christel Schmidt, European Court of Justice of 14 April 1994 (Case C-329/92) Christel Schmidt vs. Spar- und Leihkasse der früheren Ämter Bordesholm, Kiel and Cornshagen, ECR I 1994, p. 1311

The Spar- und Leihkasse dismissed Ms. Schmidt who was employed as only cleaning worker in the subsidiary in Wacken. The Spar- und Leihkasse stated as reason for the dismissal that a constructional enlargement of the subsidiary in Wacken was planned and that after that, the necessary cleaning work should be transferred to an independent cleaning business.

The cleaning company offered Ms. Schmidt the conclusion of a labor contract after termination of construction works in the Spar- und Leihkasse Kiel und Cornshagen. There should not be any essential changes with regard to the working tasks of Ms. Schmidt. Ms. Schmidt was meant to fulfill the same cleaning tasks in the premises of the Spar- and Leihkasse Kiel and Cornshagen for her new employer.

Ms. Schmidt rejected the offer, because the cleaning area in the subsidiary Wacken had grown after the reconstruction and turned against her former employer Spar- und Leihkasse because of her dismissal. Ms. Schmidt referred to the prohibition of dismissal in connection with transfers of undertakings.

The European Court of Justice assumed a transfer of undertakings if (only) partial functions (here: cleaning tasks) which were carried out by an individual employee are contractually transferred to another employer.

---

[878] OJ 1975, N° L 48, p. 29

[879] OJ 1992, N° L 245, p. 3

[880] OJ 1977, N° L 61, p. 26

[881] OJ 1980, N° L 283, p. 23

[882] OJ 1977, N° L 61, p. 26

[883] Comp. Krimphove, 2001, p. 363 et seq., 373 et seq. (with further references)

As the "transfer" of Ms. Schmidt to the cleaning company was a case of transfer of undertakings in the sense of Art. 1 of Directive 77/187/EEC, the European Court of Justice affirmed the prohibition of dismissal in favor of Ms. Schmidt.

Ms. Schmidt's former employer, Spar- and Leihkasse, was not allowed to dismiss Ms. Schmidt because of the transfer of cleaning work to a cleaning company in accordance with Art. 4 subs. 1 Directive 77/187/EEC.

With this judgment the European Court of Justice recognized the so called "**functional succession**" as element of change of company identity and therefore as individual form of transfer of undertakings in the sense of Directive 77/187/EEC.[884]

This jurisprudence of the European Court of Justice would have the consequence that any company transfer of tasks to other, individually managed and organized company units, would constitute a transfer of undertakings (outsourcing).

Now the European Court of Justice limits this broad interpretation in its decision in the case Ayse Süzen again:

Ayse Süzen, European Court of Justice 11 March 1997 (Case C-13/95) Ayse Süzen vs. Zehnacker-Gebäudereinigung GmbH Krankenhausservice

The Zehnacker Gebäudereinigung GmbH used her employee Ms. Ayse Süzen for cleaning a secondary school. As the operator of the secondary school terminated the cleaning contract between the Zehnacker-Gebäudereinigung GmbH and the operator and assigned the cleaning tasks to another company instead, the Zehnacker Gebäudereinigung GmbH dismisses Ms. Süzen.

Ms. Süzen assumed her dismissal to be ineffective because the change of cleaning companies constitutes a business transfer, which would give her a special dismissal protection in the sense of Art. 4 subs. 1 of directive 77/187/EEC.

The European Court of Justice rejected a transfer of undertakings in the sense of Directive 77/187/EEC and therefore a special dismissal protection. Charging another company with a certain task would constitute a change of contractual partners. In principle, this would not affect the identity of the business. Therefore this would not be a case of transfer of undertakings.

58.6 Collective European Labor Law

Even though the European Labor Law recognizes the **right to coalition** and **collective actions** of employees and **their co-determination rights** in company issues in principle,[885]

[884] See also already: European Court of Justice 12 November 1992 (case C-209/91) Anne Watson Rask, Kirsten Christensen vs. ISS Kantinenservice A/S, ECR I 1992, p. 5755 mn 15

up to now only questions regarding European law on **collective contracts**[886] and the works constitution of European-wide operating companies[887] could be bindingly clarified:

Subject of norm	*Norm*
Formation of coalitions	Art. 11 of the Convention of the Protection of Human Rights and Fundamental Freedoms[888], Art. 5 of the European Social Charter of 18 October 1961[889], N° 11 of the Community Charter of the Fundamental Social Rights of Workers of 9 December 1989[890]. Workers from another European member state may participate in the works council activities of their place of employment actively or by making use of their voting rights. Art. 8 of Regulation 1612/67/EEC on the freedom of movement of workers in the Community.[891]
Entering into commitments from collective contracts	Art. 6 European Social Charter[892]; Art. 4 of the Agreement on Social Policy
Freedom of strike	Art. 6 N° 3 of the European Social Charter (of 18 October 1961) and N° 13 of the Community Charter of Social Fundamental Rights of workers (of 9 December 1989)

Table 58.2: Rules of the "collective" European Labor Law

885 See in detail: Krimphove, 2001, p. 387 et seq., 407 et seq., 412–414, 421–423 (each with further references)

886 Ibid, p. 390 (with further references)

887 Ibid, p. 395 et seq. (with further references)

888 BGBl (German Federal Gazette) II 1952, p. 686, 953; comp. European Court of Human Rights 27 October 1975 (file number 1/1974/12/19 Syndicat national de la police belge vs. Belgian State; Publications de la Cour Européenne des Droits de l' Homme, Série A, Vol. 19.; European Court of Human Rights 13 August 1981 and 18 October 1982 (file number 2/1980/33/49-50) Young, James and Webster vs. United Kingdom, Publications de la Cour Européenne des Droits de l'Homme, Série A, Vol. 44

889 BGBl (German Federal Gazette) II 1964, p. 1262

890 Com [89] 248 fin.

891 OJ 1968, N° L 257, p. 1; on the function of this regulation see above.

892 On the problems regarding European collective agreements comp. Krimphove, 2001, p. 390 et seq. (with further references)

58.6.1 European law on collective contracts

Art. 3. and especially Art. 4 of the Convention of Social Policy provide the social partners – in the context of European Primary Law – with comprehensive rights in the creation of European Labor Law regulations.[893] The social partners have a right to be heard in any legislative procedure of the European legislator (Art. 3 Social Policy Convention). The social partners are also entitled to submit themselves– by way of collective agreements – to labor law regulations (Art. 4 Social Policy Convention).

The legal character of these regulations, however, is controversial.[894] The wording and the teleological interpretation of Art. 4 Social Policy Convention indicate – in the author's opinion – only a contractual and not a normative effect of the agreements made between the social partners.[895] Therefore, the legal situation after the conclusion of the Social Policy Convention does not exceed the regulation of Art. 118 EC. Art. 4 Social Policy Convention has not introduced a "real" European collective contract.

58.6.2 European Works Constitution Law

In practice the "European Works Constitution Law" is far more important for the company work relations:

With the

- Council Directive 94/45/EEC of 22 September 1994 on the establishment of a European Works Council or a procedure in Community-scale undertakings and Community-scale groups of undertakings for the purposes of informing and consulting employees

the European legislator provides the possibility of employee representation in Community-scale undertakings.

---

893 See Krimphove, 2001, p. 39 et seq. (with further references); 390 et seq. (with further references)

894 See Krimphove, 2001, p. 390 et seq. (with further references)

895 Ibid, p. 390 et seq.

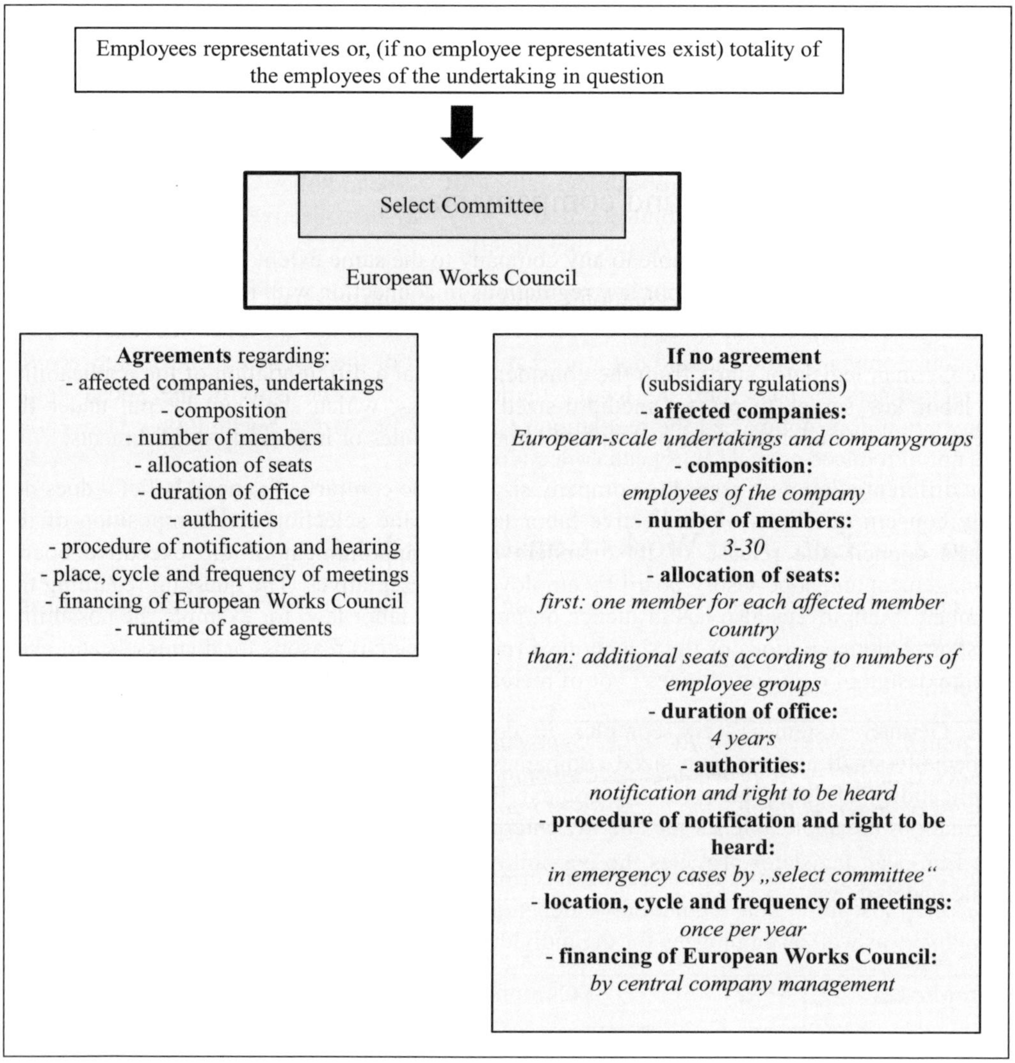

Fig. 58.1: European Works Council

The "European Works Council" is above all an information and advisory board.[896] There are no own, real co-determination rights of the European Works Council at all.

The establishment and organization of the European Works Council is primarily a matter of negotiation between the employer and the employees. If the negotiations do not come to any conclusion, the minimum requirements on the organization of the European Works Council in accordance with the European Works Council Directive apply. Often the establishment of

[896] Comp. Krimphove, 2001, p. 396 et seq. (with further references)

the European Works Council resorts to the existence of diverse national regulations.[897] Especially the combination of negotiation rules, national Works Constitution Acts and the European Works Council Regulation opens the possibility to strategically influence the functionality of the European Works Council.[898]

58.6.3 Labor Law and company size

Labor law rules are not applicable to any company to the same extent. The German legislator differentiates the validity of labor law regulations in connection with the size of the company and the number of the regularly active employees of an undertaking or a company

The German legislator starts from the consideration that a differentiation of the applicability of labor law serves to protect medium-sized business, which shall not be put under the pressure or not under the same pressure of large companies or industrial establishments.

The differentiation with regard to company size – to the contrary of general belief – does not only concern questions of "collective labor law", as the selection and composition of the works council, the release of its members or the appointment of the executive board, management and supervisory board by employee representatives. The question regarding the number of employees also has influence on individual labor law, for example the possibility to shorten notice periods or the selection of (person-related) reasons for dismissals and even on the design of company notices[899] or of recreation rooms of employees.

The German system is very complex in contrast to other European member states.[900] Especially small and medium-sized companies partially have no knowledge on the diverse applicability of labor law. In this context the question is: when and which labor law regulations are applicable for me and my enterprise?

The following overview shall close this informational gap and shall give a compact overview over the most important regulations – depending on the number of persons employed – to employers as well as employees for the individual areas of personnel management.

Employees	Area	Content of Regulation	Rule
1 to 20	Dismissal	Notice periods can be shortened	sec. 622 subs. 5, sentence 1 BGB
1 to 20	Health insurance	Duty of Continued Pay Insurance	sec. 10 subs. 1 LohnForzG

[897] See Krimphove, 2001, p. 406 (with further references)

[898] See Krimphove, 2000, p. 505 et seq.

[899] On the general comprehensive duty of notices comp. also: sec. 8 TVG (collective agreements); sec. 11 (minimum labor conditions). These obligations apply independent from the number of employees in a company

[900] Comp. Krimphove, 2001

Employees	Area	Content of Regulation	Rule
		Duty of 80 percent of refund of health insurance	sec. 10 subs. 1 LohnforzG
		employer has to provide necessary information	sec. 10 subs. 5; sec. 11 LohnforzG
3 and more (*young people*)	Notice	Notice on work time (incl. break times)	sec. 48 JArbSchG
5 to 20	Works Council	1 representative	sec. 9 BetrVG
5 to 20	Youth representation	1 representative	sec. 62 BetrVG
5 to 50	Election of works council	Simplified election process possible	sec. 14 a BetrVG
5 and more	Works council	Possibility of employees to establish works council	sec. 1 BetrVG
more than 5	Dismissal	Application of the Kündigungsschutzgesetzes (Dismissal Protection Act)	sec. 23 subs. 1 KSchG
more than 5	Dismissal	Military service as "*business-related*" reason for redundancy	sec. 2 subs. 3 ArbPlSchG
more than 5	Notice	Notice of statutory regulations with regard to sex-related discrimination under sec. 611a, 611b, 612 subs. 3, 612a BGB and 61g ArbGG	Art. 9 GleiBG
10	Tax	Company annual tax declaration possible	sec. 42 b EstG
more than 10	Breaks	Installation of recreation rooms necessary	sec. 29 subs. 1 ArbeitsstättVO
20	Severely handicapped persons	Obligatory workplaces for severely handicapped people	sec. 71 SBG IX
more than 20	Unemployment benefits	Obligation of employer to reimburse unemployment benefits paid by the social security	sec. 147 a subs. 1 SGB III)

Employees	Area	Content of Regulation	Rule
		authority for a 58 old unemployed person who was occupied in the company for at least 4 years.	
more than 20	Participation of works council	Notification of employment, classification or re-classification, relocations	sec. 99 BetrVG
21 to 50	Number of works council members	3 works council members	sec. 9 BetrVG
21 to 50	Youth representation	3 youth representatives	sec. 62 BetrVG
51 to 100	Number of works council members	5 works council members	sec. 9 BetrVG
51 to 150	Youth representation	5 representatives	sec. 62 BetrVG
more than 100	Information to employees	Establishment of Economic and Financial Committee	sec. 106 BetrVG
more than 100	Economic and Financial Committee	min. once per quarter in agreement with committee written notification of all employees over economic matters (situation and development) in the company	sec. 106 BetrVG
101 to 200	Number of works council members	7 works council members	sec. 9 BetrVG
151 to 300	Youth representation	7 representatives	sec. 62 BetrVG
200 to 500	Release	1 works council members	sec. 38 BetrVG
201 to 400	Number of works council members	9 works council members	sec. 9 BetrVG
301 to 500	Youth representation	9 representatives	sec. 62 BetrVG
401 to 700	Number of works council members	11 works council members	sec. 9 BetrVG

Employees	Area	Content of Regulation	Rule
more than 500	Participation Works Council	Works council can ask for establishment of personnel selection guidelines	sec. 95 BetrVG
more than 500	Company co-determination	In publicly listed corporations and partnerships limited by shares 1/3 employee participation in supervisory board	sec. 76 BetrVG '52
501 to 900	Dismissal	2 works council members	sec. 38 BetrVG
701 to 1000	Number of works council members	13 works council members	sec. 9 BetrVG
701 to 1000	Youth representation	13 representatives	sec. 62 BetrVG
901 to 1500	Release of works council members	3 works council members	sec. 38 BetrVG
1000 to 1500	Number of works council members	15 works council members	sec. 9 BetrVG
more than 1000	Youth representation	15 representatives	sec. 62 BetrVG
more than 1000	Company co-determination (in executive and supervisory board)	mining companies in form of stock-listed companies, GmbH (limited partnership), bergrechtl. Gewerkschaft (mining union) participation in supervisory board and executive board = labor director	sec. 1 subs. 1 and 2 Montan-MitbestG sec. 12 et seq. Montan-MitbestG
1501 to 2000	Dismissal	4 works council members	sec. 38 BetrVG
1501 to 2000	Number of works council members	17 works council members	sec. 9 BetrVG

Employees	Area	Content of Regulation	Rule
2000 und more	Company co-determination (in executive and supervisory board)	mining companies in form of stock-listed companies, GmbH (limited partnership), bergrechtl. Gewerkschaft (mining union), cooperative society participation in supervisory board and representation labor director	sec. 1, sec. 7 et seq. MitbestG sec. 33 MitbestG
		also for non mining companies (in form of stock-listed companies, limited partnerships, mining unions), if they dominate those mining companies, subject to co-determination obligations, which earn more than ½ of the group revenues	sec. 1 Montan-MitBestErgG
2001 to 2500	Number of works council members	19 works council members	sec. 9 BetrVG
2001 to 3000	Release	5 works council members	sec. 38 BetrVG
2501 to 3000	Number of works council members	21 works council members	sec. 9 BetrVG
3001 to 4000	Release	6 works council members	sec. 38 BetrVG
3001 to 3500	Number of works council members	23 works council members	sec. 9 BetrVG
3501 to 4000	Number of works council members	25 works council members	sec. 9 BetrVG
4001 to 4500	Number of works council members	27 works council members	sec. 9 BetrVG
4001 to 5000	Release	7 works council members	sec. 38 BetrVG
4501 to 5000	Number of works council members	29 works council members	sec. 9 BetrVG
5001 to 6000	Release	8 works council members	sec. 38 BetrVG
5001 to 6000	Number of works council members	31 works council members	sec. 9 BetrVG

Employees	Area	Content of Regulation	Rule
6001 to 7000	Release	9 works council members	sec. 38 BetrVG
6001 to 7000	Number of works council members	33 works council members	sec. 9 BetrVG
7001 to 8000	Release	10 works council members	sec. 38 BetrVG
7001 to 9000	Number of works council members	35 works council members	sec. 9 BetrVG
8001 to 9000	Release	11 works council members	sec. 38 BetrVG
from 9001	Number of works council members	Addition of 2 members per each additional fraction of 3,000 employees to the 35 works council members	sec. 9 BetrVG
9001 to 10.000	Release	12 works council members	sec. 38 BetrVG
from 10.000	Release	1 further member of the works council released for each additional fraction of 2,000 employees in addition to 12 released works council members	sec. 38 BetrVG

Table 58.3: Meaning of number of employees in Works Constitution law

There is no uniform way of counting the number of employees. It depends:

1. on calculation parameters of the workplace ("undertakings" or "businesses") and under certain circumstances also on the
2. employment relation which the employees have (part-time employees, home-workers, trainees etc.)

Concerning 1): In principle the number of employees of the business in question is decisive for issues concerning works constitution law and individual labor law. If questions arise with regard to organizational areas of work relations; e.g. for existence of the co-determination right of employees in the supervisory board or executive board of a legal person (corporation or association), the number of active employees is decisive.

Concerning 2): In principle the number of employees comprises also those employees who are employed as part-time employees, external workers, home workers or trainees[901] by the undertaking or business.

---

[901] Comp. sec. 5 subs. 1 BetrVG (Works Council Constitution Act)

Exceptions, however, are possible. For example sec. 622 subs. 5 sentence 2 BGB and sec. 2 subs. 3 sentence 3 ArbPlSchG limit the necessity to attribute part-time workers to the group of employees, in the sense that part-time employees with regular weekly working hours of not more than 20 hours are counted as 0.5 employees and those working not more than 30 hours are counted as 0.75 employees.

59 Conclusion

Especially labor law aspects of personnel policy confirm the coherence and consequence of the economic personnel cost approach. In this context it is less international labor law, but rather European and national labor law which has considerable influence on personnel management practice and cost advantages. Especially the jurisdiction of the European Court of Justice makes the immense influence clear which the European Labor Law has these days on managerial practice.

The significance of European Labor Law will grow because of the efforts of the European legislator to approximate the laws and harmonize the economic and competitive circumstances in order to establish a uniform European Economic Area. Also the changing structure of European companies promotes the development of European Labor Law: groups with cross-border activities need uniform law already for organizational reasons. At the same time a Europe-scale enterprise entails a Europe-wide coalition of employees to secure and exercise employee rights.

With the advancement of the above mentioned trends. the pressure for innovation of labor law increases. It has to be adjusted in order to be able to do justice to the current and future requirements. Not only questions regarding the financial and economic crisis, but also problems of growing unemployment point the necessity of a cost-economic personnel policy out. Not only the European legislator and the jurisdiction of the European Court of Justice have to offer adequate instruments for cost-efficient personnel management. Also the employers, employees and their associations need to know the instruments in order to be able to make an independent choice of the most efficient alternative.

XIV Conclusion

With the book "International Human Resource Management and International Labor Law" a trilogy in the competence and research field "Finance-oriented Personnel Management" has been developed. This trilogy has started with the book "Finance-oriented Human Resource Management" and as subsequent volume with the book "Controlling and the Berlin Balanced Scorecard Approach". The authors wanted to prove that personnel management, practiced and taught under managerial aspects, is a "functional instrument". The "theoretical crisis" of the last 25 years can neither be overcome by borrowings from economics (labor economic and human resource economic approach) nor from psychology (behavioral personnel management). It can only happen with its own strengths which are obtained from classical business management – especially from accounting, financing and investment theory – in connection with jurisprudence. Since more than 100 years the German business management theory is coined by professors like Schmalenbach, Gutenberg, Kosiol, Wöhe, Baetge, Coenenberg, Haller, Küting/Weber, Pellens, Weber/Schäffer to mention only some. On their fundaments and from their basic understanding the instruments of finance-oriented human resource-management were created. Examples are the personnel management logic, the Berlin Balanced Scorecard Approach, the Innovation Success Calculation, International Human Resource Controlling and the Berlin Human Capital Assessment Model as elements of the remuneration management systems with and without stock-option programs. The basic philosophy is the shareholder-value approach with the basic model of company assessment as an application of the decision-oriented company management theory.

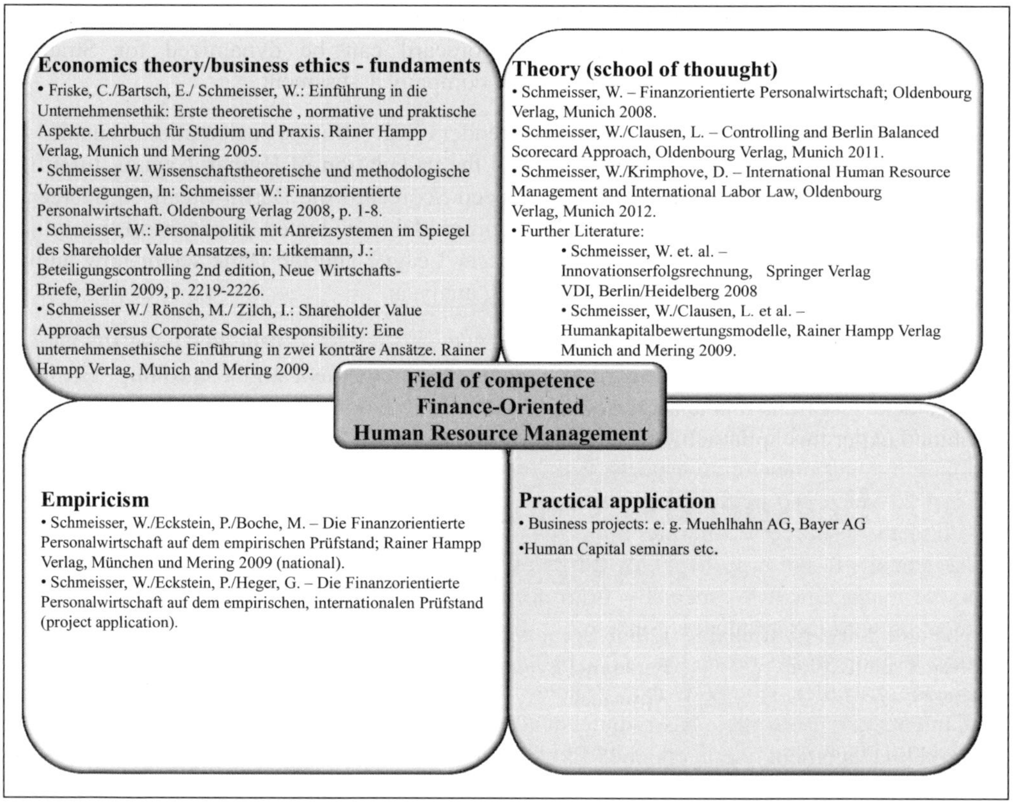

Fig. 59.1 Field of competence and research of Finance-Oriented Human Resource Management

To say it in Kant's words: each academic research and each partial discipline needs to have fundaments from normative-ethics and economic theory and to give theoretical evidence in form of stable models as well as empiric support. As business management is to be considered as applied science, it has to provide both – statistic-empirical proof and the applicability of the models in companies as organizational recommendations.

Thanks to the Muehlhan AG, as reference company with its Financial and Human Resource Manager Carsten Ennemann and the Head of HR Olaf Drewicke, we succeeded to do so.

Therefore, it is obvious that the second book of the trilogy "Controlling and Berlin Balanced Scorecard Approach" is also published in our names by the Oldenbourg Wissenschaftsverlag in English language. The book "Corporate Finance und Risk Management" shall equally provide basics for finance-oriented human resource management, because in the bank and insurance sector, in the health sector and in innovative industry such models will be more requested in future.

We have been improving the Berlin Balanced Scorecard Approach for years and prove that any perspective of the Balanced Scorecard can be calculated either with instruments of internal and/or external accountancy. Moreover we demonstrated that all perspectives can be connected and that the Berlin Balanced Scorecard can be dynamized for Strategic Management in the course of years with help of company assessment.

The Berlin Balanced Scorecard Approach renders e.g. strategy calculations equally as possible as innovation success calculations and the calculation of Human Capital. From the learning and potential perspective of the Balanced Scorecard the Berlin Balanced Scorecard Approach, the Berlin Human Capital Assessment model is deducted. It proves in theory and practice with an example from the Federal Soccer League and the Bank sector how human capital can be calculated with and without utility analysis.

The Berlin Human Capital Assessment model gives a first instruction how to manage international subsidiaries with regard to personnel management and controlling. Business management research, financial personnel aspects and analogies in practical application are combined under this approach.

Perspectives of Human Capital Assessment		
Psychological-organizational approach (motivation oriented)	**Labor Economical and Personnel Economical Approach**	**Finance-oriented or Business-Management Approach: innovation, strategy and cost and revenue orientation of HR management**
- Internal employee interview: does the company have a well-being, cuddly organization to increase motivation? - Internal competence survey within a potential form survey: personnel management as competence management? - External interviews of other companies: e.g. via "psychonomics", do you take Human Capital into consideration?	- Sum of investments into business-internal and external training activities of an economy - forced integration of quantitative/ qualitative Human Capital into the labor market (Hartz 4)	- Berlin Human Capital Assessment Model - HR reporting in Human Resource and Performance Reports - Basis: Finance-oriented logic and Berlin Balanced Scorecard Approach

Fig. 59.2: Perspectives of Human Capital Assessment

Of course, psychological-organizational as well as labor economic models will be further analyzed by us. They help us to make practical and theoretical improvements especially also in connection to legal requirements and to look for new application possibilities.

The next consequent step will be the elaboration of a human capital index which will only be calculable with the help of the Berlin Human Capital Assessment approach. The calculation of the Basel II Approach as analogy of rating approaches provides a perfect model in this context. The fundamental classic business management knowledge in accountancy and law are indispensible in this area.

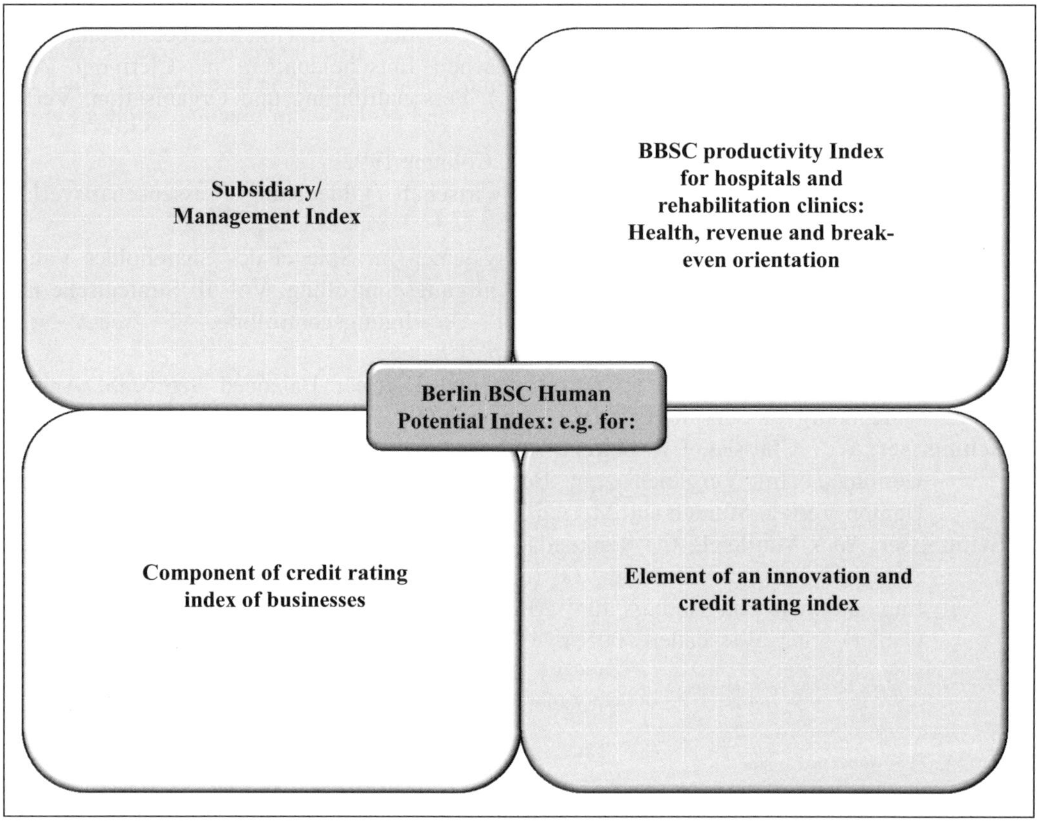

Fig. 59.3: Berlin BSC-Human-Potential-Indexes

Literature

Krimphove, D.: Das Europäische Sachenrecht, Eine rechtsvergleichende Analyse nach der Komparativen Institutionenökonomik, Lohmar, 2006.

Krimphove, D.: Europäisches Arbeitsrecht. 2nd edition, Munich 2001.

Krimphove, D.(a): Rechtstheoretische Aspekte der „Neuen Ökonomischen Theorie des Rechts". in: Rechtstheorie, 2001.

Krimphove, D.: Europäischer Betriebsrat / Betriebsräte in Europa – Strategien und Chancen der Europaweiten Vereinheitlichung der Arbeitnehmerbeteiligung an unternehmerischen und personalpolitischen Entscheidungen. in: Clermont, A. / Schmeisser, W. / Krimphove, D. (ed.): Personalführung und Organisation. Verlag Franz Vahlen, Munich 2000.

Krimphove, D.: Europäische Fusionskontrolle, Cologne 1992.

Schmeisser, W.: Finanzorientierte Personalwirtschaft. Oldenbourg Wissenschaftsverlag, Munich 2008.

Schmeisser, W.: Personalpolitik mit Anreizsystemen im Spiegel des Shareholder Value-Ansatzes. in: Littkemann, J. (ed.): Beteiligungscontrolling. Vol. II: Strategische und operative Unternehmensführung im Beteiligungscontrolling, 2nd ed., Neue Wirtschaftsbriefe, Berlin/ Herne 2009, p.219–266.

Schmeisser, W. / **Clausen**, L.: Controlling und Berliner Balanced Scorecard Ansatz. Oldenbourg Wissenschaftsverlag, Munich 2009.

Schmeisser, W. / **Clausen**, L. / **Seifert**, A. / **Stülpner**, K.: Modelle zur Humankapitalbewertung – Im Vergleich zum Berliner Humankapitalbewertungsmodell. Rainer Hampp Verlag. Munich and Mering 2009.

Schmeisser, W. / **Mauksch**, C.: Risikoadäquate Kreditzinsenkalkulation nach Basel II. in: Achleitner, A.-C. / Everling, O. (ed.): Rechtsfragen im Rating. Grundlagen und Implikationen von Ratings für Agenturen, Investoren und geratete Unternehmen. GablerVerlag, Wiesbaden 2005, p. 297–326.

Subject Index

V

W

UTB 8243

Eine Arbeitsgemeinschaft der Verlage

Beltz Verlag Weinheim und Basel
Böhlau Verlag Köln · Weimar · Wien
Wilhelm Fink Verlag München
A. Francke Verlag Tübingen und Basel
Paul Haupt Verlag Bern · Stuttgart · Wien
Verlag Leske + Budrich Opladen
Lucius & Lucius Verlagsgesellschaft Stuttgart
Mohr Siebeck Tübingen
C. F. Müller Verlag Heidelberg
Ernst Reinhardt Verlag München und Basel
Ferdinand Schöningh Verlag Paderborn · München · Wien · Zürich
Eugen Ulmer Verlag Stuttgart
UVK Verlagsgesellschaft Konstanz
Vandenhoeck & Ruprecht Göttingen
WUV Facultas · Wien

Einführungstexte Erziehungswissenschaft

Herausgegeben von Heinz-Hermann Krüger

Band 8:
Einführung in die Soziologie der Bildung und Erziehung

Martina Löw

Einführung in die Soziologie der Bildung und Erziehung

Leske + Budrich, Opladen 2003

Meinen Eltern, Ingrid und Willy Löw,
den ersten Bildungsexperten meines Lebens.

Dank

Für die Aufforderung, dieses Buch überhaupt zu schreiben, bedanke ich mich bei Heinz-Hermann Krüger. Meherangis Bürkle, Katja und Sergej Stoetzer, Brigitte Petendra, Silke Steets und Jutta Güldenpfennig haben inhaltlich und organisatorisch zugearbeitet. Sergej Stoetzer danke ich dabei besonders für die zahlreichen inhaltlichen Hinweise zu elektronischen Netzen und für unermüdliche Hilfen mit den Computer-Programmen.

Gedruckt auf säurefreiem und altersbeständigem Papier.

Die Deutsche Bibliothek – CIP-Einheitsaufnahme
Ein Titeldatensatz für diese Publikation ist bei
Der Deutschen Bibliothek erhältlich

UTB-ISBN 3-8252-8243-0

ISBN 3-8100-3400-2

Satz: Berthold Druck und Direktwerbung, Offenbach
Einbandgestaltung: Atelier Reichert, Stuttgart
Druck: DruckPartner Rübelmann, Hemsbach
Printed in Germany

Editorial zu den Einführungstexten Erziehungswissenschaft

Die Reihe Einführungstexte der Erziehungswissenschaft in sechzehn Bänden ist so konzipiert, dass sie Studierenden in erziehungswissenschaftlichen Hauptfachstudiengängen an Universitäten und Fachhochschulen im Grundstudium sowie Lehramtsstudierenden eine Einführung in Geschichte, Grundbegriffe, theoretische Ansätze, Forschungsergebnisse, Institutionen, Arbeitsfelder, Berufsperspektiven und Studienorte der Pädagogik/Erziehungswissenschaften sowie der verschiedenen Studienschwerpunkte und Fachrichtungen geben soll. Die einzelnen Bände sind so strukturiert, dass sie sich als Grundlagentexte für einführende Lehrveranstaltungen eignen.

Die Lehrtextreihe umfasst 16 Bände:

1. Einführung in die Erziehungswissenschaft
2. Einführung in die Schulpädagogik und Didaktik
3. Einführung in die Sozialpädagogik/Sozialarbeit
4. Einführung in die Erwachsenenbildung
5. Einführung in die Sonderpädagogik
6. Einführung in die Berufspädagogik/Wirtschaftspädagogik
7. Einführung in die Pädagogische Psychologie
8. Einführung in die Soziologie der Bildung und Erziehung
9. Einführung in die Interkulturelle Pädagogik
10. Einführung in die Kultur- und Freizeitpädagogik
11. Einführung in die Medienpädagogik
12. Einführung in die Frauenstudien
13. Einführung in die Vorschulpädagogik
14. Einführung in die Gesundheitspädagogik
15. Einführung in das Bildungs- und Sozialmanagement
16. Einführung in die Altenhilfe/Altenbildung

Die Bände 4 und 8 erscheinen erstmals im Sommer 2003. Alle weiteren Bände werden in den Jahren 2003 bis 2005 erscheinen. Die Autorinnen und Autoren der sechzehn Bände sind von dem Herausgeber gebeten worden, folgende Aspekte bei der Texterstellung zu berücksichtigten:

1. Geschichte des Feldes und der Teildisziplinen
2. Grundbegriffe und Theorieansätze
3. Forschungsfelder, Forschungsthemen
4. Institutionen, Arbeitsfelder, Berufschancen für Absolventen
5. Studienorte, Studiertipps
6. Thematisch sortierte Literaturempfehlungen

Es bleibt abschließend noch allen Autorinnen und Autoren, die Bände für die Reihe Einführungstexte Erziehungswissenschaft verfasst haben bzw. noch fertig stellen, für die bisherige produktive und reibungslose Kooperation zu danken. Mein besonderer Dank gilt Petra Essebier für die umsichtige und ausdauernde Mithilfe bei den vielfältigen Redaktionsarbeiten.

Heinz-Hermann Krüger
Martin-Luther-Universität Halle-Wittenberg

Inhaltsverzeichnis

Vorwort für Studierende

Die „Soziologie der Bildung und Erziehung“ ist eine von vielen Teildisziplinen der Soziologie. Sie ist besonders für jene Studierende der Soziologie relevant, welche sich auf Bildungs- und Erziehungsfragen spezialisieren wollen, und für die Studierenden der Erziehungswissenschaft und des Lehramts von Bedeutung, die mit Hilfe der Soziologie die gesellschaftlichen Grundlagen ihres Gegenstandes beleuchten möchten. Ich habe versucht, in der vorliegenden Einleitung allen drei Gruppen Rechnung zu tragen. Angesichts der Tatsache, dass bei vielen Nebenfach (oder Beifach)-Studierenden zunächst nur mit geringen soziologischen Grundkenntnissen zu rechnen ist, habe ich mich bemüht, so wenig wie möglich voraus zu setzen. Ich nehme an, dass das auch den Soziologie-Hauptfach-Studierenden zu Gute kommt.

Gleichzeitig mache ich es Ihnen nicht einfach. Dieses Buch bietet nur eine erste Orientierung. Sie lernen die Spezifika dieses Faches kennen, welches immer zwischen Erziehungswissenschaft und Soziologie hin- und hergerissen war. Sie erfahren die Grundthesen von Klassikern wie zeitgenössischen TheoretikernInnen und lernen die aktuellen Forschungen in zentralen Feldern der Bildungssoziologie kennen. Aber, das ist nur der Anfang. Sie bekommen einen Überblick und Entscheidungsgrundlagen, in welche Richtung sie ihre Kenntnisse vertiefen wollen. Dieses Buch ersetzt nicht das Studium der Primärquellen. Vielleicht lesen Sie die Klassiker ganz anders als ich. Über die Bedeutung mancher Textstellen könnten wir sicherlich Abende lang diskutieren. Überprüfen Sie also meine Lesart am Original und konfrontieren Sie sie mit anderen Interpretationen. Nur so werden Sie ein tieferes Verständnis für den Autor oder die Autorin gewinnen und vor allem nur so werden Sie zu einer eigenen Position gelangen.

Deshalb verzichte ich (vor allem in Kapitel 3 und 4) auch bewusst darauf, gängige Kritiken der theoretischen Text zu präsentieren. Was nützt es Ihnen, wenn Sie wissen, dass Pierre Bourdieu häufig vorgeworfen wird, sein Habitusbegriff sei zu deterministisch? Ein Teil der wissenschaftlichen Community ist der Auffassung, dass Bourdieu zu wenig die Handlungsfreiheit der Menschen in seiner Theoriebildung berücksichtigt und macht dies am Habitusbegriff fest. Ein anderer Teil ist umgekehrt der Überzeugung, dass es eine Illusion der Moderne ist, dass Menschen autonom handlungsfähig seien. Bourdieu beschreibt für diese AutorInnen den Zustand, wie er nun einmal ist und nicht wie wir ihn uns wünschen. Wieder andere interpretieren den Habitusbegriff gar nicht als deterministisch. Sie weisen also den Vorwurf zurück. Das heißt, Kritik bedarf der intensiven Auseinandersetzung mit der Originalquelle und den dazu erschienen wissenschaftlichen Sekundärveröffentlichungen und Debatten. Das „Nacherzählen“ von Thesen und Kritikpunkten ist noch keine wissenschaftliche Leistung, wie sie von Ihnen im Studium erwartet wird. Zu einer solchen wird es erst, wenn Sie sich mit dem aktuellen Stand der Debatte durch neue und neueste Quellen vertraut gemacht haben und in dieser Auseinandersetzung zu einem eigenen Standpunkt gelangt sind. In dieser Einführung finden Sie deshalb anschließend an die Vorstellung von TheoretikerInnen die wichtigsten Hinweise auf KritikerInnen und Folgestudien. So wissen Sie gleich, wo sie weiterarbeiten können.

Das Buch ist so aufgebaut, dass Sie selbstverständlich einzelne Kapitel für sich lesen können. Manch eine/r mag vielleicht sich nur über Untersuchungen zum Beispiel zum Thema „Schule“ informieren. Dies kann er/sie in dem speziellen Kapitel tun. Vor allem jenen aber, die Soziologie nur als Bei- oder Nebenfach studieren, empfehle ich auch das Studium der ersten vier Kapitel. Hier erfahren Sie erstens, warum es so schwierig ist, soziologisches und erziehungswissenschaftliches Wissen gegeneinander abzugrenzen, und zweitens lernen Sie, den soziologischen Blick systematisch kennen.

Im Buch werde ich stets die Perspektive wechseln. Sie erfahren jeweils etwas über historische Entwicklungen, über klassische und moderne theoretische Ansätze sowie Grundbegriffe; ich demonstriere an einzelnen Problemen verschiedene soziologische (vereinzelt auch soziologienahe) Perspektiven auf Bildungsprozesse und stelle Ihnen ausgewählte Untersuchungen ausführlicher vor. So hoffe ich, dass Sie gleichzeitig einen Überblick über die Forschungslandschaft, einen Einblick in das soziologische Denken und einen Eindruck von der Prozesshaftigkeit des wissenschaftlichen Arbeitens bekommen.

Nun stellt sich gerade für Nebenfachstudierende die Frage, die vor jeder Auseinandersetzung mit Bildungssoziologie steht, und die kaum in wenigen Sätzen zu behandeln ist: Was ist eigentlich Soziologie? Eine Annäherung an eine Antwort finden Sie im Folgenden.

Was ist Soziologie?

Max Weber definiert Anfang letzten Jahrhunderts Soziologie als „eine Wissenschaft, welche soziales Handeln deutend verstehen und dadurch in seinem Ablauf und seinen Wirkungen ursächlich erklären will“ (Weber 1980[5], original 1921, S. 1). Weber verbindet damit die geisteswissenschaftliche Tradition der Interpretation kultureller Leistungen mit der naturwissenschaftlichen Tradition, allgemeingültige Gesetze zu finden. Sein Ausgangspunkt ist der subjektiv gemeinte Sinn, mit dem Menschen ihre Handlungen intentional versehen. Die Soziologie ist nun aufgefordert, die sinnhaften Handlungssituationen verstehend zu rekonstruieren und nach Regelmäßigkeiten und Gesetzmäßigkeiten zu suchen.

Emile Durkheim verfolgt eine andere Vorstellung von Soziologie. Für ihn greift die Erforschung des subjektiven Sinnes zu kurz. Soziale Ordnung, die zu untersuchen das Aufgabenfeld der Soziologie sei, lasse sich nicht nur über individuelle Handlungen erklären, sondern existiere ein Stück weit unabhängig von diesen Handlungen und strukturiere sie vor (Durkheim 1999[3]a, S. 128; vgl. auch Durkheim 1961 und Kapitel 2.1.1 in diesem Buch). Der Gegenstand der Soziologie ist für ihn daher weniger individuelles Handeln, denn dieses ist vergänglich, als vielmehr das „Kollektivbewußtsein“ (Durkheim 1999[3]b, S. 128), welches in einer Gesellschaft für alle Gruppen gleichermaßen Gültigkeit besitzt und eine Generation mit der nächsten verbindet. Die Soziologie ist für ihn die „Wissenschaft von den Institutionen, deren Entstehung und Wirkungsart“ (Durkheim 1961, original 1895, S. 100). Die Aufgabe der Soziologie ist es, die „sozialen Tatsachen“ zu analysieren, die die individuellen Handlungen hervorbringen. Das Verständnis dessen, was genau der Gegenstand der Soziologie ist, hat sich über die Jahrzehnte immer wieder verfeinert und neuen Erkenntnissen angepasst.

Die Grundidee, dass Gesellschaft über die Deutung menschlicher Aktivität und Kommunikation sowie über verallgemeinerbare Strukturen, Institutionen und Systeme analysiert werden kann, ist geblieben. Viele Autoren und Autorinnen versuchen heute, Konzeptionen einer Soziologie vorzulegen, die die Individual- und die Kollektivebene verknüpfen.

Perspektiven der Soziologie der Bildung und Erziehung. Eine Einführung

Konjunkturen des Bildungsthemas

Auch wissenschaftliche Forschung ist Konjunkturen unterworfen. Am Thema „Bildung" zeigt sich dies in der jüngsten Vergangenheit besonders deutlich. In den 70er Jahren wird in der Erwartung, soziale Ungleichheit in Deutschland über Bildung mildern zu können, mit Euphorie an vielen Orten zu Bildungsfragen gearbeitet. Die Disziplin der Soziologie der Bildung und Erziehung zieht alle Aufmerksamkeit auf sich, als Deutschland demokratischer, egalitärer und gerechter werden will und fest davon überzeugt ist, dass dies nur über die Erziehung der Kinder zu erreichen ist. Diese Begeisterung zerschellt an den eigenen hohen Zielen, aber auch am gesellschaftlichen Zeitgeistwandel. Zuerst sickert die Erkenntnis durch, dass die subtile Reproduktion sozialer Ungleichheit über Schulreformen nur schwer zu beheben ist. Neue Institutionen wie die Gesamtschulen und Reformuniversitäten entpuppen sich zunehmend als neue Orte sozialer Ungleichheit. Dann treten die negativen Folgen der Massenbildung hervor: Schlechte Studienbedingungen, Abfall des Leistungsniveaus und vor allem eine verschobene Selektion. Die Bildungsinstitutionen übernehmen, um Chancengleichheit zu verwirklichen, immer weniger die Aufgabe, die Jugendlichen auf Berufszweige und damit in Statusgruppen zu verteilen. Dies hat zur Folge, dass die Reproduktion der sozialen Lage in den Betrieben stattfindet. Noch heute sind es die Kinder von Spitzenmanagern die Spitzenmanager werden (vgl. Hartmann 2002), aber dies liegt nicht länger an den Bildungszertifikaten, sondern an der Vertrautheit des Lebensstils von Bewerber/Bewerberin und PersonalchefInnen sowie am eingesetzten sozialen Kapital. Wenn alle Abitur und Studium aufweisen können, verlassen sich Führungskräfte um so mehr auf Empfehlungen und auf „spontane" Gefühle im Vorstellungsgespräch. Dann bekommen die den Job, die es gewohnt sind, im Milieu zu verkehren.

Soziale Ungleichheit

Die Soziologie der Bildung und Erziehung, die – wie viele andere soziologische Teildisziplinen auch – es als eine der Hauptaufgaben der Soziologie erachtet, soziale Ungleichheit in der Gesellschaft zu analysieren, scheint gescheitert zu sein. Oder anders: Keiner glaubt mehr wirklich daran, dass soziale Ungleichheit über Bildung zu beheben sei. Als dann auch noch innerhalb der Soziologie systemtheoretische Stimmen lauter werden, die die Bestimmung des soziologischen Gegenstands über die moralisch hoch besetzte Frage der sozialen Ungleichheit anzweifeln, verstärkt sich die Unsicherheit über den Gegenstand. Dabei wird häufig übersehen, dass die Bildungsreformen der 70er Jahre tatsächlich zahlreiche Benachteiligungen beseitigt haben. Die quantitative Beteiligung an

Bildungsprozessen und der Erwerb von Zertifikaten ist heute noch schichtspezifisch und vor allem interkulturell ein Problem. Kinder aus Zuwandererfamilien haben nach wie vor sehr schlechte Ausbildungschancen in Deutschland. Dafür wurden Differenzen der Bildungsbeteiligung nach Stadt/Land, katholisch/evangelisch oder nach Geschlecht behoben. Dies heißt nicht, dass Mädchen (und manchmal auch Jungen) nicht in der Schule systematisch aufgrund ihres Geschlechts benachteiligt werden. Sie verfügen jedoch zunächst über gleiche Zugangschancen zu höherer Bildung, was die Voraussetzung ist, um auch die subtilen Ausschlussprinzipien kollektiv bearbeiten und verstehen zu können.

Dennoch schläft in den 90er Jahren das Interesse am Thema „Bildung" in der Öffentlichkeit fast ein. Dementsprechend stehen zu dieser Zeit Bildungs- und ErziehungssoziologInnen auch nur wenige Ressourcen für die Arbeit zur Verfügung. Zwar wird kontinuierlich an vielen Fragen weitergearbeitet, doch motivierende Nachfrage an Ergebnissen fehlt vielfach. Heute wird Bildung wieder auf die Tagesordnungen aller politischen Parteien gesetzt. Die Journale berichten, die Forscher streiten, die Förderinstitutionen investieren. Eine neue Welle der Schul- und Hochschulreformen ist angebrochen.

Bildung als ökonomische Frage

Die Soziologie der Bildung und Erziehung beteiligt sich an diesen Debatten zunächst nur zögerlich. Als die deutsche Bildungskatastrophe immer offensichtlicher wird und die ErziehungswissenschaftlerInnen und WirtschaftswissenschaftlerInnen das Thema neu besetzen, bleiben die SoziologInnen vorerst still. Roman Herzog kündigt 1996 eine Bildungsoffensive an. Die TIMSS-Studie erhebt 1997 Leistungsdefizite von Schülern und Schülerinnen in deutschen Schulen (vgl. Baumert/Lehmann 1997). Die Wirtschaft klagt über Fachkräftemangel. So wird Bildung wieder ein gesellschaftlich relevantes Thema. Aber anders als in der Bildungsdebatte der 70er Jahren ist soziale Ungleichheit zunächst ein Randthema. Im Vordergrund der Diskussion um einen notwendigen Wandel des Bildungssystems steht in den ersten Jahren der neuen Bildungsdebatte die Kritik an der zunehmenden Verrechtlichung und Bürokratisierung der Bildungsinstitutionen sowie ein (zunächst unterstellter) informationstechnologischer Modernisierungsrückstand und Leistungsdefizite im internationalen Vergleich. Argumentativ rückt die Sorge um den Wirtschaftsstandort Deutschland in den Vordergrund. Ein rohstoffarmes Land wie Deutschland müsse in Bildung als Ressource investieren, so das zentrale Argument. Beispielhaft dafür kann auf die berühmt gewordene und in den 90er Jahren viel zitierte These von Leo A. Nefiodow (1972) verwiesen werden, dass Bildung den Konjunkturfaktor „Informationstechnologie" ablösen könne. Nefiodow baut auf der Theorie der langen Wellen von Nikolai Kondratieff auf, welche besagt „daß die wirtschaftliche Entwicklung Westeuropas und der USA nicht nur durch das Auftreten kurzer und mittlerer Konjunkturschwankungen gekennzeichnet sei, sondern in den kapitalistischen Ländern auch lange Phasen der Prosperität und Rezession periodisch auftreten" (Nefiodow 1997, S. 2).

Theorie der langen Wellen

Nefiodow untersucht, welche Entwicklungen sich derzeit wirtschaftlich abzeichnen, die die Qualität des nächsten Konjunkturfaktors erreichen können. Er kommt zu dem Ergebnis, dass der Bildungsmarkt ein möglicher Aspirant eines großen, neuen Marktes des 21. Jahrhunderts ist (vgl. zur Diskussion z. B. Glotz 1998). Lebenslanges Lernen sei inzwischen für fast alle Erwerbstätigen zum Muss geworden. 2000 Milliarden US-Dollar würden weltweit in Weiterbildung investiert und eine Sättigungsgrenze sei nicht abzusehen. Neben der lebenslangen Beschulung sieht Nefiodow Produktionsreserven im sich verändernden Qualifikationsprofil. Der Umgang mit nicht klar definierbarem oder gar paradoxem Wissen sei in der Dienstleistungs- und Informationsgesellschaft unabdingbare Voraussetzung und könne zum Innovationsfaktor ausgebaut werden. Schulen und Universitäten seien jedoch bislang zu bürokratisiert und staatlich kontrolliert, um auf diese Entwicklung adäquat reagieren zu können.

Innovative Bildungsunternehmen

Ob Peter Glotz oder Roman Herzog, ob die Berichterstattung der Zeitschrift „Zeit" oder die des „Sterns", ob eine virtuelle Konferenz zu Bildung im Internet oder der Kongress der Deutschen Gesellschaft für Erziehungswissenschaft (Krüger/Olbertz 1997) im Vordergrund des Diskurses steht die Frage, wie Massenuniversitäten und Schulen zu innovativen Unternehmen umgebaut werden können oder zumindest die geeigneten Menschen für diese Unternehmen ausbilden können. Die vorgeschlagenen Strategien unterscheiden sich dabei vehement: Während die einen einen normierten Bildungskanon für alle fordern, sich an internationalen Leistungsstandards orientieren und traditionelle Primärtugenden definieren sowie den LehrerInnen mehr Autorität zugestehen wollen (exemplarisch dafür das von Dietrich Schwanitz verfasste Buch „Bildung", aber auch die Rede von Roman Herzog 1997), plädieren die anderen für die Orientierung an Schlüsselqualifikationen, für eine Bildungspolitik die Wagemut, Beweglichkeit, unprätentiöse Wurstigkeit und nicht zuletzt den Mut zum Fehler fördert (Glotz 1998; Preuss-Lausitz 1997).

Konkurrenz der Felder

Tatsächlich sind es (Bildungs)Soziologen wie Niklas Luhmann und Pierre Bourdieu, mit denen sich die neuen Bildungskämpfe gut begreifen lassen. Zu beobachten ist eine neue Version eines alten Kampfes zwischen dem ökonomischen Feld und dem Feld der Bildung. Die Ökonomie versucht ihren Einfluss auf das Bildungssystem zu verstärken, indem sie deren Kompetenz, leistungsstarke Mitarbeiter und Mitarbeiterinnen zu produzieren, in Frage stellt. Insbesondere durch die Drohung und die teilweise Umsetzung, eigene, vom Unternehmen geleitete Hochschulen zu etablieren, versuchen sie die Machtpotentiale der Bildungsbranche, nämlich Titel vergeben zu können, zu beschneiden. Die Erziehungswissenschaft reagiert darauf, indem sie sich erneut auf eine Definition ihrer Tätigkeit über „Leistung" einlässt. Wohl wissend, dass Leistung nicht messbar ist, antwortet sie in ihrer Logik, indem Leistung durch Verbesserung der Organisation von Bildung angestrebt wird. So agieren die Akteure des ökonomischen und des pädagogischen Feldes in den für sie typischen Muster (und können auch nicht anders handeln). Sie verhandeln über Geld und über Lernfähigkeiten. Es zeigt sich, dass die Ökonomie starken Einfluss auf die Bildung gewinnen kann, letztlich jedoch nicht in der Lage ist, Bildung und Erziehung im eigenen Feld zu verorten. Die Soziologie – ihrer eigenen Logik folgend – beobachtet und analysiert die Prozesse. Zu Fragen der sozialen Ungleichheit äußern sich jedoch zunächst nur wenige Fachleute. Chancengleichheit im Bildungswesen scheint ein überholtes Thema zu sein. Die Bildungsdebatte der späten 90er Jahre findet daher weitgehend ohne die Soziologie statt.

PISA

Die Veröffentlichung der PISA-Studie (vgl. Baumert/Schümer 2001) durchbricht die Ruhe unwiederbringlich. Das zentrale Ergebnis der Studie, dass nämlich in Deutschland beschulte Kinder nicht nur leistungsschwach sind, sondern darüber hinaus in Deutschland die Benachteiligung der einkommensschwachen

Haushalte besonders groß ist (und damit wiederum Leistungspotentiale vergeudet werden), ruft die Soziologie erneut auf den Plan. Plötzlich verzeichnet zum Beispiel die Sektion Bildungssoziologie in der Deutschen Gesellschaft für Soziologie wieder Neueintritte. Nun ist wieder von öffentlichem Interesse, dass die Kinder der Arbeiterschicht nach wie vor die schlechtesten Bildungschancen haben (vgl. Hradil 1997; Shavit/Blosfeld 1993) bzw. dass die erweiterte Bildungsbeteiligung großer Gruppen der Bevölkerung sogar den Nebeneffekt hat, dass die nun zur Minderheit gewordene Gruppe der HauptschülerInnen in einem bislang ungekannten Maß abgewertet werden. Die Soziologie der Bildung und Erziehung tritt in eine neue Phase der Selbstverständigung und Neubestimmung ein. Dazu gehört auch das Verhältnis von Bildung und Ökonomie erstens unter Bedingungen von Globalisierung und zweitens in einer als Wissensgesellschaft deklarierten Welt neu zu bestimmen.

Globalisierung

Spätestens seit Claus Offes Überlegungen zum Verhältnis von Bildung, Beschäftigungssystem und Bildungspolitik (Offe 1975) ist bekannt, dass Bildungsprozesse zwar strukturell zielorientiert sind und dieses Ziel in der spätindustriellen, modernen Gesellschaft maßgeblich der Eintritt in das Wirtschaftssystem ist, doch der zukünftige Bedarf der Wirtschaft nie zuvor bekannt ist. Jede Forderung nach leistungsfähigeren Individuen steht somit vor dem Problem, definieren zu müssen, welche Leistungen das Wirtschaftssystem in den nächsten Jahrzehnten benötigt. In der modernen Wirtschaft verändert sich nun die Bestimmung von Wissen erstens in sich schnell abwechselnden Zyklen und zweitens abhängig von internationalen Verflechtungen. Nicht zufällig werden internationale Bildungstest in einer historischen Epoche der finanziellen und kulturellen Globalisierung durchgeführt. Unter dem Druck internationaler Konkurrenz wird Bildung als messbares Gut behandelt, welches internationalen Standards unterliegt. Befürchtet wird deshalb, dass marktrelevantes Wissen zunehmend der Vergleichsfaktor wird und damit langsam ein komplexes Verständnis von Bildung ersetzt (vgl. z.B. Brock/Suckow 2001).

Wissensgesellschaft

Dies schließt an prinzipielle Überlegungen zur Bedeutung von Wissen in der modernen Gesellschaft an. Nico Stehr zufolge (2000) ist Wissen ein zentrales Integrationsprinzip der modernen Gesellschaft (vgl. dazu grundlegend auch Bell 1975). Unter Wissen versteht er die „Fähigkeit zum sozialen Handeln (Handlungsvermögen)" (Stehr 2000, S. 81). Im Unterschied zu stereotypen Verhaltensmustern setze soziales Handeln Spielräume und Entscheidungsmöglichkeiten voraus. Wissen schaffe die nötigen Voraussetzungen. Zum Wissen zählt er sowohl wissenschaftliches Wissen als auch Alltagwissen, deklariertes wie prozedurales Wissen (Stehr 2000, S. 78). Seine zentrale These ist, dass Gesellschaften, die über Wissen integrieren, zerbrechliche Gesellschaften sind, weil mit dem Wissenszuwachs der Einzelnen die großen gesellschaftlichen Institutionen einen Herrschaftsverlust hinnehmen müssen. Institutionen des Staates, der Wissenschaft, der Kirche, des Parlaments, der Justiz oder der Wirtschaft erschienen zunehmend fragil, weil ihnen nicht mehr geglaubt würde und alle mitbestimmen wollten. Soziale Bewegungen und Individuen erhielten im Vergleich zu den Institutionen einen Machtgewinn. Er bezeichnet es als paradox, dass mit dem Wachstum des Wissens, größere gesellschaftliche Unsicherheit, Meinungsverschiedenheit und Kontingenz erzeugt würde. In der Wissensgesellschaft zerfalle die intellektuelle Autorität von Experten und wachse die Skepsis gegenüber der

Unparteilichkeit und der Objektivität wissenschaftliches Wissen. Die mangelnde Durchsetzungsfähigkeit der Institutionen durch das mangelnde Zutrauen in politische oder technologische Lösungen lasse die gesellschaftliche Stabilität prekär werden.

Weitgehend akzeptiert und seit langem diskutiert wird die These, dass der Zuwachs an Wissen die Hochachtung gegenüber und Akzeptanz von Wissenschaft schmälert. Umstrittener ist die inhaltliche Bestimmung des gefragten Wissens. Matthias Wingens (1999) argumentiert, dass nur eine spezifische Wissensart, nämlich wissenschaftlich-technisches Wissen, gesellschaftliche Bedeutung erlangt und diese erzielt sie keineswegs allgemein, sondern als Kapital der Wirtschaft (Wingens 1999, S. 433).

Themen der Soziologie der Bildung und Erziehung

Bildungsprozesse wandeln sich unter dem Druck der Wissensproduktion in der Moderne. Diese Kämpfe um die Definition von Bildung, um die Durchsetzung von Wissenstypen, um die Inhalte der Bildungszertifikate, um die Strukturen der Ausbildung, um die KlientInnen der Bildungsprozesse, um die heimlichen Lehrpläne der Bildung analysiert die Soziologie der Bildung und Erziehung. Der Zeitpunkt ist günstig, um durch die öffentliche Darstellung der Beobachtungen verändernd einzugreifen. Der heimliche Leitfaden beim Schreiben dieser Einführung ist daher eine Standortbestimmung durch kurzfristiges Innehalten und Zusammentragen der Kenntnisse. Diese Einführung orientiert sich in den Kapiteln an den hier nur kurz angesprochenen Facetten des Bildungsthemas. Sie behandelt das zentrale Thema der „Bildung und Ungleichheit“ ausführlich (Kapitel 4), zeigt Erklärungsansätze für aktuelle Bildungskonflikte zwischen ökonomischer Orientierung, didaktischer Aufbereitung und gesellschaftlicher Analyse durch moderne Theorien (Kapitel 3), reflektiert der Grundlagen in klassischen Ansätzen und verfolgt die Geschichte der Soziologie der Bildung und Erziehung (Kapitel 2), klärte die grundlegenden Begriffe (Kapitel 1), widmet sich dann ausführlich den drei zentralen Institutionen des Bildungsgeschehens, der Vorschule und der Schule (Kapitel 5) sowie der Hochschule (Kapitel 6), um dann an zwei grundlegenden Dimensionen der Konstitution sozialer Welt, nämlich der über den Lebenslauf organisierten Zeitlichkeit (Kapitel 7) und der auch die Materialität der Welt berücksichtigenden Räumlichkeit (Kapitel 8) das Aufwachsen in der modernen Welt exemplarisch zu verdeutlichen. Dass diese Zusammenstellung des bildungssoziologischen Kenntnisstandes subjektiv geprägt ist, erwähne ich nur für die Studierenden der ersten Semester. Alle anderen wissen das.

1. Soziologische Grundbegriffe

1.1 Bildung

Bildung hat „wer nicht mit der Hand arbeitet, sich richtig anzuziehen und zu benehmen weiß, und bei allen Dingen, von denen in Gesellschaft die Rede ist, mitreden kann" (Paulsen 1903, S. 658). Nein, möchte man einwenden, Bildung ist mehr, etwas Höheres, Allgemeingültiges. Bildung hat doch mit Selbstverwirklichung des Menschen zu tun. Und dann schleichen sich die Zweifel ein: Ist Bildung hier nicht doch ganz richtig beschrieben als ein kulturelles Kapital, dass es einem ermöglicht, die eigene soziale Position zu halten oder aufzuwerten? Ist Bildung nicht in erster Linie ein Mittel, um sich von anderen absetzen zu können?

Ideal und Kapital

Wer, soziologisch kundig, sich mit dem Bildungsbegriff beschäftigt, begibt sich notgedrungen in ein Spannungsfeld. Der Bildungsbegriff entstammt einem normativ-idealistischen Umfeld, das auch bei der heutigen Nutzung stets mitschwingt. Gleichzeitig ist Bildung in der modernen Gesellschaft eine Ressource. Jenseits normativer Vorstellungen, was einen gebildeten Menschen ausmacht, gibt es einen ganz bodenständigen Verwertungsprozess von Bildungsinhalten und Bildungszertifikaten, welcher für die moderne Gesellschaft konstitutiv ist. So ist Bildung gleichermaßen Ideal und Kapital. Dies führt dazu, dass GesellschaftskritikerInnen sich zu dem Spagat gezwungen sehen, gleichermaßen die idealistischen, distinktiven Züge von Bildung freilegen zu wollen und dabei doch selbst auf Bildung als emanzipatives Potenzial und als Ressource der Unterdrückten zu vertrauen. Um diese grundsätzliche Ambivalenz und damit auch die Schwierigkeiten, den Bildungsbegriff zu definieren, zu verstehen, muss die historische Entwicklung betrachtet werden.

Klassisches Bildungsideal

Die Karriere des Bildungsbegriffs als theoretisch gefüllte Idee, nicht einfach nur im Sinne der Wortgeschichte „behauen" oder „einer Sache Gestalt und Wesen geben", beginnt im ausgehenden 18. Jahrhundert, also im Zuge der Industrialisierung der Gesellschaft. Gerichtet gegen den Utilitarismus, das Effektivitätsdenken und die Berufsorientierung der Ausbildung setzt das Bürgertum einen emphatischen Bildungsbegriff. Das klassische Bildungsideal des Neuhumanismus ist eine Vorstellung von der Einverwandlung der Welt durch das Individuum, welches dadurch zum Höchsten gelangt. Vordenker sind zum Beispiel Schiller, Herder und W. von Humboldt. Der Gebildete sucht nach Wilhelm von Humboldt „soviel Welt als möglich zu ergreifen und so eng als er nur kann mit sich zu verbinden" (Humboldt 1960, original 1794, S. 235). Bildung soll Perfektibilität im Sinne der Selbständigkeit im Denken und Handeln sowie die Freiheit zu Urteil und Kritik ermöglichen.

Antike Wurzeln

Insbesondere Humboldt, aber auch Schiller, beziehen sich dabei auf das Ideengut der Antike. Daher werden die Wurzeln des abendländischen Bildungsverständnisses in der griechisch-hellenistischen Antike gesehen (vgl. z.B. Blankertz 1969; Lichtenstein 1970; Heydorn 1970). Das griechische Konzept der Paideia, der Formung des Menschen zur Vollkommenheit an Leib und Seele, wird heute mit Bildung übersetzt. Die Paideia ist der Akt der Menschenformung, welcher jedoch schon damals nicht nur als individuelle Prägung, sondern auch als Notwendigkeit für das Gemeinwesen betrachtet wird. Über Bildungsprozesse lernt das Individuum die gesellschaftlich relevanten Inhalte kennen. Es bekommt das notwendige Wissen, um sich in der Gesellschaft zu Recht zu finden. In der Bildungsidee steckt gleichzeitig die Vorstellung von individueller Entfaltung durch Wissens- und damit auch Entscheidungszuwachs sowie von der Begrenzung durch Beeinflussung auf gesellschaftliche Ziele hin.

Humboldtsche Bildungsreform

Der preußische Direktor für Kultus und Unterricht Wilhelm von Humboldt greift zu Beginn des 19. Jahrhunderts auf die antiken Ideen zurück und prägt damit nicht nur theoretisch den Bildungsbegriff, sondern er verankert auch die bürgerliche Bildungsvorstellung institutionell. Humboldt reformierte das Bildungssystem und ist maßgeblich an der Legitimierung und Durchsetzung eines dreigliedrigen Bildungssystems (Elementarschule – Gymnasium – Universität) beteiligt. 1810 gründet er die Berliner Universität. Hochschullehrer sollen von nun an Forschung und Lehre nicht mehr trennen. Die Studierenden werden nicht als Lernende, sondern als Mitforschende betrachtet. Diese Freiheit der Lehre aufbauend auf der eigenen Forschung prägt noch heute das universitäre Bildungsverständnis.

Humboldt wendet sich in seinem Reformkonzept vor allem gegen die Berufs- und Standesschulen (vgl. z.B. Fuhrmann 1999, S. 29ff.). Er setzt sich für eine Trennung von Allgemeinbildung und Berufsbildung ein. Jeder, auch der Ärmste, soll, so die Theorie, eine allgemeine Menschenbildung erhalten (Humboldt 1964, S. 175). Aufbauend auf dem Elementarunterricht sollen spezielle Kenntnisse einzelner Gewerbe in einer Berufsausbildung erworben werden oder aber der Bildungsprozess zunächst im Gymnasium, dann in einer universitären Laufbahn vertieft werden. Deshalb werden die Gymnasien streng allgemein bildend (und damit persönlichkeitsbildend) und nicht berufsorientiert konzipiert.

Mädchenbildung

Indem jedoch die Gymnasial- und vor allem die Universitätsbildung prinzipiell nur den Männern und – aufgrund der finanziellen Anforderungen – nur den Reicheren vorbehalten ist, entwickelt sich Bildung zu einer begehrten Ressource (Bollenbeck 1994). Die Bildung der Mädchen dieser Zeit orientiert sich in keiner Weise an den von Humboldt entworfenen Zielen der Selbständigkeit im Denken und Handeln. Bürgerliche Mädchen werden auf ihre Rolle als Hausfrau, Gattin und Mutter vorbereitet, für die Urteilsfähigkeit und Selbstreflexivität als eher schädlich erachtet wird (vgl. dazu auch Kapitel 7.1). Da Mädchen weder auf Berufstätigkeit vorbereitet werden sollen noch durch Allgemeinbildung vor Einseitigkeit und Entfremdung durch die Industrialisierung geschützt werden müssen (Häuslichkeit wird als Gegensatz zur Erwerbstätigkeit und damit nicht-entfremdend konstruiert), erfolgt die Ausbildung mehr schlecht als recht in Klöstern und Pensionen (vgl. Schmid 1986). Über Bildung setzt sich das Bürgertum vom Proletariat und vom Adel, die Männer von den Frauen ab.

Mit Bildung verbindet sich zunächst die Idee von der Freiheit des Individuums und der Selbstverwirklichung des Menschlichen im Menschen. Bildung bezeichnet dabei sowohl den Prozess des Bildens als auch das Resultat, das Gebildetsein (Fuhrmann 1999, S. 27). Im 19. Jahrhundert gewinnt aber zunehmend neben dieser subjektorientierten Idee von Bildung auch die gesellschaftsintegrative Vorstellung vom Zusammenwachsen einer deutschen Nation durch Bildung an Bedeutung. Die einzelnen industriell zurückgebliebenen Staaten sollten über eine gemeinsame Bildungsidee symbolisch zu einer Kulturnation vereinigt werden (Bollenbeck 1994). Mit dem Nationalismus, der zunehmenden Industrialisierung und der Militarisierung der Gesellschaft Ende des 19. Jahrhunderts kehrt auch der vom Bildungsbürgertum verworfene Verwertungsanspruch von Bildung zurück. Bildung wird wieder zunehmend auch über die ökonomische, technische und industrielle Brauchbarkeit des Erlernten definiert.

Gleichheit und Differenz

Zusammenfassend heißt das, dem Bildungsbegriff haftet erstens die Vorstellung an, dass im Prozess des Sich-Bildens und durch das Resultat des Gebildet-Seins ein Kulturverständnis und somit Entscheidungsgrundlagen und Urteilsvermögen geschaffen werden. Zweitens dient dieser Wissenszuwachs auch der Selbstformung und Persönlichkeitsentwicklung. Bildung ist drittens in einem Egalitätsdiskurs verankert. Da Bildung gesellschaftliche Bedeutung gewinnt, indem über Allgemeinbildung alle Menschen die Chance erlangen (sollen), Kultur zu begreifen und sich selbst zu entfalten, haftet dieser Bildung ein Moment von Emanzipation und Demokratie an. Aber gleichzeitig hat die Benennung einer Gruppe als gebildet immer auch Bevölkerungsschichten und -gruppen produziert, die nun als ungebildet galten. Diese Gruppen sind selten freiwillig ungebildet, sondern ihnen wird aus ideologischen und finanziellen Gründen der Zugang zu höherer Bildung verwährt. Diese höhere (gymnasiale/universitäre) Bildung verankert sich im öffentlichen Bewusstsein als „eigentliche“ Bildung. So trägt „Bildung“ von Anbeginn an auch distinktive Züge. Bildung trennt die Schichten und Geschlechter voneinander. Über Bildungszugang wird auch der Zugang zu Geld und Macht geregelt. Damit ist Bildung also nicht nur in einem Gleichheitsdiskurs, sondern gleichermaßen in einem Differenzdiskurs verankert. Bildung verweist zudem nicht nur auf Selbstentfaltung, sondern auch auf gesellschaftliche Notwendigkeiten. Auch gegen den Willen des Bildungsbürgertums gibt es immer wieder gesellschaftliche Kräfte, die an der ökonomischen Verwertbarkeit von Bildungsinhalten und -zielen interessiert sind (und diese gewinnen trotz des hehren Bildungsverständnisses, welches sich in den Köpfen festsetzt, im historischen Verlauf zunehmend gesellschaftlichen Einfluss). Bildung dient schließlich auch dazu, die Deutschen von den Fremden zu trennen. Eingebunden in einen nationalistischen Diskurs erfüllen die deutschen „Dichter und Denker“ sowie die „deutsche Kultur“ nicht nur die Aufgabe einer Unterscheidung nach innen, sondern auch der Abgrenzung nach außen.

Bildung verweist also auf Selbstreflexivität und ökonomische Funktionalität, auf Urteilsfähigkeit und staatliche Beeinflussung, auf Gleichheit und auf Hierarchie. Bildung ist das Produkt der bürgerlichen Gesellschaft. Eine soziologische Arbeit mit dem Bildungsbegriff muss die dialektische Spannung, die dem Begriff innewohnt, reflektieren, kann aber auch gerade in dieser Reflexion den widersprüchlichen Bedingungen der modernen Gesellschaft nachspüren.

1.2 Erziehung und Sozialisation

Die Soziologie tritt erst Ende des 19. bzw. Anfang des 20. Jahrhunderts auf die Bühne der Wissenschaften. Sie interessiert sich, gerade wegen der ideologischen Implikationen, zunächst wenig für den Bildungsbegriff. Stattdessen rücken der Erziehungs- und der Sozialisationsbegriff in den Mittelpunkt der Auseinandersetzung. Der französische Soziologe Emile Durkheim zum Beispiel definiert Erziehung als Einwirkung der Erwachsenen auf die Kinder (vgl. ausführlich das folgende Kapitel). Ziel von Erziehung ist die Gesellschaftlichkeit des Menschen. Das heißt, Erziehung ist sowohl die Einweisung eines Menschen in die Normen und Werte der Gesellschaft als auch das Erlernen der schichtspezifisch speziellen Fähig- und Fertigkeiten. Aufgrund dieser vergesellschaftenden Aspekte des Erziehungsprozesses spricht Durkheim auch von Sozialisation.

Gezielte Beeinflussung

Die Begriffe Erziehung, Bildung und Sozialisation überlappen sich bis heute in ihrer Bedeutung. Jeder Begriff setzt einen anderen Akzent und verweist gleichzeitig auf einen gemeinsamen Sachverhalt. Am klarsten ist der Erziehungsbegriff definiert. Mehrheitlich wird unter Erziehung die geplante Beeinflussung Heranwachsender verstanden (vgl. z.B. Helsper 2002^5). Dabei ist diese Beeinflussung selbstverständlich gesellschaftlich vorstrukturiert (indem z.B. die Erziehenden selbst vergesellschaftete Wesen sind) und erfolgt mit dem Ziel, die Kinder und Jugendliche zu vollwertigen Mitgliedern einer Gesellschaft zu machen. Dieser Aspekt der Integration in die Gesellschaft wird jedoch weniger über den Erziehungsbegriff als über den Sozialisationsbegriff problematisiert. Die Soziologie interessiert sich für Erziehung, indem sie zum Beispiel die gesellschaftliche Wirkung von Erziehungsstilen diskutiert (Adorno 1977), die Schichtspezifik des Erziehungshandelns untersucht (z.B. Müller 1998), die Strukturen von Erziehungsinstitutionen analysiert (z.B. Arbeitsgruppe Bildungsbericht am Max-Planck-Institut für Bildungsforschung 1994) oder den Erziehungsbereich als gesellschaftliches System ins Verhältnis zu anderen Systemen setzt (Luhmann 1996). Um die pädagogischen und psychologischen Aspekte von Erziehung, also um die Individualentwicklung des Kindes, kümmert sie sich nicht.

Lebenslange Formung

Zentraler als der Erziehungsbegriff ist für die Soziologie daher der Sozialisationsbegriff. Er fokussiert weniger als Erziehung auf die Beziehung zwischen Lernendem und ErzieherIn, sondern stärker auf den lebenslangen Prozess, in dem das Soziale das menschliche Handeln formt. Während Erziehung die absichtsvolle Einwirkung bezeichnet, umfasst Sozialisation auch die ungeplanten und unbeabsichtigten Einflüsse. Peter L. Berger und Thomas Luckmann definieren Sozialisation als „grundlegende und allseitige Einführung des Individuums in die objektive Welt einer Gesellschaft oder eines Teiles einer Gesellschaft" (1972^3, S. 140f.). Menschen eignen sich im Handeln und durch Kommunikation gesellschaftliche Regeln, Deutungsmuster und Wissensbestände aktiv an und bilden so kognitive, emotionale und sprachliche Fähigkeiten, ein Verständnis der eigenen Person und individuelle Handlungsfähigkeit aus. Dieser lebenslange

Prägungs- und Aneignungsprozess wird als „Sozialisation" oder auch als „Vergesellschaftung" bezeichnet (vgl. dazu auch Geulen 1989, original 1977; Hurrelmann 1993; Grundmann 1999). Sozialisation ist also im Vergleich zu Erziehung der umfassendere Begriff.

1.3 Ein Vergleich der Begriffe

Vergleicht man nun den klassischen Bildungsbegriff und den Sozialisationsbegriff, so fallen zahlreiche Parallelen auf. Während über Bildung in aktiver Auseinandersetzung mit den Kulturgütern Reflexivität und Handlungsfähigkeit erzielt werden sollen, beschreibt Sozialisation den Vorgang der aktiven Aneignung der gesellschaftlichen Güter ebenso mit dem Effekt der Reflexivität und Handlungsfähigkeit. Beide benennen also den Prozess der Auseinandersetzung mit der sozialen Umwelt, mit dem Unterschied, dass der Bildungsbegriff mehr auf Wissen und Bewusstseinsprozesse zielt, der Sozialisationsbegriff dagegen stärker auf unbewusste Aneignung jeglicher Form des Gesellschaftlichen fokussiert. Der Bildungsbegriff impliziert in seiner klassischen Form jedoch ein normatives Moment, das heißt, Bildung soll erfolgen, ist aber nicht notwendige Voraussetzung der Existenz. Sozialisation dagegen ist unumgänglich. Damit wird Bildung im Sinne eines Besitzes zu einem gesellschaftlichen Unterscheidungsmerkmal; Sozialisation erfolgt schicht- und geschlechtsspezifisch, somit sind Inhalte Unterscheidungskriterium, nie aber der Prozess selbst.

Trotz der immensen Widersprüche, die in der Idee von Bildung implizit enthalten sind, und des ideologischen Gehalts der Bildungsvorstellung bleibt der Bildungsbegriff ein theoretischer Grundbegriff der Soziologie, da kein anderer Begriff die Entwicklung und die gezielte Förderung geistiger Fähigkeiten zur Artikulation, Wahrnehmung und Reflexion der Wissensbestände aber auch der emotional-körperlichen Vorgänge in adäquater Weise ausdrückt. Damit ist Bildung als Begriff für die Tätigkeit der Vermittlung und Aneignung von Inhalten im Kindergarten, in den allgemein bildenden und beruflichen Schulen, in den wissenschaftlichen Hochschulen und in den Einrichtungen der Erwachsenenbildung unumgänglich.

National-sozialismus

Der Bildungsbegriff hat sich darüber hinaus in der Soziologie in der kritischen Auseinandersetzung etabliert. Die klassischen Soziologen haben, wie bereits erwähnt, den Bildungsbegriff tendenziell vermieden und statt dessen zunächst einmal versucht, ein Verständnis für die sozialen (und eben nicht natürlichen), auch für die sozial notwendigen, Prägungen eines Menschen zu erreichen. Nach dem zweiten Weltkrieg wird in den Sozialwissenschaften das Scheitern des Projekts „Bildung" angesichts der Verbrechen der Nationalsozialisten und ihrer Duldung oder Förderung durch zahlreiche BürgerInnen konstatiert und Bildung als idealistische Konstruktion kritisiert. (Tatsächlich war die praktizierte Bildung Anfang des 20. Jahrhunderts, wie oben beschrieben, längst nicht mehr im idealistischen Sinn Persönlichkeitsbildung in Auseinandersetzung mit den deutschen Dichtern und Denkern, sondern vielfach zweckorientierte, funktionale Ausbildung. Die Kritik reproduziert damit die Vorstellung von hehrer Bildung jenseits der realisierten Bildung.) In den sechziger und siebziger Jahren der BRD

Sozialisations-Euphorie

schließlich gelangt die Rezeption von Sozialisationstheorien als Kritik an einem normativ-idealistischen Bildungsbegriff zu ihrem Höhepunkt (vgl. Rolff 1967; Fend 1969; Habermas 1973; Oevermann 1976). Die Produktion sozialer Ungleichheit in den Bildungsinstitutionen wird mit Hilfe von Sozialisationstheorien untersucht. Dabei ist ein bekanntes Phänomen, dass Kritik einen Gegenstand erst populär macht. So wie ein Musiker durch schlechte Presse berühmt werden kann, so ergeht es auch dem Bildungsbegriff in der Soziologie. In kritischer Auseinandersetzung wird der Begriff und das Feld der Soziologie der Bildung und Erziehung als Kernbereich der Soziologie etabliert (vgl. ausführlich das folgende Kapitel).

Bildung als Resource

Ein weiterer Grund, auch in der Soziologie am Bildungsbegriff festzuhalten, ist politischer Natur. Nach wie vor definiert sich die deutsche Gesellschaft als Nation über Bildung. Man kann es täglich in den Medien vernehmen. Der ehemalige Bundespräsident Roman Herzog verkündet zum Beispiel: „Wissen ist heute die wichtigste Ressource in unserem rohstoffarmen Land. Wissen können wir aber nur durch Bildung erschließen. (...) Bildung ist der Schlüssel zum Arbeitsmarkt und noch immer die beste Prophylaxe gegen Arbeitslosigkeit. Sie hält die Mechanismen des sozialen Auf- und Abstiegs offen und damit unsere offenen Gesellschaften in Bewegung. Und sie ist zugleich das Lebenselixier der Demokratie in einer Welt, die immer komplexer wird, in der kulturelle Identitäten zu verschwimmen drohen und das Überschreiten der Grenzen zu anderen Kulturen zur Selbstverständlichkeit wird“ (Herzog 1997, S. 1f.). Der Bildungsbegriff, so diffus er auch sein mag, ist aus dem Alltagsbewusstsein nicht fort zu denken. Diese Betonung der Bedeutung der Wissensvermittlung und der Auseinandersetzung mit als Wissen klassifizierter Inhalte bedarf einer kritischen soziologischen Reflexion, welche gezielt die widersprüchliche Spannung des Begriffs aufgreifen muss.

Collège de France

Bildung wird daher heute in der Soziologie (wie in den meisten anderen wissenschaftlichen Disziplinen) nicht mehr auf die aktive Auseinandersetzung mit höheren auf die griechische Antike oder deutsches Kulturgut bezogenen Inhalten begrenzt, sondern als ein Prozess gefasst, der „geistige Aufgeschlossenheit zum Ziel haben muss, Einstellungen und Kenntnisse von der Art, wie man sie braucht, um sich immer neues Wissen anzueignen und mit immer neuen Situationen zurechtzukommen“ (Collège de France 1987, S. 253). In einer solchen Definition gibt es nicht mehr den Zustand des Gebildet-Seins, sondern Bildung benennt die Aneignung der Kenntnisse und der kognitiv-emotionalen Motivation, um flexibel gesellschaftlich notwendiges Wissen immer neu zu erwerben. Die Professoren des Collège de France (darunter Pierre Bourdieu), die auf „Wunsch des Präsidenten der Republik“ Vorschläge für das Bildungswesen der Zukunft erarbeiten, akzeptieren, dass Bildungsinhalte abhängig vom gesellschaftlichen Geschehen sind und ihnen damit immer etwas Willkürliches anhängt. Bildung ist demzufolge ein reflexives Geschehen, dessen Inhalte immer neu bestimmt werden müssen: „in Abhängigkeit vom Wandel der Wissenschaft, die eine fortwährende Neudefinition der Vorstellungen von Natur und Gesellschaft bedeutet, aber auch in Abhängigkeit vom Wandel der wirtschaftlich-sozialen Umwelt und insbesondere von den Veränderungen am Arbeitsmarkt, die auf technologische Innovationen und auf Umstrukturierungen der Betriebe in

Industrie, Handel und Landwirtschaft zurückgehen" (Collège de France 1987, S. 253).

Gegenstand der Soziologie der Bildung und Erziehung

Der Schwerpunkt einer Soziologie der Bildung und Erziehung liegt folglich in der theoretischen und empirischen Untersuchung des Bildungssystems und der in den Bildungsinstitutionen (Kindergarten, Schule, Universität, außerschulische Bildungsträger) vollzogenen Bildungsprozesse. Erziehung im Sinne der geplanten Beeinflussung der Heranwachsenden beobachtet die Soziologie der Bildung und Erziehung vor allem als professionelles Handeln der Lehrenden und ErzieherInnen. Grundannahme einer soziologischen Herangehensweise an Bildungs- und Erziehungsprozesse ist, dass diese nicht nur durch gezielte Einwirkungen, sondern auch durch unbeabsichtigte Folgen des Handelns, durch strukturelle Bedingungen, durch familiäre Beeinflussung und durch Prägung in der Peer Group (also unter Gleichaltrigen) zustande kommen. Bildungs- und Erziehungsprozesse sind demnach immer auch Sozialisationsprozesse. Die Soziologie der Bildung und Erziehung fokussiert in dem lebenslang andauernden und alle Lebensbereiche durchziehenden Verlauf der Sozialisation auf jene Aspekte, die das Bildungswesen betreffen (Hochschulsozialisation, Schulsozialisation usw.). Durch die Konzentration auf die Analyse von Bildungsphänomenen beschäftigt sich die Soziologie der Bildung und Erziehung mehrheitlich (aber nicht ausschließlich) mit Kindheit und Jugend als Lebensphase. Die Soziologie der Bildung und Erziehung untersucht aber auch das Zusammenwirken und die Widersprüche zwischen der Bildungs- und Erwerbssphäre und damit die Auswirkungen von Bildungsprozessen auf die Biographien Erwachsener.

Ziel

Ziel der Soziologie der Bildung und Erziehung ist es, Prozesse der Bildung und Erziehung sowie deren Institutionalisierung im historisch-gesellschaftlichen Kontext zu betrachten und die Bedeutung von Bildung für eine moderne Gesellschaft zu rekonstruieren.

2. Bildung und Erziehung – ein klassisches, soziologisches Feld

Die Soziologie hat sich von Anbeginn an (auch) über die Erforschung von Bildungs- und Erziehungsprozessen profiliert. So gilt zum Beispiel Lester F. Ward nicht nur als klassischer Autor der Soziologie der Bildung und Erziehung, sondern auch als Begründer der amerikanischen Soziologie. Der erste Lehrstuhl für Sozialwissenschaften ist gleichzeitig ein Lehrstuhl für Pädagogik und wird mit Emile Durkheim besetzt. Da die Pädagogik ihr Erkenntnisinteresse im ausgehenden 19. Jahrhundert in erster Linie auf das Individuum und kaum auf die gesellschaftliche Bedeutung von Erziehung und Bildung richtet, findet die Soziologie hier eines ihrer Tätigkeitsfelder.

2.1 Die erste Phase der Institutionalisierung der Soziologie der Bildung und Erziehung

Leitwissenschaft

Die amerikanischen Forscher Lester F. Ward und Franklin H. Giddings betonen schon früh die zentrale gesellschaftliche Bedeutung des Erziehungssystems und die soziale Prägung des Erziehungsgeschehens. Ihre Einsicht in den gesellschaftlichen Charakter der Erziehung führt in den USA zu einer öffentlichen Diskussion darüber, ob nicht die bestehende Erziehungswissenschaft in eine Soziologie der Erziehung und Bildung überführt werden solle. Auch in Frankreich leitet insbesondere Emile Durkheim mittels eines Gesellschaftsvergleichs Erziehung als gesellschaftliches Produkt und damit als soziologischen Gegenstand her. Durkheim nimmt an, über soziologische Erkenntnisse auch Erziehungsziele und optimales pädagogisches Handeln bestimmen zu können und deklariert damit implizit die Soziologie zur Leitwissenschaft für die Pädagogik.

Hilfswissenschaft

Die deutsche Debatte läuft dagegen langsamer an und nimmt auch, da sie mehrheitlich von PädagogInnen geführt wird, eine andere Richtung. Zwar veröffentlicht zum Beispiel Paul Barth im „Encyklopädischen Handbuch der Pädagogik" bereits 1896 einen Artikel zum Verhältnis von Pädagogik und Gesellschaft, aber erst in den 20er Jahren des 20. Jahrhunderts beginnt eine ausführliche Diskussion über das Verhältnis der beiden Disziplinen zueinander (vgl. Geiger 1974, original 1930). Der Pädagoge Paul Luchtenberg plädiert 1925 in den Kölner Vierteljahresheften für Soziologie für eine klare Trennung zwischen der

Pädagogik mit ihrem Ganzheitlichkeitsanspruch und der Soziologie mit ihrer Orientierung auf soziale Beziehungen. Andere ErziehungswissenschaftlerInnen, zum Beispiel Aloys Fischer und Carl Weiß, tendieren dazu, die Soziologie der Erziehung, häufig auch als „Pädagogische Soziologie" bezeichnet, als „Hilfswissenschaft" der Pädagogik zu verankern. Mit Hilfe soziologischer Methoden solle die Frage nach der gesellschaftlichen Prägung und nach dem Einfluss des sozialen Umfelds auf Erziehung in die Erziehungswissenschaft integriert werden (vgl. Plake 1987). Dem halten Soziologen wie Theodor Geiger entgegen, dass die Relevanz der Soziologie der Erziehung nicht nur in der Bereitstellung von Erkenntnissen für die PädagogInnen liege, sondern sich aus dem Erziehungsgeschehen Wirkungsweisen der Gesellschaft herausfiltern lassen, die von allgemeinem Interesse seien (vgl. Geiger 1974, original 1930). Er entwickelt eine Konzeption für die Soziologie der Erziehung als Nebenzweig der allgemeinen Soziologie bzw. als besondere Soziologie. In der deutschen Debatte Anfang des letzten Jahrhunderts geht es also weniger als in den USA oder Frankreich um die Frage, ob die Erziehungswissenschaft durch die Soziologie abgelöst werden kann, als vielmehr um das Problem der Abgrenzung zwischen den Fächern oder gar um eine Funktionalisierung der Soziologie für die Erziehungswissenschaft.

Doppelleben

Die Auseinandersetzungen zwischen Soziologie und Erziehungswissenschaft brechen mit der Machtergreifung der Nationalsozialisten und der Emigration zahlreicher SoziologInnen ab. Geblieben ist bis heute die doppelte Bedeutung der Soziologie der Erziehung und Bildung sowohl als eigenständige soziologische Teildisziplin als auch als Bestandteil der erziehungswissenschaftlichen Ausbildung. In Deutschland sind die Professuren für dieses Fach teilweise den soziologischen Instituten, aber häufig auch den erziehungswissenschaftlichen Fachbereichen zugeordnet. Im letzteren Fall übernehmen die Lehrenden dann in erster Linie Aufgaben in der Ausbildung der ErziehungswissenschaftlerInnen sowie LehrerInnen und erst in zweiter Instanz sind sie in die soziologische Disziplin eingebunden. Es existieren nach wie vor vereinzelte Versuche, eine eigenständige pädagogische Soziologie neben der Soziologie der Bildung und Erziehung durchzusetzen (z.B. Böhnisch 1996). Dabei geht es darum, die pädagogische Transformation des soziologischen Wissens stärker in den Vordergrund der fachlichen Diskussion zu rücken. Umgekehrt hat sich durch die geschilderten Auseinandersetzungen die geisteswissenschaftliche Pädagogik mit ihrer individualistischen und idealistischen Bildungstradition weitgehend in eine sozialwissenschaftlich orientierte Erziehungswissenschaft gewandelt (vgl. Krüger/ Rauschenbach 1994), so dass die Erziehungswissenschaft selbst „soziologischer" geworden ist. Die Grenzen zwischen der Soziologie der Bildung und Erziehung und einer sozialwissenschaftlich orientierten Erziehungswissenschaft können daher nicht immer klar gezogen werden.

Soziologisch orientierte Erziehungswissenschaft

Im folgenden sollen exemplarisch für die Fülle klassischer Analysen in der ersten Phase der Institutionalisierung der Soziologie der Bildung und Erziehung zwei Theoretiker ausführlich vorgestellt werden: Emile Durkheim, weil er das Gesellschaftliche am Erziehungsgeschehen als einer der ersten und bis heute besonders prägend expliziert hat, und Karl Mannheim, da dieser durch die Erfahrung mit dem deutschen Faschismus besonders pointiert hergeleitet hat, wie notwendig es ist, sich mit der gesellschaftlichen Dimension von Erziehung zu beschäftigen. Wer sich mit weiteren KlassikerInnen der Erziehungs- und Bildungssoziologie beschäftigen will, dem sei von Klaus Plake „Klassiker der Erziehungssoziologie" (1987) empfohlen, weil dort nicht nur Auszüge aus zahlreichen klassischen Texten abgedruckt, sondern auch erläuternde Einführungen zu den Texten zu finden sind.

2.1.1 Emile Durkheim

Kritisch-moralische Perspektive

Emile Durkheim wird am 15. April 1858 im lothringischen Epinal geboren. Er gilt nicht nur als einer der entscheidenden Gründungsväter der Soziologie der Bildung und Erziehung, sondern überhaupt als erster Fachwissenschaftler der Soziologie. Durkheims kritisch-moralische Perspektive auf Gesellschaft ist sowohl durch sein Elternhaus, als auch durch den Zeitgeist beeinflusst. Schon in den frühen Jahren seiner wissenschaftlichen Karriere betont er die Notwendigkeit, eine auf der Soziologie basierende Morallehre für die LehrerInnenausbildung und in der Folge für die Schulausbildung zu entwickeln. Durkheim verfolgt das Ziel, über das Erziehungssystem den moralischen Zusammenhalt der Gesellschaft zu erheben. Mit einem extra auf ihn zugeschnittenen Lehrstuhl für Pädagogik und Sozialwissenschaft an der Universität von Bordeaux wird 1896 zum ersten Mal in der Geschichte „Sozialwissenschaft" als Universitätsfach verankert. Später vertritt Durkheim die beiden Fächer Pädagogik und Soziologie als Ordinarius an der Sorbonne in Paris. Bemerkenswert ist, dass auf Durkheims Betreiben die Teilbezeichnung des Lehrstuhls von Pädagogik in Erziehungswissenschaft gewandelt wird und damit die Notwendigkeit empirischer Forschung statt normativer Begründungszusammenhänge betont wird.

Durkheims Werke

Durkheim zeichnet für mehrere bildungssoziologische Werke verantwortlich. Die bekanntesten sind „Die Entwicklung der Pädagogik. Zur Geschichte und Soziologie des gelehrten Unterrichts in Frankreich" (1977, original 1938), in dem er sich historisch mit den Erscheinungsformen von Bildung und Erziehung auseinandersetzt, „Erziehung und Soziologie" (1972, original 1911), in welchem er sich mit den wissenschaftlichen Grundlagen von Erziehung aus soziologischer Perspektive beschäftigt und „Erziehung, Moral und Gesellschaft" (1999[3]a, original 1902/03), welches Moral als gesellschaftliche Verhaltensregeln im Gegensatz zu dem Druck innerer Triebe entfaltet.

Definition von Erziehung

Erziehung ist für Durkheim „die Einwirkung, welche die Erwachsenengeneration auf jene ausübt, die für das soziale Leben noch nicht reif sind. Ihr Ziel ist es, im Kinde gewisse physische, intellektuelle und sittliche Zustände zu schaffen und zu entwickeln, die sowohl die politische Gesellschaft in ihrer Einheit als auch das spezielle Milieu, zu dem es in besonderer Weise bestimmt ist, von ihm verlangen" (Durkheim 1972, S. 30). Durkheim versteht demzufolge Erziehung als Vergesellschaftungsprozess, d.h. als Prozess der Eingliederung eines Individuums in die Gesellschaft. Diese Definition von Erziehung entspricht seinem soziologischen Grundverständnis. Durkheim bestimmt in seinem Grundlagenwerk „Regeln der soziologischen Methode" die Soziologie als die „Wissenschaft von den Institutionen, deren Entstehung und Wirkungsart" (Durkheim 1961, original 1895, S. 100). Institutionen sind Durkheim zufolge soziale Phänomene, die dem Menschen äußerlich sind, d.h. sie sind ihm nicht angeboren, sondern anerzogen. Sie üben auf den Willen des einzelnen sozialen Druck aus. Sie haben den Status

Institutionen

der Allgemeingültigkeit innerhalb einer Gesellschaft. Obwohl sie das Verhalten jedes einzelnen beeinflussen, sind Institution mehr als die Summe der individuellen Handlungen. Sie sind eigenständige „soziale Tatsachen“. In wissenschaftlichen Arbeiten gilt es, sie wie Dinge zu beschreiben und zu analysieren. Die Sprache zum Beispiel wird dem Menschen anerzogen, zwingt ihm Strukturen auf, gilt weitgehend allgemein innerhalb einer Gesellschaft und ist doch, obwohl sie nur über die Handlungen der Individuen existiert, eine eigenständige Institution.

Soziales durch Soziales erklären

Daraus leitet sich ab, dass das Gesellschaftliche nicht über psychologische, biologische oder physikalische Theorien zu erklären ist, sondern nur über soziologische. Durkheims erster methodologischer Satz lautet daher, dass Soziales sich nur durch Soziales erklären lässt. Erziehung ist der Vorgang, bei dem das Soziale in den Menschen kommt. Soziologie ist die Wissenschaft, die diesen Vorgang untersucht. Bei seiner Definition nimmt Durkheim eine bedeutende Einschränkung vor. Anders als in der modernen Sozialisationstheorie untersucht er explizit weder den Einfluss der Gleichaltrigen noch die Prägung des Menschen durch die dingliche Umwelt (vgl. Durkheim 1972, S. 26ff.), sondern fasst unter Erziehung einzig und allein den Eingriff der Erwachsenen in das Aufwachsen der Kinder.

Differenzierung

Dabei muss diese Erziehung gleichzeitig auf gesamtgesellschaftliche und auf gruppenspezifische Interessen ausgerichtet sein. Ausgehend von dem Befund einer zunehmenden gesellschaftlichen Differenzierung, welche er ausführlich in seinem Buch „Über soziale Arbeitsteilung“ (1999³b, original 1893) herleitet, entwickelt er die These, dass nur prähistorische Gesellschaften, die nicht arbeitsteilig und spezialisiert organisiert sind, eine einheitliche Erziehung gewährleisten können. Differenzierte Gesellschaften (nach Kasten, Klassen, Milieus usw.) dagegen müssen sowohl einen gemeinsamen Kodex als auch gruppen-/kasten-/milieuspezifische physische und geistige Zustände vermitteln. Das Kind muss auf die spezielle Funktion, die es in der Gesellschaft zu erfüllen hat, ausgebildet werden. „Da das Kind im Blick auf die Aufgabe, die es einmal erfüllen soll, vorbereitet werden muss, kann die Erziehung von einem gewissen Alter an nicht mehr für alle, an die sie sich wendet, die gleiche sein“ (Durkheim 1999³a, S. 40).

Spezialisierung

Daraus leitet sich die Vielgestaltigkeit von Erziehungspraktiken und -inhalten ab. Diese Spezialisierung kann nicht, wie Durkheim explizit betont, mit der Natur des Menschen, sondern nur über die Notwendigkeit der Arbeitsteilung legitimiert werden. Der Durchschnittsmensch ist zu vielen Berufen fähig. Die Spezialisierung leitet sich nur aus der Notwendigkeit ab, die gesellschaftlichen Aufgaben zwischen den Mitgliedern verteilen zu müssen. Gleichzeitig müssen alle Kinder dieser Gesellschaft die schichtübergreifend spezifischen Normen einer Gesellschaft internalisieren. Da Vorstellungen und Praktiken von Erziehung und Bildung gesellschaftsabhängig sind, ist es die Aufgabe von Erziehung neben der Spezialisierung auch das Gemeinsame einer Kultur zu vermitteln (vgl. sowohl Durkheim 1972, S. 26ff als auch Durkheim 1977). „Es gibt kein Volk, in dem nicht eine gewisse Anzahl von Ideen, von Gefühlen und von Praktiken existiert, die die Erziehung unterschiedslos allen Kindern beibringen muss, welcher sozialen Kategorie sie auch angehören“ (Durkheim 1999³a, S. 42). Damit ist Erziehung nach Durkheim weit mehr als die Prägung eines Individuums. Es ist das

Mittel, mit dem eine Gesellschaft ihre Existenzgrundlagen immer wieder erneuert. Zusammenfassend betont er: „Der Mensch, den die Erziehung in uns verwirklichen muss, ist nicht der Mensch, den die Natur gemacht hat, sondern der Mensch, wie ihn die Gesellschaft haben will; und sie will ihn so haben, wie ihn ihre innere Ökonomie braucht" (Durkheim 1999^3a, S. 44).

Zum Weiterlesen empfehle ich: Walford 1998; Luhmann 1999^3, Müller 1983; Giddens 1978; König 1978.

2.1.2 Karl Mannheim

Karl Mannheim wird am 27. März 1893 in Budapest geboren. Er entstammt einer bürgerlichen, jüdischen Familie. Die ersten 26 Jahre, in denen er hauptsächlich in Ungarn studiert und 1917 an der Universität Budapest im Fachbereich Philosophie über die „Strukturanalyse der Erkenntnistheorie" promoviert, werden nur von einem mehrsemestrigen Studienaufenthalt in Berlin unterbrochen. Dort hört er vor allem die Vorlesungen von Georg Simmel. Im Dezember 1919, als die kommunistische Regierung in Ungarn gestürzt wird, flüchtet Mannheim über Österreich nach Deutschland. Als Habilitand von Alfred Weber findet er in Heidelberg eine neue Wirkungsstätte.

Wissenssoziologie

Mannheims zentrale Frage wird der Zusammenhang zwischen dem Sein und dem Bewusstsein. Mit Arbeiten wie „Konservatismus: Ein Beitrag zur Soziologie des Wissens" (1984, original 1925) oder „Ideologie und Utopie" (1929a) trägt Mannheim wesentlich zur Etablierung einer speziellen Wissenssoziologie bei. Mannheim vertritt die These, dass die Vielfalt der Ideensysteme auf drei Typen reduziert werden könne: Liberalismus, Konservatismus und Sozialismus. Jede dieser als „Ideologien" bezeichneten Richtungen deutet er als Funktion einer besonderen Seinsart, welche durch sich wandelnde Klassen- und Generationsstrukturen bestimmt sei. Daraus leitet Mannheim ab, dass das soziologische Wissen immer in Weltauslegungen eingebettet ist (Mannheim 1929b, S. 45). Aufgabe der Wissenssoziologie sei es nun, die Zeit- und Ortsgebundenheit des Denkens immer wieder neu zu untersuchen.

Totalitäre Bedrohung

1929 erhält Mannheim einen Lehrstuhl für Soziologie an der Universität Frankfurt/Main, doch schon kurz nach der Machtergreifung der Nationalsozialisten muss Mannheim erneut fliehen. Im englischen Exil findet er eine neue Anstellung an der London School of Economics. Hier findet er zu der Überzeugung, dass weder seine persönliche Situation noch die politischen Bedingungen seiner Zeit es zulassen, weiter Wissenssoziologie zu betreiben. Er sieht es nun als seine Aufgabe an, die globale Krise, die seines Erachtens den deutschen Faschismus möglich gemacht hat, zu analysieren. Seine Perspektive ist auf die sozialen Bedingungen gerichtet, unter denen demokratische Gesellschaften angesichts der totalitären Bedrohung überleben können. Fragen der Erziehung rücken in den Vordergrund seines Werkes.

Allerdings kann Mannheim die verschiedenen Manuskripte zur Erziehungssoziologie bis zu seinem Tod im Jahre 1947 nicht mehr selbst für die Veröffentlichung fertig stellen. Sein Assistent William A. C. Stewart übernimmt diese Aufgabe posthum (Mannheim/Stewart 1973, original 1962). Ungeklärt ist nun, welche Passagen originär von Mannheim stammen und an welchen Stellen Stewart in die Struktur der Texte eingegriffen hat.

Erziehung als Vergesellschaftung

Bereits in seinem 1935 erschienenen Werk „Mensch und Gesellschaft" definiert Mannheim als soziologische Perspektive auf Erziehung: „Die Soziologie betrachtet die Erziehung nicht nur als Mittel zur Erreichung abstrakter Bildungsideale, wie zum Beispiel des Humanismus oder der technischen Spezialisierung, sondern als einen Teil der schrittweisen Eingliederung des Menschen in die Gesellschaft. Man kann eine Erziehung nur dann verstehen, wenn man weiß, für welche Gesellschaft und für welche soziale Stellung die Schüler erzogen werden" (Mannheim 1967, S. 316f.). Erziehung wird hier demzufolge als Vergesellschaftung begriffen und zwar als eine langsame Eingliederung, da erstens die Individuen Entwicklungen durchlaufen müssen und zweitens Gesellschaft sich ebenfalls entwickelt und wandelt (vgl. Mannheim/Stewart 1973, S. 34). Erziehung braucht Zeit. In dieser Definition impliziert ist eine Kritik an der klassischen Pädagogik, die Bildungsideale allgemein festzulegen versucht, ohne den gesellschaftlichen Kontext zu berücksichtigen. Unterstellt werde dort, dass die Wirkung der Erziehung nur an den Individuen sichtbar werde (Mannheim/ Stewart 1973, S. 68). Diese Kritik bedeutet jedoch keineswegs, dass Mannheim sich nun einer kollektivistischen Perspektive anschließt. Diese kennzeichnet er als jene Formen der Pädagogik, welche Erziehung nur als Methode begreifen, mit der jede neue Generation auf zuvor festgelegte Funktionen und vor allem auf Staatsbürgerschaft hin ausgebildet wird (Mannheim/Stewart 1973, S. 69).

Soziales Selbst

Mannheim schlägt einen dritten Weg vor. „In dieser Sicht ist ein Individuum nicht eine abstrahierte Persönlichkeit, sondern entwickelt sich als ein soziales Selbst in der Gesellschaft, die zu einer bestimmten Zeit in der Geschichte existiert" (Mannheim/Stewart 1973, S. 69). Es gibt also für Mannheim keine menschliche Natur an sich und keine erbliche Vorbestimmung der individuellen Entwicklung, sondern das Selbst entwickelt sich im gesellschaftlichen Kontext, welcher jedoch keineswegs kontinuierlich gleich, sondern generations- und schichtspezifisch ist. Das heißt, in den Erziehungsbedingungen unterscheiden sich Kinder von ihren Eltern und Kinder von Kindern anderer Schichten. Dabei versteht Mannheim unter Erziehung, anders als Durkheim, auch den Einfluss der Primärgruppen Gleichaltriger oder der Nachbarschaft (Mannheim/ Stewart 1973, S. 141ff.). Sekundäre soziale Gruppen, zum Beispiel für Mannheim die Schule, erziehen ebenfalls die Kinder jeder Generation. Wichtig ist Mannheim und Stewart dabei, dass stets die Menschen in einer konkreten, geschichtlich-gesellschaftlich gewordenen und sich immer weiter wandelnden Gesellschaft (und für diese) erzogen werden. Will man also die Demokratie vorantreiben und sich gegen totalitäre Tendenzen zur Wehr setzen, dann muss Erziehung die „Mitglieder der Gesellschaft darauf vorbereiten (...), sich einerseits anzupassen und andererseits (...) Möglichkeiten und Spielraum für eine Individualität" (Mannheim/Stewart 1973, S. 28) zu finden. Die soziologische und gesellschaftliche Aufmerksamkeit muss demzufolge darauf gerichtet werden, Menschen zu bilden und die gesellschaftlichen Strukturen hierfür zu schaffen, dass sowohl individuelle Freiheit als auch kollektive Fähigkeiten ausgebildet werden. Das bedeutet, mit dem Gebrauch von Verboten und Hemmungen so sparsam wie möglich zu sein, zwischen humanen und schädlichen Verboten zu unterscheiden sowie die Erziehungseinrichtungen so zu konzipieren, dass sie helfen, sich einzufügen (vgl.

Sekundäre soziale Gruppen

Verbote

Mannheim 1951, original 1943, S. 117). Deshalb, so folgern Mannheim/Stewart (1973, S. 117), müsse jede erzieherische Tätigkeit und Forschung (auch) als Soziologie der Erziehung konzipiert werden.

Wenn Sie Mannheim vertiefend studieren wollen, lohnt es sich, folgendes zu lesen: Blomert 1999; Huke-Dedier 1995; Kettler u.a. 1989; Longhurst 1989; Boris 1971

2.2 Die zweite Phase der Institutionalisierung der Soziologie der Bildung und Erziehung

Soziologie im NS-Staat

Die SoziologInnen, die während der nationalsozialistischen Herrschaft in Deutschland bleiben wollen und können, beschäftigen sich – soweit bekannt ist – kaum mit Bildungsfragen. Es gibt zwar Parallelen zwischen dem Primat des Volksbegriffs in den soziologischen Arbeiten von zum Beispiel Max Rumpf, Hans Freyer oder Gunther Ipsen und der nationalsozialistischen Politik, welche darauf zielt, alle (deutschen, arischen) Schüler und Schülerinnen zur Pflichterfüllung und gehorsamen Einordnung in das politische System zu erziehen, die Soziologie ist jedoch in dieser Zeit stärker mit einer nationalsozialistischen Selbst- und Standortbestimmung als mit bildungssoziologischen Fragen beschäftigt (zur Soziologie und Nationalsozialismus siehe zum Beispiel Jaeggi u.a. 1983; Paulsen 1988). Die Nationalsozialisten selbst wenden sich explizit gegen alle liberalen Bildungsvorstellungen von individueller Förderung und verfolgen die Idee einer Gruppenerziehung im Sinne der Anpassung an gesellschaftspolitisch bestimmte Ziele.

Instrumentalisierung der Erziehung

Außerhalb der Grenzen Deutschlands jedoch wächst das Interesse an einer Soziologie der Bildung und Erziehung, wie man am Beispiel von Karl Mannheim sieht, durch die Offensichtlichkeit, mit der Faschisten Erziehung für die politischen Ziele der Regierung instrumentalisieren. Der soziale Charakter von Erziehung und die Notwendigkeit intellektuellen Engagements tritt so deutlich wie selten zuvor hervor. In Deutschland dauert es noch bis in die 1960er Jahre hinein, bis die Soziologie der Bildung und Erziehung flächendeckend an den Hochschulen und Forschungszentren institutionalisiert wird. Es bedarf noch einer tief greifenden Krise des Bildungssystems.

Sputnik-Schock

In den 1950er Jahren nimmt der Anteil der Studierenden eines Jahrgangs kontinuierlich zu. Da es sich gleichzeitig um geburtenstarke Jahrgänge handelt, entstehen an den Universitäten bald räumliche, personelle und finanzielle Engpässe (Friedeburg 1992, S. 336). Die Unzufriedenheit an den Hochschulen wächst. 1957 ereignet sich dann noch der so genannte Sputnik-Schock, als die UdSSR als erstes Land einen Satelliten in eine Umlaufbahn der Erde schickt. Die Forderungen nach einer Bildungsreform, auch um reale oder wahrgenommene Technologiedefizite aufzuholen, werden immer dringlicher. Die Lage spitzt sich weiter zu, als durch eine gute Konjunktur und anhaltende Vollbeschäftigung die AbsolventInnen der gymnasialen Oberstufe abnehmen, da viele Jugendliche die guten Beschäftigungsmöglichkeiten sofort nutzen wollen. 1964 veröffentlicht Georg Picht in der konservativen Wochenzeitschrift „Christ und Welt“ eine Analyse der „deutschen Bildungskatastrophe“: Geringere Abiturientenzahlen als

Deutsche Bildungskatastrophe

alle Nachbarländer, sinkende öffentliche Bildungsausgaben bei gestiegenem Bruttosozialprodukt etc. Modernisierungsrückstände im Bildungswesen werden nun endgültig zu einem viel diskutierten Topos (vgl. zusammenfassend auch Goldschmidt 1991).

Bildungsreform

Seit Bildung existiert, gibt es auch Bildungsreformen (vgl. zusammenfassend Friedeburg 1992; zu neueren Entwicklungen auch Reinhardt 1999). Zu Beginn des 20. Jahrhunderts entsteht im Zusammenspiel von sozialen Bewegungen (Arbeiterbewegung, Frauenbewegung), Reformpädagogik und Regierungspolitik eine allgemeine Grundschulbildung (vgl. z.B. Parsons 1997[5], S. 180), höhere Mädchenschulen (vgl. Kleinau/Opitz 1996) und das duale System der Berufsausbildung (Friedeburg 1992). Das Neue der Bildungsreform der 1960er Jahren ist nun, dass die Soziologie erstmalig die Reformen mit beeinflusst. 1958 gründet sich, unter der Leitung von Helmuth Plessner, der Fachausschuss für Soziologie der Bildung und Erziehung in der Deutschen Gesellschaft für Soziologie. 1962 ruft Hellmut Becker das Berliner Institut für Bildungsforschung in der Max-Planck-Gesellschaft (das spätere Max-Planck-Institut für Bildungsforschung) ins Leben, welches unter der Leitung von Dietrich Goldschmidt einen wichtigen soziologischen Schwerpunkt hat.

Durch die Einsicht in die Notwendigkeit einer Bildungsreform wird die Soziologie der Bildung und Erziehung mit den notwendigen Forschungsgeldern ausgestattet, die es dieser wiederum ermöglichen, den öffentlichen Diskurs um Bildungsfragen weiter anzuregen. So paart sich der Gedanke, dass Deutschland als Bildungsland leistungsfähiger werden muss, mit dem zentralen soziologischen Thema, der Frage nach sozialen Ungleichheiten bestehender Strukturen. Über „Chancengleichheit" soll gleichermaßen der Bedarf an Fachkräften gedeckt und Nachteile an Lebenschancen abgebaut werden. Das katholische Mädchen vom Lande, welches nun auch die Chance auf ein Studium erhält, wird zur Kunstfigur, die immer wieder zitiert und diskutiert wird, weil hier alle Benachteiligungen, nämlich weibliches Geschlecht, aus ländlichen Regionen und aus den eher bildungsfernen katholischen Haushalten stammend, zusammengefügt werden.

Bildung als Bürgerrecht

In empirischen Untersuchungen wird die Funktion der Auslese (Selektion) und der Zuteilung von Bildungs- und somit Lebenschancen (Allokation) im Erziehungssystem analysiert (z.B. Grimm 1966; Peisert 1967; Rolff 1967; Goldschmidt/Sommerkorn 1970; Ortmann 1971; vgl. zusammenfassend Krais 1996b). Ralf Dahrendorf (1965) begreift Bildung als Bürgerrecht und fordert eine aktive Bildungspolitik. Denn nur sie ist, Dahrendorf zufolge, in der Lage, traditionelle Verhaltensweisen aufzubrechen und somit nicht nur die Praxis der Schulen, sondern auch Entscheidungen der Eltern über die Schullaufbahn der Kinder zu verändern.

Studentenbewegung

Wichtige Impulse, insbesondere für die Hochschulbildung, kommen auch von der Studentenbewegung, welche sich Ende der 1960er Jahre in Auseinandersetzung mit der nationalsozialistischen Vätergeneration, der Diskriminierung von Minderheiten und den imperialistischen Bestrebungen, insbesondere der USA (Vietnamkrieg), bilden. Mit Sprüchen wie „Unter den Talaren – der Muff von 1000 Jahren" (wobei sich „1000 Jahre" auf den Anspruch der Nationalsozialisten, das 1000jährige Reich zu errichten, bezieht) werden die autoritären Strukturen der Hochschulen und der Fortbestand der faschistischen Tradition an den Universitäten angeprangert. Dies führt zu einer grundsätzlichen Hochschulreform, in deren Folge nun freie HochschullehrerInnen-Stellen öffentlich ausgeschrieben werden und Studierende ein größeres akademisches Mitspracherecht bekommen (Friedeburg 1992, S. 387f.). Auch dieser Prozess wird unterstützend und analytisch von Bildungssoziologen begleitet (z.B. Friedeburg 1969; Habermas 1969). Im Folgenden werden deshalb exemplarisch einige Gedanken und Texte, die die Bildungssoziologie nachhaltig beeinflusst haben und bis heute prägen, kurz vorgestellt.

2.2.1 Talcott Parsons

Revolutionärer Strukturwandel

Der amerikanische Soziologe Talcott Parsons, 1902 in Colorado Springs geboren, hat die Soziologie insbesondere in den fünfziger und sechziger Jahren so nachhaltig beeinflusst, dass seine strukturell-funktionale Theorie gelegentlich für die einzig mögliche Form der Soziologie gehalten wird. Dementsprechend einflussreich sind auch seine Ausführungen zur Schule als soziales System (Parsons 1997[5], original 1964). Noch heute wird das Erziehungssystem, angeregt durch Parsons, häufig unter der Frage seiner Funktionalität für die Gesellschaft behandelt (vgl. Kapitel 5.2).

Parsons geht von der Annahme aus, dass die moderne Gesellschaft durch drei Prozesse „revolutionären Strukturwandels" (Parsons 1990, original 1972, S. 11) gekennzeichnet sei: die industrielle Revolution, die demokratische Revolution und die Revolution des Bildungswesens. Durch die Bildungsrevolution, das heißt durch die Universalisierung der Schulbildung und die Differenzierung des Hochschulwesens, wird Wissen – Parsons zufolge – zur Voraussetzung der Handlungsfähigkeit jedes einzelnen und der Gesellschaft. Wissen werde eingesetzt, um menschliche Ziele und Werte zu verwirklichen (Parsons 1990, S. 13ff.).

Schule als soziales System

Durch die allgemeine Aufwertung von Bildung wächst die Bedeutung des Erziehungssystems. Die Schule wird neben der Familie zur entscheidenden Sozialisationsinstanz. Dementsprechend zentral sieht Parsons auch die Rolle der Soziologie der Bildung und Erziehung in der Soziologie: „Die Struktur des öffentlichen Schulsystems und die Analyse der Formen, in denen es sowohl zur Sozialisation der Individuen als auch zu ihrer Verteilung auf gesellschaftliche Rollen beiträgt, ist meiner Ansicht nach für alle, die sich als Wissenschaftler mit der amerikanischen Gesellschaft befassen, von entscheidender Bedeutung" (Parsons 1997[5], S. 192f.).

Rollen

Prozesse der Erziehung und Bildung zu untersuchen, ist dem zufolge nicht nur deshalb von Interesse, weil Wissen die Handlungsfähigkeit stärkt, sondern auch und vor allem, so verdeutlicht das Zitat, weil Erziehung sowohl Rollenübernahme als auch Rollenverteilung gewährleistet. Wie Durkheim geht auch Parsons davon aus, dass Erziehung der Differenzierung der Gesellschaft Rechnung tragen und die Individuen dementsprechend auf spezielle Rollen bzw. Berufe vorbereiten muss. Für Parsons wird diese Aufgabe nicht durch Erziehung allgemein bewältigt, sondern durch Schulerziehung im Speziellen.

Funktionen

Die Schule muss gleichermaßen Persönlichkeiten ausbilden, die fähig und motiviert sind, Erwachsenenrollen zu übernehmen sowie dieselbe Institution die menschlichen Ressourcen innerhalb der Rollenstruktur der Erwachsenengesellschaft verteilen muss. Anders gesagt, hat die Schule zwei Funktionen: Sie sozialisiert für das Erwachsenenleben und sie organisiert die Verteilung auf zukünftige Berufssparten.

Normen und Werte

Wie erfüllt sie diese Funktionen? In seinem ersten Grundlagenwerk „The Structure of Social Action" (1968, original 1937) gelangt Parsons zu der Erkenntnis, dass soziale Ordnung nur auf der Basis eines Wert- und Normkonsens möglich ist. Dabei reicht es nicht, dass Normen und Werte verpflichtend festgelegt werden, sondern der einzelne Akteur muss darüber hinaus motiviert sein, die Normen und Werte in seinem Handeln zu verwirklichen. Damit nicht nur die Fähigkeit, sondern auch die Bereitschaft zur Normübernahme entfaltet wird, bedarf es der Identifikation.

In der Familie wird sowohl die Bereitschaft, mit den Normen konform zu gehen, als auch die Bereitschaft, kognitive Leistung zu erbringen, ausgebildet. Dennoch sind die Kinder – so Parsons – zu Beginn der Grundschulphase noch wenig „determiniert" (Parsons 1997[5], S. 166). Wirklich festgeschrieben ist zu diesem Zeitpunkt nur die Geschlechtsrolle. Da die Grundschule ihr Klientel, die Kinder, aus der unmittelbaren Nähe akquiriert, sei die Sozialstruktur, die sich in der Klasse abbilde, sehr homogen. Dementsprechend würden sich die Ausgangsbedingungen für alle Kinder, mit Ausnahme der Geschlechtszugehörigkeit, gleichen. Die Grundschule ist „eine Verkörperung des fundamentalen amerikanischen Werts der Chancengleichheit" (Parsons 1997[5], S. 180). Die Startbedingungen seien weitgehend gleich. Erfolg könne durch Leistung erzielt werden.

Leistung

Dabei mischen sich gerade in den unteren Klassen zwei verschiedene Formen von Leistung: kognitive Leistung (Informationen, Fertigkeiten, Wissen) und moralische Leistung (Betragen, Respekt, Rücksichtnahme etc.). Sehr gute Schüler und Schülerinnen erbrächten beide Leistungen, welche dann von der Schule durch Noten und Beurteilungen bewertet werden. Parsons sieht zwar sehr wohl, dass der Status des Berufs des Vaters mit dem Schulerfolg der Kinder korreliert, er führt dies jedoch auf unterschiedliche intellektuelle Fähigkeiten zurück. So bleibt er bei seiner These, dass die Schule über Noten und Beurteilungen auf der Basis von Leistungen ihre Selektionsfunktion erfülle.

Identifikation

Die Bereitschaft zur Leistung, und zwar sowohl zur kognitiven Leistung wie zur moralischen Leistung, fördere die Schule durch die Identifikation mit der Lehrerin. Die Lehrerin, in der Grundschule in der Regel eine Frau, biete sich als mutter-ähnliche Identifikationsperson an. Sie ist zunächst kontinuierlich und in vielen Fächern gleichzeitig für die Kinder da. Später unterrichten verschiedene Lehrerinnen und darauf folgend Lehrer wie Lehrerinnen. So lerne das Kind, sich von der individuellen Persönlichkeit zu lösen und mit der Rolle zu identifizieren. Damit werde die Voraussetzung geschaffen, universelle Muster zu übernehmen. Wenn das Kind nun im Elternhaus gelernt hat, sich mit den Eltern zu identifizieren, wird es in der Schule auch lernen, um der Lehrerin zu gefallen. Wenn das Kind sich in der Familie jedoch eher von den Eltern abgegrenzt hat, wird es nicht in gleichem Maße effektiv lernen wie das identifizierte Kind. Es wird das Bildungssystem früher verlassen und sich einen Beruf suchen (vgl. zum Identifikationsprozess auch ausführlich Parsons u.a. 1955).

Zusammenfassend heißt das, Parsons zufolge wird die Verteilung (Allokation) und die Auslese (Selektion) über Identifikation als Prozess und über Beurteilung von Leistung organisiert. Ein geringerer Intelligenzquotient verweist auf unterschiedliche intellektuelle Fähigkeit je nach Schicht, unterschiedliche Erziehungspraxen auf differente Muster der Identifikation und damit auf unterschiedliche Leistungsmotivation. Für die Soziologie der Bildung und Erziehung folgt aus der Erkenntnis, dass die Herstellung von sozialer Schichtung eine Funktion des Bildungssystems ist, dass Schichtungstheorie (bzw. Klassentheorie) und Bildungstheorie eng zusammen gehören.

Zur kritischen Auseinandersetzung mit Parsons Werk sei empfohlen: Münch 1982; Münch 1989; Wenzel 1990; Joas 1992; Alexander 1993; Schwinn 1993; Barber/Gerhardt 1999.

2.2.2 Helmut Schelsky

Skeptische Generation

Der 1912 in Chemnitz geborene Helmut Schelsky gehört zu den großen deutschen Nachkriegssoziologen. 1957 erweckt er die Aufmerksamkeit auch größerer Teile der Bevölkerung, indem er in seinem Werk „Die skeptische Generation" aufzeigt, dass die Jugend der 1950er Jahre durch die Erfahrung mit dem Nationalsozialismus pragmatisch und distanziert gegenüber jeglichen politischen Ideen geworden ist. Sie suche ihr Glück nun in der Familie und im beruflichen Fortkommen. Schelskys These ist, dass sich durch diese Grundeinstellung der Jugendlichen die Differenz zwischen der Jugend und den Erwachsenen angleicht. Schelsky betrachtet die Nachkriegsgesellschaft ferner als „nivellierte Mittelstandsgesellschaft". Durch den Aufstieg der IndustriearbeiterInnen und der technischen bzw. Verwaltungsberufe sowie dem gleichzeitigen Abstieg der ehemaligen Oberschichten habe sich eine sich wenig unterscheidende Mittelschicht herausgebildet (z.B. Schelsky 1965, S. 134).

Nivellierte Mittelstandsgesellschaft

Auf diesen beiden Grundannahmen (nivellierte mittelständig-kleinbürgerliche Gesellschaft und Rückzug ins private Glück) baut Schelsky seine erziehungssoziologischen Überlegungen auf. Seine Grundannahme ist – auf den ersten Blick der von Parsons ähnlich – dass die Schule unter diesen gesellschaftlichen Bedingungen zur „entscheidenden zentralen sozialen Dirigierungsstelle" (Schelsky 1965, S. 136) wird. Jenseits der Klassengefüge individualisiert und auf Beruf und Familie hin orientiert, werde die Schule zur entscheidenden Instanz für den zukünftigen sozialen Rang und das Ausmaß der Konsummöglichkeiten. Nicht nur die Kinder und Jugendlichen selbst, sondern vor allem die Eltern versuchten, durch den schulischen Erfolg der Kinder ihren Lebensstandard zu sichern oder zu verbessern. Gleichzeitig formalisiere die Wirtschaft ihre Zugangsbedingungen, so dass Schule tatsächlich Rang, Stellung und Lebenschancen dirigiere. Im Unterschied zu Parsons jedoch, der das Schulsystem als Verteilungsinstanz für den zukünftigen gesellschaftlichen Status unproblematisch als gegebene Funktion beurteilt – bewertet Schelsky die Schule mit dieser exponierten gesellschaftlichen Steuerungsfunktion als gänzlich überfordert. Erstens fänden gewünschte Kooperationen nicht statt, da Eltern mit ihren „privaten Ansprüchen und Egoismen" (Schelsky 1965, S. 138) den Sachstrukturen einer bürokratischen Organisation gegenüber stünden. Zweitens verändere sich die Bedeutungsstruktur der Schule. Sie sei nicht länger die fördernde Institution, die Begabte auslese, sondern sie ist die verweigernde Institution, die mit ihren Leistungsurteilen als

Überforderte Schulen

berechtigt empfundene Ansprüche zurückweise. Schelsky schlägt vor, um der Erziehungsaufgabe willen die Dirigierungsfunktion abzubauen oder abzuschaffen.

Zur Vertiefung des Ansatzes von Helmut Schelsky empfehle ich: Baier 1986; Boll 1997.

2.2.3 Theodor W. Adorno

Institut für Sozialforschung

Die Person des 1903 in Frankfurt am Main geborenen Theodor Wiesengrund-Adorno, der sich später Theodor W. Adorno nennt, ist auf das Engste mit dem 1920 gegründeten, marxistisch orientierten Institut für Sozialforschung verknüpft. Tatsächlich entsteht die enge Zusammenarbeit mit dem ehemaligen Leiter des Instituts, Max Horkheimer, und damit das soziologische Profil des Philosophen und Musikkritikers/Komponisten Adorno erst im Exil. Adorno, der 1931 die Lehrbefugnis an Universitäten erhält, verliert diese schon 1933 mit der Machtergreifung der Nationalsozialisten wieder. Wie auch bei Mannheim wächst bei Adorno der Wunsch, Gesellschaft zu begreifen, mit der Erfahrung des Faschismus und des Exils. Nach einigen Jahren am Oxforder Merton College nimmt Adorno die Einladung Horkheimers an, Mitarbeiter am Institute of Social Research an der Columbia University in New York zu werden. Die enge Zusammenarbeit mündet in der gemeinsam verfassten „Dialektik der Aufklärung" (Horkheimer/Adorno 1988, original 1947), die auch Adornos erziehungssoziologisches Denken stark beeinflusst. Die Autoren problematisieren, dass der Vernunftglauben totalisierende Tendenzen annehmen kann. Einerseits sei Freiheit ohne Aufklärung nicht denkbar, andererseits bringe der Rationalitätsglaube jedoch auch die Probleme blinder Naturbeherrschung und der Versachlichung der Verhältnisse mit sich. Gemeinsam mit Horkheimer kehrt Adorno 1950 in seine Heimatstadt Frankfurt am Main zurück und übernimmt dort eine Professur für Philosophie und Soziologie.

Dialektik der Aufklärung

Erziehung nach Auschwitz

Hier greift er in den gesellschaftlichen Diskurs u.a. ein, indem er die Frage aufwirft, welche Lehren Erziehung aus der Existenz von Auschwitz ziehen kann. „Die Forderung, dass Auschwitz nicht noch einmal sei, ist die allererste an Erziehung" (Adorno 1977, original 1966, S. 674). Auschwitz ist – so Adorno – der Rückfall in die Barbarei. Diese Barbarei bestehe fort, solange die Bedingungen, die diesen Rückfall möglich machten, nicht beseitigt sind. Da mit objektiven, gesellschaftlichen Mitteln faschistische Vernichtungsfeldzüge nicht verhindert werden könnten, müsste auf der subjektiven Seite, bei der Erziehung des Menschen, angesetzt werden. Dabei reichten Informationen oder Wertappelle, die von außen an die Individuen herangetragen werden, nicht aus, um Veränderung denkbar werden zu lassen. Erziehung nach Auschwitz, so Adorno, müsse vielmehr Erziehung in Form kritischer Selbstreflexion sein. Diese müsse ergänzt werden durch allgemeine Aufklärung (Adorno 1977, S. 677).

Autonomie

Adorno sieht die zentrale Ursache dafür, dass Konzentrationslager und Terror im Sinne von Auschwitz möglich sind, im Zerfall etablierter Autoritäten des Kaiserreichs. Die Menschen seien psychologisch noch nicht in der Lage gewesen, ohne Rückgriff auf Autoritäten über sich selbst zu bestimmen. Um nun Auschwitz ein für alle mal unmöglich werden zu lassen, bedarf es – so Adorno im

Geist der Aufklärung – menschlicher Autonomie. „Die einzig wahrhafte Kraft gegen das Prinzip von Auschwitz wäre Autonomie, wenn ich den Kantschen Ausdruck verwenden darf; die Kraft zur Reflexion, zur Selbstbestimmung, zum Nicht-Mitmachen“ (Adorno 1977, S. 679). Gegen die Appelle für soziale Bindungen wendet Adorno ein, dass Bindungen leicht Gesinnungsterror und das Ausnutzen von Schwächen anderer mit sich bringen. Betont man zunächst die Bindungen, so legt man leicht den Akzent auf das Sich-Abhängig-Machen, anstelle der eigenen Vernunft vertrauen zu lernen. Nur ein autonomer Mensch kann auch freiwillig Bindungen eingehen. Erziehung nach Auschwitz müsse gegen die „blinde Vormacht aller Kollektive“ (Adorno 1977, S. 681) gerichtet sein.

Männer und Schmerz

Teil der Selbstreflexion sei auch, Angst und Schwäche anzuerkennen. Gerade Männer würden mit der Gleichgültigkeit gegenüber dem Schmerz Masochismus und damit potentiell ihr Gegenteil, nämlich Sadismus, einüben. „Dieser Mechanismus ist ebenso bewusst zu machen wie eine Erziehung zu fördern, die nicht, wie früher, auch noch Prämien auf den Schmerz setzt und auf die Fähigkeit, Schmerzen auszuhalten“ (Adorno 1977, S. 683). Gerade die Gefühlskälte und die Unfähigkeit zu lieben, die Auschwitz möglich machen, entstünden aus dieser angst- und schmerzverleugnenden Erziehung. Eine weitere Komponente der Emotionslosigkeit, gleichzeitig eine Folge der Aufklärung, sei ein „verdinglichtes Bewusstsein“ (Adorno 1977, S. 684). Erst wird das (männliche) Kind selbst zum Ding gemacht, dann behandelt das erwachsen gewordene Kind die anderen wie Dinge und macht sie „fertig“.

Erziehung nach Auschwitz ist also – Adorno zufolge – eine psychoanalytisch orientierte, autonomiefördernde Kleinkinderziehung. Es ist eine Schulerziehung, die ohne Angst vor Mächtigen, Diskurse und Analysen zulässt. Es ist auch die selbstreflexive Kultur der Erwachsenen, die ihr eigenes Geworden-Sein kritisch beleuchten. Hintergrund dieser Erziehung ist nach Adorno eine Soziologie, die die gesellschaftlichen Kräfteverhältnisse hinter der Oberfläche öffentlich macht.

Ich empfehle zur weiteren Lektüre: Ritsert 1995; Dubiel 1997; Hilbig 1997; Kunnemann/de Vries 1989.

2.2.4 Michel Foucault

Michel Foucault wird 1926 in Poitiers (Frankreich) geboren. Der studierte Philosoph arbeitet ein Leben lang an der Schnittstelle zwischen Philosophie, Geschichtswissenschaft und Soziologie. Foucaults Vorgehensweise kann als diskursanalytisch bezeichnet werden. Das bedeutet, er untersucht wer, wo, wann und vor allem wie über was redet. Der Diskursbegriff bezeichnet also die Formationsregeln des Redens. An historischen Texten erforscht Foucault, wie Wirklichkeiten gemacht werden. Foucault betrachtet, ganz anders als Adorno, die moderne Konstruktion eines Subjektes, welches qua Vernunft Entscheidungen treffen kann, mit größter Skepsis und richtet sein Augenmerk stattdessen auf die historisch-gesellschaftlichen Bedingungen, in denen Bedeutungen generiert werden, die dann allen Beteiligten selbstverständlich erscheinen. Dabei geht er keineswegs von einer historisch linearen Entwicklung aus, sondern nimmt an, dass verschiedene Bedeutungskontexte nebeneinander existieren, Entwicklungen abreißen, sich wiederholen etc. können. Das heißt, Geschichte verläuft ihm zufolge diskontinuierlich und in Sprüngen.

Macht

Foucault interessieren vor allem die Machtverhältnisse, die die Formierung von Wissen und Verhalten durchziehen. Seine These ist, dass es weder ausreicht, sich Macht als eine Dynamik vorzustellen, die von oben nach unten wirkt, noch Macht ausschließlich als negative Kraft, nämlich zensierend, unterdrückend etc., zu imaginieren. Stattdessen betont er: „In Wirklichkeit ist die Macht produktiv; und sie produziert Wirkliches" (Foucault 1977, original 1975, S. 250; ausführlich erläutert Foucault seinen Machtbegriff 1983, original 1976, S. 113ff.).

Überwachung und Kontrolle

Die Bedeutung und Durchdringung der Gesellschaft von Macht eruiert und erläutert Foucault u.a. an der Geschichte des ärztlichen Blicks (1973, original 1963), an den Institutionen der Psychiatrie (1969, original 1961) und des Gefängnisses (1977, original 1975) sowie – stärker verinnerlicht – an den Praktiken der Sexualisierung (1983, original 1976). Mit den Praktiken der Schule beschäftigt sich Foucault nicht zufällig in seinen Studien zur Geburt des Gefängnisses (1977, original 1975; zur Schule insbesondere S. 220ff.). Überwachungs- und Kontrollverfahren in der Pädagogik und Justiz, vielfach aber auch in der Medizin und im Militär, weisen übereinstimmende Grundstrukturen auf. Schule, Armee, Gefängnis, Krankenhaus sind seit dem 19. Jahrhundert – so der Autor – expandierende Institutionen, die mit neuen Mitteln der Abrichtung wie hierarchische Überwachung, normierende Sanktion und Prüfung arbeiten.

Der typische Klassenraum ist zunächst mit einem erhöhten Pult, später durch den stehenden Lehrer, so ausgerichtet, dass dieser das Klassengeschehen mit einem Blick überwachen kann. Die Zimmer werden entlang des Korridors wie kleine Zellen angeordnet. Im Speisesaal des 19. Jahrhunderts steht der Tisch der Studieninspektoren auf einem Podest. Die Toiletten sind mit Halbtüren ausgestattet, so dass der Aufseher Kopf und Beine der SchülerInnen sehen kann. In den Schlafsälen der Internate trennen dichte Stellwände die Betten, der Blick für die Lehrer auf das Geschehen in den Betten ist jedoch frei. Macht, so Foucaults These, entfaltet sich nicht mehr, indem der Monarch sie – und damit sich – zeigt, zum Beispiel indem er durch die Menge schreitet. Sie entwickelt sich, indem die „Schützlinge" zur Beobachtung vorgeführt werden. Macht entfaltet sich durch die gegliederte und detaillierte Kontrolle und Sichtbarmachung der Insassen. Sie verwirklicht sich in den vielen Prüfungen, die jede/r Einzelne für alle sichtbar zu absolvieren hat. Auf diese Art und Weise bleibt die Macht den Menschen nicht äußerlich, sondern sie durchdringt sie. Die Schule lehrt die Kinder, sich in einer normierten Art zu verhalten, mit der Konsequenz, dass Macht sich voll entfalten kann. Jeder Einzelne muss unter Androhung von Strafe dafür Sorge tragen, aufrecht zu sitzen, Verspätung zu vermeiden etc. und damit kontrollierbar zu bleiben.

Individualisierung und Normierung

Gleichzeitig wird das System der Strafen verfeinert. Es gibt nicht Strafe oder Lob, sondern viele feine Möglichkeiten, unterschiedliche Strafen auszusprechen. Die Menge der Schüler oder Schülerinnen kann damit differenziert werden: Jede/r erhält seine eigene Strafe. Es ist ein Prozess, so Foucault, in dem sowohl der Einzelne/das Individuum konstituiert wird, weil er/es als Einzelperson angesprochen, beurteilt und bestraft wird, als auch gleichzeitig alle einer gemeinsamen Normierung unterliegen. Alles, was abweicht, wird bestraft. Zum „Fehler" wird schon, das vorgeschriebene Niveau nicht zu erreichen. Die Disziplinarstrafe dient dazu, Abweichungen zu reduzieren. „Das Normale etabliert sich als Zwangsprinzip im Unterricht zusammen mit der Einführung einer standardisierten Erziehung und der Errichtung der Normalschulen" (Foucault 1977, S. 237). Nur vor dem Hintergrund der imaginären Normalität kann die Abweichung verfolgt und bestraft werden. Diesen Prozess zu etablieren, weiterzuführen und in jeden Einzelnen einzupflanzen, leistet nach Foucault auch die Schule.

Michel Foucault können Sie vertiefen mit folgenden Büchern: Dane u.a. 1985; Weedon 1990; Marshall 1996; Hekman 1996; Kneer 1996; Ashenden 1999; McWhorter 1999.

3. Soziologische Bildungs- und Erziehungstheorie heute

Soziologinnen und Soziologen, die sich umfassend mit der Theorie der Bildung und Erziehung beschäftigen, sind heute selten geworden. Der französische Soziologe Pierre Bourdieu und der deutsche Soziologe Niklas Luhmann gehören zu den wenigen, die Bildungs- und Erziehungsfragen in ihrem Werk systematisch berücksichtigen.

3.1 Pierre Bourdieu

Quelle: http://www.suhrkamp.de/autoren/bourdieu/bourdieubio.htm

Der französische Soziologe Pierre Bourdieu wird 1930 in Béarn (Pyrénées Atlantiques) geboren. Als Junge vom Land schafft er nicht nur den Sprung in die französischen Eliteschulen und -hochschulen, sondern später auch als Professor an das renommierte Collège de France in Paris berufen zu werden. Dieser für das französische Bildungssystem spektakuläre Werdegang weckt nicht nur Bourdieus Interesse an Bildungsprozessen, sondern schärft auch seinen Blick für die Reproduktionsmechanismen der „herrschenden Klasse“ über Bildung. Bourdieu bearbeitet in seinem Werk eine Vielzahl von Themen, deren Kern aber immer wieder die Frage nach den Bildungsarrangements und -verhältnissen ist.

Im Laufe seiner Berufsbiographie erstellt er ein theoretisches Gerüst, welches eine Basis für seine Analysen bietet. In einem gemeinsam mit Loïc D. J. Wacquant verfassten Band mit dem Titel „Reflexive Anthropologie“ werden diese theoretischen Grundlagen sowie die Abgrenzungen und Bezüge zu anderen Theorien sehr treffend erläutert (vgl. Bourdieu/Wacquant 1996; insbesondere Wacquant 1996). Am 23. Januar 2002 starb Pierre Bourdieu nach langer Krankheit.

3.1.1 Zum Werk

Soziale Strukturen

Bourdieu geht von der soziologischen Basisannahme aus, dass Gesellschaft von „sozialen Strukturen", das heißt von einer systematischen und statistisch erhebbaren Verteilung der materiellen Ressourcen und der Aneignungsmöglichkeiten seltener und begehrte Güter und Werte durchzogen ist. Diese Verteilungsformationen bilden die Matrix, welche Handlungsmöglichkeiten kanalisiert. Dabei nimmt er an, und das unterscheidet ihn von vielen anderen SoziologInnen, dass Strukturen ein „Doppelleben" (Wacquant 1996, S. 24) führen. Sie existieren nicht nur als Verteilungskonstellation, sondern auch inkorporiert als symbolische Matrix des praktischen Handelns. Das heißt Strukturen bleiben dem Menschen nicht äußerlich, sondern verwirklichen sich im Verhalten, in Gedanken, Gefühlen, Wahrnehmungen und Urteilen sozialer Akteure. Strukturen existieren also sowohl als Gesellschaftsstrukturen als auch als mentale Strukturen. Die Grundhypothese der Bourdieuschen Soziologie ist, dass zwischen den gesellschaftlichen Strukturen und den inkorporierten Strukturen eine Korrespondenz besteht. Handeln richtet sich nach den verinnerlichten Strukturen und reproduziert auf diese Weise die gesellschaftlichen Strukturen erneut.

Sozialer Raum

Um diesen zentralen Gedanken besser entfalten zu können, schlägt Bourdieu ergänzend zur Kategorie der Struktur das Begriffspaar „Sozialer Raum/Feld" und „Habitus" vor. Ein Feld, von Bourdieu auch „sozialer Raum" genannt, ist das Ensemble der Relationen zwischen Gruppen von Menschen, das auf bestimmten Formen von Macht bzw. von Kapital basiert. Es ist ein Spannungsfeld, welches sich zwischen (Berufs-)Gruppen und deren Kämpfe und Konkurrenzen aufspannt:

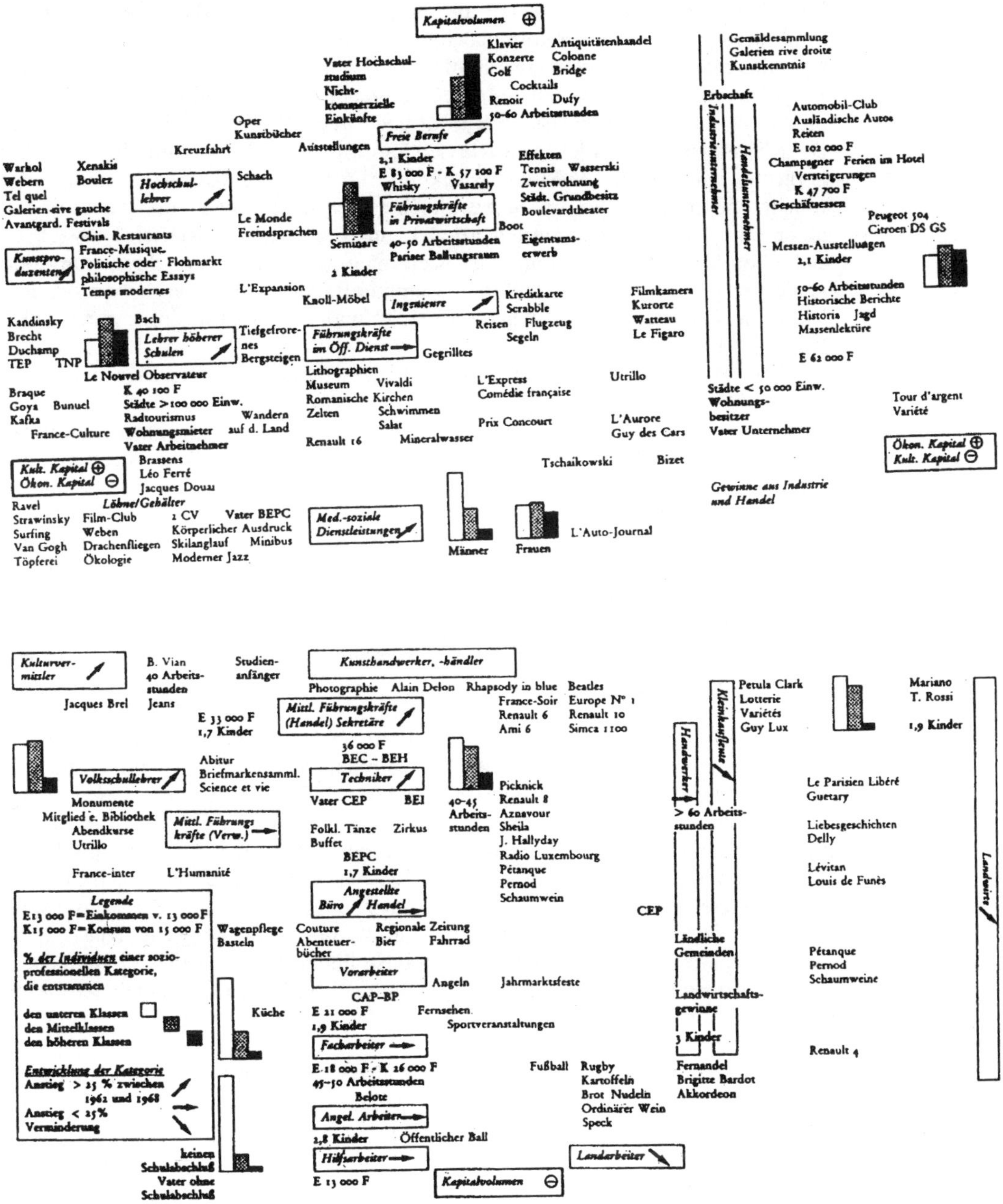

(Quelle: Bourdieu 1983², S. 212f.)

Klassen

Auf der Abbildung sieht man, wie Bourdieu die soziale Welt in Form eines Raums darstellt, welcher sich primär aus den Verteilungen von Berufsgruppen (und damit auch über Einkommen und Ausbildungsniveau) ergibt. In Erweiterung des marxistischen Gedankenguts (vgl. Kapitel 4.1.1 dieses Buches) werden Gruppen von Menschen, die unter homogenen Lebensbedingungen leben, zu „Klassen" zusammengefasst (vgl. Bourdieu 1983[2], S. 175). Dabei unterscheidet Bourdieu zwischen „objektiven Klassen" im Sinne von Menschen, die analytisch aufgrund homogener Merkmale (strukturell im Sinne von Besitz an Gütern oder Macht und inkorporiert im Sinne gleicher Wahrnehmungen und Handlungen) unter eine Klasse subsumiert werden können, und „mobilisierter Klasse", die sich ebenfalls aufgrund homogener Merkmale zum Kampf um Bewahrung oder Änderung der Verteilungsstruktur zusammenfinden (vgl. ebda. sowie bei Karl Marx die Unterscheidung „Klasse an sich" und „Klasse für sich", dargelegt in Kapitel 4.1.1).

Die Akteure und Gruppen pflegen Unterscheidungen. Die Merkmale, über die sie sich von anderen abgrenzen, sind in der Graphik dargestellt. Während die einen Fußball lieben, präferieren die anderen die Oper. „Akteure oder Gruppen von Akteuren sind anhand ihrer relativen Stellung innerhalb dieses Raums definiert" (Bourdieu 1991[2], S. 9f). Das bedeutet, dass Klassen nicht in erster Linie über *ein* Merkmal, welches sie aufweisen, definiert werden können, sondern durch die Beziehung, in der sie zu anderen Klassen stehen. In dem gleichzeitigen Prozess des Unterscheidens von anderen und des Angleichens an sozial Nahestehende bilden sich homogene Gruppen, die zu „Klassen" zusammengefasst werden können. „Eine soziale Klasse ist vielmehr definiert durch die *Struktur der Beziehungen zwischen allen relevanten Merkmalen*" (Bourdieu 19832, S. 182; Hervorhebung im Original).

Geschlecht

Geschlecht ist für Bourdieu ein Merkmal, welches klassenbildend wirkt. Die Art und Weise, wie Geschlecht gelebt wird, welchen Wert den jeweiligen Geschlechtern zugewiesen wird und wie Weiblichkeit oder Männlichkeit definiert wird, ist klassenspezifisch. „Die geschlechtsspezifischen Merkmale sind ebenso wenig von den klassenspezifischen zu isolieren wie das Gelbe der Zitrone von ihrem sauren Geschmack: eine Klasse definiert sich wesentlich auch durch Stellung und Wert, welche sie den beiden Geschlechtern und deren gesellschaftlich ausgebildeten Einstellungen einräumt. Darin liegt begründet, warum es ebenso viele Spielarten der Verwirklichung von Weiblichkeit gibt wie Klassen und Klassenfraktionen, und warum die Arbeitsteilung zwischen den Geschlechtern auf der Ebene der Praxis wie der Vorstellungen innerhalb verschiedener Gesellschaftsklassen höchst unterschiedliche Ausprägungen annimmt" (Bourdieu 1983[2], S. 185).

Felder

Bourdieu benutzt den Begriff des sozialen Raums häufig als Alternative zum Gesellschaftsbegriff, um damit die Beziehungsgefüge innerhalb einer Gesellschaft zu verdeutlichen. Die ausdifferenzierten Teilbereiche der Gesellschaft bezeichnet er als Felder. Er unterscheidet zum Beispiel zwischen dem künstlerischen, dem ökonomischen, dem akademischen Feld. Jedes Feld weist eine eigene Logik auf. Bourdieu denkt Felder als Kräftefelder. In ihnen entsteht im sozialen Handeln unterschiedlich kapitalstarker und sozial positionierter Akteure eine Dynamik, welche das Feld in einen mehr oder minder autonomen Mikrokosmos verwandelt. Da Bourdieu jedoch für „offene Begriffe" plädiert, welche wie Werkzeuge helfen sollen, Probleme zu verstehen, benutzt er Begriffe auch graduell verschieden (vgl. Wacquant 1996, S. 55). So kommt es vor, dass die Gesellschaft als Feld und ein Teilbereich als sozialer Raum bezeichnet wird, was die inhaltliche Nähe der beiden Begriffe verdeutlicht.

Habitus

Die wissenschaftliche Einteilung von Klassen basiert für Bourdieu auf einer empirisch beobachtbaren Handlungs- und Beurteilungspraxis der Akteure: dem „Habitus". Der Habitus sind die „geistigen und körperlichen Denk- Wahrnehmungs- und Handlungs-Schemata" (Wacquant 1996, S. 37). Es ist eine Matrix, die immer wieder gleiche Gefühle, Handlungen, Wahrnehmungen, Gedanken, Geschmackspräferenzen nach Klasse, Milieu oder Geschlecht reproduziert. Durch den gruppenspezifischen Habitus finden zum Beispiel Mitglieder einer Klasse Plastikblumen einen hervorragenden und schönen Dekorationsgegenstand, während Mitglieder einer anderen Klasse dieselben Blumen als hässlich verwerfen; finden Kinder der einen Klasse, dass JuristIn ein nahe liegender Berufswunsch ist, wohingegen andere sich bestenfalls Pädagogik als akademisches Fach vorstellen können; finden Mädchen Liebesfilme spannend und Abenteuerfilme langweilig, während Jungen genau umgekehrt werten usw. Mit dem Habitusbegriff richtet Bourdieu den Blick auf das handelnde Subjekt.

„freiwillige" Reproduktion

Da Bourdieu in seinem Werk vielfach belegen kann, dass es eine Übereinstimmung von Feld und Habitus gibt, folgert er daraus, dass über die Übereinstimmung, also quasi über eine unbewusste Unterwerfung unter gesellschaftliche Verteilungsprinzipien, Gesellschaft und vor allem Herrschaft reproduziert wird. Indem, um beim Beispiel zu bleiben, die Mädchen (und Jungen) der unteren Mittelschicht „freiwillig" Pädagogik und nicht Jura studieren, reproduzieren sie ihre soziale Position und das Herrschaftsverhältnis zwischen den Klassen. Indem vor allem Frauen Erziehungswissenschaft studieren stellen sie gleichzeitig das Geschlechterverhältnis mit der Zuständigkeit von Frauen für Kindererziehung immer wieder neu her. Bourdieu spricht daher sowohl vom Klassenhabitus als auch vom vergeschlechtlichten Habitus (Bourdieu 1997, S. 167).

Durch die Entstehung von klassifizierbaren, habituell gleichen Praxisformen zum einen und die gesellschaftliche Beziehungsstruktur zwischen diesen entsteht der soziale Raum auch als ein „Raum der Lebensstile" (z.B. Bourdieu 1983[2], S. 277ff). Es sind die gruppenspezifisch homogenen aber gesellschaftlich unterschiedlichen, relationalen Praxisformen. Während der Raum der sozialen Positionen, also die Klassenlage, im Schwerpunkt die objektiven sozialen Strukturen erfasst, beschreibt der Raum der Lebensstile, welche durch den Habitus generiert werden, die subjektive Lebenspraxis. Bourdieu zeigt für Frankreich, dass zwischen den milieuspezifischen Praktiken und der Klassenlage eine systematische Beziehung existiert (vgl. für Deutschland z.B. Schulze 1997[7], Vester u.a. 2001, Keim 1997).

Kapital

Wie man am oben abgebildeten Schema erkennen kann, differenziert Bourdieu zwischen einer hierarchischen Klassenstruktur mit mehr oder weniger Kapital und einer vertikalen Struktur je nach der Art des eingesetzten Kapitals und weicht damit die Vorstellung antagonistisch sich gegenüber stehender Klassen auf. Er entwickelt ein differenziertes Bild von vielfältigen Klassenrelationen. Bourdieu übernimmt aus der marxistischen Tradition nicht nur den Klassen-, sondern auch den Kapitalbegriff, meint damit aber nicht nur ökonomisches Kapital (materieller Besitz an Geld und Eigentum), sondern auch kulturelles Kapital

(Bildung und Ausbildung) und soziales Kapital – auch symbolisches Kapital genannt – (soziale Beziehungen). Macht entwickelt sich ihm zufolge durch Kapitalakkumulation. Da niemand kein Kapital hat, ist Macht immer eine Bezeichnung für ein Verhältnis. Es gibt nicht einfach Mächtige und Opfer, sondern Bourdieu betrachtet die Relationen, also das Machtverhältnis zwischen Menschen und Menschengruppen. Dennoch spricht er von Herrschaft im Sinne einer Handlungspraxis von Privilegierten (Reiche, Intellektuelle, Männer, Weiße), welche andere Klassen (oder Frauen, Schwarze) von Ressourcen und Aneignungsmöglichkeiten abhalten. Die Bourdieusche Soziologie besteht maßgeblich aus Studien über die Reproduktions- (Wiederherstellungs-) und Konversionsstrategien (die Umwandlungen, Veränderungen), die Gruppen entwickeln, um ihre soziale Position zu erhalten oder zu verbessern. Soziologie hat ihm zufolge die Aufgabe, die verborgensten Strukturen der sozialen Welt, ihre Reproduktion und Transformation, aufzudecken. Soziologie zerstört die Mythen, mit denen Machtausübung verschleiert und Herrschaft reproduziert wird.

3.1.2 Das akademische Feld

Zu den wesentlichen bildungssoziologischen Werken von Bourdieu gehören „Homo academicus" (1998[2], original 1984), „Titel und Stelle. Über die Reproduktion sozialer Macht" (1981), eine Zusammenstellung von Aufsätzen, welche er in den 1970er Jahren zusammen mit seinen MitarbeiterInnen Luc Boltanski, Monique de Saint Martin und Pascale Maldidier verfasst hat sowie auch der Text „Vom Gebrauch der Wissenschaft. Eine klinische Soziologie des wissenschaftlichen Feldes" (1998). Bourdieu und sein MitarbeiterInnenkreis interessieren sich, entsprechend der oben skizzierten gesellschaftstheoretischen Einsichten, für den Stellenwert von Bildung im Prozess der sozialen Reproduktion. Sie untersuchen die Strategien mittels derer Individuen ihre soziale Position in einer Gesellschaft behaupten oder verbessern wollen. Dazu nutzen diese Bildungsinhalte und vor allem Bildungsinstitutionen. Bildungszertifikate, also Schulabschlüsse, Diplome, Titel etc., begreift Bourdieu als kulturelles Kapital, über welches Zugangschancen zu sozialen Positionen und Lebensformen „erkauft" werden.

Relative Autonomie der Bildung

Das Feld der Bildung ist „relativ autonom". Einerseits versucht das ökonomische Feld Einfluss auf das Bildungssystem zu nehmen und durch eigene Ausbildungskonzepte dem Bildungssystem Konkurrenz zu machen, andererseits ist das Bildungssystem in der privilegierten Lage, Titel vergeben zu können, die einen universellen und relativ zeitlosen Wert haben (vgl. Bourdieu u.a. 1981, S. 89ff). Durch die Titel werden Zugangsberechtigungen und Lohnerwartungen (z.T. sogar Lohngarantien) geschaffen, die auch dann noch gelten, wenn sich Arbeitsplatzmerkmale längst verändert haben. Die Ökonomie hat, so Bourdieu u.a., das Interesse, dass Titel an Stellen gebunden sind; das Bildungssystem dagegen beharrt auf der eigenen Kontrolle über die Vergabe von Titeln. So bleibt das Bildungssystem und seine Titel ständig Gegenstand politischer Auseinandersetzungen.

Titel und Zertifikate

Aus dem Anwachsen der Zahl der Diplomierten folgt, das problematisieren Bourdieu und seine MitarbeiterInnen schon in den 1970er Jahren und das bekommt heute angesichts der Bildungsentwicklungen (siehe Kapitel 4.2.1) neue Bedeutung, erstens die völlig Abwertung Nicht-Diplomierter (heute in Deutschland vor allem der HauptschülerInnen), zweitens der Bedeutungsverlust früherer Aufstiegsmechanismen und drittens die Abwertung jedes einzelnen Titels und damit Vergrößerung der Spielräume des ökonomischen Systems. Man könne, so führt er aus (vgl. Bourdieu/Champagne 1997, S. 529), nicht Zeugnisse und Diplome an breite Schichten vergeben, ohne den Wert diese Titel zu schmälern. Dabei seien es dann vor allem die Neuankömmlinge, die als erstes die Opfer der Entwertung werden.

Klassenspezifische Effekte

Daher sind seine Arbeiten von einer großen Skepsis gegenüber der Politik der Bildungsexpansion geprägt. Viele PolitikerInnen, insbesondere in den 1970er Jahren (siehe Kapitel 4.2.1) vertreten die Auffassung, dass über die Zulassung breiter Schichten zu höherer Bildung soziale Ungleichheit gemindert, vielleicht sogar beseitigt, werden könne. Bourdieu betont dagegen, dass das Bildungssystem prinzipiell selektiv arbeite und sich „feine Unterschiede" zwischen den Bildungsinstitutionen entwickeln, die die soziale Differenz perpetuieren. Ökonomisches, soziales und kulturelles Kapital ergänzen sich gegenseitig. Wer über ökonomisches Kapital verfügt, muss zum Beispiel seine Kinder nicht auf das nächstbeste Gymnasium bringen, sondern kann das „beste" auswählen. Bourdieu zufolge stellt das Universum der schulischen Einrichtungen ein Kontinuum zwischen zwei Extremen dar: „Auf der einen Seite die Noteinrichtungen, die man hastig in den verwahrlosten Vorstädten vervielfacht hat, um die zahlenmäßig ständig anwachsenden und mit immer geringerer kultureller Mitgift versehenen Schülergruppen aufzunehmen, die auch nicht mehr viel mit dem Gymnasium, wie es sich bis in die 1950er Jahre gehalten hat, zu tun haben; auf der anderen Seite die in höchstem Maße behüteten Einrichtungen, wo Gymnasiasten aus guter Familie auch heute noch eine Schulzeit durchlaufen können, die sich nicht radikal von der unterscheidet, die ihre Väter oder Großväter gekannt haben" (Bourdieu/Champagne 1997, S. 527). Die dort erworbenen Zertifikate und sozialen Beziehung werden zum Schlüssel zu sozialen Positionen, die andere Kinder nie erreichen werden. SchülerInnen aus „gutem Hause" hätten von ihren Familien einen Platzierungssinn mitbekommen, der es ihnen ermöglicht, zum richtigen Zeitpunkt die richtige Entscheidung (für welche Fremdsprache, für einen Auslandsaufenthalt, für ein Praktikum, für die Studienwahl etc.) zu treffen, die aus benachteiligten Familien stammenden Kinder, häufig aus Einwandererfamilien, investieren ihr eh schon geringes Kapital dann häufig an der falschen Stelle. Nach einer Phase der Euphorie begriffen die Kinder der Kleinhändler, Handwerker und Industriearbeiter, so Bourdieu u.a., dass ihre Fertigkeiten entweder nicht ausreichen, um in der Schule erfolgreich zu sein, oder dass das Abitur nicht ausreicht, um beruflich erfolgreich zu sein. Wer nur über ein Kapital, in diesem Fall kulturelles Kapital im Sinne der Bildungszertifikate verfügt, hat immer schlechtere Chancen als jene, die verschiedene Kapitalsorten akkumulieren. Allerdings wird das Scheitern zeitlich hinausgezögert. Es ist nicht mehr als klassenspezifischer Effekt erkennbar, sondern geschieht schleichend in verschiedenen Lebensphasen.

Die Schule und die Hochschule wiederum geraten durch die Widersprüche und Konflikte in eine Krise. Wie Demonstrationen und Proteste zeigen, können sie weder den Massenandrang noch die Unzufriedenheit und Aggression der

SchülerInnen und StudentInnen verarbeiten. Die Kinder kulturell benachteiligter Familien, deren Eltern die größten Opfer für die Ausbildung erbringen, deren schulischer Erfolg bzw. beruflicher Aufstieg unsicher ist, erleben schließlich statt des gesellschaftlichen „Aufstiegs“, dass ihre Stigmatisierung sich erhöht. Sie hatten „ihre Chance“ und haben sie vertan.

Bauern

Dass die steigende Bedeutung nun, die Bildung in den Reproduktionsstrategien aller Klassen einnimmt, für einzelne Gruppen nicht nur nicht hilfreich, sondern sogar schädlich sein kann, zeigt sich zum Beispiel an der Situation von Bauern und Bäuerinnen (vgl. Bourdieu u.a. 1981, S. 62ff). Durch die Sekundarschulerziehung der Kinder unterwerfen sich die Bauern bzw. die Bäuerinnen der symbolischen Herrschaft der StädterInnen. Da die Schule verspricht, auch die auf einem modernen Bauernhof dringend notwendigen Fertigkeiten wie Hochsprache und Wirtschaftsrechnung zu vermitteln, widersetzen sich die Bauern/Bäuerinnen nicht der fremden Kultur, sondern akzeptieren diese als einen auch für ihre Kinder notwendigen Schritt. Mit dieser „freiwilligen“ Unterwerfung werden jedoch auch sukzessive die Werte der städtischen Kultur übernommen: „Der gesellschaftliche Raum des Dorfes, ein geschlossener Mikrokosmos mit eigenen Hierarchien (...), wird dabei durch ein weiteres gesellschaftliches Feld ersetzt, in dem die Bauern und Bäuerinnen sich allesamt »irgendwo weit unten« wieder finden“ (Bourdieu u.a. 1981, S. 64). Die lokale Anbindung, auch deutlich im Dialekt, geht durch das Studium verloren. Die Bauern und Bäuerinnen begeben sich in eine Konkurrenzbeziehung zu Angestellten und Arbeitern. Sie lernen die gleichen Dinge wie LohnempfängerInnen und sie unterwerfen sich den gleichen zeitlichen Rhythmen. Damit entfremden sie sich von der Bauernklasse und verlieren das Interesse am Bauernberuf. Die Klasse der Bauern eliminiert sich auf diese Weise selbst.

Hochschule

Besonders ausführlich beschäftigt sich Bourdieu mit dem (französischen) Hochschulsystem. „Universitäres Kapital erhält und behält, wer Positionen innehat, mit denen sich andere Positionen und deren Inhaber beherrschen lassen“ (Bourdieu 1998², S. 149). Bourdieu interessiert an der Bildungsinstitution „Universität“, wie diese sich selbst und damit die Positionen der Akteure im sozialen Raum reproduziert. Dazu gehört zum ersten die Beobachtung, dass Studierende und DozentInnen je nach Herkunft sich im gleichen Fach wieder finden (vgl. für die philosophischen und humanwissenschaftlichen Fächer Bourdieu 1998², S. 140, siehe auch Engler 1993) sowie die Rekrutierung des wissenschaftlichen Nachwuchses. Bis ein Mensch in die Position des Professors oder der Professorin ernannt wird, vergeht eine lange Zeit. Diese oft über zwanzig Jahre andauernde Zeit der Abhängigkeit ist – so Bourdieu – gleichzeitig die Phase, in der der universitäre Habitus entwickelt wird. Die Gesetze des Systems werden schleichend verinnerlicht, weil sie die Voraussetzung für den universitären Erfolg sind. Jene, die Positionen in der Universität zu vergeben haben, überlassen diese in der Regel Menschen, die das gleiche Eintrittsalter haben wie sie selbst hatten (vgl. Bourdieu u.a. 1981, S. 135) bzw. Menschen, welche glaubhaft machen können, dass sie entsprechend der Regeln des Feldes mitspielen, was vor allem beim gleichen Geschlecht (sprich bei Männern) erwartet wird (vgl. Bourdieu 1998², S. 155ff). Zu diesen Spielregeln gehört auch, sich den Ritualen von Promotion und Habilitation (bzw. vergleichbaren Abschlüssen in anderen Ländern) zu unterwerfen, welche Bourdieu u.a. drastisch auch als „Ritus der Selbstverstümmelung" (Bourdieu u.a. 1981, S. 147) bezeichnen.

Strategien zu Reproduktion

Einmal HochschullehrerIn geworden, finden die Strategien der Reproduktion jedoch nicht ihr Ende. Erstens ist der „Doktorvater" (Doktormutter) nun selbst für Zöglinge und deren Eingliederung zuständig und kann über diese selbst Reputation erlangen. Dabei sind die berühmtesten Ordinarien, wie er im Einzelfall für Frankreich belegen kann, in der Regel diejenigen, welche die wenigsten DoktorandInnen betreuen. Umgekehrt sind aber jene Universitätskarrieren besonders erfolgreich, wenn derjenige (diejenige) sich auf einen einflussreichen Ordinarius berufen kann und zwar unabhängig davon, ob man ihm (oder ihr) inhaltlich nahe steht. Mit anderen Worten: Bourdieu zeigt auf, dass die Universitätslaufbahn von der Wahl des/der BetreuerIn, der eingesetzten Zeit und der „freiwilligen" Selbstunterwerfung abhängt. Die Themenstellung ist insofern interessant, als mit der Schwierigkeit der gestellten Aufgabe, der Kandidat/die Kandidatin dokumentiert, zu welchen Weihen er/sie sich berufen fühlt. Es geht also, so lautet Bourdieus Fazit, nicht um hehre Bildung, sondern um die Strategien zur Reproduktion oder zur Verbesserung der sozialen Position und dabei haben jene, deren Familie selbst bereits kulturelles und soziales Kapital angehäuft hat, womöglich auch noch über ökonomisches Kapital verfügt, immer eine bessere Ausgangsposition.

Wettbewerb

Dabei entwickeln sich in einem relativ autonomen Feld wie die Wissenschaft Mechanismen der Selbstkontrolle (vgl. ausführlich Bourdieu 1998, S. 28ff.). Das wissenschaftliche Feld bringe gleichzeitig das Bestreben der Subjekte hervor, zu den Besten, Außergewöhnlichsten zu gehören und kontrolliere die Antriebe. So könne ein Mathematiker andere im Feld eben nur durch mathematische Leistungen ausstechen. Je heteronomer ein Feld sei, desto leichter falle es den Mitwirkenden, Kräfte aus anderen Feldern einzuschleusen. Je autonomer es sei, desto „reiner" sei der Wettbewerb. Im Feld der Wissenschaft mischen sich „reines" wissenschaftliches Kapital, das durch anerkannte Beiträge zur Wissenschaft angehäuft wird, und institutionelles wissenschaftliches Kapital, welches auf spezifischen politischen Strategien basiert (Mitgliedschaft in Kommissionen, Ausschüssen etc.). Die gleichzeitige Akkumulation beider Kapitalsorten ist – nach Bourdieu – aus zeitökonomischen Gründen heraus äußert schwierig. So lassen sich WissenschaftlerInnen auch über die soziale Position im wissenschaftlichen Feld und ihrer Strategie der Anhäufung des jeweiligen Kapitals einteilen. „So sind all diese Universen aufgrund der Tatsache, dass ihre Autonomie gegenüber äußeren Mächten niemals vollständig ist und sie gleichzeitig von zwei Herrschaftsprinzipien, einem weltlichen und einem spezifischen bestimmt werden, durch eine strukturale Zwiespältigkeit gekennzeichnet: intellektuelle Konflikte sind in gewisser Hinsicht immer auch Machtkonflikte. Jede Strategie eines Wissenschaftlers hat gleichzeitig eine (spezifisch) politische und eine wissenschaftliche Seite, und Erklärungsversuche müssen immer beide im Auge behalten" (Bourdieu 1998, S. 36f.).

Diese pessimistische Analyse der Lage bedeutet nun nicht, dass Bourdieu die Schule und Hochschule perspektivisch nur für die Reichen und bereits Gebildeten geöffnet wissen will, vielmehr geht es ihm um eine Entschleierung der Machtverhältnisse und um eine Neuformulierung schulischer und universitärer Grundlagen.

Bildungswesen der Zukunft

Auf Wunsch des französischen Präsidenten hat Bourdieu zusammen mit seinen KollegInnen vom Collège de France Vorschläge für das Bildungswesen der Zukunft vorgelegt (Collège de France 1987). Als ersten Grundsatz formulieren sie, dass das Bildungssystem in der Lage sein muss, „den dem (natur)wissenschaftlichen Denken eigenen Universalismus mit dem Relativismus der Humanwissenschaften, denen es um die Pluralität der kulturgebundenen Lebensweisen, Erkenntnisformen und Arten des Empfindens geht, in Einklang zu bringen" (Collège de France 1987, S. 255). Damit gemeint ist, dass Menschen gleichzeitig Mittel gegen Formen des Irrationalen als auch gegen eine übersteigerte Rationalität brauchen. Die Sozialwissenschaft lehrt die historischen und kulturspezifischen Wurzeln kognitiver und emotionaler Leistungen, die Naturwissenschaft zwingt zu einem rationalen auf das Universale gerichteten Blick. Es bedürfe einer über Bildungsprozesse organisierten rationalen Auseinandersetzung mit der Mannigfaltigkeit und Historizität von Kulturleistungen. Nur so könne gleichermaßen der Wissenschaft vertraut gemacht und dennoch immer die Vielfältigkeit kulturspezifischer Rationalitäten vor Augen geführt werden.

Ferner, so der zweite Grundsatz, müsse die monistische Vorstellung von „Intelligenz" verabschiedet werden, weil sie Leistung hierarchisiert, stattdessen muss das Bildungssystem für eine größere Vielfalt sozial anerkannter Formen kultureller Leistung sorgen. Weiter sollen die negativen Urteile und Stigmatisierungen abgebaut werden, indem eine breitere Auswahl von Bildungsgängen und eine stärkere Durchlässigkeit zwischen ihnen ermöglicht würde. Autonome Bildungseinrichtungen sollten in den Wettstreit um die beste Bildung treten, wobei gleichzeitig benachteiligte Personen und Institutionen vor ungezügelte Konkurrenz geschützt werden. Der Lehrstoff sei periodisch zu revidieren. Die innere Einheit der angebotenen Bildungsinhalte müsse deutlich gemacht werden. Vorgeschlagen wird eine kontinuierliche Bildung im Wechsel mit Berufstätigkeit. Mit Hilfe moderner Technologien könnten Standortnachteile ausgeglichen werden. Schließlich solle die Schule ein offener Ort eines Gemeinschaftslebens sein, der Außenstehende in die Diskussion über Entscheidungen einbezieht. Dies setze auch voraus, dass die Autonomie der Lehrerschaft durch Steigerung ihrer fachlichen und pädagogischen Kompetenzen gesteigert wird.

Zum weiteren Studium der Bourdieuschen Thesen und Theorien empfehle ich folgende Werke: Joppke (1986); Krais (1989); Wiley (1990); Ecarius (1996); Krais (1996b); Schimank/Volkmann (1999, S. 23ff). Um die Erklärungspotentiale und Anwendungsmöglichkeiten der Theorie abzuschätzen, ist es hilfreich, sich mit Folgestudien zu beschäftigen. Hier empfehle ich: Wacquant (1995a/b); Frerichs (1997); Frerichs/Steinrücke (1997); Engler (1997), Vester u.a. (2001).

3.2 Niklas Luhmann

(Quelle: Horster 1997, Umschlagseite)

Niklas Luhmann wird am 8. Dezember 1927 in Lüneburg geboren. Er studiert Rechtswissenschaft in Freiburg und arbeitet einige Jahre in verschiedenen Funktionen der öffentlichen Verwaltung. 1960/61 wird er für das Studium der Verwaltungswissenschaft und der Soziologie an der amerikanischen Harvard University beurlaubt. Dort trifft er mit Talcott Parsons zusammen, dessen Theorie ihn nachhaltig beeinflusst. Zurück in Deutschland absolviert er den Aufbaustudiengang „Verwaltungswissenschaft" in Speyer. Darauf aufbauend promoviert und habilitiert Luhmann in Münster und wird 1968 Professor für Soziologie an der Universität Bielefeld, wo er bis zu seinem Tod (1998) bleibt.

Die Luhmannsche Systemtheorie geht aus der kritischen Auseinandersetzung mit den Parsonschen strukturell-funktionalen Systemtheorie (vgl. Kapitel 2.2.1) hervor und bezieht die wesentlichen Anregungen für Novellierungen und Neudefinitionen der Parsonschen Theorie aus modernen interdisziplinären (vor allem biologisch/neurophysiologischen) systemtheoretischen Entwürfen. Gedanklicher Ausgangspunkt ist ein kybernetisches Modell, demzufolge gleiche Schemata von Systemen, Steuerung, Kontrolle, Selbstreferenz und Selbstorganisation in der Technik, der Biologie und in sozialen Prozessen unterstellt werden können (vgl. Maturana 1982; Varela 1982).

3.2.1 Zum Werk

Soziale Systeme

Im Unterschied zu Parsons, in dessen Theorie Systeme klare Strukturen aufweisen, nach deren Funktionen er dann fragt, interessiert Luhmann in seiner, nun funktional-strukturell genannten, Theorie wie System funktionieren und wie einzelne Funktionen sich historisch verändern. Unter einem sozialen System versteht Luhmann zunächst eine Konfiguration sinnhaft aufeinander verweisender Handlungen. Jedes System besitzt eine Umwelt, von der es abgegrenzt ist. Zu der Umwelt gehören alle Handlungen, die nicht in den Sinnkontext des Systems gehören (Luhmann 1991[4], S. 9). Auch hier unterscheidet sich Luhmann von Parsons, da er nicht länger davon ausgeht, dass kollektiv geteilte Normen und Werte die gemeinsame Grundlage von Systemen oder gar von Gesellschaften sind. Stattdessen wird der Sinnzusammenhang zum konstituierenden Element. Systeme leisten in einer hyperkomplexen Welt – so Luhmann – die Funktion, Komplexität zu reduzieren (Luhmann 1970, S. 116). In seinem späteren Werk (z.B. Luhmann 1986) trennt sich Luhmann zur Bestimmung von Systemen vom

Handlungsbegriff und betont – gerade um den Sinnzusammenhang hervorzuheben – Kommunikationen als das verbindende Element.

Autopoisis

Es lassen sich die verschiedensten Systeme unterscheiden: lebende, neuronale, psychische, technische, soziale Systeme. Luhmann interessiert sich als Soziologe in erster Linie für soziale Systeme. Diese versteht er, wie auch psychische oder neuronale Systeme in der neueren Systemtheorie, als autopoietisch Systeme. Das bedeutet, dass Systeme operativ geschlossen sind und durch ihre eigenen Elemente sich immer wieder neu reproduzieren. Systeme sind in diesem Sinne autonom. Allerdings kann die Umwelt eines Systems irritierend auf dieses einwirken, d.h. sie irritiert die Kommunikation eines Systems.

Kommunikation

Ein System besteht – nach Luhmann – nicht aus Menschen, weil Menschen selbst zu komplexe Gebilde sind. In ihnen steckt ein Immunsystem, ein neurophysiologisches System, ein psychisches System, ein organisches System usw. Luhmann konzentriert sich daher in seiner Begriffswahl auf den Kommunikationsbegriff. Dabei versteht er Menschen nicht als Urheber der Kommunikation, sondern Kommunikation wird als Produkt sozialer Systeme gefasst (Luhmann 1990b, S. 31). Das soziale Geschehen wird demnach von Luhmann als eigenständige analytische Dimension jenseits der kognitiven oder psychischen Operationen von Menschen begriffen. Kommunikation ist die elementare Einheit von Systemen. Sinnproduktion unterscheidet soziale Systeme von anderen Systemen. Nur psychische Systeme sind ebenfalls sinnproduzierend.

Jede Kommunikation ist selektiv, das heißt, es besteht die Auswahl aus verschiedenen Möglichkeiten. Sie ist – Luhmann zufolge – sogar in dreifacher Weise selektiv. Die „Information“ ist eine Auswahl aus den vielfältigen Möglichkeiten zu kommunizierender Informationen; auch zur „Mitteilung“ stehen verschiedene Varianten zur Verfügung (schreien, flüstern, schreiben, sprechen etc.); schließlich gibt es zahlreiche Möglichkeiten die mitgeteilte Information zu „verstehen“ (Luhmann 1993[4], S. 191ff). Da Kommunikation also die Synthese einer dreifachen Selektionsleistung ist, schlussfolgert Luhmann, dass Kommunikation theoretisch nicht als das Resultat des Handelns eines Individuums begriffen werden kann. Sie ist eine eigene analytische Dimension.

Diese hier notgedrungen sehr knapp geschilderten theoretischen Grundlagen nutzt Luhmann, und dies wird in der Fachöffentlichkeit am ausführlichsten diskutiert, zur Analyse gesellschaftlicher Prozesse. Prinzipiell unterscheidet Luhmann drei Typen sozialer Systeme: Interaktion, Organisation und Gesellschaft (vg. Luhmann 1991[4], S. 10ff). Interaktionssysteme entstehen in der gegenseitigen Wahrnehmung von Anwesenden. Ihre Wirksamkeit ist auf die Dauer der Anwesenheit beschränkt. Organisationen (oder Organisationssysteme) sind über Mitgliedschaft und Entscheidungstechniken formalisierte Sozialtypen. Gesellschaft schließlich ist das umfassendste Sozialsystem, dem Luhmann das meiste Interesse schenkt.

Differenzierungstypen

Die auf der Theorie autopoietischer sozialer Systeme aufbauende Gesellschaftstheorie geht von drei evolutionären Typen gesellschaftlicher Differenzierung aus. Einfachstes Differenzierungsprinzip ist die segmentäre Differenzierung, wie sie archaischen Gesellschaften zugeschrieben wird. Die Gesellschaft teilt sich hier in gleiche Teile, zum Beispiel Familien, Stämme, Dörfer. Mit einer Zunahme an Komplexität entwickelt sich die stratifikatorische Differenzierung. Es entwickelt sich eine Teilung in ungleiche Schichten. Sie gilt für Luhmann als

Kennzeichen der Vormoderne und wird bis in das 16. Jahrhundert hinein veranschlagt. Für sie gilt, dass es eine gesamtgesellschaftliche Grundsymbolik der Hierarchie gibt (Luhmann 1980, S. 29). Jede/r hat durch eine religiös fundierte Weltanschauung seinen Platz in der Geschichte. Das Ganze kann durch eine Person repräsentiert werden (z.B. den Kaiser). Ab dem 16. Jahrhundert lässt sich Gesellschaft – so Luhmann – nicht länger als schichtförmig gegliederte Einheiten mit einem gemeinsamen normativen Kontext beschreiben. Es entsteht eine funktionale Ausdifferenzierung, indem sich die Politik von der Religion trennt, die Erziehung sich als eigenes Feld etabliert, die familiäre Privatsphäre sich herauskristallisiert, das Recht sich von der Politik trennt, Wissenschaft und Wirtschaft als eigene Sinnkontexte entstehen. Mitte bis Ende des 19. Jahrhunderts habe sich die funktionale Differenzierung endgültig zur primären Differenzierungsform entwickelt. Dabei ist entscheidend, dass Luhmann diese Differenzierung als „primär" kategorisiert, was nicht bedeutet, dass es keine stratifikatorische Differenzierung mehr gäbe, sie ist ihm zufolge jedoch – ganz anders als im Konzept von Bourdieu – sekundär geworden (Luhmann 1985, S. 130).

Beobachtungseinheit		Funktion	Leistung	Medium	Code	Programm
Funktionale Teilsysteme	Religion	Kontingenzausschaltung	Diakonie	Glaube, Gott-Seele-Diff.	Immanenz/ Transzendenz	Offenbarung, Heilige Schrift, Dogmatik
	Recht	Ausschaltung der Kontingenz normativen Erwartens	Erwartungserleichterung, Konfliktregulierung	Recht (Rechtsprechung)	Recht/Unrecht	Konditionalprogramme, Rechtsnormen, Gesetze
	Erziehung	Selektion für Karrieren	Ermöglichung unwahrscheinlicher Kommunikationen	Lebenslauf (Kind)	besser lernen/ schlechter lernen, Lob/Tadel	Bildung, Lehr- und Lernpläne
	Politik	Ermöglichung kollektiv bindender Entscheidungen	Umsetzung kollektiv bindender Entscheidungen	Macht	Macht haben/ keine Macht haben, Regierung/ Opposition	Regierungs-, Parteiprogramme, Ideologien
	Wirtschaft	Knappheitsminderung	Bedürfnisbefriedigung	Geld	Zahlung/ Nichtzahlung	Zweckprogramme, Budgets
	Wissenschaft	Erzeugung neuen Wissens	Bereitstellung neuen Wissens	Wahrheit	Wahrheit/ Unwahrheit	Theorien und Methoden

(Quelle: Krause 1999², S. 36)

Binäre Codes

Die einzelnen Systeme, Wirtschaft, Politik, Recht, Religion, Erziehung, Wissenschaft, Kunst etc. operieren autonom und können durch keine gemeinsame Symbolik ausgedrückt werden. Die Systeme operieren auf der Basis von binären Codes: Recht-Unrecht, wahr-falsch etc. Die Systeme beobachten einander und reagieren aufeinander, aber immer in ihrer eigenen Logik. Das Wirtschaftssystem spricht in Erziehungsfragen über wirtschaftlichen Erfolg, das Wissenschaftssystem über wahre und falsche Aussagen und die Religion über den moralischen Standart. Die Systeme operieren also in ihrem eigenen Sinnhorizont. Das bedeutet, dass über den Code sowohl festgelegt wird, welchen Ausschnitt der Welt die Systeme beobachten, als auch Wirklichkeit selbst geschaffen wird. Das Rechtssystem zum Beispiel findet nicht nur Recht und Unrecht vor, sondern definiert und erzeugt auf diese Weise Recht und Unrecht. Gleichzeitig ist genau dieser Code auch der blinde Fleck des Systems. Das Rechtssystem kann selbst nicht entscheiden, ob es rechtens ist, Recht und Unrecht zu unterscheiden.

Multiinklusion

Die Inklusion von Personen erfolgt in der funktional differenzierten Gesellschaft nicht mehr über den gemeinsamen Ideenhorizont, sondern als Multiinklusion. Fast alle sind gleichzeitig in die verschiedenen Teilsysteme mit ihren spezifischen Kommunikationen integriert (Luhmann 1994), dabei sind Menschen nie gänzlich in ein System integriert, sondern immer nur rollenspezifische Teilaspekte einer Person.

3.2.2 Luhmanns soziologische Erziehungsreflexion

Reflexionsprobleme

Soziale Systeme beobachten sich selbst. Sie haben dabei das Problem des „blinden Flecks“. So wie das Rechtssystem nicht durch Beobachtung entscheiden kann, ob es rechtens ist, zwischen Recht und Unrecht zu unterscheiden, so steht auch das Erziehungssystem bei der Selbstbeobachtung vor enormen Problemen. Niklas Luhmann und Karl Eberhard Schorr (1988) analysieren explizit als Soziologen die Reflexionsprobleme des Erziehungssystems. Sie nutzen dabei den fremden Blick derjenigen, die nicht in pädagogische Prozesse involviert sind. Die von ihnen in Augenschein genommenen Probleme des Erziehungssystems, sich selbst zu reflektieren, untersuchen sie exemplarisch für die Schwierigkeiten vieler Systeme. Es handelt sich – so Luhmann/Schorr – um verallgemeinerbare Schwierigkeiten dabei, die gewandelte Ausdifferenzierung und Autonomie der Systeme sozialwissenschaftlich zu begreifen.

Pädagogik und Erziehungswissenschaft

Luhmann/Schorr gehen von der Annahme aus, dass sich im Zuge der funktionalen Differenzierung das Erziehungssystem (wie auch das Wissenschaftssystem) als autonomes Teilsystem etabliert hat. „Funktionale Differenzierung bedeutet in Hinsicht auf »Erziehungswissenschaft«, dass für wissenschaftliche Forschung und für Erziehung *verschiedene* Sozialsysteme ausdifferenziert worden sind“ (Luhmann/Schorr 1988, S. 8, kursiv im Original). Die „Erziehungswissenschaft“ sei im Hochschulsystem verankert, während die „Pädagogik“ Teil des Erziehungssystems sei. Die Pädagogik stehe unter der Anforderung, Situationsdeutungen, wenn nicht gar Handlungsanleitungen, zu liefern (Luhmann/Schorr 1988, S. 7). Wie jedes System reflektiere die Pädagogik ihre eigene Praxis. Diese Reflexion werde häufig mit „Wissenschaft“ gleichgesetzt, tatsächlich mache die gleichzeitige Erwartung von Handlungsanleitung und Selbstlegitimation als System Wissenschaft unmöglich. Das System brauche Reflexion zur Weiterentwicklung, die Wissenschaft jedoch müsse und werde in einem anderen System geleistet, was es anzukennen gilt. Die Erziehungswissenschaft ist eine „besondere Art der Sichtung und Produktion von Wissen für das Geschäft der Erziehung“ (Luhmann/Schorr 1988, S. 368).

Luhmann und Schorr stellen sich also die Aufgabe, Reflexionsprobleme des Erziehungssystems aus soziologischer Perspektive zu diskutieren. Sie wählen dazu drei Felder: Die Autonomie des Erziehens, die Ambivalenz zwischen Technik und Reflexion sowie die Frage der sozialen Selektivität. Für das Verständnis der Ausführungen ist es wichtig, dass es den Autoren nie um Lösungen der pädagogischen Probleme, sondern um Reflexion, d.h. um ein komplexes, differenziertes Verständnis der Probleme auf der Basis der Systemtheorie, geht.

Leitmotive

Soziale Systeme weisen prinzipiell – nach Luhmann – einen Bezug auf die Gesamtgesellschaft auf, indem sie eine Funktion erfüllen. Diese Funktion wird durch ein Leitmotiv bzw. eine Kontingenzformel zum Ausdruck gebracht (Luhmann/Schorr 1988, S. 36). Hier zeigen Luhmann/Schorr, und damit kommen sie zum ersten Reflexionsproblem, in Bezug auf das Erziehungssystem, dass sich das Selbstverständnis der Funktion historisch wandelt und konkretisiert. Während zunächst „humane Perfektion“ im Mittelpunkt der Selbstbeschreibung gestanden habe, wäre darauf folgend „Bildung“ zum Leitmotiv geworden und vielfach bis heute geblieben. Luhmann/Schorr bezweifeln nun, dass es angemessen ist, über die Idee von Bildung, Erziehung zu definieren. Sie schlagen stattdessen

„Lernfähigkeit“ als eine adäquate Kontingenzformel vor. Humane Perfektion sei eine Formel, die sowohl für Familienerziehung als auch für Schulerziehung greife und daher noch sehr weit gefasst sei. Bildung definiere das Erziehungssystem in Hinblick auf das Hochschulsystem und gewährleiste daher die für ausdifferenzierte Systeme typische Autonomie noch nicht. Man wähle mit dem Leitmotiv „Bildung“ den einfachen Weg, da nur Pädagogen sagen könnten, was Bildung sei und sichere sich so gegen Eingriffe von Priestern, Politikern etc. ab (Luhmann/Schorr 1988, S. 366). Über das Konzept von Lernfähigkeit werde die Autonomie anderen Systemen gegenüber substanzieller. „Lernfähigkeit schließlich formuliert eine Kondition, die vor allem auf spätere Verwendung im beruflichen Leben blickt und besagt, dass die Erziehung, was immer sie sonst mitgibt, vor allem zum Lernen von Lernfähigkeit führen müsse, um für jede spätere Lebenslage zu wappnen“ (Luhmann/Schorr 1988, S. 62).

Technologiedefizit

Systeme stellen zweitens nicht nur zur Gesamtgesellschaft einen Bezug her, sondern sie stehen auch in Beziehung zu anderen Systemen. Im Fall des Erziehungssystems wird diese Beziehung über Produktion von „Leistung“ beschrieben. Daraus ergibt sich als zweites Reflexionsproblem des Erziehungssystems, dass das eigene Technologiedefizit nicht hinreichend diskutiert werde. Der Begriff des Technologiedefizits bezieht sich auf die Erkenntnis, dass eine direkte und in ihren Folgen klar umgrenzbare Einflussnahme auf die Zöglinge nicht möglich ist. Weder ist kontrollierbar, was in das Bewusstsein der zu Erziehenden dringt, noch wie es dort verarbeitet wird. Die Pädagogik könne weder die Selbstreferenz im Menschen noch im Erziehungssystem leugnen und fühle sich doch gehalten, die Erziehungsaufgabe zu operationalisieren sowie als „Leistung“ nachvollziehbar und in andere Systeme vermittelbar zu machen. Daher verlasse es sich immer stärker auf „Organisation“ von Erziehung. Statt das eigene Dilemma ernst zu nehmen, werde Erziehung immer neu reformiert und als organisatorisches Problem definiert (vgl. Luhmann/Schorr 1988, S. 177ff). Didaktik und Methode dienen dazu, Ziele festzulegen, die dann überprüfbar werden. Messbare Leistung ersetzt die Einsicht in die Unmöglichkeit, erzieherisches Handeln in ihren Folgen zu bestimmen.

Chancengleichheit

Aus der Zuspitzung auf Leistung ergibt sich das dritte Reflexionsproblem. Soziale Systeme beziehen sich, wie schon festgestellt, nicht nur auf andere Systeme und die Gesamtgesellschaft, sondern auch auf sich selbst in Form von Selbstreflexivität. Als zentrales Problem entziffern Luhmann/Schorr hier einen endlosen Diskurs um das Problem der Chancengleichheit. Während das Erziehungssystem notwendig, und gerade über die Differenz von gebildet/ungebildet sowie über die Fokussierung auf Leistung, unterschiedliche Karrierechancen eröffnet, wird gleichzeitig jede Selektionsfunktion abgelehnt. Pädagogen reagieren darauf, indem sie die Probleme an den Rand des Systems verschieben (in den Übergang zum Beruf) und sich damit der Verantwortung entziehen. Durch Massenbildung, so argumentieren Luhmann/Schorr ähnlich wie Bourdieu, sei die Zahl der Absolventen mit höheren Abschlüssen gestiegen, ohne dass für adäquate Beschäftigung gesorgt sei. Die Selektion fände nun nicht mehr direkt durch das Bildungssystem statt, sondern indirekt im Übergang in den Beruf. Luhmann/Schorr zufolge habe dies das Selektionsproblem „erheblich verschärft“ (Luhmann/Schorr 1988, S. 274). Statt Chancengleichheit schlagen sie „Karriere“ als Reflexionsbegriff vor, welcher es erlaubt mitzubedenken, dass die Kinder aus den Familien bereits mit unterschiedlichen Fertig- und Fähigkeiten in das Erziehungssystem kommen.

Grundlegende Paradoxien

Später stellt Luhmann (1996) darüber hinaus die These auf, dass das Erziehungssystem von grundlegenden Paradoxien durchzogen sei. Eine formuliert er folgendermaßen: „Das Erziehungssystem behandelt Ungleiches gleich, um die Auflösung dieser Paradoxie, die Erzeugung schulleistungsbedingter Ungleichheiten, sich selbst zurechnen zu können" (Luhmann 1996, S. 25). Die ungleichen Kinder werden also als gleich eingestuft, um dann Leistungsdifferenzen (scheinbar) hervorbringen und messen zu können. Voraussetzung – so die Selbstbeschreibung – sei eine „gerechte", d.h. „im System kontrollierte Herstellung von Ungleichheit" (Luhmann 1996, S. 26).

Eine andere Paradoxie ergibt sich aus der über den Bildungsgedanken vermittelten Orientierung am Wissenschaftssystem. Nach der Ablösung des Religionssystems stütze sich das Erziehungssystem auf wissenschaftlich bewährte Wahrheiten und wandle diese in der Transformation zu Schulwissen in Wahrheitsverzerrungen: „Es gibt schon nichteuklidische Geometrien, aber der Schulunterricht hält sich an Euklid. Die Atome werden (noch heute) als kleine Kügelchen oder Hütchen veranschaulicht. Vor allem aber schafft das Erziehungssystem sich «Klassiker«, also Fünf-Sterne-Helden der Tradition, an denen die in Schulklassen verwendeten Bücher (...) sich orientieren können" (Luhmann 1996, S. 36). Das Erziehungssystem verkenne, dass Wissenschaft (auch Erziehungswissenschaft) sich von der Pädagogik grundlegend unterscheide.

Gesellschaftliche Funktion der Erziehung

In seinem letzten Werk „Das Erziehungssystem der Gesellschaft" – posthum von Dieter Lenzen herausgegeben – (vgl. Luhmann 2002) wechselt Luhmann die Perspektive und fragt nach der gesellschaftlichen Funktion von Erziehung. Das Erziehungssystem trage zum Erhalt der Gesellschaft bei, so argumentiert er, durch die Erziehung der Menschen zu Personen. Es produziere Eigenschaften der Menschen, die für die Übernahme von Pflichten und Aufgaben, auch für die statusadäquate Positionierung notwendig seien. In diesem Sinne sind „Personen" durch Sozialisation und Erziehung entstandene Menschen. Als Personen sind Menschen in der Lage, an der sozialen Interaktion teilzunehmen. Bildungsinstitutionen stellen den Rahmen für die Aktivitäten zur Herstellung von Personen dar. Sie bieten den Ort, an dem Erziehung in von der Gesellschaft geregelten Bahnen stattfindet, und stellen sicher, dass Erziehung jeden und jede erreicht.

Zur Einführung in die Luhmannsche Systemtheorie empfehle ich: Kneer/Nassehi 1997[3] (original 1993) sowie Schimank/Volkmann 1999. Die Theorie diskutieren: Knorr-Cetina 1989; Bude 1990; Münch 1991; Schimank 1996, S. 135ff; Schwinn 1998; Nassehi 1999 u.a. Die Erziehungsthesen diskutieren: Ruhloff 1996; Schäfer 1983; Lenzen 1996. Lesenswerte Folgestudien sind: Drewek/Harney 1986; Stichweh 1987. Einen Glossar zur Theorie haben Baraldi u.a. (1997) zusammengestellt.

4. Soziale Ungleichheit

4.1 Definition sozialer Ungleichheit

Ungleichartigkeit

Das Phänomen der sozialen Ungleichheit als strukturierende gesellschaftliche Dimension gilt als „theoretisches Schlüsselthema" (Kreckel 1992, S. 21) der Soziologie. Soziale Ungleichheit meint in der Soziologie in der Regel nicht einfach nur Ungleichartigkeit. Die Heterogenität der Menschen zum Beispiel in der Ohrengröße oder eine Präferenz für fleischlose bzw. fleischhaltige Speisen interessiert die Soziologie erst dann, wenn Gruppen von Menschen deshalb privilegierten Zugang zu Ressourcen gewinnen oder aber vom Zugang ausgeschlossen bleiben. Die soziologische Ungleichheitsforschung konzentriert sich auf gesellschaftsstrukturierende und Personengruppen benachteiligende Lebensbedingungen. Diese entstehen insbesondere durch Geschlechts- oder Klassenzugehörigkeit, aber auch durch sexuelle Präferenzen, körperliche und mentale Möglichkeiten (Behinderung, Körpergröße, Haarfarbe), durch das Alter sowie durch ethnische Zuschreibung etc.

Zugangschancen

„Soziale Ungleichheit im weiteren Sinne liegt überall dort vor, wo die Möglichkeiten des Zuganges zu allgemein verfügbaren und erstrebenswerten sozialen Gütern und/oder zu sozialen Positionen, die mit ungleichen Macht- und/oder Interaktionsmöglichkeiten ausgestattet sind, dauerhaft Einschränkungen erfahren und dadurch die Lebenschancen der betroffenen Individuen, Gruppen oder Gesellschaften beeinträchtigt bzw. begünstigt werden" (Kreckel 1992, S. 17). Der Begriff der strukturierten sozialen Ungleichheit meint „langfristig wirksame, die Lebenschancen ganzer Generationen prägende Ungleichheitsverhältnisse" (Kreckel 1992, S. 19). Reinhard Kreckel unterscheidet in seinem Buch „Politische Soziologie der sozialen Ungleichheit" zwei Aggregatzustände sozialer Ungleichheit: asymmetrische Beziehungen zwischen Menschen und ungleiche Verteilung von Gütern. Ersteres ist eine relationale Form sozialer Ungleichheit, letztgenanntes eine distributive Form. Analytisch können die distributiven Aspekte sozialer Ungleichheit unterschieden werden in eine „Reichtums-Dimension" und eine „Wissens-Dimension" (Kreckel 1992, S. 78ff.). Unter der Reichtums-Dimension versteht Kreckel die Zugangschancen zu primär materiellen Produkten und Bedingungen, unter der Wissens-Dimension die Zugangschancen zur primär symbolischen Kultur. Da Handelnde sich nicht nur an den materiellen und symbolischen Vergegenständlichungen orientieren, sondern auch an den anderen Handelnden, prägen relationale Formen ebenfalls die Entstehung und Reproduktion sozialer Ungleichheit – und zwar sowohl als Beziehung zwischen Gleichen als auch zwischen Ungleichen. Kreckel unterscheidet

Reichtum und Wissen

daher bei den relationalen Aspekten zwischen „hierarchischer Organisation" und „selektiver Assoziation". In dem einen Fall sind Verfügungschancen aufgrund der sozialen Position innerhalb einer Hierarchie gemeint, im anderen Fall symmetrische Beziehungen zwischen Gleichen, die für andere Ausschlusscharakter haben bzw. die Integrierten begünstigen, zum Beispiel studentische Verbindungen, politische Loyalitäten etc. Diesen vier Dimensionen Reichtum, Wissen, Organisation und Assoziation lassen sich institutionalisierte Tauschmittel zuordnen: Geld, Zeugnis, Rang und Zugehörigkeit als „Währungen sozialer Ungleichheit" (Kreckel 1992, S. 86).

In der soziologischen Ungleichheitsforschung kommt den Kategorien „Klasse" (oder Schicht, Milieu etc.) und „Geschlecht" eine besondere Bedeutung zu. Im Unterschied zu anderen Feldern sozialer Ungleichheit kann sehr umfassend nachgewiesen werden, dass Klasse und Geschlecht jede Handlung durchziehen und gleichzeitig als Ordnungsmuster des Gesellschaftlichen fungieren. Deshalb spricht man von ihnen auch als „Strukturkategorien". Diese sehr grundlegenden soziologischen Begriffe werde ich im Folgenden erläutern, bevor ich den Gesamtzusammenhang auf Bildung beziehe.

4.1.1 Klasse

Karl Marx

Der Begriff der sozialen Klasse bezieht sich in der Soziologie in besonderem Maße auf die Theorien Karl Marx und Max Webers. Marx versteht unter einer Klasse eine Gruppe, die eine gemeinsame Beziehung zu dem Produktionsmittel, also zu den Mitteln, mit denen sie ihren Lebensunterhalt verdient, hat. In der Industriegesellschaft sind die beiden wichtigsten Klassen die Arbeiterklasse bzw. das Proletariat auf der einen Seite und die Kapitalisten bzw. Industrielle auf der anderen Seite. Sie unterscheiden sich vor allem darin, dass die einen (die Minderheit) über die Produktionsmittel verfügen, wohingegen die anderen (die Mehrheit), da sie keine Produktionsmittel (mehr) besitzen, gezwungen sind, ihre Arbeitskraft zu verkaufen. Das Verhältnis zwischen den Klassen ist Marx zufolge eines von Ausbeutung, da der Kapitalist mehr Geld für die Produkte erzielt als er für deren Produktion aufwendet.

Mehrwert

Der so genannte Mehrwert ist der Profit des Kapitalisten, den dieser zur Stabilisierung des asymmetrischen Machtverhältnisses nutzen kann, welchen er jedoch nur durch die Ausbeutung der Arbeit des Proletariats erzielt. Das Klassenverhältnis ist somit sowohl ein ökonomisches Verhältnis als auch ein politisches Herrschaftsverhältnis (vgl. auch Giddens 1984). Der Begriff der Klasse ist für Marx folglich gleichzeitig eine analytische Kategorie und ein politischer Begriff. Das zusammen mit Friedrich Engels verfasste Manifest der Kommunistischen Partei beginnt mit der Feststellung: „Die Geschichte aller bisherigen Gesellschaft ist die Geschichte von Klassenkämpfen" (Marx/Engels 1984, original 1848, S. 44). Seit dem Verlassen der archaischen Gemeinschaften stünden sich antagonistische Klassen gegenüber, wobei immer die eine über die erforderlichen ökonomischen Ressourcen verfüge und somit die andere unterwerfe. Zum Klassenkampf kommt es jedoch nur, wenn die Menschen ein „Klassenbewusstsein" entwickeln und ihre soziale Lage begreifen lernen. Durch die Stellung zu den Produktionsmitteln bildet das Proletariat eine „Klasse an sich". Allein daraus leite sich jedoch keine Handlungsfähigkeit ab.

Erst durch das „Klassenbewusstsein" und die Einsicht in die Widersprüche der Klassengesellschaft kann sich das Proletariat zu einer „Klasse für sich" formen: (MEW Bd. 4, 1964, S. 180ff.).

Max Weber

Max Weber bezieht sich auf den Klassenbegriff bei Marx und geht mit diesem soweit konform, dass auch er die Klassenbildung auf wirtschaftliche Bedingungen zurückführt. „Wir wollen da von einer ‚Klasse' reden, wo 1. einer Mehrzahl von Menschen eine spezifische ursächliche Komponente ihrer Lebenschancen gemeinsam ist, soweit 2. diese Komponente lediglich durch ökonomische Güterbesitz- und Erwerbsinteressen und zwar 3. unter den Bedingungen des (Güter- oder Arbeits-)*Markts* dargestellt wird" (Weber 1980[5], S. 531; Hervorhebung im Original). Jedoch erweitert Weber das Klassenkonzept von Marx erheblich. In seinen unter dem Titel „Wirtschaft und Gesellschaft" (Weber 1980[5]) versammelten Arbeiten argumentiert er, dass Klassenunterschiede nicht nur auf der Verfügungsgewalt über Produktionsmittel basieren, sondern im Besonderen auch auf Leistungsqualifikationen und damit auf Bildungsprozessen (Weber 1980[5], S. 177ff.). Weber unterscheidet zwischen der Erwerbsklasse und der Besitzklasse. Menschen, die zur Erwerbsklasse gezählt werden können, sind auf Erwerb angewiesen und müssen folglich arbeiten gehen. Die anderen verfügen über Besitz und können davon leben. Im Unterschied zu Marx ist der Wechsel von einer Klasse zur anderen jedoch in der theoretischen Konzeption von Weber durchaus möglich. Darüber hinaus besteht für die Erwerbsklasse die Möglichkeit, durch Qualifikationserwerb die eigenen Chancen auf dem Arbeitsmarkt zu erhöhen. Somit ist die Klassenlage bei Weber nicht nur durch die Verhältnisse geprägt, sondern auch durch die „Lebensstellung" und das „Lebensschicksal" der Menschen.

Erwerbs- und Besitzklassen

Aufgrund dieser Betonung von Bildungsprozessen betrachtet Weber es nicht als das Interesse der Arbeiterklasse/Erwerbsklasse, die Spielregeln des Marktes durch Revolution umzugestalten, sondern die Interessen sieht er in der Verbesserung der eigenen Chancen durch Spezialqualifikationen, Standortvorteile etc.

Stände

Weber differenziert in Klassen, Ständen und Parteien. Die Klassenlage wird ihm zufolge durch die Ständebildung überlagert. Stände basieren auf Gemeinschaftsbildungen von Gruppen (z.B. durch Religion, Ethnizität, Hautfarbe, Berufe). Sie implizieren im Unterschied zur Klasse ein subjektives Zusammengehörigkeitsgefühl und basieren nicht notwendig auf ökonomischen Faktoren. „Im Gegensatz zur rein ökonomisch bestimmten ‚Klassenlage' wollen wir als ‚ständische Lage' bezeichnen jene typische Komponente des Lebensschicksals von Menschen, welche durch eine spezifische positive oder negative Einschätzung der ‚Ehre' bedingt ist, die sich an irgendeine gemeinsame Eigenschaft vieler knüpft" (Weber 1980[5], S. 534).

Parteien

Parteien wiederum sind auf ein planvoll angestrebtes Ziel gerichtet, welches sowohl sachlicher als auch persönlicher Natur ist. Parteien sind Zusammenschlüsse, um Einfluss auf das Gemeinschaftshandeln auszuüben. Insofern sind Parteien nicht wie Stände oder Klassen Dimensionen der sozialen Ungleichheit, sondern vielmehr Zusammenschlüsse, die sich aus Stand oder Klasse ergeben, allerdings auch auf ganz anderen Interessen basieren können (vgl. zum Stand z.B. Weber 1980[5], S. 539).

Zusammenfassend heißt das, die Klassenlage bestimmt nach Marx und Weber die Position eines Individuums oder einer Gruppe in der Gesellschaft. Nach

Marx bestimmt sich die Position primär über die Teilhabe bzw. den Ausschluss von Eigentum an dem Produktionsmittel. Bei Weber ist die Klassenlage über die Chance zur Versorgung mit Ressourcen bestimmt, welche sowohl auf die äußeren Lebensumstände als auch auf die subjektiven Bewältigungen zurückzuführen ist. Klasse ist demnach ein analytischer Begriff, um die Lebensumstände von Personengruppen in der Regel auf der Basis ökonomischer Faktoren zu analysieren.

Wandel der Erwerbsstruktur

Ein wesentlicher Einwand, der gegen die Marxsche Klassentheorie vorgetragen wird, ist, dass der soziale Wandel in der Erwerbsstruktur in diesem theoretischen Modell nicht erfasst werden kann (vgl. Gottschall 2000, S. 71). Auf Weber trifft diese Kritik nicht in gleichem Maße zu, weil er in seiner Konzeption nicht auf die Produktion, sondern auf den Markt fokussiert, Bildungsqualifikationen als strukturierenden Faktor berücksichtigt sowie sich durch die Unterscheidung von Klasse und Stand einem mehrdimensionalen Modell sozialer Ungleichheit annähert. Dadurch hat Weber die deutsche Ungleichheitsforschung schließlich stärker geprägt als Marx.

Soziale Schichten

Der Klassenbegriff wird vielfach durch den Begriff der „sozialen Schicht" ersetzt, um die politischen Implikationen der Klassentheorie zu vermeiden, aber auch aufgrund einer auf eine quantitative Bestimmung von Status verkürzte Rezeption der Weberschen Theorie (vgl. Kreckel 1982). „Schicht" wird zu einem rein deskriptiv genutzten Ordnungsbegriff. Dabei werden die Schichten eingeteilt, indem die Berufsstruktur als Indikator für die soziale Position genutzt wird (vgl. Hradil 1987).

Allerdings muss man berücksichtigen, dass der Schichtbegriff so in das Alltagsvokabular eingegangen ist, dass auch jene Theorien, die nach wie vor dem marxistischen Grundgedanken, dass Industriegesellschaften sich in einem Zustand von Widersprüchen und sozialen Konflikten befinden, folgen, den Begriff synonym mit „Klasse" verwenden. Letztlich folgen beide Richtungen und damit die soziologische Ungleichheitsforschung einem gemeinsamen „Hintergrundkonsens" (Kreckel 1983, S. 3), dass eine einheitliche und dominante gesellschaftliche Struktur auf der Basis der Erwerbsarbeit herausgefiltert werden kann, die eine – alle gesellschaftlichen Bereiche durchziehende – Struktur sozialer Ungleichheit darstellt.

Stabile Gattenfamilie

Diese ausschließliche Akzentuierung auf Erwerbsarbeit erwies sich jedoch als reduktionistische Herangehensweise, da hierbei die Frauen bruchlos der Position des erwerbstätigen Mannes zugeordnet werden. Reinhard Kreckel (1992) kritisiert an der Zuordnung der Frauen zum Status der Männer, dass eine stabile Gattenfamilie in modernen westlichen Gesellschaften nicht länger unterstellt werden könne. Sie existiere zwar weiterhin als Norm, könne jedoch als durchgängige Praxis in der Forschung nicht mehr angenommen werden. Man müsse daher parallel zur klassenspezifischen Vergesellschaftung und zur klassenspezifischen Ungleichheit auch die institutionalisierte Zweigeschlechtlichkeit (vgl. Kapitel 4.1.2 im Folgenden) und die daraus folgende geschlechtsspezifische Vergesellschaftung sowie die geschlechtsspezifische Ungleichheit betrachten.

Petra Frerichs und Margareta Steinrücke (1997) zeigen darüber hinaus, dass sich Paare zwar mehrheitlich innerhalb der gleichen Schicht finden, jedoch ist der Status der Frau in der Regel etwas niedriger als der des Mannes. So beziehen Frauen klassenübergreifend auf der gleichen sozialen Stufe bzw. in der gleichen beruflichen Stellung ein geringeres Einkommen als Männer, was, wie Ursula Rabe-Kleberg (1987, 1993) und Reinhard Kreckel (1992) belegen, nicht länger mit Qualifikationsunterschieden begründet werden kann (vgl. Kapitel 4.2.3 in diesem Buch). Mit Bezug auf Michael Mann (1986) argumentiert Frerichs (1997), dass Frauen eine Art „Pufferzone" zwischen der Klasse ihres Mannes und der nächst niedrigeren Klasse bilden und damit eine wesentliche gesellschaftliche Vermittlungsleistung erbringen.

Frauen als „Pufferzone"

Zusammengenommen bedeutet dies, dass das Phänomen der sozialen Ungleichheit zum einen an die Klassenlage (oder unpolitischer formuliert an die Schicht) geknüpft ist, aber parallel dazu sich die Geschlechtszuweisungen auf den sozialen Status auswirken. Die Klassenlage wird über die Struktur der Erwerbsarbeit aber, wie sich in der eingangs referierten Position von Kreckel und durchaus mit Bezug auf Weber zeigt, auch über die Bildungsverhältnisse bestimmt (welche, wie später noch genauer zu zeigen sein wird, in Wechselwirkung mit dem Arbeitsmarkt wirksam werden). Neben den Dimensionen Wissen und Reichtum existieren quer liegende Vergesellschaftungsformen (bei Weber „Stand", bei Kreckel „hierarchische Organisation" und „selektive Assoziation"), welche Zugangschancen zu gesellschaftlichen Ressourcen eröffnen oder verschließen und damit an der Entstehung, Aufrechterhaltung oder Veränderung von sozialer Ungleichheit mitwirken.

Lebensstile

Diese quer liegenden Vergesellschaftungsformen haben unter dem Begriff der „sozialen Milieus" oder der „Lebensstile" in den letzten Jahren den Klassen- und Schichtbegriff in Teilen der Soziologie, vor allem aber im öffentlichen Diskurs, in den Hintergrund gedrängt. Insbesondere Ulrich Beck vertritt die These, dass sich moderne Industriegesellschaften „jenseits von Klasse und Schicht" (Beck 1986, S. 121) entwickeln. Der Prozess der Individualisierung und Diversifizierung von Lebenslagen und Lebensstilen ließe sich in einem starren Modell hierarchischer Klassen und Schichten nicht ausreichend erfassen. Die Becksche Überlegung trifft in einer Gesellschaft, die sich selbst gern als egalitär und voll von gleichen Chancen für alle Menschen beschreibt, auf fruchtbaren Boden. Vergessen wird dabei schnell, dass Beck nie die nach wie vor existierende soziale Ungleichheit geleugnet hat. „Auf der einen Seite sind die Relationen sozialer Ungleichheit in der Nachkriegsentwicklung der Bundesrepublik weitgehend *konstant* geblieben. Auf der anderen Seite haben sich die Lebensbedingungen der Bevölkerung radikal verändert. Die Besonderheit der sozialstrukturellen Entwicklung in der Bundesrepublik ist der ‚*Fahrstuhl-Effekt*': die ‚Klassengesellschaft' wird *insgesamt* eine Etage höher gefahren" (Beck 1986, S. 122; Hervorhebung im Original).

Dass die Ausformung unterschiedlicher Milieus und Lebensstile, die Zuordnung dieser zu Klassen und die parallele Analyse von Geschlechterverhältnissen nicht in einem Widerspruch stehen muss, hat Pierre Bourdieu in seiner Theorie bewiesen (vgl. ausführlich Kapitel 3.1.1 in diesem Buch). Mit Bourdieu gesprochen kann abschließend Klasse als „Ensembles von Akteuren, die homogenen Lebensbedingungen unterworfen sind" (Bourdieu 1983[2], S. 175) definiert werden. Die Klassenlage ist systematisch mit der Geschlechtslage verknüpft.

4.1.2 Geschlecht

Soziale Konstruktion

Der Begriff des Geschlechts ist ein ähnlich komplexer Begriff wie der der Klasse, akzeptiert als Strukturkategorie der Gesellschaft wird er jedoch erst seit ca. zehn Jahren. Die meisten Menschen glauben zu wissen, dass es zwei Geschlechter gibt, die aufgrund biologischer Merkmale voneinander unterschieden werden können. Die Soziologie verfolgt eine andere Perspektive: Es ist ein weithin akzeptierter Konsens, dass jener soziologisch denkt, „wer Wirklichkeit als soziale Konstruktion, also als sozial konstruiert, aber eben als prinzipiell sozial konstruiert begreift" (Hitzler 1997, S. 11). Das heißt, es geht u.a. darum, die Herstellung sozialer Tatsachen im gesellschaftlichen Prozess zu betrachten. Untersucht man die scheinbare Tatsache einer geschlechtsspezifischen Differenz (einer scheinbar objektiven Natur in Form von Körpern, Charaktermerkmalen oder spezifischen Emotionen) auch als soziale Konstruktion, so zeigt sich ein Herstellungsprozess auf zwei Ebenen. Es handelt sich zunächst um einen sozial überaus komplexen *Sortierungsvorgang*. Bei Neugeborenen ausgehend von den Genitalien, bei LeistungssportlerInnen ausgehend von den Chromosomen, im Alltag durch Kleidung, Gestik, Mimik oder Sprache werden Menschen in zwei Gruppen eingeteilt und müssen sich, selbst handelnd als das richtige Geschlecht zu erkennen geben.

Sortierungsvorgang

Dieser Sortierungsvorgang basiert zunächst auf unterschiedlichen biologischen Gegebenheiten, die aber sozial vereinheitlicht werden müssen, um eine eindeutig zweigeschlechtliche Aufteilung zu ermöglichen. Das Kontinuum zwischen weiblich und männlich, welches sich durch die Variationsbreite von Anatomie, Chromosomen und Hormonen ergibt, wird zu einer zweipoligen Einteilung vereinfacht (vgl. Gildemeister/Wetterer 1992). Die eigentliche Aufgabe jedes einzelnen besteht darin, das eigene Geschlecht angemessen darstellen zu lernen und die Inszenierungen anderer entschlüsseln zu müssen, ohne dass die Kleidungsstücke entfernt werden können. Kinder lernen diese Aufgabe zu bewältigen, ohne auf die Erkennungsmerkmale der Erwachsenen wie Lippenstift oder Krawatte zurückzugreifen. Gerade hierin liegt die Begründung, warum Mädchen und Jungen oft die einem Geschlecht zugewiesenen Eigenschaften wie Aggressivität oder Koketterie über die Maßen betonen (vgl. Hagemann-White 1984). Die Irritation, die Menschen auslösen, die nicht klar zuzuordnende Signale senden, oder die Unfähigkeit fast jedes Menschen mit einem Baby zu kommunizieren, dessen Geschlecht man zuvor nicht erfragt hat, zeigt die unterschwellig stets aktuelle Bedeutung geschlechtsspezifischer Zuordnung.

Schule und Zweigeschlechtlichkeit

Die zweigeschlechtliche Sortierung ist nicht nur unvermeidbar, sie ist auch Teil des Bildungsprozesses. Gerade in der Schule lassen sich die Sortierungsprozesse gut beobachten und zwar nicht nur von den Lehrern und Lehrerinnen in Bezug auf die Schüler und Schülerinnen, sondern auch unter den Kindern und Jugendlichen. Georg Breidenstein (1997) analysiert, dass Kinder andere Kinder des gleichen Geschlechtes oft namentlich ansprechen, wohingegen sie Kinder des anderen Geschlechts über die Genusgruppe definieren. Typische Aussprüche sind: „Ich fänd´s gut, wenn zwei Mädchen mitspielen" oder „Nee, Jungen wollen wir nicht dabei haben". Die Konstruktion einer grundlegenden Differenz schafft Fremdheit und damit Neugier oder Abwehr. „Die Mädchen" oder „die Jungen" werden in schulischen Interaktionen als grundlegend Andere etabliert (ausführlich zur Studie Breidenstein/Kelle 1998).

Der Akt der Rede und der Handlung

Der Sortierungsvorgang entsprechend den Prinzipien der Zweigeschlechtlichkeit wird durch Strukturen vorgegeben (z.B. Namensrecht, Geschlechtszuweisung bei der Geburt). Diese Strukturen werden, so betonen diskurstheoretisch orientierte AutorInnen (z.B. Butler 1991), im performativen Akt der Rede stets aufs Neue wiederholt. „Es ist ein Mädchen", sagt die Krankenschwester, und gibt dabei nicht nur die Geschlechtszugehörigkeit wieder, sondern konstruiert sie auf diese Weise. Stärker handlungstheoretisch argumentierende AutorInnen untersuchen die Reproduktion im Alltagshandeln. Gerade in der Transsexuellenforschung zeigt sich, welchen Aufwand Menschen betreiben (müssen), um als das „richtige" Geschlecht erkannt zu werden (vgl. z.B. Lindemann 1993). Dies beginnt mit der Art, die Zigarette zu halten, geht über die Wahl der Getränke und Speisen, den Gang, die Körperhaltung, die Kleidung, Mimik (die Art des Lächelns), Gestik bis hin zur Wahl des Autos oder der Lieblingslektüre.

Auf diese Weise geht das Geschlecht in Fleisch und Blut – oder mit Bourdieu gesagt (vgl. Kapitel 3.1) – in den Habitus über. Dieser ist jedoch nicht für alle Menschen einer Genusgruppe gleich. Wir erkennen Frauen unabhängig von der Schicht als Frauen und dennoch sind die Prinzipien der Darstellung ihrer Geschlechtszugehörigkeit unterschiedlich – sie unterscheiden sich aber prinzipiell von denen der Männer (vgl. z.B. zur Klassen- und Geschlechtsspezifik im Kochen: Frerichs/Steinrücke 1997).

Hierarchische Differenzierung

Über diese Praxis des Handelns wird jedoch nicht nur die zweigeschlechtlichte Sortierung reproduziert, sondern auch darüber hinaus – und das ist die zweite Ebene – eine hierarchische *Differenzierungsform* geschaffen und bestätigt (vgl. zum Differenzierungsbegriff Löw 2002). Ein typisches Beispiel hierfür ist der geschlechtsspezifisch segmentierte Arbeitsmarkt. Arbeit selbst wird vergeschlechtlicht und verselbständigt sich zu weiblich und männlich assoziierten Bereichen (Wetterer 1995), welche in Bezahlung, Rang und Aufstieg etc. unterschiedliche Chancen und Verhinderungen mit sich bringen (ausführlich Kapitel 4.2.3).

Geschlecht als soziale Tatsache

Die Geschlechterdifferenz ist also ein Kategorisierungsprinzip und ein Muster der Vergesellschaftung. Geschlecht ist dabei strukturell gesehen sowohl Strukturierungs- als auch Differenzierungsprinzip. Diese Strukturen verwirklichen sich im Handeln und Sprechen. Die Soziologie versucht demzufolge nicht zu bestimmen, was ein Mann und was eine Frau ist, sondern untersucht, wie Mann und Frau, Männlichkeit und Weiblichkeit als soziale Tatsache geschaffen werden. Die Kategorien Frau und Mann werden als soziale Konstruktionen begriffen, die nicht aus der Biologie abgeleitet werden können, da der Mensch selbst das Biologische betrachtet und dann in zwei Kategorien einteilt.

Prinzip der Zweigeschlechtlichkeit

Vielmehr formt das Prinzip der Zweigeschlechtlichkeit mit der Anforderung, Geschlecht eindeutig darzustellen und zu erkennen, und das Prinzip der geschlechtsspezifischen Differenzierung im Sinne einer hierarchischen binären Gegensatzkonstruktion von Mann und Frau eine „asymmetrische Geschlechterkultur" (Müller 2000). Diese Kultur begünstigt strukturell Männer als Gruppe. Auch Sexualität ist in diese Geschlechterkultur eingebunden und gliedert sich entsprechend den Kategorien Frau – Mann in Heterosexualität – Homosexualität auf. Sexualität wird wie Geschlecht in einem binären System verankert und basiert auf der genauen Zuordnung zu Geschlechtern (Frau mit Mann, Frau mit Frau, Mann mit Mann).

Öffentlichkeit und Privatheit

Ursachen für diese Geschlechterkultur werden meistens in der Spezifik der bürgerlichen Gesellschaft gesehen. Diese ist durch die Trennung von Öffentlichkeit und Privatheit sowie durch die Zuweisung von Frauen in die Sphäre des Privaten gekennzeichnet (vgl. Kapitel 7.1). Entwickelt hat sich so eine Organisationsform für Erwerbsarbeit und Generativität (die Regelung der Nachkommenschaft) (vgl. z.B. Hausen 1976), durch die Männer strukturell bessere Zugänge zu gesellschaftlichen Ressourcen haben und Männlichkeit gesellschaftlich höher bewertet wird als Weiblichkeit. Historische Unkenntnis und Ethnozentrismus führen häufig dazu, dieses spezifische moderne Muster für ein universelles und ahistorisches zu halten. Tatsächlich weiß man sehr wenig zum Beispiel über die verschiedenen (!) Kulturen der so genannten Jäger und Sammler. Die Interpretation zum Beispiel von Höhlenmalereien im Sinne der dem Hause nahen weiblichen Sammlerinnen und der in die Welt ziehenden männlichen Jäger geschieht auch auf der Basis der heute für natürlich erachteten kulturellen Konstruktion von Zweigeschlechtlichkeit und einer binären Differenzierung von Geschlecht.

4.2 Bildung und soziale Ungleichheit

Wie man an den soziologischen Theorien gut sehen kann, gilt Bildung als ein Feld, über das soziale Ungleichheit in besonderem Maße hergestellt wird. Über die Differenz von gebildet/ungebildet bzw. über den Erwerb/Ausschluss von Bildungszertifikaten werden Bildungsinstitutionen als Horte der Reproduktion sozialer Ungleichheit und als Schaltstellen für mögliche Veränderungen angesehen (vgl. z.B. Kolbe/Sünker/Timmermann 1994, S. 19). Vor allem drei Problemkontexte, die im Folgenden genau betrachtet werden, beschäftigen die Fachleute: Die soziale Ungleichheit quantitativ in der Bildungsbeteiligung, die soziale Ungleichheit qualitativ im Bildungsgeschehen, die soziale Ungleichheit im Übergang vom Bildungs- in das Berufssystem.

4.2.1 Die soziale Ungleichheit in der Bildungsbeteiligung

Besuch der weiterführenden Schulen

Rein auf der Ebene der Bildungsbeteiligung lässt sich feststellen, dass sich die Unterschiede zwischen den Geschlechtern und zwischen den Konfessionen nivelliert haben (vgl. Eigler/Hansen/Klemm 1980). Auch die Stadt-Land-Disparitäten nehmen ab (vgl. Henz/Maas 1995). In der Literatur herrscht ferner Einigkeit darüber, dass durch die Bildungsreform alle Kinder und Jugendlichen insgesamt mehr in Bildungsprozesse integriert sind. Ob sich durch dieses Mehr an Beschulung auch die schicht- und klassenspezifische Ungleichheit reduziert, darüber herrscht keine Einigkeit. Diese Diskrepanz ist erklärungsbedürftig: Zunächst kann man feststellen, dass die Zahl der Kinder, die weiterführende Schulen besuchen, gestiegen ist. Während 1955 gerade mal 10% der 13jährigen eine weiterführende Schule besuchten, waren es 1970 bereits 40% und 1995 betrug der Anteil schon 70% (vgl. Köhler 1992). Heute liegt der Anteil derjenigen, die schließlich erfolgreich eine weiterführende Schule abschließen, bei 65,4 %.

Das heißt, die Mehrzahl der deutschen Kinder und Jugendlichen ist heute wesentlich besser ausgebildet als noch vor dreißig oder gar 50 Jahren. Das gilt ebenfalls für die Bildungsentwicklung in der DDR. Diese Entwicklung führt dazu, dass auch wesentlich mehr Kinder aus Arbeiterfamilien das Gymnasium besuchen. Bei den siebzehn- bis achtzehnjährigen Arbeiterkindern ist der Anteil der Gymnasiasten von 1976 bis 1989 von 7,1% auf 11,8% gestiegen. Allerdings ist die gymnasiale Beteiligung bei den Kindern von Angestellten im gleichen Zeitraum von 30,7% auf 40% gestiegen (vgl. Köhler 1992). Prinzipiell kann man festhalten, dass die Differenz von 11,8% zu 40% nach wie vor hoch und somit soziale Ungleichheit keineswegs beseitigt ist. Auch kann man betonen, dass fast 9% mehr Angestelltenkinder, aber nur knapp 5% mehr Arbeiterkinder durch die Bildungsreform dem Gymnasium zugeführt wurden. Dabei muss man jedoch beachten, dass die Steigerung um fast 5% bei den Arbeiterkindern den Ausgangswert um gut zwei Drittel erhöht, während der Ausgangswert der Angestelltenkinder nur um etwa ein Drittel gestiegen ist (vgl. zu den Interpretationen der Daten auch Krais 1996b). Je nach Deutung wird in der Literatur mehr Bildungsgewinn der Arbeiterkinder oder die Bildungsdifferenz zwischen den Schichten betont (zur unveränderten sozialen Ungleichheit vor allem Köhler 1992; Meulemann 1992; Blossfeld 1993; Büchner/Krüger 1996; zur verringerten sozialen Ungleichheit Müller/Haun 1994; Henz/Maas 1995; Schimpl-Neimanns 2000).

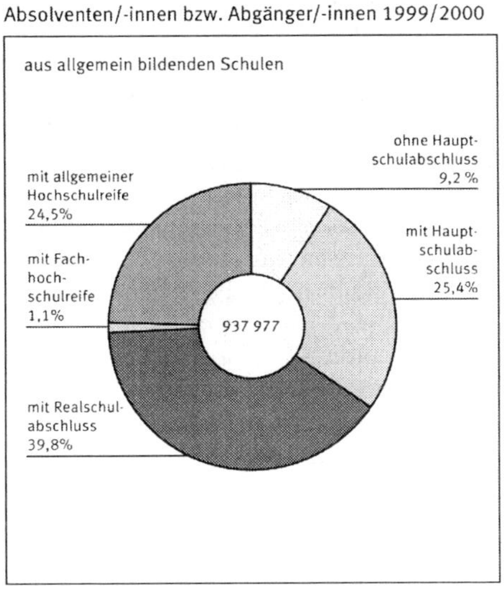

(Quelle: Statistisches Bundesamt, Stat. Jahrbuch 2002, S. 373)

Bildungsarmut

Insgesamt bezweifelt niemand, dass es nach wie vor klassenspezifische soziale Ungleichheit gibt, fraglich ist nur, „ob es im Zeitablauf eine Angleichung im Gymnasial- und Hochschulbesuch der verschiedenen sozialen Schichten gegeben hat, die es wert ist, als solche bezeichnet zu werden“ (Krais 1996b, S. 133). Dazu noch ein paar Zahlen: Unter den siebzehn- bis achtzehnjährigen SchülerInnen eines gymnasialen Jahrgangs kamen 1989 immerhin 16,9% aus Arbeiterfamilien (Köhler 1992) und 25 – 29% der Arbeitereltern streben das Abitur als Schulabschluss für ihre Kinder an (vgl. Rolff u.a. 1994). Andererseits reproduziert sich auch Bildungsarmut von Generation zu Generation in vielen Familien. Immerhin haben 15% der Söhne und 26% der Töchter, deren Väter keinen Schulabschluss haben, selbst keinen Schulabschluss. Hat der Vater keinen Berufsabschluss, dann haben 21% der Söhne und 54% der Töchter keinen beruflichen Abschluss (worin sich auch zeigt, dass selbst auf der quantitativen Ebene in Bildungsfragen das Geschlechterverhältnis nicht irrelevant wird) (vgl.

Allmendinger 1999). Bedenkt man dabei, dass von den SchulabsolventInnen ohne Schulabschluss lediglich ein Viertel einen Ausbildungsplatz findet (Klemm 2000), dann wird offensichtlich, dass Bildungsarmut in der Regel finanzielle Armut nach sich zieht und umgekehrt.

„Kellerkinder" der Bildungsexpansion

Jutta Allmendinger (1999) zeigt darüber hinaus auf, dass fehlende Bildung neben der finanziellen Situation auch die Familienplanung, das politische Interesse und die Lebenserwartung beeinflusst. Somit werden diejenigen, die – in der Regel aufgrund ihrer familiären Situation – mit der gestiegenen Bildungserwartung nicht mithalten können, zu den Verlierern der Bildungsreform, sie sind, wie Klaus Klemm (1991) es formuliert, die „Kellerkinder der Bildungsexpansion".

PISA

In der als PISA bekannt gewordenen Studie (vgl. Baumert/Schümer 2001, S. 323ff., insbesondere S. 359) wird deutlich, dass nicht Leistung über den Besuch einer Hauptschule oder einer Realschule entscheidet, sondern die soziale Herkunft: „So beträgt zum Beispiel die Chance eines Jugendlichen aus einem Facharbeiterhaushalt, ein Gymnasium anstelle einer anderen Schulform zu besuchen, ungefähr 3:17. (...) Für Jugendliche, die aus Familien der oberen Dienstklasse stammen, betragen die Chancen, ein Gymnasium statt einer anderen Schulform zu besuchen, in etwa 1:1" (Baumert/Schümer 2001, S. 356). Den ForscherInnen zufolge gelingt es der Schule nicht, Bildungsnachteile abzubauen, sondern im Gegenteil, in einem kumulativen Prozess wird die Spirale von Leistung und Schicht verstärkt (vgl. Baumert/Schümer 2001, S. 372). Dies ist, so bemerken die AurorInnen im internationalen Vergleich, für alle Schüler und Schülerinnen von Nachteil, da fehlende Fähigkeiten einiger, insbesondere in der Lesekompetenz, die Leistungen aller schmälern. „Eher deutet sich eine Tendenz an, dass bei einer Verminderung sozialer Disparitäten auch das Gesamtniveau steigt, ohne dass in der Leistungsspitze Einbußen zu verzeichnen wären" (Baumert/Schümer 2001, S. 393).

4.2.2 Die soziale Ungleichheit im Bildungsgeschehen unter besonderer Berücksichtigung der Schule

Wie im Detail wird soziale Ungleichheit in der Schule, im Kindergarten, in der Hochschule reproduziert? Welche Rolle spielen Inhalte, Sprachstil, Habitus? Mit welchen Vorurteilen sind Lehrende behaftet, wie positionieren sich Lernende in der Institution? Mit solchen Fragen beschäftigt sich häufig (aber nicht ausschließlich) die qualitative Forschung zu sozialer Ungleichheit in den Bildungsinstitutionen. Die Forschungsliteratur wird im Verlauf des Buches noch anhand der einzelnen Institutionen vertiefend dargestellt werden, nun geht es zunächst, vor allem am Beispiel der Schule, um ein Verständnis für die Problemlage.

Qualitative Forschungsergebnisse

Sprachstil

Bereits in den 1970er Jahren wurden sehr grundlegende Untersuchungen zu den Auswirkungen des Bildungsgeschehens auf die soziale Ungleichheit durchgeführt. Der Soziolinguist Basil Bernstein (1973) zum Beispiel argumentiert auf der Basis breiter empirischer Untersuchungen, dass sich je nach Klasse unterschiedliche Sprachstile und Codes entwickeln, die später den schulischen Erfolg nachhaltig beeinflussen. Kinder aus der Arbeiterklasse verfügten, so Bernstein, über einen *restringierten Code* (auch „public language" genannt). Kurze apodiktische Feststellungen bestimmen den Sprachstil. Die Aussagen sind deskriptiv und kaum verallgemeinernd. Die Jugendlichen zeichnen sich durch einen

Sprachgebrauch aus, bei dem bei den Zuhörern permanent vorausgesetzt wird, dass Kontextkenntnisse bekannt sind. Betont werden die emotionalen Aspekte eines Inhalts und weniger die logischen. Dieser Sprachstil entwickle sich in den familiär und nachbarschaftlich organisierten Arbeitervierteln. Hier, wo erstens eher praktische Erfahrungen diskutiert würden und zweitens ein kollektives Wissen um Geschehnisse vorausgesetzt werden könne, sei eine Sprache, die Normen und Werte nicht expliziert, sondern implizit transportiert, zweckdienlich. In der Schule jedoch, in der die Diskussion abstrakter Gedanken, Prozesse oder Beziehungen verhandelt würde, scheiterten die Arbeiterkinder mit ihren implizierenden Sprachmustern.

Im Unterschied dazu entwickelten Kinder aus der Mittelschicht einen *elaborierten Code* (auch als „formal language" bezeichnet), das heißt einen Sprachstil, bei dem die Bedeutung der Worte und das Kontextwissen der Zuhörenden beim Reden reflektiert werden. Die Rede wird der jeweiligen Situation angepasst, gleichzeitig können Prinzipien eloquent erläutert werden. Diese Kinder haben weniger Schwierigkeiten zu verallgemeinern oder abstrakte Gedanken zu formulieren. Sie sind in der Lage, sprachlich zwischen Logik, individuellen Intentionen und Gefühlen zu differenzieren. Bernstein begründet dies mit dem Erziehungsstil der Mütter, den Kindern das eigene Erziehungsverhalten zu begründen. Die akademische Ausbildung basiere, so Bernstein, auf einer vergleichbaren Begründungskultur, die Kinder aus der Mittelschicht besser reproduzieren können als Kinder aus der Arbeiterklasse.

Sprachkulturen

Mittlerweile ist erwiesen, dass diese zwei Klassen kontrastierende Sicht von Bernstein zu einfach ist. In empirischen Folgeuntersuchungen (vgl. zusammenfassend Steinkamp 1991) konnte der Nachweis eines Kausalzusammenhangs zwischen Sprachcode und Schicht eindeutig gegliedert in zwei Klassen nicht überzeugend erbracht werden. Dennoch zeigt sich in Untersuchungen immer wieder, dass es unterschiedliche Sprachkulturen nach Klassen und Milieus gibt, die auf unterschiedliche Akzeptanz und differentes Verständnis der mehrheitlich aus der Mittelschicht kommenden Lehrer und Lehrerinnen treffen (z.B. Tough 1976; Tizard/Hughes 1984). Hans-Günter Rolff (1967) wertet eine Vielzahl empirischer Studien zur sozialen Herkunft der LehrerInnenschaft aus. Es zeigt sich, dass die Mehrzahl der LehrerInnen aus kleinbürgerlichen Milieus kommt, bei GymnasiallehrerInnen darüber hinaus fast die Hälfte aus Beamtenfamilien. Die Kultur einer Schule wird durch deren schichtspezifisches Berufsverständnis, Gesellschaftsbild und Sozialisationsvorstellung maßgeblich geprägt.

Migration und Sprache

Sprache ist auch ein umkämpfter Gegenstand in den Auseinandersetzungen zwischen MigrantInnnen und Einheimischen. Nach wie vor sind in der Gruppe der SchulabgängerInnen ohne Hauptschulabschluss fast doppelt so viel Jugendliche aus Migrantenfamilien wie aus deutschen Familien (17,6 % zu 8,1%) (vgl. Bellenberg u.a. 2001, S. 106). In der so genannten PISA-Studie (Baumert/Schümer 2001, S. 374) wird deutlich, dass die Sprachkompetenz die entscheidende Schwierigkeit der eingewanderten Jugendlichen im deutschen Bildungssystem ist. Dementsprechend wird in Deutschland darüber gestritten, ob zum Beispiel der islamische Religionsunterricht in Deutsch oder in Türkisch abgehalten werden soll. Während die einen die Unvermeidbarkeit der kompetenten Anwendung des Deutschen für den Bildungserfolg betonen, streben andere danach, (auch) die Herkunftssprache zu fördern. In den USA wird bereits über die Akzeptanz verschiedener Varianten des Englischen diskutiert. Unter dem Begriff Ebonics wird gefordert, das Englisch der BewohnerInnen von Vierteln mit vorwiegend schwarzen Bevölkerungsgruppen als eigene, regelgeleitete Sprache anzuerkennen und nicht länger als „falsches Englisch" abzuwerten. Ebonics soll als eigenes Schulfach institutionalisiert werden (vgl. zusammenfassend Perry/Delpit 1998).

Paul Willis

Kulturkonfrontationen zeigt auch Paul Willis (1982[2]) in seiner Studie. Willis führt in den Jahren 1972 bis 1975 in Großbritannien eine Untersuchung über die Statuspassage von Jungen aus der Arbeiterklasse ohne höhere Schulbildung in die Erwerbsarbeit durch. Mittels Fallstudien, Interviews, Gruppendiskussionen und teilnehmender Beobachtung bei Jungengruppen während der letzten beiden Schuljahre und der ersten Monate im Arbeitsleben will er die Bildungs- und Aufstiegschancen dieser Jungen erforschen. Willis schreibt eine „Ethnographie der Gegenschulkultur der weißen männlichen Arbeiterjugend" (Willis 1982[2], S. 13). So ist es ihm möglich zu belegen, wie sich Arbeiterjungen durch ihre Kultur – auch und gerade durch ihre Widerstandskultur – effektiv darauf vorbereiten, unqualifizierte Fabriktätigkeiten auszuüben.

Oppositionskultur

Die zentrale Untersuchungsgruppe, die „Lads", so der selbstgewählte Name der Gruppe, umfasst zwölf nicht akademisch geschulte Arbeiterjungen, die als Clique zusammen agieren. Willis wählt diese Gruppe aufgrund ihrer oppositionellen Kultur in einer überwiegend aus der Arbeiterklasse stammenden Jungenschule. Willis zeichnet die permanente Rebellion der Arbeiterjungen im kleinbürgerlichen Schulsystem nach. Er zeigt dabei, dass sich die Jungen in ihrem Widerstand gegen das Autoritätssystem der Schule selbst in die Lage bringen, ohne Abschluss die Schule zu verlassen und somit – wie die Eltern – unqualifizierte, schlecht bezahlte Jobs annehmen zu müssen, dass also Widerstand zur Anpassung führen kann. Der Schule gelingt es demnach nicht, die schichtspezifische Ungleichheit an Lebenschancen und Zugang zu Ressourcen aufzuheben. Trotz Schulbeteiligung reproduzieren die Jugendlichen ihre Klassenlage.

Diese Rebellion gegen die Schule, so führt Willis weiter aus, ist weder konditioniertes Verhalten noch hilfloses Aufbegehren, sondern zielt so geschickt auf die Schwächen des Systems, dass genaue Kenntnisse der Funktionsweise der Schule bei den Jungendlichen vorausgesetzt werden können. Weder sind die Jugendlichen nicht gescheit genug, um die Schule zu bewältigen, noch verweigern sie den Abschluss, weil sie ihre Unfähigkeit zu akademischer Bildung akzeptieren, sie nutzen vielmehr ihre Intelligenz, um die als fremd und diskriminierend erlebte Kultur zu bekämpfen.

Humor als Rebellion

Diese Ergebnisse bieten sich für Anthony Giddens (1988, S. 343ff.) an, um an der Studie von Willis die Grundzüge seiner Theorie der Strukturierung zu illustrieren. Er hebt zunächst hervor, dass Willis die Jungen als Akteure thematisiert, die über ein praktisches, zum Teil auch diskursives Wissen über ihre Schulumwelt verfügen. So kann er die reflexive Steuerung des Handelns selbst in einer Kultur, deren Sprachcode nur als gering diskursiv gilt, erläutern. Die sprachlichen Fähigkeiten der Lads werden aber, so Willis und Giddens, vielmehr in Humor, Hänseln und aggressivem Sarkasmus deutlich. Gerade in der Witzkultur offenbare sich ein komplexes Verständnis der Institution Schule. Weiter bietet sich die Rebellion der Lads an, um zu zeigen, dass die reflexive Steuerung nicht mit den Handlungsfolgen identisch ist. Die Widerstandskultur führt in der Konsequenz zum Leben als ungelernter Arbeiter. Ihre „Durchdringung" der Institution Schule und das daran orientierte Aufbegehren bringt die Zementierung ihrer Lebensbedingungen mit sich. Im Handeln der Jungen werden wesentliche Züge industriekapitalistischer Arbeitsorganisation reproduziert. Gleichzeitig entfalten die strukturellen Bedingungen der Gesellschaft ihre Wirkung im begründeten Handeln der Einzelnen. Die Strukturen werden im Handeln aufrechterhalten und durch es reproduziert, aber nicht als einfache Widerspiegelung, sondern im komplexen Handlungsverlauf (vgl. zu Willis auch Kapitel 8 in diesem Buch).

Schulische Belastungen

Eine die Bundesländer Hessen und Sachsen-Anhalt vergleichende Studie (Büchner/Krüger 1996) zeigt, dass Kinder und Jugendliche aus der Arbeiterklasse deutlich stärker von schulischen Belastungen betroffen sind als Kinder und Jugendliche aus bildungsnahen Schichten. Als schulische Belastungen definieren die Autoren Klassenwiederholung, Nichtaufnahme in die Wunschschule, Schulwechsel, Versetzungsgefährdung und Nachhilfeunterricht. Die Kinder aus Familien mit niedrigem sozialen Status sagen häufiger von sich, dass sie weniger zustande bringen als andere (in den neuen Bundesländern 7 Prozent mehr, in den alten Bundesländern sogar 12 Prozent mehr) und dass die LehrerInnen nie richtig mit ihnen zufrieden sind (in den neuen Bundesländern 15 Prozent mehr, in den alten Bundesländern 14 Prozent mehr) (Büchner/Krüger 1996, S. 25). Zu einer Kumulation von Problemen kommt es, wenn Klasse und Geschlecht bzw. Klasse und Ethnizität zusammenkommen. Zum Beispiel haben Mädchen aus der Arbeiterklasse mit dem Phänomen zu kämpfen, dass sie sowohl als Mädchen wie auch als Arbeiterkinder relative Neulinge im höheren Bildungswesen sind. Verschiedene Untersuchungen (z.B. Bublitz 1980; Theling 1986) können aufzeigen, dass Arbeitertöchter subjektiv große Schwierigkeiten mit dem über Bildung vermittelten sozialen Aufstieg haben. Belastungen, Ängste und Versagenserfahrungen werden entweder in Form von Selbstzweifeln gegen die eigene Person gerichtet („Vielleicht wäre ich als Verkäuferin glücklicher", vgl. Theling 1986) oder in Wut über die diskriminierenden Anderen, zu denen man gar nicht gehören möchte, übersetzt (Schlüter/Metz-Göckel 1989). Umgekehrt tragen gerade Jungen aus unterprivilegierten Einwandererfamilien ein besonders großes Risiko, ohne Abschluss die Schule zu verlassen (Baumert/Schümer 2001, z.B. S. 401).

Heimlicher Lehrplan

Die Schule ist eine Institution, die nicht nur die offiziell festgelegten Inhalte, sondern darüber hinaus einen so genannten *heimlichen Lehrplan* vermittelt. Ivan Illich (1995[4], original 1972) argumentiert, dass Kinder in der Schule durch die dort herrschende Disziplin und Ordnung einen passiven Konsum, d.h. eine unkritische Akzeptanz der existierenden gesellschaftlichen Strukturen erlernen und so auf ein entfremdetes Leben unter Leistungs- und Konsumdruck vorbereitet werden. Illich geht soweit, dass er eine Abschaffung der Schulpflicht zugunsten eines Netzes freiwilliger, kommunikativer Institutionen vorschlägt. Kinder, Jugendliche oder Erwachsene jeden Alters sollten jederzeit die Möglichkeit haben, Bildung zu erwerben, dies sollte jedoch auf dem Freiwilligkeitsprinzip basieren.

Koedukation

Besonders ausführlich ist der heimliche Lehrplan in den letzten Jahren im Zusammenhang mit der Koedukationsdebatte untersucht worden. Kinder durchlaufen in Schulen geschlechtsspezifische Sozialisationsprozesse, in denen sie sich neben dem offiziell vermittelten Lehrstoff auch Wissen über als angemessen geltendes geschlechtsspezifisches Handeln aneignen. Neben den Prinzipien der Selbstsortierung entsprechend der Zweigeschlechtlichkeit werden in der Koedukationsforschung vor allem Differenzierungsprobleme untersucht. Eine wesentliche Frage ist, wie Mädchen schleichend im Unterricht lernen, ihre geschlechtsspezifische (meist gesellschaftlich abgewertete und in Bezug auf beruflichen Erfolg häufig benachteiligte) Position einzunehmen. Die Diskussion setzt an dem Problem an, dass quantitativ betrachtet die Mädchen mittlerweile ähnliche, z.T. auch bessere, formale Qualifikationen (Schulabschlüsse) als Jungen aufweisen, gleichzeitig sie jedoch – so der durch verschiedene Studien abgesicherte Befund (zuerst: Horstkemper 1987) – trotz des schulischen Erfolgs ein geringeres Selbstwertgefühl und ein negativeres Selbstbild haben als Jungen. Ursachen hierfür werden u.a. in den geschlechtsspezifischen Fachpräferenzen, die gleichzeitig gesellschaftlich unterschiedlich bewertet werden, gesehen. So gelten die von Jungen favorisierten Fächer wie Naturwissenschaften und Technik als „schwere" Fächer, wohingegen Geisteswissenschaften und Sprachen, in denen Mädchen mehrheitlich ihre Kompetenzen entwickeln, als „weiche", „einfache" oder „Redefächer" gelten.

Wissensreservate

Es gelingt Schule offensichtlich nicht, geschlechtsspezifische Kompetenzentwicklung, die sich später in einer geschlechtsspezifischen Berufs- oder Studienwahl verstärkt und mit unterschiedlichen Chancen auf dem Arbeitsmarkt einhergeht, abzubauen. Im Gegenteil muss angenommen werden, dass schulische Lehr- und Lernformen geschlechtsspezifisch angeeignete Wissensreservate eher verstärken, denn kompensieren (z.B. Faulstich-Wieland 1991).

Computerkompetenz

Geschlechtsspezifisch unterschiedliche Lernvoraussetzungen und Zugangsweisen finden im Unterricht keine Berücksichtigung. Besonders deutlich wird dies am Informatikunterricht. Obwohl bekannt ist, dass Mädchen weniger private Zugriffsmöglichkeiten auf Computer haben, über weniger technische Vorkenntnisse verfügen und mit anderen Interessen an die Computerarbeit herangehen, nämlich den Gesamtablauf, nicht nur einzelne Operationen verstehen zu wollen, wird im Unterricht nur ein an Jungen orientiertes didaktisches Konzept verfolgt (Faulstich-Wieland 1991). Mädchen, die sich für Computer interessieren, werden von anderen Mädchen, die dies nicht tun, als „männliche Mädchen" betrachtet. Die Grenzüberschreitung löst damit Verunsicherungen in der Identität aus, die im Unterricht kaum bearbeitet werden (vgl. Ritter 1994). Weibliche „Computerfreaks" müssen demnach die paradoxe Anforderung bewältigen, gleichzeitig ihr Geschlecht zu neutralisieren, um als Kompetenz anerkannt zu werden, und doch als Mädchen erkennbar zu bleiben, um die geschlechtsspezifische Sortierung nicht zu verwirren (vgl. zur grundsätzlichen Problematik auch Heintz u.a. 1997).

Christine Roloff kommt in einem Projekt zum Thema „Geschlechtsspezifische Umgangsformen mit dem Computer" zu dem Ergebnis, dass Mädchen im koedukativen Computerunterricht maßgeblich Zuschauerinnen und Zuarbeiterinnen sind (Roloff 1990). Während der Computerunterricht, ebenso wie naturwissenschaftliche Fächer, an der Interessenlage der Jungen orientiert wird, werden umgekehrt jedoch nicht Mädcheninteressen in deren leistungsstarken Fächern wie zum Beispiel Deutsch und Fremdsprachen stärker berücksichtigt. Wie Mädchen in einer SchülerInnenbefragung selbst kritisieren, werden auch hier männliche Autoren mit ihrem geschlechtsspezifischen Blick auf die Welt mehrheitlich gelesen (Faulstich-Wieland/Horstkemper 1995).

Mädchenschulen

Bessere Zugangsweisen zu mathematisch-naturwissenschaftlichen und technischen Fächern finden Mädchen offensichtlich in Mädchenschulen. Eine Untersuchung zum Studienverlauf von Chemie- und Informatikstudentinnen in der Bundesrepublik kommt zu dem Ergebnis, dass immerhin 35% der Studentinnen Absolventinnen von Mädchenschulen waren (Roloff u.a. 1987). Eine Folgeuntersuchung gelangt zu der Erkenntnis, dass sogar 47% der Informatikstudentinnen Mädchenschulen besucht hatten. Da insgesamt im Untersuchungsland Nordrhein-Westfalen nur 14% der Mädchen monoedukative Schulen besuchen, zeigt sich sehr deutlich, dass Mädchenschulen erheblich erfolgreicher für das Studium von naturwissenschaftlich-technischen Fächern qualifizieren.

Pubertät

Sozialkonstruktivistische Herangehensweisen verfolgen andere Deutungsmuster für die geschlechtsspezifische Kompetenzentwicklung. Da Mädchen und Jungen in der Kindheit noch vergleichbare Schulleistungen erbringen und erst in der Adoleszenz die Leistungen der Mädchen in den mathematisch/technischen Fächern, die der Jungen in Sprachen abfallen, wird dieser Phase der Identitätsbildung besondere Bedeutung zugemessen. Die Pubertät ist der Zeitraum, in dem spätestens von den Mädchen und Jungen erwartet wird, dass sie sich ihrem Geschlecht angemessen verhalten (Hagemann-White 1984; Sobiech 1991). Um also für alle sichtbar, aber eben ohne nackte Körper zu zeigen, ein Geschlecht repräsentieren zu können, müssen die Jugendlichen bewusst und unbewusst auf Stereotype zurückgreifen. Mädchen beginnen, als männlich assoziierte Verhaltensweisen abzulehnen, Jungen verweigern gesellschaftlich als weiblich verstandene Handlungsmuster. Mit der Ausbildung eines geschlechtsspezifischen Habitus sinken die Leistungen in den jeweils als männlich oder weiblich assoziierten Fächern. So lassen zum Beispiel die Leistungen der Mädchen in Tests nach, die mit Mathematik assoziiert werden (ausführlich Löw 1995; Rabe-Kleberg/Löw 1998).

Biographische Engführung

Beide Geschlechter erfahren so in schulischen Interaktionsprozessen eine biographische Engführung. Mädchen lernen durch das brave, hilfsbereite Verhalten nicht, sich zur Wehr zu setzen. Dies bezieht sich sowohl auf ihre verbale Durchsetzungsbereitschaft, so erklären sie sich zum Beispiel mit Noten schneller einverstanden als ihre männlichen Mitschüler, als auch auf ihre Fähigkeiten, sich bei körperlichen Angriffen zur Wehr zu setzen. Mädchen werden als schutzbedürftig wahrgenommen und begreifen sich selbst als schutzbedürftig. Dies führt jedoch nicht dazu, dass sie von Jungen geschützt werden, sondern prädestiniert sie zum Objekt von Gewalt. Im Gegensatz dazu lernen Jungen, Aggressivität und Gewaltbereitschaft als Teil männlicher Identität wahrzunehmen. „Jungen dagegen neigen stärker als Mädchen zu individualistischen und aggressiven Orientierungen, und sie tendieren dazu, auf strukturelle Ungerechtigkeiten mit Verantwortungsabwehr zu reagieren“ (Stanat/Kunter 2001, S. 311). Dieser Prozess ist für die Jungen selbst schmerzhaft und gefährlich. Die einseitige Sozialisation führt dazu, dass Jungen weniger soziale Kompetenzen erwerben und demzufolge in sozialen Beziehungen von Frauen abhängig werden oder bleiben.

Schulbuchanalysen

In der Koedukationsforschung werden neben der fachlich-geschlechtsspezifischen Kompetenzentwicklung auch Verfestigungen sozialer Zuständigkeiten und Eigenschaften nach Geschlecht untersucht. Breite Aufmerksamkeit wird der Schulbuch- und Medienanalyse geschenkt, welche, von wenigen Ausnahmen abgesehen, einhellig zu folgenden Resultaten kommt (vgl. Brehmer 1982; Glumper 1995): Mädchen und Frauen werden in Abbildungen und Texten seltener repräsentiert als Jungen und Männer; Frauen werden überwiegend als Hausfrau und Mutter dargestellt; Darstellungen berufstätiger Frauen beschränken sich auf sorgende und pflegende Berufe; dargestellte Frauen verfügen über geringere Entscheidungskompetenzen und materielle Ressourcen als dargestellte Männer; bereits Kinder werden mit scheinbar typisch weiblichen und männlichen Persönlichkeitsattributen dargestellt, zum Beispiel Mädchen als lieb und sorgend, Jungen als aggressiv und durchsetzungsfähig.

Vorbilder

Die lerntheoretisch begründete Annahme ist, dass Mädchen durch die stereotype Zuweisung lernen, sich in der Familie und in pflegenden oder sozialen Berufen zu verorten, Jungen dagegen eindeutig und einseitig auf die gesamte Berufs- und Erwerbssphäre sozialisiert werden. Dieser Prozess werde, so die weiterführende Argumentation, durch fehlende weibliche Vorbilder unter den Lehrerinnen verstärkt (vgl. zu Lehrerinnen z.B. Händle 1998). Durch die quantitativ nur geringe Besetzung von Funktions- und Leitungsstellen in der Schule durch Frauen und durch teilweise falsche, teilweise fehlende Verweise auf weibliche Gelehrte im vermittelten Unterrichtsstoff lernen beide Geschlechter Männer als klüger, mächtiger und durchsetzungsfähiger wahrzunehmen. Auch dies trägt zur Schwächung des Selbstbewusstseins von Mädchen bei.

Interaktionsstudien

Interaktionsstudien aus den USA, England, Skandinavien und der Bundesrepublik Deutschland kommen übereinstimmend zu dem Ergebnis, dass geschlechtsspezifische Differenzen im Bewertungs- und Kommunikationsverhalten von Lehrkräften vorliegen. Diese beziehen sich insbesondere auf die unterschiedliche Verteilung von Aufmerksamkeit und Bestätigung. Ein unter Leitung von Angelika Wagner durchgeführtes DFG-Projekt zu „Unterrichtsstrategien und ihre Auswirkungen auf Schülerverhalten" kam bereits in den 1980er Jahren zu dem Ergebniss, „dass Jungen im Unterricht deutlich häufiger drangenommen, gelobt und getadelt werden als Mädchen" (Frasch/Wagner 1982, S. 272; Stanat/Kunter 2001, S. 299ff.). Jungen erhalten demzufolge mehr positive wie negative Aufmerksamkeit und fühlen sich in ihrem Verhalten bestätigt. Neuere Untersuchungen (z.B. Kaiser 1994) zeigen, dass sich das LehrerInnenverhalten geringfügig verändert hat, jedoch immer noch eine signifikante Benachteiligung der Mädchen aufweist. Mädchen werden in schulischer Interaktion als brav, sozial vermittelnd, hilfsbereit, verantwortungsvoll und fleißig beschrieben. Sie werden im Unterrichtsverlauf von Lehrern und Lehrerinnen eingesetzt, um Konkurrenz und Aggression von Jungen zu mildern (vgl. z.B. Faulstich-Wieland 1991). Gerhard Amendt führt dies exemplarisch aus: „Viele Mädchen scheinen für die Schule psychodynamisch produktiv zu sein und darin einen nicht unerheblichen Teil ihrer Produktivität zu verschleudern. Sie bändigen Konflikte, die andere nicht lösen können oder lösen wollen, indem sie den Konfliktstoff neutralisieren. Zu ihrem Leistungserfolg trägt das wenig und zu ihrer Konfliktfähigkeit absolut nichts bei" (Amendt 1996, S. 377). Da Jungen in ihrem Verhalten durch Aufmerksamkeit bestätigt werden und gleichzeitig erleben, dass die Sozialkompetenz der Mädchen zwar gebraucht, aber nicht positiv sanktioniert wird, führt dies zu einer Verstärkung der Gewaltsozialisation von Jungen.

Selbstvertrauen

Der oben beschriebene Prozess wird durch den geschlechtsspezifisch unterschiedlichen Umgang mit Leistung und Erfolg – einer der für schulisches Lernen wichtigsten Kategorien – weiter verstärkt. Gute Schulleistungen führen nicht selbstverständlich zu hohem Selbstvertrauen, dies die überraschende Beobachtung einer bedeutenden Differenz zwischen Mädchen und Jungen. Marianne

Horstkemper unterscheidet unterschiedliche Rückmeldungen durch die Lehrkräfte in Bezug auf die Leistungen und Erfolge bei Jungen und Mädchen. Während Jungen bei Versagen oder mäßiger Leistung mit Hinweisen nach dem Muster „Du könntest, wenn Du wolltest" auf ihre nur noch nicht geweckten Potenzen aufmerksam gemacht werden, werden Mädchen eher getröstet („Du hast Dir Mühe gegeben"). Allerdings werden auch Mädchen, die gute Noten haben, eher wegen ihres Fleißes nicht ihrer hohen Intelligenz gelobt. Während also Jungen den Glauben an ihre Intelligenz auch bei schlechten Leistungen nicht aufgeben müssen, fällt es Mädchen schwer, trotz guter Leistungen an eben diese zu glauben (Horstkemper 1987).

Zudem kommt es vor allem in der Adoleszenz zur Verschiebung der Relevanz von schulischer Leistung bei den Mädchen. Während es zum männlichen Selbstbild gehört, durch Leistung zu imponieren, kommt es bei den Mädchen vor allem in den als „männlich" gedeuteten Fächern eher zu einer Leistungszurückhaltung. Hier geht es nicht nur darum, die Konkurrenz mit den Jungen zu meiden, sondern auch darum, Anerkennung durch körperliche Attraktivität zu gewinnen. Darin liegt u.U. eine Erklärung für die höhere Effektivität von Mädchenschulen in den naturwissenschaftlich-mathematischen Fächern.

Die latente Reproduktion sozialer Ungleichheit in Bildungsinstitutionen bezieht sich nicht nur auf Klasse und Geschlecht. In der Forschungscommunity entwickelt sich – noch langsam und vorsichtig, aber wahrnehmbar – eine Diskussion über die Reproduktion heterosexuell fixierter oder ethnozentristischer bzw. nationaler Normen und Werte.

Homosexualität

Heterosexualität gilt als Norm. In Schulbüchern, Unterrichtsgestaltung, Lektüren etc. finden sich selten Hinweise auf Homosexualität. In der Sexualpädagogik wird zunächst über Sexualität aufgeklärt und dann, häufig am Ende, noch eine Einheit über Homosexualität angefügt. Allein die Thematisierung von Homosexualität kann man als Fortschritt werten, die zugrunde liegende Konstruktion ist jedoch, dass Heterosexualität das Allgemeine und somit die Norm ist (nämlich „Sexualität") und Homosexualität das Besondere, das in einer Extra-Einheit behandelt wird (vgl. Holzkamp 1996, S. 9). Jutta Hartmann (1996) zeigt, dass die Konstruktion von Zweigeschlechtlichkeit (siehe Abschnitt 4.1.2 dieses Buches) auf einer heterosexuellen Matrix basiert. Die Geschlechter müssen eindeutig und unverwechselbar voneinander zu unterscheiden sein, um sich dann in der Logik der romantischen Liebe ohne Zweifel und Missverständnisse wieder aufeinander beziehen zu können.

Das bedeutet, das weitgehende Verschweigen von Homosexualität erschwert Mädchen und Jungen den Zugang zu dieser Lebensform. Die Art, wie Homosexualität vereinzelt als besondere Form von Sexualität abgehandelt wird, reproduziert die Vorstellung, Heterosexualität sei die Norm. Insgesamt wird in der Schule (wie in der gesamten Gesellschaft) eine dualistische Vorstellung, dass Menschen entweder homo- oder heterosexuell seien, reproduziert. Dabei gibt es keine wissenschaftlichen Beweise für die unterstellte Zweiteilung und die Vielzahl bisexuell lebender Menschen deutet in eine andere Richtung. Es sind jedoch keine institutionalisierten Angebote bekannt, Jugendlichen beide Optionen bzw. die eigene Homo- *und* Heterosexualität zu vermitteln.

Ethnische Fiktionen

Etienne Balibar (1990, S. 118ff.) problematisiert die Reproduktion *ethnischer Fiktionen* im Schulunterricht. Dem modernen Nationalstaat als historisch junge Gesellschaftsformation gelänge es nur dann, als „natürlich" zu erscheinen, wenn eine ethnische Fiktion aufgebaut würde. Diese basiere auf zwei Konstruktionsprinzipien, nämlich „Sprache" und „Rasse". Über eine Nationalsprache, die in Schulen eingeübt werde, imaginierten die Individuen eine kollektive Identität. Mit der Idee, das Erlernen der Nationalsprache ermögliche allen Menschen, ZuwanderInnen wie Einheimischen, an der fiktiven Ethnizität zu partizipieren, wird Egalität signalisiert. Tatsächlich verfügen jedoch nicht alle über die gleiche Sprachkompetenz und die gleichen sozialisatorischen Bedingungen zum Erlernen der Nationalsprache, so dass die Schule über die Sprachdominanz Spaltungen produziere.

Da jedoch verschiedene Völker eine gemeinsame Sprache sprechen können, ohne eine Nation zu bilden, reiche Sprache zur Konstruktion von Nation nicht aus. Daher werde über die Idee von „Rasse" gleichzeitig noch eine biologische und geistige Generationenabfolge konstruiert. Die Vorstellung eines gemeinsamen rassischen Ursprungs ergänze meist schleichend und implizit die Fundierung eines nationalstaatlich organisierten Gesellschaftsmodells, in dem es Fremde und Heimische, Eindringende und Hierhin-Gehörende gäbe.

Zusammenfassend ist festzuhalten, dass in einer Reihe von empirischen Untersuchungsergebnissen für Schulen Strukturen und Prozesse zu benennen sind, die – trotz formaler und subjektiv gewünschter Gleichheit – soziale Ungleichheit herstellen. Die Schule „fädelt die Heranwachsenden in unterschiedlich privilegierte Lebensläufe ein" (Krappmann 1999, S. 238). Es handelt sich um Strukturen und um Handlungsprozesse, die Lehrkräften kaum oder gar nicht bekannt oder bewusst sind.

4.2.3 Die soziale Ungleichheit im Spannungsverhältnis von Arbeit und Bildung

Beschäftigungssystem

Pierre Bourdieu und Niklas Luhmann kommen trotz völlig divergierender theoretischer Ansätze zu dem gemeinsamen Schluss, dass eine unbeabsichtigte Folge der Bildungsreform der 1970er Jahre die Verlagerung der Selektion in das Beschäftigungssystem ist (vgl. Kapitel 3 dieses Buches). Der Befund gilt mittlerweile in weiten Kreisen der Bildungssoziologie als gesichert (vgl. z.B. Weymann 1987; Krais 1996b). Dieser Tatbestand zeigt noch einmal die traditionell eng verwobene Verzahnung der Felder „Bildung" und „Arbeit" (vgl. die Einleitung dieses Buches) unter dem Aspekt der Reproduktion sozialer Ungleichheit. Zeugnisse, insbesondere das Abiturzeugnis, gelten einerseits als notwendige Voraussetzung für den beruflichen Erfolg, andererseits zeigt sich, dass die nun massenhaft erworbenen Dokumente als eine alleinige Voraussetzung für einen erfolgreichen Start ins Berufsleben nicht mehr ausreichen (vgl. Weymann 1987). Konstant bleibt, dass HochschulabsolventInnen insgesamt ein höheres Einkommen erzielen als AbsolventInnen anderer Bildungsgänge (vgl. Müller 1998, S. 96), also darüber die Klassenrelation erhalten bleibt. Sehr unterschiedlich ist jedoch die Bezahlung innerhalb der jeweiligen Ausbildungsgruppe. Um dies zu verstehen, muss man sich die Herausbildung von Berufen und Professionen genauer ansehen.

Jürgen Kocka (1991) hat darauf hingewiesen, dass sich das Bürgertum nicht nur über den Begriff der Bildung konstituiert hat (vgl. Kapitel 1 dieses Buches), sondern dass auch der „Beruf" eine selbstkonstituierende Rolle einnimmt. Für Teile des Bürgertums war der Beruf die Möglichkeit, sich vom müßiggängerischen Adel abzusetzen und sich über den Wert der Arbeit zu identifizieren, für andere Teilc dcs Bürgertums erfolgte die Abgrenzung sozusagen „nach unten". Sie setzen sich von anderen Formen der Arbeit ab, nämlich der Lohnarbeit des Proletariats, der Handarbeit und dem Handel des Kleinbürgertums und der privaten Hausarbeit der Frauen. Ihr Berufsverständnis grenzt alle Tätigkeiten aus, die ohne (Aus)Bildung durchgeführt werden können, repetitive Handarbeiten oder diffuse Hausarbeiten beinhalten und als unehrenhaft klassifiziertem Gewinnstreben folgen (vgl. Beckmann 1991). Die heutige Differenzierung und Bewertung zwischen Berufen, über die auch Einkommensunterschiede legitimiert werden, basiert auf diesem bürgerlichen Arbeitsethos, demzufolge über Bildung berufliche Selbstverwirklichung ermöglicht wird sowie Leistung und Erfolg auf Bildungsformen zurückgeführt werden. Diese Konstruktion der Erwerbsarbeitssphäre, bei der Frauenarbeit, Fabrikarbeit und Handel aufgrund historisch gewachsener Interessen des Bildungsbürgertums abgewertet werden, ist an sich schon ein Feld sozialer Ungleichheit (vgl. dazu auch Rabe-Kleberg 1993), allerdings basiert diese Konstruktion auf einem breiten sozialen Konsens und wird daher selten als ungerecht erfahren.

Beruf

Gebrochen wird dieser soziale Konsens jedoch, wenn gleiche Qualifikation nicht zu gleichen Chancen auf dem Arbeitsmarkt führen oder ungleicher Lohn für gleiche Arbeit gezahlt wird. Dies ist vor allem bei der geschlechtsspezifischen Segregation (räumliche bzw. feldspezifische Aufteilung) des Arbeitsmarktes der Fall. Reinhard Kreckel zeigt, dass diese Form der geschlechtsspezifischen Diskriminierungen im Vergleich zum Beginn des 20. Jahrhunderts zugenommen haben (Kreckel 1992, S. 228). Während damals die den Arbeitsmarkt strukturierenden, sozialen Ungleichheiten den Qualifikationsdifferenzen entsprachen, sind Frauen heute „besser gebildet – und doch nicht gleich" (Rabe-Kleberg 1990). Frauen haben trotz ihrer im Vergleich zu Männern sogar besseren allgemeinen und beruflichen Bildungsabschlüsse nicht die vergleichbar guten Positionen im Beruf. Die einflussreichen und hoch dotierten Positionen sind zum Beispiel überproportional mit Männern besetzt (Kreckel 1992, S. 212) und die Einkommensdifferenzen sind erheblich größer als die Qualifikationsunterschiede (Frauen verdienen in allen Klassen weniger als Männer) (Kreckel 1992, S. 235; Frerichs 1997, S. 213ff.). Dies ist zum einen auf direkte Diskriminierung zurückzuführen (zum Beispiel, indem unterstellt wird, dass der Mann der Ernährer ist oder dass Frauen nicht so viel leisten können), zum anderen liegt es jedoch darin begründet, dass Frauen und Männer in unterschiedlichen Berufen arbeiten, die sich einem direkten Vergleich entziehen. Frauen erhalten dann in der Fabrikfertigung von Textilien weniger Lohn als Männer in der Autoindustrie, obwohl beide „unqualifizierte" Fließbandarbeit ausführen. Diese Segregation der Berufe und die damit einhergehenden Differenzen in Status und Bezahlung lässt sich auch für die DDR nachweisen (Rabe-Kleberg 1995 und 1999[3]).

Geschlechtsspezifische Segregation

Frauen- und Männerberufe

Die geschlechtsspezifische Segregation des Arbeitsmarktes ist eine Grundstruktur der modernen Gesellschaft. Man kann klar zwischen Frauen- und Männerberufen bzw. zwischen Frauen- und Männerarbeitsplätzen unterscheiden. Dabei ist es keineswegs so, dass Frauenberufe oder Männerberufe inhaltlich irgendeinem gemeinsamen Tätigkeitsfeld zugeordnet werden können. So basieren zum Beispiel Frauenberufe keineswegs, wie man anfänglich angenommen hat, alle auf hausarbeits-verwandten Tätigkeiten. Vielmehr teilen alle Frauenberufe nur ein Schicksal, nämlich dass sie als vergleichsweise gering qualifiziert gelten und einen niedrigen Status aufweisen, was sich in Aufstiegschancen und Bezahlung niederschlägt. Für Frauenberufe ist kennzeichnend, wie Ursula Rabe-Kleberg (1993) zeigt, dass es kaum Aufstiegsmöglichkeiten gibt, die Verdienstmöglichkeiten gering sind und dass das Qualifikationsprofil zwischen beruflicher und Laienqualifikation angesiedelt ist (vgl. dazu ausführlich Kapitel 5.3). Dies zeigt sich zum Beispiel sehr deutlich in der Schule. Aufgrund der doppelten Aufgabenstellung der Grundschule von Kindorientierung und Basisausbildung (vgl. auch Hempel 1996) ist das Aufgabenprofil diffuser als in der Sekundarstufe. Aufstiegschancen und Bezahlung sind darüber hinaus geringer. Typischerweise findet man mehr männliche Lehrer, je klarer die Wissensvermittlung in den Vordergrund rückt und je höher die Position in der schulischen Hierarchie ist.

Erwerbstätige im April 2001 nach Wirtschaftsbereichen [1]

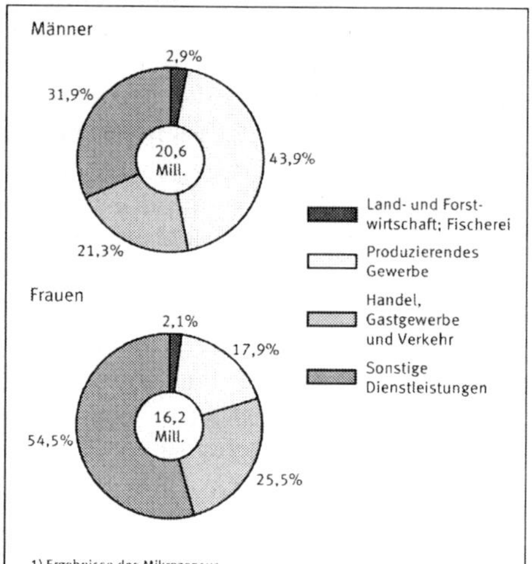

1) Ergebnisse des Mikrozensus.

(Quelle: Statistisches Bundesamt, Statist. Jahrbuch 2002, S. 103)

Hierarchische Verteilung

Die Segregation ist also nicht nur eine geschlechtsspezifische Sortierung des Arbeitsmarktes, sondern auch eine hierarchische Verteilungspraxis (vgl. Wetterer 1993, S. 52ff.; Rabe-Kleberg 1999[3], original 1996, S. 277). Angela Willms-Herget (1985) vergleicht Frauenarbeit von 1880 bis 1980 und zeigt, dass Frauen immer dann Chancen in als Männerberufe klassifizierten Branchen haben, wenn Männer sich aus der Berufssparte zurückziehen. Auch in expandierenden Bereichen haben Frauen gute Chancen, werden jedoch nach der Konsolidierung der Sparte schnell wieder verdrängt. Dieses Muster gilt bis heute und lässt sich ebenfalls für die New Economy beobachten.

Geschlechtswechsel eines Berufes

Dringen Frauen in einen Beruf vor, so erfährt der ganze Beruf einen „Geschlechtswechsel". Tätigkeiten, die vorher als männlich galten, werden plötzlich als typisch weiblich wahrgenommen. Mit der „Verweiblichung" eines Feldes geht stets ein Statusverlust einher, mit der „Vermännlichung" ein Statusgewinn. Dies kann man sehr gut beim Geschlechtswechsel von der Sekretärin zum Sekretär oder von der Putzfrau zum Lehrberuf des Gebäudereinigers beobachten (vgl. Wetterer 1993). Regine Gildemeister und Angelika Wetterer (1992) zeigen dies sehr prägnant am Beispiel der Setzerinnen. Den Einsatz von Maschinen zum Satz von Texten und Büchern erlebten die Männer in der zweiten Hälfte des 19. Jahrhunderts als Bedrohung ihrer Arbeitsplätze und verweigerten zunächst die Mitarbeit. So kam es, dass bürgerliche Frauen (die lesen und schreiben konnten) als Setzerinnen angeworben wurden. Maschinensetzerin etablierte sich nach kürzester Zeit als typischer Frauenberuf, was in dem folgenden längeren Zitat erläutert werden soll: „Um den Frauen die Sache schmackhaft zu machen, wurde den Setzmaschinen ein dem ‚Pianotyp' ähnliches Aussehen verliehen und die Fingerfertigkeit der klavierspielenden Bürgerstöchter bot ein breit gefächertes Repertoire für Analogiebildungen, in denen die ‚besondere Eignung der Frauen' für den Maschinensatz beschworen wurde. So gut wie alle ersten Modelle der Setzmaschinen wurden von Frauen eingeführt und wurden in Werbebildern mit anmutig an der Tastatur sitzenden Bürgerstöchtern abgebildet. Erst mit der sukzessiven Konsolidierung der Maschinensetzerei, erst in der darauf folgenden Phase der (neuerlichen) Verberuflichung verschwanden die Setzerinnen ebenso aus den Druckereien wie aus der sozialen Erinnerung. Das neu entstandene Berufsbild des Männerberufs Maschinensetzer fußte fortan auf anderen Analogien. Die Assoziationskette ‚Maschine-Männlichkeit' wurde in Szene gesetzt: Der Lärm und Dreck, der die Setzmaschine (jetzt plötzlich?) umgibt, trat in den Vordergrund, das beim Satz verwendete Blei galt (jetzt plötzlich) als Gefahr für die Gebärfähigkeit der Frauen, und die Maschinen selbst wurden nicht mehr so konstruiert, dass sie dem Klavier möglichst ähnlich sind" (Gildemeister/Wetterer 1992, S. 224f.). Die Frauen konzentrierten sich nach ihrer Verdrängung aus der Setzerbranche auf die strukturell ähnliche Arbeit an der Schreibmaschine, welche jedoch mit Argumenten von Höher- bzw. Niedrigerqualifikation geringer entlohnt wurde.

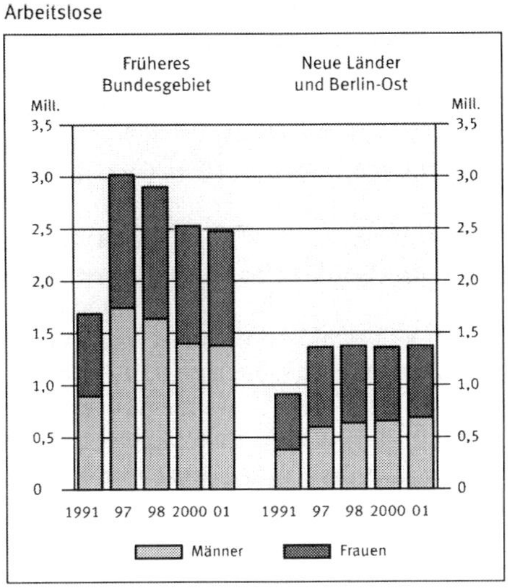

(Quelle: Statistisches Bundesamt, Statist. Jahrbuch 2002, S. 125)

Zusammenfassend heißt das, soziale Ungleichheit entsteht durch eine hierarchisierende Sortierung des Arbeitsmarktes. Innerhalb einer Berufssparte haben Frauen nicht die gleichen Chancen, Leitungspositionen zu erreichen, wie Männer; meist verteilen sich die Geschlechter jedoch „freiwillig" auf unterschiedliche Berufssparten, nämlich auf die Berufe, die in einem gesellschaftlichen und historisch wandelbaren Zuweisungsprozess als „männlich" oder „weiblich" klassifiziert werden. Die Sparten, die als weiblich gelten, zeichnen sich durch ein niedrigeres Lohnniveau und durch einen geringeren sozialen Status aus. Trotz gleicher formaler Bildungsabschlüsse haben Frauen also weniger gute Chancen auf dem Arbeitsmarkt. Frauen sind darüber hinaus stärker von Arbeitslosigkeit betroffen als Männer (Engelbrech 1991).

Der Bezug auf „Bildung" wird dabei eingesetzt, um Berufe, die entweder an Reproduktionsarbeiten, an Handarbeiten, an Handel oder an unqualifizierte Tätigkeiten erinnern, abzuwerten. Proletarische Männer sind von dieser Abwertung auch betroffen, aber strukturell sind es vor allem die Frauen, die im Dienstleistungsgewerbe (darunter auch der Handel) arbeiten. Darüber hinaus basieren viele Berufe, die junge Frauen gerne ausüben, nicht auf einer dualen (betrieblichen und berufsschulischen Ausbildung), sondern auf einer schulischen Berufsbildung (die z.T. noch von den Frauen selbst bezahlt werden muss), was wieder zu einem ökonomischen und einem Statusverlust der Frauen führt (vgl. Krüger 1992).

Berufswünsche

Warum wählen Mädchen und Frauen dann diese Berufe? Eine Studie zu Berufswünschen von Mädchen in der Hauptschule (Rettke 1987) zeigt, dass diese in der 7. Klasse noch relativ geringe geschlechtsspezifische Ausprägungen zeigen. Sie sind sich jedoch des Arbeitslosigkeitsrisikos, welches mit ihrem Schulabschluss einhergeht, bewusst und beginnen, ihre Chancen auf dem Arbeitsmarkt auszuloten. Bis zur 9. Klasse und damit dem Herannahen des Abschlusses sind sie zu der Einschätzung gelangt, dass sich ihre Chancen noch mehr verschlechtern, wenn sie einen geschlechtsuntypischen Beruf anstreben. Durch erste Bewerbungsbemühungen und durch Ratschläge der Familie oder der Freunde, aber auch vom Arbeitsamt werden sie in der Regel in der Einschätzung bestätigt, dass ihre Ausgangsqualifikation so defizitär sei, dass sie jedes Risiko vermeiden müssten. Ihre Berufseinmündung, so Rettke, ist also eine rationale Entscheidung. Dies ist jedoch nur eine Erklärung für das Phänomen. Eine andere ist die häufig offen diskriminierende Stimmung in Betrieben und Berufsschulen, denen Mädchen in so genannten Männerberufen ausgesetzt sind, oder die Freude, in einem von einer Frauenkultur geprägten Beruf zu arbeiten sowie der Wunsch, die in der geschlechtsspezifischen Sozialisation erworbenen Fähigkeiten beruflich zum Einsatz zu bringen.

5. Bildung und (Vor)Schule

5.1 Bildung im Kindergarten

„Sie leben mitten unter uns. Wir sehen sie auf dem Bürgersteig, wo sie zwischen den Passanten auf die Knie gehen und, Nase am Boden, die Fugen zwischen den Betonplatten inspizieren. Es sind jene Wesen, die sich auf einer Autoeinfahrt verstohlen Kies in den Mund stopfen, um den Geschmack zu prüfen. Sie zerrupfen Spinnen, statt zum Abendessen zu kommen, sie wälzen sich schreiend am Boden, wenn jemand versucht sie daran zu hindern, in Wasserlachen zu springen, um die Verdrängung der Materie zu testen. Und nehmen dabei keinerlei Rücksicht auf den Preis der neuen Schuhe. Die Rede ist von Kindern" (Mayer 2001, ohne Seitenangabe).

Bewahranstalten

Kinder sind, wie Susanne Mayer hier lebhaft verdeutlicht, wissenshungrig und bildungsbegierig, was – langsam – dazu führt, dass auch in (West) Deutschland die Vorschulerziehung als Bildungsprozess begriffen wird. Lange Zeit galt der „Kindergarten" in erster Linie als soziale Institution. Der Kindergarten (oder die Krippe, die Kindertagesstätte, der Hort) wurden vornehmlich als Bewahr-Anstalt begriffen, welche insbesondere die Defizite der Familie ausgleichen sollte. Im Zuge der Industrialisierung und Urbanisierung führten die langen Arbeitszeiten der Eltern außerhalb des Hauses (vgl. dazu Kapitel 7.1) dazu, dass viele Kinder auf sich allein gestellt in den elenden Wohnungen und auf der Straße aufwuchsen. Die Bewahranstalten wurden als schlechtere, aber notwendige Alternative zur Familienerziehung gedacht. Der Staat (oder die Kirche) nahmen sich derer an – so die Idee – die in der Familie nicht ausreichend betreut wurden (vgl. Heinsohn/Knieper 1975; Großmann 1990; Colberg-Schrader/Krug 1999).

Friedrich Fröbel

Dieser soziale Auftrag mischte sich mit der pädagogischen Idee in der Tradition der Pädagogik Friedrich Fröbels. Fröbel verfolgte die Vorstellung, dass pädagogische Prozesse sich nicht einfach aus dem praktischen Handeln von allein entwickeln, sondern sich in Gestalt einer Kunstlehre, d.h. angeleitet, ergeben. Dabei ging es Fröbel stets darum, durch den Lehrenden angeleitet, die Lernenden in die Lage zu versetzen, Ereignisse und Erfahrungen in einen biographisch konsistenten Zusammenhang zu setzen (vgl. zusammenfassend zu Fröbel Giel 1979).

DDR-Kindergarten

In der jüngeren Entwicklung wurde dem Kindergarten in den beiden deutschen Staaten unterschiedliche Bedeutung zugemessen. In der DDR galten Kindergärten als Teil des Volksbildungssystems und dienten sowohl der Entfaltung allgemein gebildeter Persönlichkeiten als auch der Absicherung der Frauenerwerbstätigkeit. Mit zentralistisch verabschiedeten Erziehungszielen und -plänen wurde der Kindergarten zur Regeleinrichtung (vgl. Hildebrand/Musiol 1995). In der BRD setzte der Kindergarten sich erst langsam durch. Kirchliche, staatliche und alternative Kindereinrichtungen (z.B. antiautoritär organisierte Kinderläden) prägten ab den 1960/1970er Jahren im Kontext der Bildungsreformen und sozialen Bewegungen eine vielfältige Landschaft. In einer Gemengelage aus Frühförderungsbestrebungen, Freistellung der Mütter und institutionalisierten Gleichaltrigenkontakten erlangte die institutionalisierte Vorschulerziehung alltagspraktische Bedeutung.

Lesekompetenz

Heute wird der Kindergarten von Fachleuten aus West- wie Ostdeutschland als Bildungseinrichtung verstanden (vgl. z.B. Krappmann 2001). Gerade die Ergebnisse der PISA-Studie zeigen, dass Deutschland zu den Staaten gehört, „in denen die potentielle Risikogruppe schwacher und extrem schwacher Leser relativ groß ist“ (Baumert/Schümer 2001, S. 401). Da Lesekompetenz als besonders entscheidend für den Bildungserfolg betrachtet wird, ist ein Anteil von 23% potentiell leseschwacher Kinder einer Alterskohorte alarmierend. Dementsprechend – (und) da sich Bildungsdefizite im Durchgang durch die Institutionen immer mehr verstärken (vgl. Kapitel 4.2.1 und 4.2.2 dieses Buches) – wird der Frühförderung zunehmend Bedeutung geschenkt. Kinder sollen dort kognitive Anregungen bekommen, die die familiäre Prägung ergänzen. Der Kindergarten soll – so die einhellige Position der PädagogInnen und BildungssoziologInnen, die jedoch in der Praxis nicht immer umgesetzt wird – Kinder „in ihren eigenen Sichtweisen und Interessen unterstützen, eigene Erfahrungen fördern, ihnen helfen, verschiedene Möglichkeiten der Einflussnahme auszutesten und das eigene Handeln zu reflektieren“ (Krappmann 2001, S. 5). Die Kinder sollen befähigt werden, diese Fähigkeit zur Eigenständigkeit in ein balanciertes Verhältnis zur Teamfähigkeit zu bringen. Mit anderen Worten: Der Kindergarten wird als eine Institution konzeptualisiert, in der bereits kleine Kinder die Fähigkeiten und Fertigkeiten erlernen, die Menschen in einer modernen Gesellschaft benötigen. In diesem Sinne – und nicht einfach als Vorbereitungskurs auf die Schule – wird der Kindergarten als Bildungsinstitution verstanden. Lothar Krappmann formuliert dies folgendermaßen: „Demgegenüber möchte ich unterstreichen, dass der Kindergarten als Bildungseinrichtung weder dadurch ausreichend beschrieben ist, dass er kognitive Anregungen ersetzt, die überlastete Eltern in vielen Fällen nicht leisten, noch dadurch, dass er Kinder zum Stillsitzen abrichtet, um es einmal scharf zu pointieren, sondern weil er jungen Kindern ein sachliches und soziales Erfahrungsfeld eröffnet, das eigene Anforderungen an die Entwicklung von Fähigkeiten und Problemlösungsmustern der Kinder stellt“ (Krappmann 2001, S. 6).

Eigenständigkeit und Teamfähigkeit

Entlastung der Eltern

Die Unterbringung von Kindern in einem Kindergarten oder einer Kindertagesstätte ermöglicht es beiden Elternteilen, ihren beruflichen, sozialen, kulturellen oder politischen Interessen nachzugehen und damit den Anforderungen nach Individualisierung und subjektiver Sinnsuche wie auch nach finanzieller Reproduktion nachzukommen. Die Kinder erfahren auf diese Weise beide Elternteile als gesellschaftlich integrierte Persönlichkeiten. Darüber dass die Eltern, Vater und Mutter, institutionell abgesichert für einige Zeit des Tages freigestellt werden, ihre gesellschaftlich notwendigen Aufgaben zu übernehmen und für die Reproduktion der Familie Sorge zu tragen, können Kinder Eltern als Vorbilder für eine eigenständige und verantwortungsvolle gesellschaftliche Existenz erfahren (vgl. auch hier Krappmann 2001, S. 5).

Praxis des Kindergartens

All dies bezieht sich jedoch auf die Bildungsidee, wie sie im pädagogisch-bildungssoziologischen Diskurs entworfen wird. Die Praxis des Kindergartens sieht vielerorts anders aus. Allerdings gibt es kaum empirische Studien zum Kindergarten. Was tagtäglich im Kindergarten passiert, ist ein dunkler Fleck in der Forschungslandschaft. Diese Unkenntnis hat u.a. zur Folge, dass zwar normativ darüber gestritten wird, wie ErzieherInnen ausgebildet werden sollen, jedoch niemand empirisch gesichert weiß, welche Qualifikationen Erzieherinnen in der Praxis brauchen bzw. wie sie ihre Qualifikationen verwerten können (vgl. Kapitel 5.3 sowie Rabe-Kleberg 1997).

Frühförderung

Gleichzeitig werden gerade die frühen, grundlegenden Bildungsprozesse gesellschaftlich vernachlässigt. Einer Studie von Tietze u.a. (1998) zufolge lassen sich bei vergleichbaren Rahmenbedingungen Unterschiede in der Qualität der pädagogischen Betreuung beobachten, die erhebliche Kompetenzdifferenzen bei Kindern verursachen. Im Extremfall sind dann Gruppen von Kindern in ihrer Entwicklung anderen – gemessen am Durchschnitt – über ein Jahr voraus oder hinterher. Auch in den USA wurde eine Reihe von Studien durchgeführt, die die Langzeiteffekte der Frühförderung aufzeigen (Clarke/Campbell 1998). Das bisherige Forschungsdefizit, aber auch der Mangel an öffentlicher Diskussion über die Geschehnisse im Kindergarten lassen sich nur mit einer doppelten Abwertung erklären. Da es sich beim Kindergarten erstens um einen typischen Frauenbereich und zweitens um Kinder (d.h. in einer kapitalistischen Logik um nichtproduktive Gesellschaftsmitglieder) handelt, wurde dem Feld keine soziologische und kaum öffentliche Aufmerksamkeit beigemessen. Nur vereinzelt liegen Studien vor, die nicht die Vorschulerziehung selbst zum Gegenstand haben, sondern andere Themen wie beispielsweise das Geschlechterverhältnis und dieses im Feld der Kleinkinderziehung untersuchen. Martin D. Lampert (1996) zum Beispiel kommt in einer ethnographischen Humoruntersuchung zu dem Ergebnis, dass Jungen im Spiel mit anderen Jungen nonverbalen Humor entwickeln (z.B. Herumkaspern), wohingegen Mädchen mehr verbalen Humor ausprägen (Wortspiel, lustige Lieder etc.). Auch erzählen die Mädchen häufiger zu Handlungen Geschichten, die lustig sein sollen. Wenn nun Mädchen und Jungen zusammenkommen, dann passen sich die Mädchen dem von Jungen entwickelten Humor an, das heißt, sie inszenieren ebenfalls nonverbalen Humor. Dies deutet darauf hin, dass auch im Kindergarten (oder schon im Kindergarten) Mädchen lernen, die Welt sowohl aus ihrer eigenen als auch aus der Perspektive von Jungen zu betrachten, und dass sie ihre Handlungen auf die Interessen von Jungen ausrichten. Pädagogische Intervention bezüglich der Aneignung geschlechtshierarchischer Muster scheint dabei selten zu sein.

Märchen

Zu ähnlichen Ergebnissen kommt eine australische Untersuchung. Die Forscherin Bronwyn Davies (1992) arbeitet mit teilnehmender Beobachtung im Kindergarten und analysiert die Reaktion von Kindern im Kindergarten auf Märchen. Sie zeigt, dass Mädchen lernen, sich sowohl mit weiblichen als auch mit männlichen Helden zu identifizieren. Ein Phänomen, welches auch aus der Produktion von Spielfilmen bekannt ist, ist, dass Filme mit Heldinnen in der Regel nur von Mädchen angeschaut werden, wohingegen sich für Filme mit Helden beide Geschlechter begeistern können. So lernen Kinder, wie Davies zeigt, schon im Kindergarten die Einteilung der Welt in weiblich und männlich. Jungen lernen in diesem Prozess eine Positionierung als männliches Geschlecht einzunehmen, Mädchen aber müssen – da sich die Alltagspraxis latent an Jungenperspektiven orientiert – sowohl die für ihr eigenes Geschlecht relevante als auch die von Jungen akzeptierte Positionierung einnehmen (ein Phänomen, welches auch für das Handeln von Frauen und Männern bekannt ist, vgl. Bourdieu 1997). Auf Grund der Latenz des Prozesses und im Interesse der Gewährleistung eines reibungslosen Ablaufs wird dieser Vorgang im Kindergarten kaum pädagogisch gestaltet (vgl. dazu auch Kapitel 4.2.3). Gleichzeitig betont Davies, dass Mädchen und Jungen mit ihrem – gerade im Kindergartenalter – häufig überbetont klischeehaft weiblichen und männlichen Verhalten versuchen, sich in das System der Zweigeschlechtlichkeit einzugliedern (vgl. dazu Kapitel 4.1.2). „Man kann von Kindern nicht verlangen, sich als erkennbar männlich oder weiblich zu zeigen, und ihnen gleichzeitig die Mittel, mit deren Hilfe sie dies tun könnten, vorenthalten" (Davies 1992, S. 7).

5.2 Schule und Gesellschaft

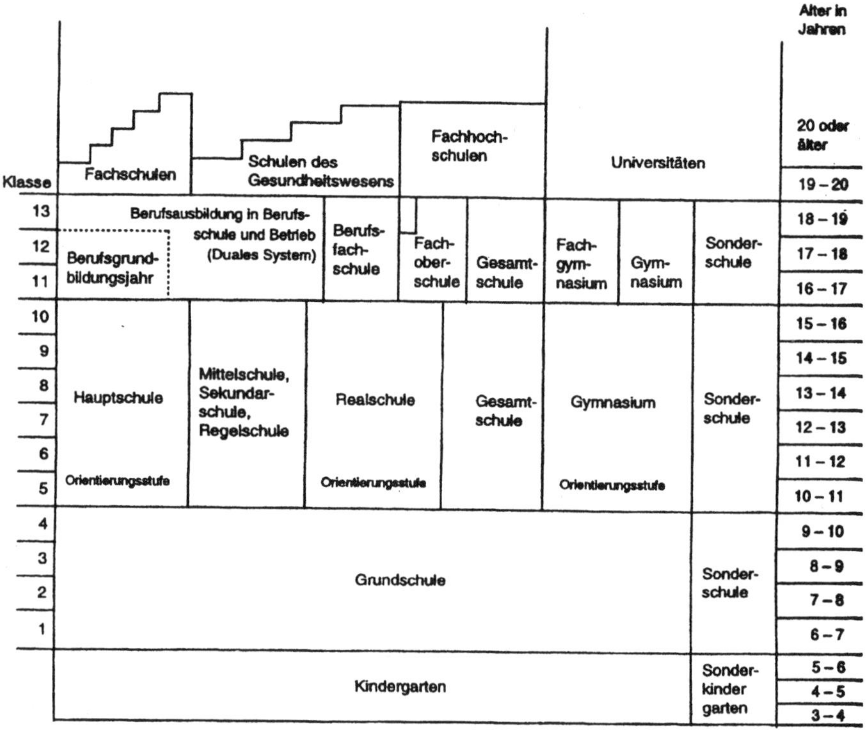

(Quelle: Reinhardt 1999, S. 312)

Wie an der Graphik abgelesen werden kann, ist das deutsche Bildungssystem stark in verschiedene Leistungs- und Fachbereiche aufgegliedert. Nur Kindergarten und Grundschule richten sich an (fast) alle Kinder. Während die institutionalisierte Vorschulerziehung zwar massenhaft genutzt wird, aber auf Freiwilligkeit basiert, endet die Teilnahme am Bildungssystem auf der Basis eigener Entscheidung mit dem 6. oder 7. Lebensjahr. Der Schulbesuch ist Pflicht. Eltern werden bestraft, wenn sie Kinder von der Schule fernhalten. Die Schule ist demzufolge in die Struktur des modernen Staates eingebunden. Die Schulpflicht bzw. das Recht auf Beschulung für alle ist grundlegender Bestandteil der bürgerlichen Gesellschaft (vgl. Kapitel 1 und 7.1; zur Geschichte der Schulpflicht vgl. Herrlitz 1981). Der Staat bestimmt (bislang) die Kernelemente der Schule: die allgemeinen Inhalte, die Prüfungen, die Zugangs- und die Berufsqualifikationen. Legitimiert ist der Staat zu dieser Gestaltung der schulischen Praxis dadurch, dass es in einer Demokratie als die Pflicht des Staates angesehen wird zu gewährleisten, dass die Menschen Wahl- und Mitwirkungsrechte ausüben können, was auch bedeutet, über Lesen und Schreiben eine eigene Meinung entfalten zu können, darüber hinaus auch geschäftsfähig zu werden und so ein administratives wie ökonomisches Grundverständnis zu erlangen. Schulische Bildung bezieht sich demnach auf die Vermittlung kultureller Wissensbestände, auf die Entwicklung kognitiver, sozialer und symbolischer Fähigkeiten und damit auch auf die Ausbildung von Autonomie in einem Beziehungsgefüge.

Schulpflicht

In Anlehnung an Helmut Fend (1974, S. 68ff), welcher sich wiederum auf Talcott Parsons bezieht (vgl. Kapitel 2.2.1), werden in der Forschungsliteratur häufig drei verschiedene Funktionen der Schule unterschieden (vgl. z.B. auch Büchner 1985; Grimm 1987):

Funktionen der Schule

1. Die Qualifikationsfunktion
   Die Schule hat die Aufgabe, im schulischen Unterricht die Kinder und Jugendlichen für das Beschäftigungssystem wie für die biographische Berufsfähigkeit zu qualifizieren. Dazu soll sie die für die Reproduktion und Wettbewerbsfähigkeit einer Gesellschaft notwendigen Fähigkeiten und Fertigkeiten, das über Generationen gesammelte, aber immer wieder neu ausgewählte Wissen, vermitteln.
2. Die Allokations- und Selektionsfunktion
   Durch Leistungsvergleich mittels Prüfungen sollen die Zuweisung der Schüler und Schülerinnen auf die verschiedenen Berufssparten und damit eine bedarfsgerechte Verteilung von Wissen erfolgen.
3. Die Integrations- und Legitimationsfunktion
   Die Schule soll die Schüler und Schülerinnen mit jenen sozialen oder politischen Kompetenzen ausstatten, die sie oder ihn befähigen, als loyal-kritische StaatsbürgerInnen zu handeln und vollwertige Mitglieder der Gesellschaft zu werden. Dazu soll die Schule die dominanten Normen, Werte und Interpretationsmuster vermitteln.

Das Schulsystem soll durch unterschiedliche Bestandteile (Unterricht – Prüfungen – „Schulleben“) die Verwirklichung der jeweiligen Funktion ermöglichen. Die folgende Graphik verdeutlicht dies exemplarisch:

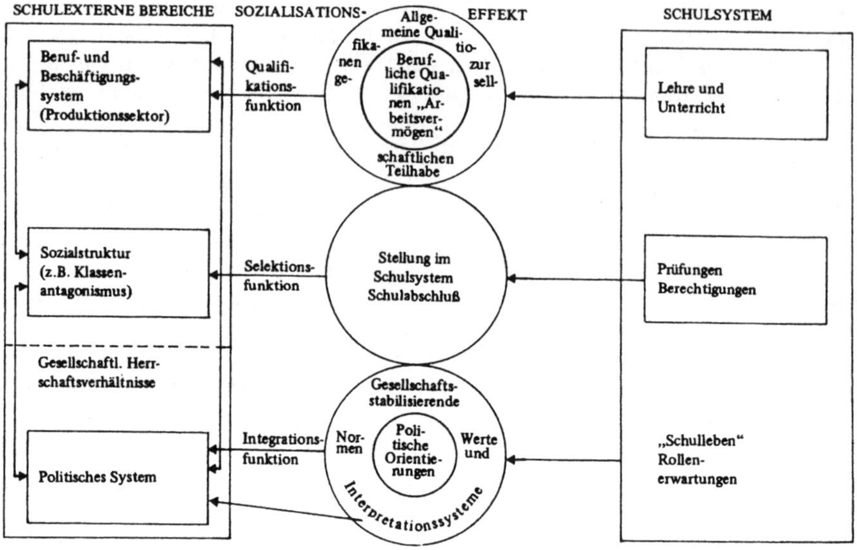

(Quelle: Fend 1974, S. 17)

Kritik am Funktionsgedanken

In der Literatur finden sich viele Belege dafür, dass die Schule diese Funktionen nur bedingt erfüllen kann. Fend selbst weist darauf hin, dass die Qualifikationsfunktion dadurch eingeschränkt wird, dass Bildungsprozesse langfristige Investitionen sind, die Anforderungen des Wirtschaftssystems sich jedoch häufig kurzfristig wandeln. Die Allokations- und Selektionsfunktion wird durch die Ausarbeitungen von Schelsky (vgl. Kapitel 2.2.2) und Bourdieu (vgl. Kapitel 3.1), aber auch aufgrund der aktuellen empirischen Daten (vgl. Kapitel 4.2.1) in Frage gestellt. Ausführlich dargelegt wird dort, dass die Zuweisung auf Berufe in klassen- und geschlechtsspezifischen Bahnen erfolgt, so dass die Prüfungen stark legitimatorischen und nur bedingt objektiv differenzierenden Charakter haben. Gegen die Integrations- und Legitimationsfunktion schließlich ließe sich einwenden, dass Bildungserfolg die Vertrautheit mit der Mittelschichtskultur ist. Ein vollwertiges Mitglied wird hier entsprechend der spezifischen kleinbürgerlichen Normen und Werte entworfen. Die Pluralisierung der Lebensformen wird zu wenig berücksichtigt.

Aufgrund dieser Einwände stellt sich die Frage, ob die Schule adäquat über die Funktionen beschrieben wird. Welchen Sinn macht es, eine Institution über Aufgabenbereiche zu beschreiben, die sie strukturell nicht in der Lage ist, angemessen zu erfüllen. Die Schule wird mit einer Erwartungshaltung betrachtet, welche Funktionen sie sinnvoller Weise übernehmen sollte, und gleichzeitig wird ihr Versagen diagnostiziert.

Reproduktion der bürgerlichen Gesellschaft

Peter Büchner kommt daher zu dem Schluss, dass „die Schule heute eine Institution (ist), deren Hauptaufgabe auf die Reproduktion der bürgerlichen Gesellschaft ausgelegt ist" (Büchner 1985, S. 140). Er stellt sich damit in die Tradition reproduktionslogischer Argumentationen, wie sie vor allem unter dem Begriff „new sociology of education" als gesellschaftskritische Bildungssoziologie in den USA und Großbritannien in den 1960er Jahren entstanden sind (vgl. zum

Beispiel die Studien von Willis in Kapitel 4.2.2 und 8), aber auch in Frankreich durch Autoren, die sich an Pierre Bourdieu anlehnen, vertreten werden. Im Unterschied zu strukturfunktionalistischen Thesen, in denen die Möglichkeiten des individuellen Aufstiegs hervorgehoben werden, betonen diese Ansätze die soziale und kulturelle Reproduktion (auch und im Besonderen von sozialer Ungleichheit) in der Schule. Statt konsensfähigen Werten und Normen, welche durch die Schule vermittelt werden, rücken Konflikte, Opposition und Widerstand in den Blick (vgl. zusammenfassend Kolbe/Sünker/Timmermann 1994). Parelius/Parelius (1987[2]) unterscheiden deshalb bildungssoziologische Analysen in jene, die die Notwendigkeit sozialer Stabilität durch (schulische) Bildung betonen (Normreproduktion), und andere, die die Möglichkeit sozialen Wandels durch Bildung hervorheben (Analyse der Reproduktion und Skandalisieren sozialer Ungleichheit). In reproduktionslogischer Perspektive scheitert die Schule nicht an den ihr gestellten Aufgaben, sondern sie erfüllt die Anforderung der stetigen Wiederherstellung der gesellschaftlichen Strukturen, welche geschlechts- und klassenspezifisch sind. Dass sich dabei LehrerInnen in grundsätzlichen Antinomien bewegen müssen, ist ein anderes Problem (mehr dazu in Kapitel 5.3).

Funktionale Differenzierung

Aber auch Ansätze, welche sich in der gedanklichen Weiterentwicklung des Parsonsschen Ansatzes verstehen und sich auf Niklas Luhmann beziehen, sind mit der einfachen Funktionszuschreibung von Schule heute unvereinbar. Armin Nassehi (1999) zufolge ist Integration in die Gesellschaft ein Modus, mit dem die Internalisierung sozial integrierender Norm- und Wertmuster in früheren segmentären und stratifizierten Gesellschaften beschrieben werden kann. Funktional differenzierte Gesellschaften, seiner Position nach die treffende Charakterisierung der derzeitigen modernen Gesellschaft, inkludieren nach einer völlig anderen Logik. Gesellschaft differenziert sich in operativ autonome funktionale Teilsysteme, die wechselseitig aufeinander reagieren, aber doch primär Differenzen (z.B. systemspezifische Logiken, Kommunikationen über Werte) hervorbringen. Die Schule oder das Bildungssystem können demnach weder eine allgemeine Integrationsfunktion erfüllen, noch kann sie/es in die Logik der Ökonomie eingreifen (vgl. dazu Kapitel 3.2). Moderne Gesellschaften, so Nassehi, inkludieren über die Zugehörigkeit zu verschiedenen Teilbereichen.

Dividuum

In der funktional differenzierten Gesellschaft führe die gleichzeitige Zugehörigkeit zu verschiedenen Systemen zur Multiinklusion. Das Individuum wird zum „Dividuum", das die verschiedenen Inklusionen biographisch zusammenführen muss. Das heißt, die Biographisierung individueller Perspektiven (damit auch die Verunsicherung in der Lebensplanung und den sozialen Beziehungen etc.) ist eine Folge der funktionalen Differenzierung. In dieser Perspektive (die wie Parsons nicht den Konflikt in den Vordergrund stellt) ermöglicht die Schule Karrieren oder sie organisiert die Lernfähigkeit, vielleicht vermittelt sie sogar maßgeblich die Fähigkeit zur Biographisierung, aber sie kann nicht in alle Systeme integrieren und somit eine allgemeine Integrationsfunktion erfüllen.

Freude am Schulbesuch

Jenseits dessen, ob man die Aktivitäten in der Schule als Funktionen beschreibt, die nicht erfüllt werden, oder die soziale Ungleichheit hervorhebt, fest steht, dass das Scheitern der Schulen in den letzen Jahren besonders deutlich wahrgenommen wird. Die PISA-Studie (vgl. z.B. Baumert/Schümer 2001) bescheinigt den Schulen ungenügende Kompetenzvermittlung sowie die Verstärkung sozialer Ungleichheit (letzteres wird ausführlich dargestellt in Kapitel 4, insbesondere in Kapitel 4.2.2 zum heimlichen Lehrplan). Die Schüler und Schülerinnen selbst gehen, nach Einschätzung ihrer Eltern, immer weniger gern zur Schule. Nach einer repräsentativen Untersuchung des Instituts für Schulentwicklung (IFS) geben nur noch 40% der Schülereltern im Westen und 37% der Schülereltern im Osten an, ihr Kind gehe gern zur Schule (1993 waren es noch 62% und 54%). Das Vertrauen in die LehrerInnen ist so gering wie nie zuvor. Die Kompetenzen der Schüler und Schülerinnen nach Schulabschluss werden zunehmend schlechter bewertet. Zeitgleich steigt die Anforderungserwartung der Eltern. Schule soll auch soziale Kompetenzen und Teamfähigkeit stärker fördern (alle Ergebnisse IFS 2000, S. 14ff). Da der Staat mit der Aufsicht der Schulen gescheitert zu sein scheint, liegt es nahe, darüber nachzudenken, die Schulen in die Autonomie zu überantworten.

5.2.1 Schulautonomie

Entscheidungsstruktur

Die Frage, ob der Staat sich aus den Entscheidungen der Schulpolitik zurückziehen soll, wird seit vielen Jahren kontrovers diskutiert. Dabei stehen verschiedene Modelle von „Autonomie“ und „Teilautonomie“ zur Diskussion. Prinzipiell geht es um die Frage, welche Entscheidungen vom Staat auf die Ebene der Einzelschule verlagert (z.B. Curricula, die Art des Prüfungssystems, die Organisation der Schule, dabei auch Zeitrhythmen oder die Altersstrukturierung der Klassen oder die Verwendung von Ressourcen, seien es Geldmittel, Räume oder Personal) und welche Entscheidungseinheiten gebildet werden. Wer bestimmt Inhalte und Verwendung: Lehrer? Eltern? Schüler? Schulleitung?

Für Schulautonomie werden eine Reihe von Argumenten und Gegenargumenten vorgetragen (vgl. Timmermann 1996):

Finanzen

1. Autonomie entlastet die öffentlichen Haushalte und schafft finanzielle Selbstbestimmung für die Schulen. Das Problem hierbei ist, dass finanzielle Autonomie der Schulen in einem Atemzug mit Kürzungen diskutiert wird. Der Staat bewegt sich, weil die Schulbildung zu teuer ist. Nun soll die schwierige Übergangsphase in die Autonomie ausgerechnet bei schlechterer finanzieller Ausstattung vollzogen werden. Die Aufgabe scheint für viele Schulen nicht zu bewältigen zu sein und die Frage, warum der Staat ausgerechnet bei der Bildung spart, bleibt unbeantwortet.

Wahl

2. Die Wahlmöglichkeiten von Eltern und Schule werden erweitert. Die Basisannahme ist, dass die Konkurrenz zwischen den Schulen eine Vielfalt der Bildungsangebote hervorbringt und somit den Eltern wie SchülerInnen eine Wahl nach individuellen Interessen ermöglicht. GegnerInnen stellen hier die Rationalität der elterlichen Entscheidung in Frage. Die meisten Eltern bringen ihre Kinder in nahe gelegene Schulen und viele Eltern/SchülerInnen sind nicht in der Lage, die Qualität einer Schule zu beurteilen. Das gesellschaftliche Gesamtinteresse werde aufgegeben zugunsten zufälliger und häufig ungerecht verteilter Schulzugänge (schlechte Wohngegenden haben schlecht ausgestattete Schulen). Andererseits wird jedoch auch betont, dass eine als reflexiv bezeichnete Moderne, Eltern und SchülerInnen schon früh

und vielfältig Entscheidungen abverlange. Eltern müssen also nicht zwangsläufig überfordert sein. In Bezug auf das scheinbar existierende gesellschaftliche Gesamtinteresse wird eingewandt, dass derzeit schulpolitische Entscheidungen von kurzfristigen Stimmenmaximierungen abhängen und insofern nicht das Gesamtinteresse berücksichtigen.

3. Schulautonomie ist effizienter.
   Staatliche Behörden arbeiten – so die Annahme – ineffizient. Staatliche Entscheidungsstrukturen gelten als „zeitaufwendig, inflexibel, ressourcenaufwendig und innovationshemmend“ (Timmermann 1996, S. 73). Fehlentscheidungen werden nicht sanktioniert. Das Problem einer solchen Argumentation ist, dass statt pädagogischem HandelnWirtschaftlichkeit/Effizienz im Vordergrund der Diskussion steht. Die Gefahr einer Schulautonomie ist, dass Bildungsentscheidungen nun unter Zeit- und Wirtschaftlichkeitsdruck von den Einzelschulen getroffen werden müssen und dabei zwar weniger zeitaufwendig, aber auch weniger durchdacht, gefällt werden.

Effizienz

4. Staatsschulen sind ungerecht.
   Einkommensschwache Haushalte finanzieren über Steuerabgaben die höhere Bildung mit, die hauptsächlich von einkommensstärkeren Haushalten nachgefragt wird. Im marktgesteuerten System autonomer Schulen könnten – dies ist das Argument – einkommensstärkere Haushalte für Bildung zuzahlen. Als Gegenargument wird eingewandt, dass sich die existierende Bildungsdistanz der einkommensschwachen Haushalte durch eine Einführung von Schulgeld – selbst wenn gleichzeitig Stipendien vergeben werden – noch verstärkt werden könnte. Insgesamt betonen die GegnerInnen der Schulautonomie die Sorge, dass Kinder aus einkommensschwachen Familien unter Konkurrenzbedingungen zu Verlierern der Umstrukturierung werden, die BefürworterInnen dagegen sehen Gestaltungschancen für Bildungsprozesse.

Gerechtigkeit

5.2.2 Schulkulturen und Schulmythen

Profile

Unabhängig von der Autonomiefrage lassen sich jedoch bereits jetzt Tendenzen feststellen (und werden auch staatlich gefördert), Schulen unter Konkurrenzdruck und damit unter Profilierungszwang zu bringen. Die sinkenden SchülerInnenzahlen forcieren diesen Prozess. Im Werben um Schüler und Schülerinnen bzw. in der gegenseitigen Abgrenzung bemühen sich Schulen, spezifische Profile zu gewinnen, indem sie Zusatzangebote unterbreiten und besondere pädagogische Programme auflegen. Diese Bestrebungen werden in neueren Forschungen als Schulkulturen zusammengefasst. Hervorgehoben wird hierbei, dass Schulen nicht formalisierte, abstrakt-bürokratische Organisationen sind, sondern dass sie komplexe symbolische und sinnhafte Gebilde sind. Werner Helsper/Jeanette Böhme/Rolf-Torsten Kramer/Angelika Lingkost (2001) plädieren in einer theoretischen und empirischen Studie zu „Schulkultur und Schulmythos“ für einen deskriptiv-analytischen Schulbegriff, der Schulkultur nicht über ein normatives Ideal bildet, sondern in der Tradition kulturanthropologischer Studien über die Erschließung der Alltagspraxen, Rituale, Symbole und Interaktionsformen rekonstruiert. Demzufolge gibt es nicht Schulen, die „Kultur“ haben und andere, denen sie fehlt, sondern jede Schule hat eine Schulkultur – es fragt sich nur welche.

Schulkultur

Schulkultur ist den AutorInnen zufolge „als die symbolische Ordnung der einzelnen Schule in der Spannung von Realem, Symbolischem und Imaginärem zu verstehen“ (Helsper u.a. 2001, S. 25). Das Reale definieren sie als die vorstrukturierenden und rahmenden gesellschaftlichen Strukturierungen und die damit einhergehenden Antinomien. Mit dem Symbolischen sind die Interaktions- und Kommunikationsprozesse erfasst. Das Imaginäre wiederum bezeichnet das Selbstverständnis der Schule. Schulkultur wird durch die „handelnde Auseinandersetzung der schulischen Akteure mit systemischen Vorgaben, bildungspolitischen Strukturentscheidungen vor dem Hintergrund historisch spezifischer Rahmenbedingungen und sozialer Aushandlungen um die Durchsetzung und hierarchisierende Distinktion pluraler kultureller Ordnungen generiert“ (Helsper u.a. 2001, S. 25). Schulkultur entsteht demnach in einer Wechselwirkung zwischen Handeln und Strukturen. In Aushandlungsprozessen zwischen verschiedenen AkteurInnen und mit den strukturellen Bedingungen wird die symbolische Ordnung der Einzelschule generiert.

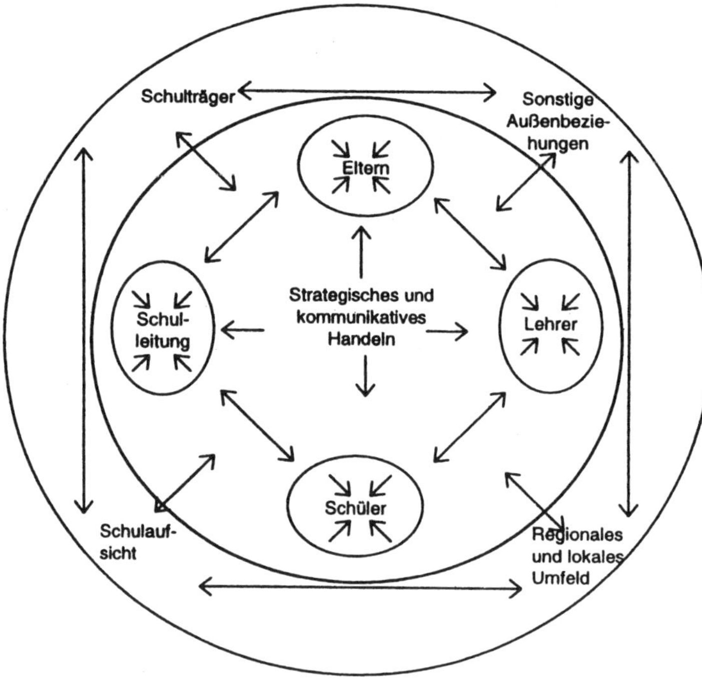

(Quelle: Helsper u.a. 2001, S. 28)

Kulturelle Ordnung

Die plurale kulturelle Ordnung bildet die Folie, vor der erst spezifische Kulturen zur Wirkung gelangen können. Auf dem allgemeinen Bedingungsnetz treten LehrerInnen, SchülerInnen, Eltern und Schulleitung in einen Aushandlungsprozess, der auch von regionalen, bürokratischen und sozialen Akteur-Innen beeinflusst wird. Die gemeinsame kulturelle Ordnung, auf die sich dieser Aushandlungsprozess bezieht, löst die gesellschaftliche Ungleichheit nicht in eine gleichberechtigte Existenz unterschiedlicher Formen auf. Vielmehr entwickeln sich in den verschiedenen Schulen unterschiedliche kulturelle Ordnungen, denen verschiedene soziale Milieus und Lebensformen unterschiedlich nah oder fern (stehen) und die in einem hierarchischen Verhältnis zueinander stehen (vgl. dazu auch Helsper u.a. 2001, z.B. S. 17).

Mythen

Schulkulturen sind spezifische Varianten der Auseinandersetzung mit den strukturellen Vorgaben des Bildungssystems. Insofern basieren sie auf verallgemeinerbaren Grundlagen, stellen jedoch spezifische kulturelle Ausformungen dar. Auf der Ebene des Imaginären folgen aus den spezifischen Schulkulturen auch unterschiedliche Schulmythen. Trotz fortschreitender Rationalisierung sei die Moderne weiter auf Mythen zur Bewältigung der Lebenspraxis angewiesen. „Mythen werden als kreative Konstruktionen von Sinn gefasst, in denen die Ambivalenzen und Widersprüche der Moderne imaginär überbrückt werden" (Helsper u.a. 2001, S. 69; vgl. auch Böhme 2000). Ein Schulmythos ist ein dominanter, legitimatorischer Sinnentwurf einer Schule; zum Beispiel die Konstruktion einer wie „heilig" behandelten schulischen Tradition. Unsicherheit und Ungewissheit sowie das prekäre Verhältnis von Kollektivität und Individualität werden durch Mythen überspielt und geglättet.

In ihrem Werk rekonstruieren die AutorInnen die Schulkultur von drei ostdeutschen Gymnasien. Bei aller Differenz der jeweiligen Ausgestaltung zeigt sich in den drei Schulen doch eine vergleichbare Handlungspraxis. Innerschulische Strukturprobleme werden als außerschulische Einflussfaktoren wie Gefährdungen konstruiert und damit innerschulische Problemlagen entthematisiert oder negiert. An die Stelle tritt der Schulmythos. Das Imaginäre ersetzt das schulische Reale (vgl. Helsper u.a. 2001, S. 165ff sowie S. 555). Hierfür ein Beispiel: Während einerseits der gesellschaftliche Wahn der Fortschrittsgläubigkeit und Umweltzerstörung kritisiert wird, wird andererseits von den Schülern und Schülerinnen selbst höchste Leistung erwartet. Die Idee des „Immer-Schneller-Immer-Mehr" erscheint als Umweltproblem und muss daher als schulisches Problem nicht thematisiert werden. Gleichzeitig entsteht ein Schulmythos von der Schule bzw. ihren SchülerInnen als Heilsbringer für eine von Zerstörung bedrohten Welt, die zum Beispiel in Baumpflanzungen ihren Ausdruck findet. Das Imaginäre richtet sich gegen das Reale, das – da es als „Außen" definiert wird – nicht mehr in seinen den Schulalltag konstituierenden Facetten berücksichtigt wird.

5.3 Erziehen als Profession

Pluralisierung der Lebensbedingungen

Die Krise der Schul- und Vorschulerziehung rückt die Ausbildung der ErzieherInnen und LehrerInnen in den Mittelpunkt des Interesses. LehrerInnen und ErzieherInnen stehen vor dem Problem, dass sich das Anforderungsprofil an ihre Arbeit gewandelt hat, sie jedoch keine verallgemeinerbaren Strategien erlernt haben, diesen Anforderungen zu begegnen. Als Veränderungen werden an erster Stelle stets die Pluralisierung der Lebensbedingungen genannt. Dazu gehört, dass Kinder heute in sehr unterschiedlichen Familienkonstellationen aufwachsen. Nach Schätzungen wachsen in der BRD ca. 1,5 Mio. Kinder mit schwulen Vätern oder lesbischen Müttern auf. Der Anteil von Alleinerziehenden macht ca. 20% der Erwachsenen mit Kindern aus. Viele Kinder leben nach der Scheidung in zwei Wohnungen. Die unterschiedlichen Erfahrungen, die die Kinder in die Bildungsinstitutionen mitbringen, müssen im gemeinsamen Unterricht oder Spiel koordiniert werden. Die Verinselung der Lebensbedingungen (vgl. Kapitel 8.2) und die prinzipiell gestiegenen Wahlmöglichkeiten und Entscheidungszwänge, d.h. zum Beispiel die Wahrnehmung vielfältiger Bildungsangebote in der Freizeit, führen dazu, dass auf die spezifischen Interessen einzelner Kinder unter Bedingungen eines gemeinsamen Unterrichts/Spiels eingegangen und mit den Kindern ausgehandelt werden muss (vgl. zu Veränderungen z.B. Fölling-Albers 1992, zum Anforderungsprofil an professionelle Akteure vgl. auch Paschen/ Wigger 1996; Daschner/Rolff/Stryck 1995). Diese Pluralisierung wird von Universalisierungstendenzen konterkariert, indem allgemeine Richtlinien für Notengebung etc. erlassen werden, die vom einzelnen Schüler/Schülerin abstrahiert werden. Diese werden jedoch durch individuelle Zertifikate beantwortet. Werner Helsper (1999[3], original 1996, S. 538) verdeutlicht, dass Lehrer und Lehrerinnen mit verschiedenen Antinomien zu kämpfen haben. Der Widerspruch zwischen Generalisierung und Pluralisierung ist die eine, eine andere Antinomie ergibt sich aus dem Widerspruch zwischen den zweckrationalen Organisationsstrukturen einer Organisation einerseits und gestiegenen kommunikativen Erfordernissen andererseits. Diese kommunikativen Anforderungen beziehen sich nicht nur auf die Aushandlungen im Unterricht, sondern auch auf Veränderungen in Richtung Schulautonomie mit den selbststeuernden Verständigungsprozessen und dem Kommunizieren schulischer Aktivitäten und Profile außerhalb der Schule. Eine weitere Antinomie besteht aus der gleichzeitigen Anforderung der Moderne, Affekte zu kontrollieren (vgl. Elias 1976) und LehrerInnen-SchülerInnen-Beziehungen als intime, auf Vertrauen basierende Verhältnisse zu realisieren. Für die Lernenden hat dies Vor- und Nachteile: Die Emotionalisierung schulischer Kontakte ermöglicht es den Lernenden, sich als konkrete Personen mit subjektiven Hintergründen einzubringen. Gleichzeitig ermöglicht Distanz auch den Schutz des eigenen Selbst. Schlechte Noten werden schnell zur Infrage-Stellung der gesamten Person (vgl. Helsper 1990). Auch Aushandlungschancen und die pädagogische Zuschreibung der Eigenverantwortlichkeit bringt gleichzeitig weniger Zwang in einem egalitäreren Verhältnis von LehrerIn und SchülerIn und mehr *Selbst*zwang hervor. Die Anreicherung der Schule mit sozialpädagogischen Elementen wie der Förderung von sozialen Kompetenzen, Teamfähigkeit etc., (Entscholarisierung der Schule) bei gleichzeitiger Nutzung vielfältiger Bildungsangebote in der Freizeit (Scholarisierung der Freizeit), bringt den Kindern/Jugendlichen mehr spielerische, weniger leistungsorientierte Elemente in der Schule (Projektphasen, Abschaffung der Ziffernoten in der Grundschule) und gleichzeitig eine Verschulung der Freizeit.

Antinomien

Ausbildung

Dieses Handeln unter Bedingungen struktureller Widersprüche führt zu einem häufig unverstandenen und damit nicht artikulierbaren Gefühl von Überlastung (vgl. Combe 1999[3], original 1996) und nicht selten zu dem notwendig scheiternden Versuch, die Widersprüche einseitig aufzulösen (dazu z.B. Schütze 1999[3], original 1996), auf die in Ausbildungskonzepten bislang kaum reagiert wird.

Für die Neugestaltung der ErzieherInnen-Ausbildung stellt sich das Problem noch komplexer dar, da es kein anerkanntes Selbstverständnis über die Arbeit und den Beruf der ErzieherIn gibt. Ursula Rabe-Kleberg (1993; 1997) kritisiert, dass die Debatte um die Professionalisierung sozialer Berufe bzw. allgemein um Profession zumeist unter Außerachtlassung des Geschlechts geführt wird. Dabei sind es gerade die Frauenberufe wie zum Beispiel Erzieherin, Grundschullehrerin oder Krankenschwester, denen das Etikett „Semi-Professionalität" anhaftet. In der berufssoziologischen Literatur werden nur jene Berufe als Professionen bezeichnet, die einen langen formalisierten, in der Regel akademischen Bildungsprozess voraussetzen und fast ausschließlich von Männern ausgeübt werden, insbesondere Richter oder Pfarrer. Rabe-Kleberg überprüft daher, „wie weit der (positiv besetzte) Begriff selbst bereits den Prozess der Segregation und Monopolisierung von Machtstellungen legitimiert, die mit dem Verständnis von Profession verbunden sind, statt ihn kritisch zu untersuchen" (Rabe-Kleberg 1997, S. 64, vgl. zur geschlechtsspezifischen Segregation auch Kapitel 4.2.3). Sie führt die wenig oder ungeordneten Bildungswege und das wenig oder gar nicht strukturierte Verhältnis von Wissenschaft und Praxis in sozialen Berufen auf die geschlechtsspezifische Aufteilung des Arbeitsmarktes zurück. Daher betrachtet sie Profession nicht als ein besonders entwickeltes Verhältnis zur Arbeit, sondern in seiner bisherigen Form als soziale Schließung und als Instrument der Monopolisierung von Macht- und Marktpositionen.

Professionen

In Anlehnung an Andrew Abbott (1988) schlägt Rabe-Kleberg statt dessen vor zu untersuchen, wie „Professionen und Professionals in spezifischer Weise mit ihrer Arbeit umgehen und wie sich diese von der Art, wie Frauen mit Frauenarbeit umgehen, unterscheidet" (Rabe-Kleberg 1997, S. 65). Nach diesem Perspektivwechsel, der hier eingenommen wird, ist nicht länger davon auszugehen, dass Frauen per se im professionellen Kampf um Privilegien unterliegen, sondern dass Professionen über Zuständigkeiten („jurisdiction" bei Abbott) für Bereiche auf der Basis von Wissen und Fähigkeiten definiert werden. Verschiedene Professionen konkurrieren dann um das Recht, für bestimmte Probleme in der Gesellschaft zuständig zu sein (Rabe-Kleberg 1993, S. 96ff). Voraussetzung hierfür ist die Kontrolle über die eigene Ausbildung und das generierte Wissen, was genau bei vielen Frauenberufen aufgrund eines (angenommenen) Qualifikationsprofils zwischen beruflicher und Laienqualifikation wie zum Beispiel beim Erzieherberuf und bei der Grundschullehrerin nicht der Fall ist (vgl. auch hier Kapitel 4.2.3). Die durchaus verbreitete Vorstellung, jede halbwegs gebildete Frau könne diese Tätigkeiten eigentlich auch ohne Ausbildung ausüben, wie es Arno Combe (1999[3]) analysiert – ohne dabei das Geschlechterverhältnis zu berücksichtigen – ist hierfür beispielhaft.

Diffusität

Nun kommt es jedoch durch den gestiegenen Bildungsgrad der Bevölkerung insgesamt und durch eine reflexive Kultur dazu, dass die traditionellen Professionen mit gebildeteren Laien verhandeln müssen, und somit – wie Rabe-Kleberg (1993) oder auch Schütze (2000) aufzeigen – die Differenz zwischen Laien- und professionellem Wissen auch dort diffuser wird. Wie oben an den Antinomien des LehrerInnenberufs dargestellt, unterliegt das professionelle Handeln einer zunehmenden Kontrolle durch die Laien und die Professionen sind gezwungen zu kooperieren, so dass nicht immer deutlich ist, welches Wissen zur Problemlösung beiträgt (vgl. zum Verhältnis von Lehramt und Profession auch Dewe u.a. 1992; Wagner 1998).

Ungewissheit

Dies hat zur Konsequenz, dass Erziehen als professionelle Arbeit in mehrfacher Hinsicht als „Arbeit unter Ungewissheitsstrukturen" (Rabe-Kleberg 1997, S. 66) charakterisiert werden kann. Ungewiss sind sowohl das Ziel als auch die Mittel. Man muss nur eine aktuelle Definition von Bildung als einen Prozess betrachten, der „geistige Aufgeschlossenheit zum Ziel haben muss, Einstellungen und Kenntnisse von der Art, wie man sie braucht, um sich immer neues Wissen anzueignen und mit immer neuen Situationen zurechtzukommen" (Collège de France 1987, S. 253, vgl. auch Kapitel 1 dieses Buches), um sich zu verdeutlichen, dass weder das Ziel noch der Weg eindeutig definiert sein können. Die Spielräume für unterschiedliche Entwürfe und Konzepte wachsen, die Möglichkeiten zur Selbstgestaltung ebenfalls – damit vergrößern sich aber auch die Unsicherheiten über das „richtige" Vorgehen.

Reform

Professionalität kann daher, so Rabe-Kleberg, sowohl unter frauenpolitischen Gesichtspunkten als auch unter Berücksichtigung des Strukturwandels des Dienstleistungssektors und der Antinomien des LehrerInnenhandelns nur „als die Bereitschaft und die Fähigkeit verstanden werden, unter den Ungewissheitsstrukturen professioneller Arbeit verantwortlich zu handeln" (Rabe-Kleberg 1997, S. 67). Dazu bedarf es, wie Fritz Schütze herausstellt, einer besseren Ausbildung in der Analyse widersprüchlicher Handlungslogiken (Schütze 1999[3], S. 253) sowie einer Reformierung der Ausbildungsstrukturen für Frauenberufe, insbesondere einer größeren Gender-Sensibilität in der Ausbildung. Eine Reform der LehrerInnenbildung in Hamburg (Keuffer/Oelkers 2001) versucht immerhin die universitäre Ausbildung auf der Grundlage kultureller und sozialer Heterogenität, Eigenverantwortung und Medienkompetenz neu zu gestalten.

6. Bildung und Hochschule

6.1 Kurze Geschichte der Universitäten

Geschichte

Die Institution der Universität entsteht im hohen Mittelalter. Zu den ersten Gründungen zählen Bologna, Reggio, Vicenza, Arezzo und Padua zwischen 1158 und 1222. In Neapel wird 1224 die erste Staatsuniversität durch Kaiser Friedrich II. von Hohenstaufen gegründet. Zeitgleich entstehen vor allem in Frankreich, England und Spanien Universitäten. In deutschen Territorien ruft 1348 in Prag Karl IV die erste Universität ins Leben (vgl. Grundmann 1976[2]; Seidenfaden 1988). Von Anbeginn an ist die Geschichte der Universität mit der Geschichte der Kirche verwoben. Eine große Zahl der Lehrenden sind Geistliche oder werden Geistliche, um Universitätslehrer zu werden. Bis heute zeugt zum Beispiel die Praxis, eine Rangliste potentieller KandidatInnen für eine Professur mit drei Plätzen dem Ministerium zur Ernennung einzureichen, von dem kirchlichen Einfluss. Zur Berufung eines Bischofs werden ebenfalls so genannte Dreierlisten erstellt und nach Rom in den Vatikan gesandt.

Artistenfakultät

Die inneruniversitäre Struktur bildet damals vielerorts eine Aufteilung in Juristische, Medizinische und Theologische Fakultät als Hauptbereiche, deren Basis die Artistenfakultät darstellt. Die Künste, die unter dem Dach der Artistenfakultät, aus der die spätere philosophische Fakultät hervorgeht, versammelt werden, sind Grammatik, Rhetorik, Dialektik, Arithmetik, Geometrie, Astronomie und Musiktheorie. Die Gesamtzahl der Studierenden an deutschen Universitäten übersteigt im Mittelalter selten 700. Die meisten Studierenden sammeln sich in der Artistenfakultät, deren Lehrstoffniveau mit der heutigen gymnasialen Oberstufe verglichen wird (vgl. Seidenfaden 1988, S. 20). Dies korrespondiert mit dem Aufnahmealter an der Universität, das in der Regel bei 15 Jahren liegt. Die Vorlesungen jener Zeit dienen vor allem dazu, ein Buch öffentlich vorzulesen und zu kommentieren, da die Mehrzahl der Studierenden sich Bücher nicht leisten kann. Universitäten sind jedoch nicht nur eine kirchlich, sondern auch eine städtisch geprägte Einrichtung. Im 16. Jahrhundert zeigt sich, dass zwar flächendeckend Universitäten entstehen, sich jedoch überall dort Knotenpunkte bündeln, wo sich auch städtische Zentren herausbilden.

(Quelle: Seidenfaden 1988, S. 26)

Territorialstaaten

Im 16. und 17. Jahrhundert wird mit der Entwicklung der Territorialstaaten auch der Einfluss der Landesfürsten größer. Mit der Gründung von Wittenberg (1502) und von Frankfurt an der Oder (1506) sichert sich jeder der sieben Kurfürsten universitären Einfluss, um die Ausbildung der Beamten überwachen zu können. Die Kurfürsten übernehmen Richterfunktionen für die Universitäten, sie regulieren Ausbildungsprogramme und -zugänge, zahlen Gehälter der Professoren und vertreten in begrenztem Maße die Interessen der Universität gegenüber der Stadt.

Nationalerziehung

Wie bereits dargelegt (vgl. Kapitel 1.1), durchzieht im 18./19. Jahrhundert die Universität ein neuer Geist. Geknüpft an die Idee der Bildung und gegen reine Ausbildung gerichtet, werden freie Forschung und Diskussion zu wichtigen Bestandteilen des universitären Alltags. Die Idee der Nationalerziehung gewinnt, wie Rudolf Stichweh ausführlich begründet (vgl. Stichweh 1991; 1998), europaweit an Bedeutung. In dem Versuch, die Machtpotentiale wichtiger Einflussgruppen wie dem Jesuitenorden zu schwächen, entsteht die Leitvorstellung einer nationalspezifischen Bildung. Diese führt zum Beispiel dazu, dass Vorlesungen und Disputationen zunehmend auf Deutsch gehalten werden, um die nationale Idee zu stärken und der Erneuerung symbolisch Ausdruck zu verleihen, indem das an Kirche und Tradition geknüpfte Latein in den Hintergrund gedrängt wird. In der Gründungsphase der Soziologie wird sich diese „Nationalisierung wissenschaftlicher Kommunikationszusammenhänge" (Stichweh 1998, S. 63) in einer Gesellschaftsidee niederschlagen, deren Basis die Nation als territorial gebundenes Gebilde mit Innen-Außen-Struktur ist (vgl. Berking 2002, S. 109ff.). Zunächst führt jedoch in der Hochschulentwicklung allgemein die Leitidee der Nationalerziehung zu einer gleichzeitigen Bewegung von Globalisierung und Nationalisierung: „Überall werden Institutionen postuliert und geschaffen, die einem entstehenden nationalstaatlichen Zusammenhang verpflichtet sind. Aber die Formung dieser nationalen Institutionen vollzieht sich in einem europaweiten, später weltweiten Beobachtungs- und Imitationszusammenhang, der durch die zunehmende Häufigkeit von Interaktionen bestimmt wird" (Stichweh 1998, S. 64).

Die „Humboldtsche Universitätsidee" (vgl. Kapitel 1.1) greift vor dem Hintergrund dieser Entwicklung als deutscher Versuch, über ein emphatisches Bildungsverständnis sowohl das Bürgertum als auch die deutsche Nation zu stärken. Ab 1908 dürfen auch Frauen an diesem Projekt mitwirken. Nach langen Kämpfen werden sie zum Universitätsstudium zugelassen.

Vorkapitalistische Organisationsform

In gewisser Weise sind Universitäten heute Dinosaurier in der modernen Welt. Rudolf Stichweh (1998) macht darauf aufmerksam, dass es angesichts wirtschaftlicher und kultureller Entwicklungen der Globalisierung auffällig ist, dass Universitätsrektoren und -räte vieles nicht tun, was andernorts für selbstverständlich gilt, etwa mit dem „Unternehmen" zu expandieren oder es über Schulden zu finanzieren. Erfolgreiche Universitäten wie Harvard oder Oxford zum Beispiel richten keine Filialen im In- oder Ausland ein. Dies lasse sich nicht allein dadurch erklären, dass es sich bei Universitäten um lokale Einrichtungen des Staates handle. Vielmehr gäbe es auch zahlreiche erfolgreiche Privatuniversitäten, die nicht expandieren. Während Wirtschaftsunternehmen über Verschuldung die Ausweitung ihrer Geschäfte finanzieren, neigen Universitäten dazu, Kapital anzuhäufen, dieses jedoch nie für Kreditzwecke zu nutzen. Stichweh schlussfolgert: „Was sich hier offenbart, ist eine extreme Sicherheitspräferenz der Universitäten, die als Stellvertreter für andere Motive fungiert: Reputation und Langlebigkeit der Organisation" (Stichweh 1998, S. 68). In dieser Strategie offenbare sich sowohl der historisch anhaltende Erfolg (im Sinne ihrer fortwährenden Existenz) der Institution, als auch ihre genuin vorkapitalistische Organisationsform. Wie die religiösen Körperschaften und Orden arbeite die Universität mit der Anhäufung von Besitz, sei aber im Unterschied zu ersteren deshalb nie in die öffentliche Kritik geraten. Insofern reproduzieren die Universitäten zwar nationale Ideen, sie sind auch global im Sinne der Präsenz der Institution „Universität" in fast allen Ländern der Welt, sie sind aber auch explizit lokale Gebilde, da die jeweilige Universität sich als stadtspezifisches Einzelgebilde präsentiert. In Hochschulen, so kann man schlussfolgernd feststellen, bilden sich Widersprüche zwischen Lokalität und Globalisierung, zwischen emphatischer Bildungsidee und Rentabilitätsvorstellung, zwischen Bildung und Ausbildung ab.

6.2 Die Professorenschaft

Stereotype

Professoren (selten auch Professorinnen) verkörpern, so Daxner (z.B. 1999), die Institution „Universität". Mehr noch als die Studierenden sind sie die öffentlichen Repräsentanten des Systems und – wie das System – so hat auch der Berufsstand der Professoren die Jahrhunderte alte Stereotypenbildung relativ unbewegt überstanden. Diese bezieht sich, nach Daxner, vor allem auf drei Facetten:

1. Die Professorenschaft bildet einen eigenen Stand. Nur im Militär und in der Kirche existieren vergleichbare Hierarchien.
2. Der Professor besitzt immer noch eine recht hohe Reputation (und zwar in einer nicht-ständisch organisierten Umgebung).
3. Der Professor gilt als weltfremd und zerstreut.

Verunsicherung

Stereotype sind nach Daxner Produkte von Unsicherheit und Ergebnis von Machtkämpfen. Im Konflikt zwischen Experten- und Laienkultur, Vertrauenswünschen und Misstrauensanlässen entstünden sich widersprechende Bilder der Professorenschaft. Die Hochschullehrer/Hochschullehrerinnen hätten sich zwar an die widersprüchlichen Zuschreibungen und die Tatsache, als fremde Spezies beschrieben zu werden, gewöhnt, wären jedoch gleichzeitig als „Stand" verunsichert. Dies zeige sich, so Daxner, vor allem darin, dass sie um ihre Reputation bangen, wenn sie die Schwächen der Universität offenlegen. Ihre Macht basiere ferner nicht unwesentlich auf ihrer Prüfungsberechtigung und gleichzeitig hätten sie das Prüfen nie gelernt. Ihre Fachsprache schließlich, ebenfalls Machtmedium, würde im Studium nie einer systematischen Kritik unterzogen. So zeige sich in der Verunsicherung der Professorenschaft eine gesellschaftliche Unsicherheit bezüglich des Nutzens und der Strukturen der Universität.

Stereotype haben allerdings nie nur aktuelle Ursachen, sondern basieren auch auf historischen Entwicklungen. Max Weber setzt sich in einem 1917 in München auf Einladung des Bundes freier Studenten gehaltenen Vortrag „Wissenschaft als Beruf" mit „Persönlichkeit" auseinander. Seine Grundannahme ist: „‚Persönlichkeit' auf wissenschaftlichem Gebiet hat nur der, der rein der Sache dient" (Weber 1992, original 1917, S. 15). Er entwickelt in seinem Vortrag eine Vorstellung vom Beruf des Wissenschaftlers über die Bestimmung innerweltlicher Askese. Wissenschaftler könne nur sein, wer einer innerlichen Berufung folgend, sich völlig und ausschließlich der wissenschaftlichen Arbeit hingebe. Die wissenschaftliche Tätigkeit wird damit auf rein geistige Facetten zurückgeführt. Die Suche nach Erlebnissen, körperliche Ausschweifungen, der Versuch verschiedene Aspekte der Persönlichkeit zu entdecken, all dies verurteilt Weber als Sackgassen und Irrwege, die den Wissenschaftler von seiner eigentlichen Bestimmung abhalten.

Askese

Weber argumentiert dabei gegen eine studentische Kultur der Trinkgelage und des militärischen Kampfgeistes, die sich mit der Orientierung der Universität an Nützlichkeit und Verwertbarkeit bei der Produktion von Wissen entwikkelt.

Einsamkeit und Freiheit

Er rekurriert auf ein Jahrhunderte altes asketisches Wissenschaftsverständnis. Die Akzentuierung einer körper- und emotionsfreien Wissenschaft findet sich bereits im Zölibatsprinzip der frühen Universitäten (vgl. Janshen 1999), durch die eine materielle und immaterielle Struktur in den Universitäten gelegt wird, die auch in der säkularisierten Universität fortbesteht. Die Gründung der Universität basiert auf der Vorstellung, so arbeitet Doris Janshen in ihrer historischen Herleitung heraus, dass der Kopf erst wirksam ist, wenn der Körper nicht mehr spürbar ist (vgl. Janshen 1999, S. 16). Dazu passt auch das Tragen der klerikalen Tracht, welches den Körper unter dem Gewand versteckt. Es ist also ein Verzicht an Weltlichkeit im Sinne körperlicher Erfahrungen und Genüsse, der den „richtigen" Wissenschaftler auszeichnet. Humboldt spitzt dieses Imago im

19. Jahrhundert in der bekannten Forderung nach „Einsamkeit und Freiheit" zu. Wenn heute das Stereotyp des weltfremden, zerstreuten Professors reproduziert wird, dann ist es auf die immer wiederkehrenden Versuche zurückzuführen, Wissenschaft körper- und alltagsfremd als reine Kopfleistung zu konzeptualisieren, was in der Alltagspraxis immer wieder zu Ambivalenzen und Widersprüchen führt.

Liest man vergleichend die frühen wissenschaftssoziologischen Arbeiten von Robert K. Merton, der sich seit den 1930er Jahren des letzten Jahrhunderts mit der Wissenschaft beschäftigt und zeigt, wie wissenschaftliche Forschung in den Ideenkontext und die sozio-ökonomischen Probleme der jeweiligen Zeit eingebunden ist (vgl. auch Merton 1939), so werden auch hier Ambivalenzen der Berufsgruppe thematisiert. Das Funktionieren der Wissenschaft versucht Merton, vor allem in späteren Arbeiten, über institutionalisierte Normen und Werte des Wissenschaftlers zu erklären. Das wissenschaftliche Ethos umfasst – nach Merton (1985, original 1942, S. 90ff.) – vier affektiv getönte Komplexe bindender Normen und Werke: Ethos

a) Universalismus (Die Persönlichkeit des Wissenschaftlers spielt für die Beurteilung und den Wahrheitsanspruch keine Rolle.)
b) Kommunismus (Wissenschaftliche Erkenntnisse sind Allgemeingut.)
c) Uneigennützigkeit
d) Organisierter Skeptizismus (Glaubenssätze sollen unvoreingenommen durch empirische und logische Mittel geprüft werden.)

Wie später auch Pierre Bourdieu (vgl. Kapitel 3.1.2) betrachtet auch Merton das Streben nach Anerkennung über originelle Ideen und Erfindungen als Leitmotiv der WissenschaftlerInnen. Konkurrenz betrachtet er dabei als treibende, produktive Kraft. Er reflektiert jedoch deutlich die Ambivalenzen, in denen sich WissenschaftlerInnen dabei bewegen müssen. Am Beispiel der Verschleierung von Mehrfachentdeckungen problematisiert er: „Man kann sich den Widerstand gegen die Untersuchung von Mehrfachentdeckungen und Prioritätsfragen vorstellen als Resultate einerseits aus Kräften, die sehr intensiv in Richtung auf die öffentliche Anerkennung wissenschaftlicher Leistungen drängen, und andererseits aus entgegengesetzten, der sozialen Rolle des Wissenschaftlers inhärenten Kräften, die den Wissenschaftler zu bescheidener Anerkennung der eigenen Grenzen, wenn nicht gar zu regelrechter Demut drängen" (Merton 1985, S. 117). Der Kampf zwischen Individualität und Bescheidenheit bestimmt nach Merton das Leben eines Wissenschaftlers/einer Wissenschaftlerin. Konkurrenz

Dabei berücksichtigt er kaum, wie zum Beispiel Steffani Engler (2001, S. 139) hervorhebt, dass Wissenschaft von Macht- und Abhängigkeitsverhältnissen durchzogen ist. Wissenschaftlern/Wissenschaftlerinnen würden bei Merton und Weber, wie auch im Alltagsdiskurs, spezifische Eigenschaften als Persönlichkeitsmerkmal zugeschrieben. Steffani Engler misstraut der Erklärung, dass spezifische Persönlichkeiten sich für die Wissenschaft eben besonders eignen und bemüht sich darum, in qualitativen Interviews herauszufinden, wie das wissenschaftliche Feld (vgl. dazu die Ausführungen über Bourdieu in Kapitel 3.1.1 und 3.1.2) wissenschaftliche Persönlichkeiten hervorbringt. Persönlichkeit wird als soziales Konstrukt behandelt, das „im Geschehen der Wissenschaft, im Alltag entworfen wird und entsteht“ (Engler 2001, S. 15). In diesem Sinne versucht Engler die Funktionsprinzipien des Systems Wissenschaft zu erkunden, indem sie die Produktion ihrer RepräsentantInnen betrachtet.

Wissenschaftliche Persönlichkeit

Sie interviewt dazu ProfessorInnen, die nach 1980 ihre erste Professur erhielten (also in einer Phase von Stellenknappheit nach vorher vollzogenem Hochschulausbau), in der Regel also lange um die Professur kämpfen mussten. Zwei Fachkulturen werden berücksichtigt: Erziehungswissenschaft/Soziologie auf der einen Seite, Elektrotechnik/Maschinenbau auf der anderen Seite. Sie zeigt, dass in den Interviews alle HochschullehrerInnen den Weg zur wissenschaftlichen Persönlichkeit thematisieren. Dabei handelt es sich um einen langen, schwierigen Weg in Auseinandersetzung mit den Regeln des wissenschaftlichen Feldes. Im Gegensatz zu den gesellschaftlichen Erzählungen über einzelne aufgrund ihrer Leistung herausragende Wissenschaftlerpersönlichkeiten werten die interviewten HochschullehrerInnen ihre Berufung nicht als Ergebnis ihrer wissenschaftlichen Leistung. Obwohl alle glauben, dass Leistung wichtig ist, führen sie die Tatsache, ProfessorIn geworden zu sein, nicht auf diese Leistung zurück. WissenschaftlerInnen, so argumentiert Engler, streben zwar nach Anerkennung, aber nicht jede wissenschaftliche Leistung wird zwangsläufig anerkannt. Insofern sagt die Anerkennung nichts über Leistung im Verhältnis zu anderen Leistungen aus, sondern in der Anerkennung offenbart sich die Wahrnehmung durch die KollegInnen. „Das heißt aber auch, nicht die eigentliche Arbeit und Leistung führt zum Erfolg, sondern die in sozialen Prozessen anerkannte und zugeschriebene Leistung durch andere WissenschaftlerInnen“ (Engler 2001, S. 447). In sozialen Spielen nimmt man sich gegenseitig zur Kenntnis (oder auch nicht) und bestimmt die eigene Position in Relation zu den anderen. In diesem Prozess werde Eigenes und Originelles zugeschrieben und hervorgebracht. Das wissenschaftliche Feld funktioniert – so Engler – weder über Normen/Werte noch über Leistung, sondern über die Vorstellung, dass wissenschaftliche Persönlichkeiten agieren. Dies jedoch ist eine Illusion, da die Persönlichkeiten als Effekte der Kräfteverhältnisse hervorgebracht werden. Zu der Illusion gehört auch, dass die soziale Herkunft als Einflussfaktor auf den wissenschaftlichen Erfolg geleugnet wird.

Anerkennung

Matrix

Auch Dirk Baecker (2000, S. 57) geht in seinen Überlegungen davon aus, dass Wissenschaft als System eine eigene „Matrix“ entfaltet, die sich von den Strukturen und der Logik anderer unterscheidet. Er vertritt die These, dass die Universität mit einer Matrix von Wissenden und Unwissenden, Fragenden und Antwortenden etc. arbeiten muss. Diese Matrix jedoch, die ein Interesse an Wissenden voraussetze, sei in einer Gesellschaft, in der jeder über „eigenwilliges Wissen“ verfüge, nicht mehr interessant. Abgelehnt werde dabei nicht das Wissen an sich, sondern die Matrix. „Derselbe Vortrag desselben Vortragenden, der an Universitäten eine Handvoll Unentwegter in einen Saal lockt, füllt in einer Galerie oder in einem Kunstmuseum die Säle – einfach deswegen, weil dort die Matrix eine andere ist. Kommunikation im Kontext von Kunst führt eigenwillig Handelnde vor, führt Performanzen vor; Kommunikation im Kontext von Lehre führt eigenwillig Wissende vor. Das eigenwillige Wissen findet jedoch kein Interesse mehr“ (Baecker 2000, S. 57). Baecker zufolge ist ein zentrales Problem der traditionsbeladenen Institution Universität, dass in einer Gesellschaft, in der Wissen zum Allgemeingut geworden ist, der Wissende als Typus uninteressant wird. Dies läge sowohl daran, dass der Glanz des Besonderen nicht mehr vom Wissenschaftler ausgeht (zumindest solange er in der für Wissenschaft vorgesehenen Institution bzw. systemtheoretisch gesprochen im System Wissenschaft auftrete), als auch dass das Wissen entmystifiziert sei. Auf die Behauptung neuer Wahrheiten werde aufgrund der Erfahrung steter Revision mit Skepsis reagiert. Die WissenschaftlerInnen kennen zwar die Zufallsabhängigkeit und die schnellen Verfallszeiten ihrer Leistungen, bauen jedoch – gerade aufgrund der Akzeptanzschwierigkeiten – eine starke Argumentationskette auf, die die Fragilität ihrer Arbeit verschleiere. Für die Lehre habe dies die fatale Folge, dass man darüber erst recht lernt, „den Gründen der anderen zu misstrauen und die eigenen Gründe überprüfungsgerecht zu stilisieren" (Baecker 2000, S. 61).

6.3 Fachkulturen

Fachspezifischer Habitus

Hochschulen sind keine in sich homogenen Gebilde, sondern ausdifferenziert nach Fachkulturen. Mit dem Eintritt in die Hochschule lernen die Studierenden nicht nur ihre fachspezifischen Inhalte, sondern eignen sich auch ein Wissen um den für ihr Fach typischen Habitus an. Man erkennt es häufig auf den ersten Blick. Architekturstudierende und ProfessorInnen kleiden sich anders (häufig schwarz) als Studierende und DozentInnen der Erziehungswissenschaft (besonders leger). Die Cafeteria bei den KulturwissenschaftlerInnen sieht anders aus als bei den JuristInnen (Selbstbedienung versus Service). Die BotanikerInnen schreiben mit billigeren Stiften als JuristInnen. Eine Studie des Ludwig-Uhland-Instituts für Empirische Kulturwissenschaft der Universität Tübingen (1998) zeigt an vielfältigem Bildmaterial die symbolischen Differenzen zwischen den Fächern. Es existieren, meist angeregt durch die Arbeiten von Pierre Bourdieu (vgl. Kapitel 3.1.1), eine Reihe von Untersuchungen zu Fachkulturen an den Universitäten. Steffani Engler (1993) zeigt quantitative Differenzen auf. So wohnen zum Beispiel Studierende der Erziehungs- und der Rechtswissenschaft häufiger in Wohngemeinschaften, aber viel seltener bei den Eltern als Studierende der Elektrotechnik und des Maschinenbaus. ErziehungswissenschaftlerInnen legen ihre Matratze eher auf den Fußboden als Studierende des Maschinenbaus und sie beziehen ihre Möbel öfter vom Flohmarkt oder vom Sperrmüll. Barbara Friebertshäuser (1992) beschäftigt sich ausführlich mit der Initiation in den Studiengang „Erziehungswissenschaft". Sie beschreibt, wie alle Mitglieder des Fachbereichs auf ihre je spezifische Weise, zum Beispiel in der Einführungsveranstaltung, darauf hinwirken, dass Studierende die Geschmackspräferenzen und Bewertungen der Fachkultur kennen lernen. Der Raum, den die Studierenden am ersten Tag ihres Studiums betreten, ist auffällig bemalt und vermittelt so bereits Elemente einer Oppositionskultur. Der Raum ist zu klein für die vielen Studierenden, so dass alle durcheinander, zum Teil auch auf dem Boden, sitzen. Differenzen nach Statusgruppen und Hierarchien werden so verwischt. Die Studierenden lernen bald, dass in der Erziehungswissenschaft mit subtilen Machtstrategien gearbeitet wird. Die Lehrenden präsentieren sich in ihren Selbstdarstellungen als Menschen, die offen für Probleme der Studierenden sind und offenbaren Unsicherheiten. Sie geben Fehler öffentlich zu. So vermitteln sie den Anspruch, das

Initiation

eigene Handeln kritisch zu hinterfragen. Ältere Studierende übernehmen die Detailarbeit der Einführung. In der Orientierungswoche präsentieren sie den Novizen und Novizinnen die Fachkultur ganz alltagsnah. Man organisiert einen Brunch (Vollwertkost; kein Plastikgeschirr), zeigt die Kneipen, in denen ErziehungswissenschaftlerInnen verkehren, spricht über Politik, lädt auch zu sich nach Hause ein (Wohnkultur), trifft sich erst gegen Mittag zum Frühstück (Zeitkultur) (vgl. Friebertshäuser 1992, S. 231ff.).

Studienwahl

Als „ältere Vorbilder" vermitteln Studierende ihren neuen KommilitonInnen die Fachkultur und damit auch Abgrenzungsstrategien zu anderen Fachkulturen. Sie brauchen dazu jedoch nicht völliges Neuland zu betreten. Tatsächlich entscheiden sich bereits sozialisierte Menschen für das Studienfach. Wie Friebertshäuser darlegt, führen spezifische Werte und Erfahrungen spezifische Personengruppen (vielfach Frauen, wenig ökonomisches Kapital) in das Studium der Erziehungswissenschaft. Diese habituelle Prägung wird nun im Studium weiter entwickelt und geformt. Über die Fachkultur wird gleichermaßen interner Zusammenhalt und damit eine Ressource im Konkurrenzkampf der Fächer wie auch spezifische Perspektiven auf die Welt, die mit fachspezifischen Inhalten koalieren, vermittelt. Für eine Erziehungswissenschaftlerin, die eine Kontaktstelle für Drogenabhängige leitet, ist es zum Beispiel notwendiger, bürokratiekritisch zu agieren, als für einen Betriebswirt in der Versicherungsbranche (vgl. zu den Fachkulturen auch Huber 1991).

Fachkulturen

Beate Krais (1996a) vergleicht die sozialen Unterscheidungen und die mental differenten Orientierungen der Fächer Erziehungswissenschaft und Chemie. Über eine dichte Beschreibung der Sozialisation von Studierenden kann sie die feinen Unterschiede der beiden Fachkulturen aufzeigen. Chemie konstituiert sich als soziales Feld über experimentelle Forschung. Die Promotion ist ein überdurchschnittlich häufig angestrebter Abschluss. Das Fach selbst ist klar in Arbeitsbereiche strukturiert und international ausgerichtet. Der wissenschaftliche Alltag von Studierenden und DozentInnen findet nicht nur in der Universität, sondern auch in der Privatwirtschaft statt. Die Grenzen zwischen diesen Bereichen sind demzufolge fließend. Die Forschergemeinschaft ist eine Männerdomäne und klar hierarchisch strukturiert. Ganz anders sind die kulturellen Bedingungen, in die Studierende der Erziehungswissenschaft hineinwachsen. Hier baut das zu erwerbende Wissen kaum auf Versuchen auf. Stattdessen wird ein „Identitätsdiskurs" gepflegt. Damit meint Krais, dass einmal gewonnene Erkenntnisse und scheinbar offensichtliche Alltagserfahrungen stets wieder in Frage gestellt werden. Das Fach ist multidisziplinär ausgerichtet. Verschiedene Paradigmen finden gleichermaßen Anerkennung. Die Abgrenzung der Arbeitsbereiche ist lose strukturiert. Neue Zuschnitte sind jederzeit möglich. Das Fach ist in erster Linie national orientiert. Wie schon Steffani Engler so zeigt auch Beate Krais, dass Studierende völlig unterschiedliche Ausgangsbedingungen und Wissensvorstellungen in ihren jeweiligen Fächern vorfinden, die ihre berufliche Sozialisation prägen werden. Fachkulturen zu untersuchen, heißt demnach, Strukturen der Wissenschaft im Verhältnis zu gesellschaftlichen Strukturen zu betrachten, wie sie erstens durch die Gliederung der Welt in Berufe und zweitens durch die Gliederung der Welt in Schichten/Klassen/Milieus wie Geschlechter existiert (vgl. dazu auch Kapitel 4, Kapitel 5.2.2 und Kapitel 5.3).

6.4 Hochschule und Geschlecht

Wie selbstverständlich geht zum Beispiel Max Weber von der „Wissenschaft als Laufbahn eines jungen Mannes" (Weber 1992, original 1917, S. 3), der „Laufbahn eines Mannes der Wissenschaft" (Weber 1992, original 1917, S. 4) oder der „innere(n) Stellung des Mannes der Wissenschaft (...) zu seinem Beruf" (Weber 1992, original 1917, S. 18) aus. Während der von Studierenden als „Kommilitonen und Kommilitoninnen" (Weber 1992, original 1917, S. 36) spricht, wird der Beruf des Wissenschaftlers ganz selbstverständlich für die Männerwelt reserviert.

Erweiterung der allgemeinen Bildung

Das Zulassen junger Frauen zum Studium ist nach langen Kämpfen zunächst denkbar als eine Erweiterung der allgemeinen Bildung von Frauen oder auch als Ausbildung für bestimmte personenorientierte Berufe (Lehrerin, Ärztin). Die geschlechtsspezifische Grenzziehung verläuft im Übergang zum professionellen wissenschaftlichen Arbeiten. Auch heutigen Statistiken zufolge (vgl. zusammenfassend Krais 2000, S. 11ff.) ist der Übergang zur wissenschaftlichen Karriere – mittlerweile als Sprung auf die Professuren – geschlechtsspezifisch codiert. Trotz eines zu verzeichnenden Anstiegs der Besetzungen von Professuren mit Frauen liegt in den meisten Fächern der Anteil immer noch unter 10%, in den Natur- und Technikwissenschaften, in der Philosophie und Medizin sogar unter 5%. Die Forschungslage zur Situation von Frauen an Universitäten dagegen ist mittlerweile sehr vielfältig (vgl. zusammenfassend Müller 1999; Krais 2000). Hier können also nur einige Thesen zur Erklärung des „Verschwindens" der Frauen auf höheren und hohen Positionen genannt werden. Die wesentlichen sind:

Gabe

- Die historisch erst kurze Anwesenheit der Frauen in den Universitäten führt manche Forscherinnen zu der Annahme, dass Frauen nur selten uneingeschränkt das Gefühl entwickeln, dass die Universität selbstverständlich ihre Wirkungsstätte ist. Marianne Schuller (o.J.) hat die Wissensaneignung von Frauen an der Universität mit einer Gabe verglichen. Frauen haben das Gefühl, das Wissen wie eine Gabe von den männlichen Professoren erhalten zu haben. Nicht die Selbstverständlichkeit des Angeeigneten oder des „Einem-Zustehenden" steht im Vordergrund, sondern die Empfindung von Dankbarkeit, teilhaben zu dürfen. Wissen wird zu einem ganz und gar kostbaren Gut. Dementsprechend labil bleibt die Konstruktion der eigenen Identität. Immer wieder schleicht sich die Angst ein, die Gabe – großzügig überreicht – könnte zurückgefordert werden. Schuller schreibt: „Und manchmal, im Halbschlaf vielleicht, taucht ein undefinierbarer, zerstörerischer Schrecken auf: als würde die sorgsam gehütete und lustvoll ausstaffierte Gabe wieder genommen. Und in der plötzlich aufgerissenen Leere nistet sich eine, die ‚Begabte' noch einmal belastende Schuld ein: als habe sie sich, eine Diebin, verstohlen in den Besitz eines (ihr) anderen gebracht" (Schuller o. J., S. 16).

  Atmosphäre

  Die Männerkultur erzeugt eine Atmosphäre – so eine verbreitete Annahme –, die Ein- und Ausschlüsse organisiert. Durch den subtilen, kaum zu greifenden Charakter von Atmosphären werden die unterschiedlichen Zugriffsmöglichkeiten auf Ressourcen und Positionen in der internen Hierarchie, auf

Forschungsgelder oder auf Wissen um Strukturen verschleiert. Unter sozialen Bedingungen, die Frauen formal Zugang zu allen sozialen Positionen ermöglichen, erscheint dann der Abbruch einer akademischen Laufbahn, die Ablehnung von Forschungsförderanträgen oder auch der nicht erfolgte Ruf auf eine Professur als individuelles Scheitern.

Homosoziale Institutionen –

Anschlussfähig an die erste These sind Untersuchungen zu Hochschulen/ Forschungsinstitutionen als homosoziale Institutionen. Damit gemeint ist, dass im Wissenschaftssystem nach wie vor Männer mit Vorliebe Männer gleichen Alters, gleicher Hautfarbe und gleicher Schicht rekrutieren. Dies zeigt zum Beispiel eine Evaluierung der Deutschen Forschungsgemeinschaft und der Max-Planck-Gesellschaft. Als Ergebnis kann man zusammenfassen, dass Frauen dort marginalisiert werden, wo sie von den Entscheidungen des anderen Geschlechtes abhängig sind, bzw. umgekehrt zeigt sich, dass bei einem einmal erreichten Anteil von Frauen in Leitungspositionen eine Vermehrung des Frauenanteils zu beobachten ist (Allmendinger u.a. 1997). Die beiden schwedischen Forscherinnen Christine Wenneras und Agnes Wold (2000) untersuchen das Peer-Review-Verfahren (Statusgleiche beurteilen Forschungsanträge, Zeitschriftenartikel etc.) in Bezug auf die Geschlechterfrage. Hauptergebnis dieser quantitativen Studie ist, dass Frauen, die sich bei einer Förderinstitution um Forschungsmittel bewerben, auch bei gleichen Voraussetzungen deutlich geringere Chancen auf Zuwendungen haben. Die Autorinnen vergeben Punkte für mögliche Auswahlkriterien. Danach werden Frauen mit 100 Punkten ebenso viele Fördermittel zugesprochen wie Männern mit nur 20. Das heißt, Frauen müssen fünf Mal so gut sein wie Männer, um berücksichtigt zu werden. Auch bei der Rezeption wissenschaftlicher Schriften lässt sich belegen, dass Arbeiten von Frauen seltener gelesen werden als von Männern. Ist das Geschlecht des Autors/der Autorin unbekannt, so wird das Werk besser beurteilt, wenn unterstellt wird, dass es von einem Mann geschrieben wurde (vgl. Müller 1999, S. 148).

Angst vor Prestigeverlust –

Strukturtheoretisch betrachtet, werden Gründe für die Schließung der Universität gegenüber Frauen in der Furcht vor Konkurrenz gesucht. Bereits in historischen Untersuchungen lässt sich zeigen, dass der Kampf um die Zulassung zum Hochschulstudium vielfach auch ein Kampf um die knappen Ressourcen und eine Angst vor Prestigeverlust ist (vgl. Hausen 1986). Angelika Wetterer zeigt in ihren umfassenden Studien zur Professionalisierung (z.B. 1993, vgl. auch Kapitel 5.3), dass eine Öffnung der Institution Hochschule für Frauen stets nur eine Öffnung für einzelne Segmente (erst Studium, dann Mittelbau) ist, welcher stets vorausgeht, dass sich Männer bereits in anderen Bereichen alleinige Zugriffsmöglichkeiten gesichert haben. Es wäre in dieser Perspektive deshalb durchaus plausibel, wenn das zunehmende Vordringen von Frauen in die akademischen Spitzen mit einer gesellschaftlichen Abwertung der Universität einherginge, ein Verdacht, den derzeitige Hochschulreformen durchaus nahe legen (Angleichung von Fachhochschule und Universität, geringere Bezahlung der ProfessorInnen etc.). Das hieße, Frauen erreichten ihr Ziel, während eine bürgerliche Männermehrheit sich bereits wieder besser bezahlte und prestigeträchtigere Positionen gesichert hat.

Aufgrund der vielfältigen Befunde zum Sexismus an den Hochschulen betonen Wissenschaftlerinnen immer wieder, „dass Organisationen, in denen Wissenschaft betrieben wird, nicht geschlechtsneutral sind“ (Krais 2000, S. 20; vgl. dazu auch Wetterer 1992; 1995). Dennoch ziehen es viele Frauen und Männer im Studium wie in der Forschung vor, sich als individuell nach Leistung bewertete Menschen zu imaginieren (vgl. Engler 2001, dargestellt in Absatz 6.2). Universitäten bauen ferner auf geschlechtsspezifischer Arbeitsteilung auf. Neben offenen und versteckten Diskriminierungen versuchen Frauen stärker als Männer in einer Institution, die in ihrem Zeitmanagement und in dem geforderten Engagement, auch in der unterstellten Mobilität wenig auf die Vereinbarkeit von Familie und Beruf Rücksicht nimmt, beides in Einklang zu bringen (vgl. z.B. Zukkermann u.a. 1991; Bauer u.a. 1993). Männer können sich nach wie vor stärker auf Reproduktionsunterstützung durch (Ehe)frauen verlassen (vgl. dazu auch Bochow/Joas 1987). Sexismus

6.5 Hochschulbildung durch das elektronische Netz

Die Ursprünge des heutigen Internets gehen unter anderem auf die Arbeiten der Advanced Research Projects Agency, die 1958 gegründet wurde, zurück. Motiviert durch die Sorge der US-Airforce, dass im Kriegsfall die Datenübertragungen anfällig für Zerstörung sind, wird von Paul Baran 1964 eine Netzwerk-Topologie vorgeschlagen. Es handelt sich um ein „distributed network“, in dem jeder Computer bzw. jedes Subnetzwerk mit möglichst vielen weiteren Computernetzen oder einzelnen Computern verbunden ist. Damit ist eine zentrale Steuerung nicht mehr notwenig und das gesamte System äußeren Einflüssen gegenüber sehr robust. Entstehungsgeschichte

Vom heutigen Standpunkt aus betrachtet leistet das elektronische Netz diese Sicherungsfunktion nur bedingt. Obwohl das Internetprotokoll darauf ausgelegt ist, Hindernisse, zum Beispiel nicht antwortende Server oder zusammengebrochene Datenleitungen, selbstständig zu umgehen, ist das Internet heutzutage an einigen neuralgischen Punkten sehr verletzbar: Der Mythos der Unverwüstbarkeit des elektronischen Netzes setzt voraus, dass ein sehr homogenes Netzwerk gebildet wird. Die tatsächliche Infrastruktur ist aber stark inhomogen: Einzelne Bereiche (Zugangssysteme, „Vermittlungsstellen“ zwischen Firmen/Ländern/Kontinenten) sind stärker vernetzt als andere. Physiker der University of Notre Dame haben die Gefahr des Auseinanderbrechens des Internets in voneinander vollständig getrennte Netzwerke („Inseln“) anhand von Computersimulationen bei einem Ausfall von nur vier Prozent der am stärksten vernetzten Knotenpunkte vorausgesagt (vgl. Rötzer 2000).

Dennoch hat das elektronische Netz als Informationsquelle und als Medium der Freizeitgestaltung (Spiele, Chats etc.) wichtige gesellschaftliche Bedeutung erhalten. Sein Wert für Bildungsprozesse ist jedoch sehr umstritten und soziologische Literatur zum elektronischen Netz ist zwar in allgemeiner Form erhältlich (vgl. Shields 1996; Turkle 1998; Maresch 2001; Funken/Löw 2002; Briggs 2002), jedoch selten zu Bildungsprozessen (Ausnahme aus eher erziehungswissenschaftlicher Perspektive z.B. Marotzki 1997; Gogolin/Lenzen 1999; Marotz- E-Learning ki/Meister/Sander 2000). Es existiert stattdessen ein umfassender Markt an kurzen Texten zum Lehren und Lernen, vor allem an Hochschulen, mittels internetgestützter Medien, in deren Mittelpunkt Überlegungen entweder zu technischen Fragen der Realisierbarkeit oder zur Didaktik stehen. Viele Schlagworte wie E-Learning, Telelearning, IT-gestütztes Lernen, LAA (Learning Anytime Anywhere), Web Based Education, Virtuelles Lernen erschweren allerdings den Zugang zu Informationen über die tatsächliche Rolle oder zukünftige Potentiale neuer Medien in der Bildung.

Weltweite Vorlesungen

Bildungspotentiale mittels elektronischer Netze werden an erster Stelle in dem asynchronen Zugriff auf Vorlesungen gesehen. Berufsbegleitende Weiterbildung oder die Option, weltweit sich Vorlesungen anzuhören, seien die wichtigsten Anwendungsfelder. Die Digitaltechnik im Hörsaal könne auch dann sinnvoll werden, wenn zum Beispiel der Umgang mit Statistiksoftware demonstriert wird oder Filmausschnitte gezeigt und analysiert (schneller Wechsel zwischen Szenen möglich) oder wenn im Rahmen einer Übung die Online-Untersuchungen entwickelt werden sollen.

Vorlesungskino

Tatsächlich beschränkt sich die Nutzung „neuer Medien" häufig jedoch auf ins Internet gestellte Vorlesungsskripte, die überdies mit kostenpflichtiger Standardsoftware („Powerpoint") erstellt werden und sich in ihrer didaktischen Konzeption von „gefürchteten Monologen am Overheadprojektor" (Loviscach 2001, S. 109) nicht wesentlich unterscheiden. Deren Transformation ins Multimediazeitalter mit Hilfe von Videoprojektoren (Beamer) bringt die Gefahr mit sich, so eine häufig geäußerte Kritik, dass der Hörsaal zum „Vorlesungskino" mutiert. Während der Präsentation ist es schwer, das Material zu verändern, um zum Beispiel auf Zwischen- oder Verständnisfragen einzugehen; das Durchspielen von verschiedenen Lösungsansätzen oder die adhoc-Diskussion alternativer Erklärungsansätze ist mit herkömmlichem Material immer noch einfacher. Komplexe Diagramme oder Theorien lassen sich oft besser verstehen, wenn sie direkt an der Tafel entwickelt und nicht auf Knopfdruck abgerufen werden bzw. lediglich die Legende vorgelesen wird. Ganze Vorlesungen fertig aufbereitet, per Mausklick abspulen zu können, verleitet dazu, die Aktualität des Materials aus den Augen zu verlieren. Hinzu kommt, dass der Aufwand dafür, eine komplette Vorlesung abgefilmt ins Internet zu stellen, immens groß ist: Die Kamera muss dem Dozenten/der Dozentin folgen, der Ton muss nachgeregelt werden und die Aufnahmen hinterher geschnitten werden: Versprecher, Fehler oder „Denkpausen" wirken beim späteren Abrufen über das Internet nicht nur unprofessionell, sondern sind je nach Zugangsmodell auch teuer. Werden neue Medien in diesem Sinne eingesetzt, so verlagert sich praktisch der Ort des Lernens vom Hörsaal (ob real oder virtuell), der dann primär nur noch der Präsentation des zu lernenden Stoffes dient, auf den Privatbereich des Studierenden. Damit verstärkt sich die Befürchtung, dass der öffentliche Raum durch neue Medien an Bedeutung verliert (vgl. Turkle 1998).

Bedeutungsverlust des öffentlichen Raumes

Hemmnisse anderer Art liegen darin, dass Live-Übertragungen oder die filmische Archivierung von Vorlesungen eine zuvor in dieser Weise nicht existente psychologische Barriere schaffen. Rückfragen an die Studierenden werden ebenso wenig beantwortet wie Zwischenfragen gestellt. Schließlich möchte niemand seine vermeintlich „dumme Frage" oder eine falsche Antwort im Internet verewigt wissen: Weltweit abrufbar in Farbe und Stereo.

Ein Schwachpunkt der Nutzung neuer Medien bei der Ausbildung von Studierenden ist eng mit dem viel gepriesenen Vorteil der zeit- und räumlichen Unabhängigkeit verbunden. Der Internetzugang, den Studierende zur Nutzung von Videostreaming-Angeboten oder der Möglichkeit, Filmsequenzen der jeweiligen Vorlesung herunter zu laden, benötigen, muss ein Breitband- bzw. Hochgeschwindigkeitszugang sein. Die Gefahr einer neuen sozialen Kluft liegt auf der Hand, denn nicht alle Studierenden können sich die technischen Lösungen leisten.

Finanzielle Investitionen

Hier soll durch die Anbindung des gesamten Campus-Geländes an ein Funknetz entgegengewirkt werden. So können StudentInnen multimediale Studienangebote überall innerhalb des Universitätsgeländes kostenlos wahrnehmen, allerdings wird die Illusion der räumlichen Unabhängigkeit multimedialer Bildung im Internetzeitalter durch die Präsenz an der Universität beschränkt. Auch bei diesem Ansatz ist eine Anfangsinvestition in eine entsprechende Computerausstattung vonnöten.

Medienkompetenzen

Aufgrund der vielfältigen Schwierigkeiten läuft es meistens darauf hinaus, dass kurze Skripte und Folien zu den jeweiligen Themen der Vorlesung bzw. des Seminars im Internet verfügbar gemacht werden. Dies liegt auch an den Medienkompetenzen der Akteure. Eine im August 2001 vorgelegte Studie der Sozialforschungsstelle Dortmund im Auftrag des Bundesministeriums für Bildung und Forschung (Klatt u.a. 2001) vermittelt einen guten Einblick in die Anwendungsformen elektronischer Medien an deutschen Hochschulen. In einer repräsentativen Untersuchung wurden Dekanate, HochschullehrerInnen und Studierende zur Nutzung elektronischer Fachinformationen in der Hochschulausbildung befragt. Dabei soll die Informationskompetenz der Studierenden und der HochschullehrerInnen ermittelt sowie Potentiale und Hemmnisse der Nutzung elektronischer Fachinformationen ausgelotet werden. Die Ergebnisse der Studie sind ernüchternd: Obwohl 73% der Studierenden über einen Computer mit Internetanschluss verfügen (Erziehungswissenschaften nur 58%), ist die Kompetenz der Studierenden im Umgang mit neuen Medien bei der Recherche nach Fachinformationen nur sehr wenig ausgeprägt. Der wichtigste Zugang zu wissenschaftlicher Literatur ist nach Ansicht der Studierenden der Wissensaustausch mit KommilitonInnen (60%) und die Anfrage beim jeweiligen Dozent/Dozentin (42%). Recherchen über das Internet oder den lokalen Bibliotheksserver werden an zweiter bzw. dritter Stelle genannt, wobei die Suchstrategien mit Hilfe des Internets alles andere als effizient sind. Es werden allgemeine Suchmaschinen und kaum spezielle Fach- oder Themenangebote genutzt. Entsprechend unzufrieden sind die befragten Studierenden dann auch mit den Ergebnissen ihrer elektronischen Recherchen: Nur 21% schätzen diese als zuverlässig ein, 12% als systematisch und 37% beurteilen die gefundenen Informationen als aktuell (vgl. Klatt u.a. 2001, S. 139ff.).

Auch die Kenntnisse zur Nutzung der neuen Medien im Rahmen des Studiums wurden unsystematisch erworben. Die überwiegende Mehrheit hat diese Kenntnisse durch Versuch und Irrtum erworben (knapp 80%), lediglich eine Minderheit von fast 16% durch Betreuungspersonal der jeweiligen Bibliotheken (vgl. Klatt u.a. 2001, S. 135). Die autodidaktische Vermittlung entsprechender Kompetenzen halten allerdings nur rund 25 Prozent der HochschullehrerInnen für geeignet und plädieren eher für entsprechende Einführungsveranstaltungen der Fakultät oder Universität (57%) (vgl. Klatt u.a. 2001, S. 180). Hierdurch erlangen aber ihren eigenen Aussagen zufolge nur rund ein Achtel der Studierenden ihre Kenntnisse im Bereich der Informationsbeschaffung mit neuen Medien.

Die Haupthemmnisse bei der Nutzung neuer Medien im Rahmen des Studiums sind analog der unsystematischen Aneignung entsprechender Suchstrategien die empfundene Unübersichtlichkeit des Angebots und die Einschätzung der Qualität.

7. Bildung in lebensgeschichtlicher Perspektive

Obwohl wesentliche Bildungsprozesse im Erwachsenenleben verlaufen, basiert doch die Bildungs- und Erziehungsidee hauptsächlich auf der Vorstellung, das unmündige Kind müsse durch pädagogische Begleitung und durch Bildung in die Mündigkeit geführt werden. Im ausgehenden 18. Jahrhundert (vgl. Kapitel 1.1) profiliert sich das Bürgertum mit einem emphatischen Bildungsbegriff und der Idee der Formbarkeit des individuellen Lebens. Es entstehen eigene Lebensphasen, die speziell an die Idee des Bildungserwerbs geknüpft sind, nämlich Kindheit und Jugend.

Institutionalisierung des Lebenslaufs

Martin Kohli (1985) zufolge kommt es allmählich zu einer „Institutionalisierung des Lebenslaufs". Damit gemeint ist eine allgemein gültige Einteilung der Lebenszeit in eine Vorbereitungsphase auf die Erwerbsarbeit, eine Zeit der Produktionstätigkeit und eine Altersphase auf Rentenbasis. Die Lebensphase Kindheit/Jugend, Erwachsenenalter und Alter entsprechen diesen Phasen. „Lebenslauf als Institution bedeutet also zum einen die Regelung des sequentiellen Ablaufs des Lebens, zum anderen die Strukturierung der lebensweltlichen Horizonte bzw. Wissensbestände, innerhalb derer die Individuen sich orientieren und ihre Handlungen planen" (Kohli 1985, S. 3). Es kommt demnach zu einer Verzeitlichung des Lebens. Alter wird zu einer strukturierenden Kategorie. Lebens-, auch Bildungsentscheidungen werden vom Alter abhängig gemacht. Diese Verzeitlichung wirkt sowohl als äußere gesellschaftliche Strukturierung als auch als biographische Orientierung. Die Institutionalisierung des Lebenslaufs ist – so Kohli – um die Erwerbsarbeit zentriert. Die Erwerbstätigen sind die „vollwertigen" Mitglieder der Gesellschaft. Kinder und Jugendliche befinden sich in über Bildung organisierten Vorbereitungsphasen, die Alten werden als jene wahrgenommen, die nicht mehr in das Erwerbsarbeitssystem integriert werden können.

Normalbiographie

Denkt man die von Martin Kohli dargelegte sequenzielle Abfolge des Lebenslaufs als real existierende Abfolge für jeden Menschen, so stößt man schnell auf Gegenbeispiele. Man denke allein an die Mehrheit der Frauen, die entweder nicht erwerbstätig sind oder in viel mehr Sequenzen mit Erziehungsphasen und Wiedereingliederung in den Beruf ihr Leben organisieren (vgl. auch Hoerning/Krais 1987 zum zweiten Bildungsweg). Dennoch hat diese Vorstellung von einer Normalbiographie das gesellschaftliche Handeln maßgeblich geprägt. Sie liegt dem Rentensystem ebenso zugrunde wie dem Schulsystem und formt die (Selbst)Wahrnehmung der Menschen als Erwachsene, Kinder etc. Der Lebenslauf, im Sinne einer sequenziellen Abfolge, wie sie Kohli beschreibt, ist eine soziale Konstruktion und als solche strukturiert sie die soziale Wirklichkeit.

Normatives Ideal

Nur vor diesem Hintergrund kann heute von einer Destrukturierung des Lebenslaufs gesprochen werden. Denn einerseits folgen immer mehr Menschen durch Arbeitslosigkeit, Verzögerung des Arbeitsbeginns, später Mutterschaft und aktiver Altenarbeit nicht der Triade Ausbildung, Berufstätigkeit, Ruhestand. Andererseits verläuft die Abweichung vor dem Hintergrund einer erfolgreichen Institutionalisierung des Lebenslaufs: Frauen, die Mitte 40 das erste Kind bekommen, wissen, dass sie „spät" ein Kind in die Welt setzen. Die Arbeitslosenpolitik funktioniert vor dem Hintergrund der Annahme, dass man in der mittleren Lebensphase normalerweise arbeitet. Insofern hat Kohli Recht, wenn er davon ausgeht, dass die erfolgreiche Institutionalisierung einer Normalitätsvorstellung vom lebensgeschichtlichen Ablauf auch die Möglichkeit bietet, sich individualisierend davon abzusetzen (vgl. Kohli 1986). Wichtig ist es jedoch, nicht das normative Ideal der Normalbiographie und die gelebte Praxis gleichzusetzen, weil dann zum Beispiel die doppelte Lebensführung von Frauen in Familie und Beruf aus dem Blickwinkel gerät. Die Ausgestaltung der Lebensphasen ist vielfältiger geworden, die Länge der Lebensphasen wird immer wieder neu bestimmt, die Fixierung auf Erwerbsarbeit zur Bestimmung des Lebenslaufs wird vielfach in Frage gestellt, die ausschließliche Bindung von Bildung an Kindheit und Jugend löst sich auf, aber die Idee vom sequentiellen Lebenslauf bietet nach wie vor die Folie, vor der das Leben organisiert wird.

Altersdifferenzierung

Soziologisch bedeutet dies, dass Kindheit und Jugend nicht mehr angemessen als Übergangsphasen oder Entwicklungsaufgaben beschrieben werden können, sondern als eigene Lebensphasen ernst genommen werden müssen (vgl. Zinnecker 1991; Honig 1999). Diese eigene Lebensphase verweist nicht nur auf die biographische Ausgestaltung und damit auf die altersgebundenen Handlungen der Individuen, sondern – gesellschaftsstrukturell gedacht – auf den Tatbestand, dass Kindheit und Jugend zu Elementen der Sozialstruktur werden (zuerst beobachtet: Qvortrup 1985; 1993). Moderne Gesellschaften lassen sich als altersdifferenzierte Gebilde betrachten. Über das chronologische Lebensalter werden Zugangschancen und -zwänge organisiert. Alterstufen bilden einen differenzierenden Komplex rechtlich abgesicherter Handlungserwartungen. Michael Honig betont daher: „Die Pointe der neueren sozialwissenschaftlichen Kindheitsforschung steckt jedoch in ihrer Fragestellung: Sie ist an Kindern als ‚Personen aus eigenem Recht' interessiert, nicht an zukünftigen Erwachsenen. Sie bewegt sich also an den Grenzen der Erziehungskindheit und stellt ihre Selbstverständlichkeit in Frage" (Honig 1999, S. 158).

Kinder als Akteure

Kinder und Jugendliche werden als eigenständige KonsumentInnengruppe angesprochen; sie entwickeln eine Spiel- und Medienkultur, die nicht einfach nur auf die Erwachsenenwelt vorbereitet, sondern unabhängig von den Praktiken der Erwachsenen eine eigene Erfahrungswelt beinhaltet. Zu fragen ist auch, ob sich die Konstruktion von Kindheit oder Jugend, der Kinder/Jugendliche folgen, nicht systematisch von den Konstruktionen Erwachsener unterscheidet. Daraus folgt methodisch, dass nicht nur Jugendliche, sondern zunehmend auch Kinder als Akteure in der Forschung betrachtet werden, welche nicht ohne weiteres allein über Interviews mit Erwachsenen, zum Beispiel Eltern, erforscht werden können. Kinder werden aufgefordert, selbst über ihr Leben, ihre Erfahrungen und vor allem auch über ihren eigenen Entwurf von Kindheit zu berichten (zu den Methoden vgl. Heinzel 2000).

Von entscheidender Bedeutung für die Soziologie der Bildung und Erziehung ist es also, die Sequenzialisierung des Lebenslaufs in den Blick zu nehmen. Aus diesem gesellschaftlichen Deutungsmuster folgt, dass Kindheit und Jugend zwar nicht (mehr) als ausschließliche, aber doch als privilegierte Phasen für Bildung und Erziehung gelten, denen deshalb besondere Forschungsaufmerksamkeit gewidmet wird, wie im Folgenden noch zu erläutern ist.

7.1 Kindheit

Entdeckung der Kindheit

Die heutige Vorstellung von Kindheit als Lebensphase, als eigene Welt neben der Erwachsenenwelt, ist historisch neuartig. Phillipe Ariès (1975, original 1960) zeigt in seiner Geschichte der Kindheit, dass noch im 16. und frühen 17. Jahrhundert Kinder nicht in pädagogisch kontrollierten Institutionen aufwachsen, sondern wie kleine (defizitäre) Erwachsene behandelt werden, die – indem sie in alle Lebensbereiche der Erwachsenen integriert werden – in die Erwachsenenanforderungen einsozialisiert werden. Erst im 17./18. Jahrhundert setzt ein Prozess ein, der als „Entdeckung der Kindheit" bezeichnet wird. Hiermit ausgedrückt wird nicht, dass es vorher überhaupt keine Übergangsphase zwischen Kindheit und Erwachsenenalter bzw. keine altersgebundenen Rechte und Pflichten gab (vgl. Hardach-Pinke/Hardach 1981, S. 10), sondern dass sich mit dem Übergang von der feudalen zur bürgerlichen Gesellschaft die Einstellung zu Kindern gravierend verändert hat. Im Übergang vom Mittelalter zur Neuzeit entsteht die Wahrnehmung von der Besonderheit der Kinder. Ariès zufolge hatte sich die Bildungsidee der griechischen Antike bis zum Mittelalter völlig verloren (Ariès 1975, S. 559) und entstand erst im Zuge der Industrialisierung und Urbanisierung und damit der Entstehung der bürgerlichen Gesellschaft neu. In der feudalen Gesellschaft wuchs die große Mehrheit der überwiegend aus ländlichen Kontexten stammenden Kinder in der Erwachsenenwelt auf, ohne speziell an schulischen oder außerschulischen Bildungsangeboten zu partizipieren. Lediglich einem kleinen Kreis im Kontext der Kirche stand eine Ausbildung in Fächern wie Grammatik, Rhetorik, Arithmetik, Geometrie, Musik oder Astronomie offen. In weltlichen Kreisen, zum Beispiel im Ritterstand, wurde eher auf körperliche Fähigkeiten in der Ausbildung gesetzt (vgl. Blankertz 1982).

Ganzes Haus

Die Masse der Bevölkerung lebt jedoch, wie Heidi Rosenbaum (1990[5], original 1982) aufzeigt, als Bauern, Dorfhandwerker und Heimarbeiter auf dem Land in der Familienform des „ganzen Hauses". „So hat Brunner darauf hingewiesen, dass der Begriff ‚Familie' in Deutschland erst im Laufe des 18. Jahrhunderts in der Umgangssprache geläufig wurde. Vorher habe man vom ‚Haus' gesprochen, in dem durch die Einheit von Haushalt und Produktion, wie sie prototypisch bei Bauern und Handwerkern zu finden war, sachliche und Gefühlsbeziehungen eine Einheit gebildet hätten. Erst mit der Lösung der Produktion aus dem ‚Haus' im Zuge der kapitalistischen Entwicklung und mit der Sonderung eines Bereiches ‚Wirtschaft' und eines Bereiches ‚Familie' sei die Rationalität der Wirtschaft, das Gefühl der Familie zugeordnet worden" (Rosenbaum 1990[5],

S. 30f.). Im „ganzen Haus" lebten neben Verwandten auch Lehrlinge, Gesellen, Knechte und Mägde in einer Gemeinschaft. Die „Familie", wie wir heute sagen, bildete sich nicht allein auf der Grundlage von Blutsverwandtschaft, sondern als Arbeitsgemeinschaft unter patriarchaler Leitung durch den Hausvater. „Der Hausvater verfügt über eine nahezu absolute Autoritätsposition gegenüber den Kindern und dem Gesinde, aber auch gegenüber der Ehefrau, obwohl letztere durchaus auch eigene Arbeits- und Autoritätsbereiche hat. Aber nur der Mann und Hausvater ist rechtlich und politisch handlungsfähig in Bezug auf die Mitglieder seines Hauses und nur ihm steht die Ausübung des körperlichen Züchtigungsrechtes zu" (Büchner 1985, S. 19).

Binäre Gesellschaftsordnung

Ausgelöst durch Veränderungen in der Produktionsweise entstand eine binäre Gesellschaftsordnung, welche die Erwerbsarbeit als kalte Welt, die auf ökonomischen Kalkül, Entlohnung und Entfremdung basiere, der Familie als emotional-warmer Welt, die auf Gefühlen, unbezahlter Arbeit und sozialer Nähe Gründe, entgegenstellt. Die als Privatheit gedachte Familie wird als komplementär zur als Öffentlichkeit konzipierten Erwerbsarbeit entworfen und Zuständigkeiten werden geschlechtsspezifisch verteilt (Terlinden 1990). Die Frauen übernehmen die Haus- und Gefühlsarbeit, die Männer die außerhäusliche Erwerbsarbeit. Um diese Entwicklung sozial zu legitimieren, entsteht das ideologische Konstrukt der Geschlechtscharaktere (Hausen 1976). Männer und Frauen werden als grundsätzlich verschiedene, sich aber komplementär ergänzende Wesen beschrieben. Während dem Mann (genauso wie der Erwerbsarbeit) Rationalität zugeschrieben wird, gilt die Frau als emotional. Der Mann scheint natürlich die für das öffentliche Leben notwendigen Eigenschaften wie Tapferkeit, Willenskraft und Selbständigkeit mitzubringen, die Frau als sein Pendant wird dementsprechend als bescheiden, wankelmütig und anhänglich beschrieben.

Bürgerliche Familie

Gleichzeitig mit der Entstehung der bürgerlichen Familie als einen „Ort unabdingbarer affektiver Verbundenheit" (Ariès 1975, S. 48) unter der Zuständigkeit von Frauen wird das Kind als zu Erziehendes neu entdeckt. Die Frau gewinnt – und das muss als Grund für ihre Akzeptanz der geschlechtsspezifischen Arbeitsteilung betrachtet werden – die Aufgabe der familiären Kindererziehung. Dieses Modell der bürgerlichen Familie gilt zunächst nur für jene, deren Einkommensverhältnisse eine Emotionalisierung und eine Freisetzung der Frauen aus der Erwerbsarbeit zulassen. Es entwickelte sich jedoch zunehmend zum normativen Ideal, dem auch proletarische Familien nachstreben. Die familiäre Erziehung durch die Mütter wird ergänzt durch die Institutionalisierung öffentlicher Bildung in den Schulen. Mit der allmählichen Verbreitung des bürgerlichen Erziehungsgedankens entstehen erste Einrichtungen der „öffentlichen" Kleinkinderziehung, zunächst zur Kompensation von Erziehungsschwächen im proletarischen Milieu (vgl. Kapitel 5.1). Kindheit entsteht als eigene Lebensphase, indem Kinder als vollwertige und im Vergleich zu Erwachsenen andersartige Wesen wahrgenommen werden.

Lebensbedingungen von Kindern

Die wesentlichen Folgen dieser Entwicklung für die Lebensbedingungen von Kindern sind:

– Kinder werden von Arbeitsprozessen freigestellt. Dies hat zur Folge, dass sie Zeit gewinnen, um sich Bildungsprozessen zu widmen, aber auch dass sie aus der Erwachsenenwelt ausgegrenzt werden. Eine eigene Kinderwelt entsteht.

- Kindheit wird als Schutzraum konzipiert, wobei die Einschätzung, was gut für Kinder ist, stark variiert. Schläge und strenge Disziplin – Erziehungsmaßnahmen, die auch als „schwarze Pädagogik" bezeichnet werden (vgl. z.B. Rutschky 1977; Miller 1983), – werden nicht als Widerspruch zur Schutzidee erlebt.
- Kinder dienen zunehmend weniger der finanziellen Sicherheit des Haushalts, übernehmen durch die Emotionalisierung der Beziehungen jedoch zunehmend die Aufgabe, als „Wunschkind" Sehnsüchte der Eltern zu erfüllen.
- Man investiert in Kinder, damit sie als Erwachsene gesellschaftlich notwendige Aufgaben übernehmen; der Status von Kindern in der Kultur ist jedoch gering.

Zivilisatorischer Fortschritt

Insgesamt wird die „Entstehung der Kindheit" in der Forschung sehr widersprüchlich beurteilt. Während Phillipe Ariès stark die Einschränkungen beschreibt, die mit der Kindheit als eigener Lebensphase einhergehen, zum Beispiel strengere Disziplinierung, Kontrolle, Schulzwang und pädagogische Dressur, betont Lloyd deMause (1989[6], original 1974) den zivilisatorischen Fortschritt, der damit einhergeht, dass die Erziehungsnotwendigkeit von Kindern erkannt wird. Kindestötungen – so deMause – nehmen ab und die Willkür durch Disziplin und Aufmerksamkeit wird eingedämmt. Auch methodische Kritik wird häufig formuliert (exemplarisch Arnold 1980; Pollock 1983; Shahar 1991). All die Kritiken bezweifeln jedoch nicht die zentrale theoretische Einsicht der Arièsschen Studie, dass Kindheit ein sich historisch wandelndes soziales Phänomen ist.

Der drastische und bis heute anhaltende Geburtenrückgang verstärkt die Aufmerksamkeit, die den ein bis zwei Kindern, die heute in einer Familie aufwachsen, zukommt, immer mehr. Bildung – und damit die Hauptadressaten der Bildung, nämlich die Kinder – gelten als fundamentale Investition in die Zukunft. Insofern ist kein Ende der spezifischen Lebensphase abzusehen. Ihre Bedeutung für Erziehungs- und Bildungsprozesse hält an. Dennoch sind die Lebensbedingungen von Kindern heute gänzlich andere als noch vor 100 oder 200 Jahren. Für Deutschland wird ein gravierender Einschnitt mit dem Ende des Zweiten Weltkriegs konstatiert.

Nachkriegszeit

In den 1950er und 1960er Jahren erlebt (West)-Deutschland einen wirtschaftlichen Aufschwung, der Kindern einen exzessiven Konsum ermöglicht (vgl. Büchner 1985, S.71ff.). Noch heute ist Deutschland eines der reichsten Länder der Welt. Damit dies so bleibt, gilt Bildung als förderungswürdiges Gut (auch wenn sich das nicht immer in finanziellen Investitionen des Staates niederschlägt). Spätestens seit den 1970er Jahren in Westdeutschland und seit den 1990er Jahren in ganz Deutschland setzt sich ein individualisiertes Konzept kindlicher Förderung durch. Spezielle Fähigkeiten werden – soweit es das finanzielle Budget der Eltern erlaubt – in der Freizeit ergänzend zur Schule gefördert. Kinder beginnen an zahlreichen Kursen von Musikunterricht über Sport bis hin zu Internetkursen teilzunehmen. Darüber lernen sie ein gezieltes Zeitmanagement (vgl. Büchner 1985) sowie neue Muster der Konstitution des Raums (vgl. das folgende Kapitel).

Wertewandel

Die Erfahrungen mit dem Nationalsozialismus und im Zweiten Weltkrieg verändern aber auch die Werte der Gesellschaft. Die Erziehung zu blindem Gehorsam mit Prügelstrafe und unnachsichtiger Strenge wird für die blinde Gefolgschaft unter Adolf Hitler mitverantwortlich gemacht (vgl. Adorno in Kapitel 2.2.3). Kindern wird nun ein Mitspracherecht und langsam auch ein eigener Wille eingeräumt (Fuchs 1983). Da eine Demokratie auf Aushandlung und Abstimmung basiert, müssen Kinder die Fähigkeiten des Verhandelns, des Vertretens eigener Interessen und das Akzeptieren von Mehrheitsvoten erlernen. Die Familie wandelt sich vom „Befehls- zum Verhandlungshaushalt" (Du Bois-Reymond 1991; vgl. zu Ostdeutschland auch Krüger/Haak/Musiol 1993).

Biographisierung der Kindheit

Aufgrund dieser Veränderungen gehen viele AutorInnen von einer „Biographisierung der Kindheit" aus (Krüger u.a. 1994; Büchner u.a. 1994). Damit gemeint ist, dass Kinder heute nicht mehr traditionelle Milieus vorfinden, in die sie zwangsläufig hineinwachsen, sondern in einem reflexiven Umgang Alltag und Lebensführung selbst gestalten müssen. Zwar „wählen" Kinder häufig milieuspezifische Hobbys, geschlechtsspezifische Interessen usw., aber diese sind nicht vorgeschrieben und unterliegen scheinbar der eigenen Verantwortung. Gleichzeitig ist die Vielfalt der Angebote größer geworden, so dass zum Beispiel selbst bei milieutypischen Hobbys noch Wahlanforderungen gestellt werden.

7.2 Jugend

Feudale Gesellschaft

Historisch lassen sich in vielen Gesellschaften institutionalisierte Formen und Riten finden, die den Übergang in das Erwachsenenleben symbolisch organisieren. Doch wie auch bei der neuerlichen Etablierung der Kindheit geht man in der Jugendforschung ebenfalls davon aus, dass der Jugend in einer so hochkomplexen Gesellschaft wie der modernen bürgerlich-kapitalistischen Gesellschaft eine besondere Bedeutung zukommt (vgl. zusammenfassend Mitterauer 1986). In der feudalen Gesellschaft unterscheidet sich die Lebenslage junger Erwachsener nach Schicht und Stand. Die Mehrheit der ländlichen Bevölkerung initiiert den Übergang nach subjektiver Einschätzung der Erwachsenen, indem ab einem bestimmten Zeitpunkt die Kinder an allen wesentlichen Praktiken der Erwachsenen teilhaben. Die Darbietung der ersten Pfeife wird zum Beispiel bei Jungen als nun erfolgte Akzeptanz als Erwachsener interpretiert. Ein komplexes Verständnis von Jugend fehlt.

Flegeljahre

Ideengeschichtlich beginnt im 18. Jahrhundert mit dem Erziehungs- und Bildungsgedanken auch die Vorstellung von der Jugend als besonderer Lebensphase (Hornstein 1966; Dudek 1992[2]). Der Begriff der „Flegeljahre", der Ende des 18. Jahrhunderts gebräuchlich wird, zeigt, dass die Erprobungsphase der Jugend zunächst nur für Jungen gilt. Der männliche christliche Jüngling, bürgerlicher Herkunft und gymnasial ausgebildet, bedarf eines sozialen Moratoriums, so die Vorstellung, in dem er Zukunftspläne entwickeln, sexuell reifen und die Übernahme von Verantwortung lernen kann (vgl. z.B. Hurrelmann 1993). Im 19. Jahrhundert weitet sich diese Vorstellung auch auf die Jungen „aus dem Volk" aus (Roth 1983). Erst Ende des 19. Jahrhunderts entsteht mit dem „Backfisch" ein romantisierendes weiblich konnotiertes Jugendbild.

In der gleichen Zeit setzt sich der Begriff des Jugendlichen in der Semantik der Juristen durch. Die „Flegeljahre" und mit ihnen die potentiell kriminellen Handlungen der Jugendlichen werden nach und nach institutionalisiert in einer eigens für Jugendliche gültigen Rechtsprechung. Eine staatliche Jugendpolitik entsteht. Die Jugendphase etabliert sich als eine für die Gesellschaft notwendige Phase, um den komplexen Anforderungen des Erwachsenendaseins zu begegnen. Die Verinnerlichung der Zwänge der Erwerbsarbeit sowie die emotionale Bindung auf Lebenszeit mit Treueerwartung in der Familie scheint für Männer nur dann realisierbar zu sein – so die soziale Konstruktion – wenn vorher eine Phase des „Austobens" bereitgestellt wird. Mädchen wird zunächst nur eine Phase der unrealistischen Träumerei eingeräumt.

Rechtsprechung

Die Ambivalenz dieser Konstruktion zeigt sich, wenn man sich die Lebensumstände junger Menschen im beginnenden 20. Jahrhundert vor Augen führt. Durch die Urbanisierung der Gesellschaft im Zuge der Industrialisierung leben mehr und mehr Jugendliche in den Städten. 1933 wohnen zwei Drittel aller Jugendliche in der Stadt, jeder zweite sogar in der Großstadt (Siemering 1937, S. 81). Die Mehrzahl der Jugendlichen (1906 sind es 93,3%) besucht die Volksschule, welche mit dem 14. Lebensjahr endet. Somit sind die meisten Jugendlichen keineswegs von Erwerbsarbeit freigestellt, sondern sie werden spätestens mit dem 14. Lebensjahr in die lohnabhängige Erwerbsarbeit integriert. Dies gilt auch für einen großen Teil der Mädchen (Grossmann 1985; Benninghaus 1997). Das „Ausprobieren" wird in den Freizeitbereich verlagert und kann – sofern es finanzierbar ist – im lebendigen Nachtleben der Großstädte erfolgen. Einzig die wohlhabenden jungen Männer können im Studium die Lebensphase der Jugend bis in das dritte Lebensjahrzehnt verlängern.

Urbanisierung

Obwohl der Alltag der meisten Jugendlichen durchaus von Erwachsenen-Tätigkeiten durchzogen ist, hat sich jedoch gesellschaftlich Jugend als Institution etabliert. Jugendliche Widerstandskultur wird staatlich eingebunden und bisweilen sogar wohlwollend begleitet. Jugendbewegungen entstehen. Der Nationalsozialismus kann ohne Widersprüche an das Jugendideal anknüpfen. Jugend gilt der Partei als Garant für eine neue Gesellschaft. Die Protestbereitschaft und die Lust, Neues auszuprobieren, wird in den Jugendorganisationen der NSDAP gepflegt. Über Kameradschaftsbeschwörungen wird die Jugend nicht nur zu einem Zustand im eigenen Lebenslauf, sondern als Generationslage inszeniert (vgl. Koebner/Janz/Trommler 1985).

Widerstand

Es setzt langsam ein Prozess ein, der bis heute immer deutlichere Konturen annimmt, nämlich dass „Jugend" als schillerndes Idealbild und als Mythos kultiviert wird. Der Lebensstil der Jugend wird zum erstrebenswerten Ideal für Menschen aller Altersgruppen. Ist das Konstrukt der Jugend in seiner Entstehungsphase zunächst an einen spezifischen Werdegang im Bildungssystem geknüpft, nämlich die Teilhabe an gymnasialer Bildung, so löst sich in das 20. Jahrhundert hinein die Konzeption von Jugend von der Bildungsidee ab. Geblieben ist nur, dass sich Bildungsinstitutionen wie alle anderen gesellschaftlichen Bereiche auch auf die spezifischen Belange von Jugendlichen einstellen müssen.

Ideal

Analysen zur aktuellen Situation von Jugendlichen basieren auf unterschiedlichen theoretischen Zugängen. Überwiegt in der Kindheitsforschung eine modernisierungstheoretische Begründung der Veränderung von Kindheit, die vor allem auf die gestiegenen Wahlmöglichkeiten und die Vielfalt der Lebensformen abzielt (z.B. Krüger 1994; Büchner u.a. 1994; theoretischer Hintergrund Beck 1986), so finden sich in der Jugendforschung auch systemtheoretisch motivierte Argumentationen, welche die Veränderung der Jugendphase aus dem Prozess der funktionalen Differenzierung herleiten (vgl. zur funktionalen Differenzierung die Ausführungen zu Luhmann in Kapitel 3.2; zur Jugendforschung z.B. Dudek 1992[2]; Olk 1989). Hier unterscheidet sich die Begründung der Veränderung, nicht aber die Beobachtung. Festgestellt wird durchwegs eine zeitliche Ausdehnung der Jugendphase sowie eine Vervielfältigung der jugendkulturellen Muster (zusammenfassend z.B. Abels 1993).

Bildungsmoratorium

Die zeitliche Verlängerung der Jugend gründet auf der Verlängerung der Schul- und Ausbildungszeiten sowie der gestiegenen Partizipation von Kindern aller Schichten an höherer Schul- bzw. Hochschulbildung (Coleman 1986; vgl. die Ausführungen zur Bildungsreform, Kapitel 4.2.1). Dementsprechend betont zum Beispiel Jürgen Zinnecker die Bedeutung der Jugendphase heute als Phase des „Bildungsmoratoriums“ (Zinnecker 1990a). Parallel zur Ausweitung der Jugendphase wird es immer schwieriger, verallgemeinernde Aussagen über „die Jugend“ zu treffen. Allgemeine Tendenzen, wie sie in der 13. Shell-Jugendstudie (vgl. Fischer u.a. 2000) und der 14. Shell-Jugendstudie (vgl. Hurrelmann/Albert u.a. 2002) erhoben werden, sind, dass mehr als die Hälfte der Jugendlichen ihre Zukunftsperspektiven recht optimistisch beurteilt. Dabei sind es – wie zu erwarten ist – vor allem jene Mädchen und Jungen, die über Bildung, Unterstützung durch die Eltern und Persönlichkeitsressourcen wie Selbstvertrauen verfügen, welche ihre Lebenschancen als gut einschätzen. Überhaupt kann festgestellt werden, dass Jugendliche eine sichere Einschätzung ihrer Lebensumstände vornehmen. So wie sie um ihre Chancen und Risiken wissen, so können sie auch die Leistungserwartungen und die Ausdauer, die zu beruflichem wie privaten Erfolg führen, in der Mehrheit gut benennen. Für Mädchen und Jungen in Ost- wie in Westdeutschland gilt dabei, dass ihnen eine Vereinbarkeit von Beruf und Familie wichtig ist. „Sie wollen keineswegs einem Individualismus huldigen, der sich von diesen traditionellen Lebensformen freimacht und jenseits von Beruf und Familie ein gegenwarts- und selbst bezogenes Leben propagiert“ (Fischer u.a. 2000, S. 14). Jeder Zweite möchte heiraten, viele wollen mit dem Partner/der Partnerin zusammenleben. Die Jugendlichen zeigen jedoch auch – häufig aus pragmatischen Überlegungen heraus – eine grundsätzliche Bereitschaft zu anderen Lebensformen. Ähnlich pragmatisch ist das Verhältnis der Jugendlichen zu Politik. Während sie zwar Demokratie prinzipiell für eine gute Staatsform halten, ist das allgemeine Interesse an Politik rückläufig. Gesellschaftliches Engagement basiert nicht länger auf Selbstverwirklichungsinteressen, sondern häufig auf pragmatischen Überlegungen (vgl. Hurrelmann/Albert u.a. 2002). Institutionalisierte Religiosität wird bei christlichen Jugendlichen weitgehend abgelehnt, was allerdings nicht zu einer prinzipiellen Ablehnung von Religion führt, sondern zu einem Set privater Glaubensüberzeugungen. Ähnlich verhält es sich mit der Politik. Das ForscherInnenteam stellt fest, dass der Vertauensverlust der PolitikerInnen und der Parteien weiter zunimmt. Dabei ist das Image der Politik besser, je besser die Jugendlichen ihre eigenen Chancen auf dem Arbeitsmarkt und im Alltag einschätzen.

Vereinbarkeit von Beruf und Familie

Jugendkulturen

Um hinter der Statistik die Verschiedenheit zu erkennen, wird – ähnlich der Schulkulturforschung (Kapitel 5.2.2) – in der Jugendforschung ein Hauptaugenmerk auf die verschiedenen Jugendkulturen geworfen. Beispielhaft hierfür sind die Arbeiten des Birminghamer Centre for Contemporary Cultural Studies (vgl. z.B. Willis 1991). Jugendliche neigen in besonderem Maße dazu, unterschiedliche Lebensstile, die stets auf differente Milieus verweisen, auszuprobieren und über dort typische Symbole (Kleidung, Musik, Gruppen-Codes) gleichermaßen Gruppenzugehörigkeit zu produzieren und politische/kulturelle/soziale Inhalte gegenüber den Erwachsenen zu vertreten.

Dieter Baacke und Wilfried Ferchhoff (1992², S. 436ff.) unterscheiden 5 Formen der Jugendkulturen:

1. Religiös-Spirituelle (Grufties, Anhänger asiatischer Religionen, Pfadfinder etc.)
2. Kritisch-Engagierte (Frauen-, Friedens-, Ökologie-, Antiglobalisierungsbewegung etc.)
3. Manieristisch-Postalternative (Spielerisch-ironische Demonstration der Symbole früherer Epochen, zum Beispiel Barock-Kleidung kombiniert mit modernen Elementen)
4. Körper- und Actionorientierte (Orientierung an Sport und Abenteuer; Modellierung des eigenen Körpers)
5. Institutionell-Integrierte (Vereins- und Verbandsarbeit, zum Beispiel Freiwillige Feuerwehr, Chormitgliedschaft oder auch Sportverein).

Allen Jugendkulturen ist gemeinsam, dass über den Gleichaltrigenkontakt vier zentrale Aufgaben der Jugendphase bewältigt werden sollen:

1. Familienablösung
2. Orientierung auf politische/kulturelle/soziale Partizipation
3. Festigung einer sexuellen Orientierung
4. Entwicklung eigenständiger Konsumgewohnheiten und Freizeitplanung.

7.3 Die Ausweitung der Bildungs- und Lernprozesse auf das Erwachsenenalter

Prozessualer Bildungsbegriff

Die OEEC und die UNESCO haben mit der Idee der éducation permanente die Öffentlichkeit auf die Notwendigkeit lebenslangen Lernens hingewiesen. Sich stets fortzubilden, ist zu einer Erwartung an alle im erwerbsfähigen Alter geworden. Die Professoren des Collège de France formulieren in ihren „Vorschlägen für das Bildungswesen der Zukunft" (1987) den Bildungsbegriff selbst als einen in diesem Sinne prozessualen: „Sicher wird niemand bestreiten, dass jede Bildung geistige Aufgeschlossenheit zum Ziel haben muss, um sich immer neues Wissen anzueignen und mit immer neuen Situationen zurechtzukommen" (Collège de France 1997, S. 253). Für die Umstrukturierung des Bildungssystems schlagen sie u.a. einen kontinuierlichen Wechsel von Bildung und Berufstätigkeit vor („alternierende Bildung") (Collège de France 1987, S. 272).

Zwang

Kaum einer wird heute bezweifeln, dass zumindest Lernprozesse in einem ganz konkreten Sinn (z.B. Umgang mit neuen Computerprogrammen etc.) und stetige Teilnahme an Fortbildungsmaßnahmen auch von vielen Erwachsenen verlangt werden. Allerdings variiert die Einschätzung dieses Prozesses. Bolder/Hendrich (2000, S. 18ff.) betonen die immanente Zumutung. Der/die Erwerbstätige müsse sich permanent Entwicklungen anpassen, die weder technologisch unausweichlich noch autonom gewählt seien, sondern die einzig den privaten Renditeerwartungen des Wirtschaftssystems genügen. Dabei habe sich eine Haltung durchgesetzt, in der nicht etwa höhere Qualifikation mit besserem Lohn beglichen werde, sondern umgekehrt die permanente Weiterqualifizierung als Voraussetzung für den Arbeitsplatzerhalt entworfen werde. Mehr noch: Der/die Arbeitnehmerin müsse im zunehmend marktförmig organisierten Weiterbildungssystem für die Weiterbildung häufig sogar selbst bezahlen und Freizeit dafür aufwenden. ArbeitnehmerInnen seien durch den Anspruch auf lebenslanges Lernen und durch die Ausweitung von Bildungsprozessen auf die Lebensphasen der Erwachsenen in einen permanenten Druck geraten, so ihre Konsequenz.

Chance

Winfried Marotzki (1991) wirft einen sehr anderen Blick auf die lebensgeschichtlich andauernden Bildungsprozesse. Mit Blick auf die Erwachsenenbildung zeigt er, dass Bildungsprozesse, die nicht auf unmittelbar verwertbare Qualifikationen zielen, sondern die Lebensgeschichte selbst zum Gegenstand haben können, biographische Sinnzusammenhänge erzeugen. Unter dem Stichwort der Biographisierung betont Marotzki, dass die individuelle Lebensgeschichte ein vom Subjekt selbst hervorgebrachtes Konstrukt ist, das die Fülle der Erfahrungen und Ereignisse des gelebten Lebens zu einer Einheit zusammenfügt. In diesem Akt werden Bedeutungen und Schlussfolgerungen für die Gegenwart abgeleitet. Bildung auch im Sinne der Selbstformung ist damit als Akt der Selbstreflektion stets notwendig. Organisierte Bildungsprozesse, wie sie zum Beispiel auch in Psychotherapien angeboten werden, bieten, so Marotzki, eine Chance umzulernen, gewohnte Pfade des Handelns zu verlassen und damit den Einfleischungen des Habitus Neugestaltungen der eigenen Biographie entgegenzusetzen.

8. Die Raumdimension der Bildung

Bebauung

Der Schwerpunkt der Bildungssoziologie liegt, wie zu Beginn des Buches festgestellt, in der theoretischen und empirischen Untersuchung des Bildungssystems und der sich in den Bildungsinstitutionen vollziehenden Prozesse. Diese Institutionen, also zum Beispiel Schule oder Universität, haben nicht zufällig ein eigenes Gebäude, um Bildung zu organisieren. Raum wird systematisch in den Prozess des Bildens einbezogen. Allerdings wäre es vereinfacht anzunehmen, Raum sei auf die gebauten Anordnungen zu reduzieren. Raum entsteht auch aus der Platzierung beweglicher Objekte oder menschlicher Körper. Die Anordnungen von Jugendlichen auf dem Schulhof, die Platzierung von Stellwänden im Unterricht, die Positionierung der Tische zueinander, all das sind Elemente, die Raum hervorbringen (vgl. dazu z.B. auch die Analyse von Foucault, dargestellt in Kapitel 2.2.4). Um dies auszuführen, ist es hilfreich, noch einmal auf die von Paul Willis (1982²) durchgeführten Studien zur Statuspassage von Jungen aus der Arbeiterklasse ohne höhere Schulbildung in die Erwerbsarbeit zurückzugreifen (vgl. Kapitel 4.2.2 dieses Buches). Willis beschreibt eine Situation, die sich täglich mehrfach ereignet: „Die meisten ‚lads' rauchen und, was noch wichtiger ist, sie werden beim Rauchen gesehen. Wenn Schüler rauchen, so geschieht dies meist draußen an der Schulpforte. Die ‚lads' verbringen typischerweise viel Zeit damit, ihre nächste Zigarettenpause zu planen und Stunden zu ‚schwänzen', um einen ‚raschen Zug' zu nehmen" (Willis 1982², S. 35).

Gegenkulturelle Platzierungen

Wenn die Lads, so der Name der Jungenclique, zur Schule gehen, wechseln sie ständig zwischen zwei Orten. Morgens rauchen sie auf der Straße vor dem Schulaufgang, so dass der türöffnende Lehrer sie sehen muss. Zwischen den Stunden, aber auch während der Stunden wechseln sie zwischen Schulgebäude und Straße hin und her. Selbst bei Schulausflügen, zum Beispiel beim Besuch eines Museums, nutzen die Jugendlichen jede Gelegenheit, um auf der Straße „eine" zu rauchen.

Der Straßenraum vor der Schule gehört sowohl in den alltäglichen Handlungen als auch in der Erinnerung nach der Schulzeit für die Lads zum Schulraum. Er wird durch die ständige Bewegung zwischen Straße und Schulgebäude im Handeln hergestellt. Die Straße ist für die Lads unmittelbar mit dem Schulgebäude verknüpft. Gemeinsam bilden sie den Raum der Schule. Jeden Morgen behaupten sie mehrmals die Straße als Teil des Schulraums neu. Diese Konstruktion ist jedoch wirkungslos, wenn die Bestätigung durch die LehrerInnen fehlt.

Körper

Der Raum der Lads ist flüchtig; er ist an ihre körperliche Präsenz gebunden. Institutionalisiert ist der Schulraum als Innenraum mit dem durch Mauern oder Zäune integrierten Schulhof. Diese institutionalisierten Anordnungen sind erstens materiell festgeschrieben (Mauern, Zäune, Bebauungspläne, Eigentumsverhältnisse) und basieren zweitens auf symbolischen Verknüpfungen, zum Beispiel der Eingangspforte mit dem Beginn/Ende eines Raums. Diese Raumkonstruktion, der zufolge der Schulraum aus einer relationalen Anordnung von Pforte, Schulgebäude und Schulhof besteht, und eben nicht die Straße integriert, sondern explizit die Straße als anderen Raum durch Grenzziehung ausweist, wird von den LehrerInnen, so zeigt die Studie, mehrheitlich akzeptiert.

Cliquenzugehörigkeit

Die Lads bemühen sich nun, dieser Raumkonstruktion eine eigene entgegenzusetzen. Sie verfügen aber weder über rechtliche noch über planerische oder bauliche Mittel, ihren Raum materiell festzulegen. Daher können sie nur mit dem Einsatz ihres eigenen Körpers die gegenkulturelle Raumkonstruktion materiell und symbolisch markieren oder aber mit kurzzeitigen symbolisch/materiellen Besetzungen wie herumliegende Zigarettenstummel oder Graffitis an den Wänden arbeiten. Während die LehrerInnen über soziale Güter, Wissen und soziale Positionen verfügen und somit institutionalisierte Raumkonstruktionen durchsetzen können, besteht für die Lads die Chance, die Zugehörigkeit zur Clique dagegenzustellen. Sie sind daher darauf angewiesen, dass ihre Raumkonstruktion wahrgenommen und als different erkannt wird. Nur in der (strafenden) Anerkennung des gegenkulturellen Raums der Lads durch die LehrerInnen beginnt dieser, sich materiell und symbolisch zu verfestigen. Der Raum, den die Lads als Schulraum etablieren, ist daher ohne die handelnden Menschen, die Lads in ihrer körperlichen Präsenz und die LehrerInnen als anerkennende, nicht zu erklären. LehrerInnen und Lads sind demnach ebenfalls ein „Baustein" der relationalen (An)Ordnung.

Allerdings ist die Machtbalance zwischen Lads und LehrerInnen keineswegs einseitig als „oben-unten-Verhältnis" zu verstehen. Nicht wenige LehrerInnen reagieren auf die Provokationen der Lads hilflos ignorierend. In LehrerInneninterviews reflektieren diese, dass sie nur wenig Sanktionsmöglichkeiten haben und schließlich nicht herumlaufen und ständig Strafen aussprechen können (Willis 1982[2], S. 103). Ein Lehrer vergleicht die Lads sogar mit einer Sturmflut.

> „Es ist wie eine Sturmflut, man kann sie nicht aufhalten, wir versuchen es, wir versuchen sie zurückzudämmen ... manche lassen die Flut einfach über sich hinweg gehen" (Willis 1982[2], S. 127).

Bewegte Räume

Nur durch ein hohes Maß an Mobilität, das heißt durch den stetigen Wechsel zwischen beiden Orten, gelingt es den Lads, Schulgebäude und Straße zu einem Raum verschmelzen zu lassen. Dieses In-Bewegung-Sein wird von dem zitierten Lehrer kraftvoll wie eine Flut wahrgenommen. Sturmfluten überschreiten Grenzen, reißen die Dämme ein und nehmen das Land in Besitz. Sturmfluten sind beängstigend, aber auch anziehend. Die LehrerInnen, wie sie von Paul Willis beschrieben werden, synthetisieren entsprechend der institutionalisierten (An)Ordnungen den Raum der Schule. Die Lads synthetisieren dagegen nicht nur Pforte, Schulgebäude und Schulhof zu einem Raum, sondern integrieren auch noch die Straße in ihre Raumkonstruktion. Auch sie überschreiten die gesetzten Grenzen und integrieren „neues Land" und zwar mittels bewegten Körpereinsatzes. Die

Machtpotentiale, die in der Bewegung und in der Grenzverschiebung liegen, werden durch die Kennzeichnung als „Flut" prägnant zum Ausdruck gebracht.

Durch den Zwang zur eigenen Bewegung bleibt die Raumkonstruktion im ständigen Fluss. Der Raum der Lads ist nicht starr, sondern an die bewegten Handlungen gebunden. Aber auch die Konstitution von Raum durch die LehrerInnen kann nicht als starrer, fester Raum analysiert werden, da dieser ebenfalls – zumindest zeitweilig – von den gegenkulturellen Bestrebungen der Lads in Frage gestellt wird.

8.1 Definition von Raum

Syntheseleistung

Raum kann also nicht, wie im Alltagsdenken tief verhaftet, einfach als umschließender Behälter gedacht werden, sondern Raum wird durch Handeln hergestellt. Es gibt Räume, die sind flüchtig und strukturieren nur kurzfristig eine Handlungssituation, andere Räume sind institutionalisiert und in Bauordnungen oder Normen festgelegt und dementsprechend langlebig. Raum entsteht immer dadurch, dass Menschen Objekte zu einem Raum verknüpfen, sozusagen in einer Syntheseleistung die sozialen Güter oder Lebewesen, die sie wahrnehmen oder erinnern, zu einem Raum zusammenschließen. Manche Räume entstehen so in erster Linie auf der Basis gebauter oder gewachsener (bei Pflanzen z.B.) Güter, andere basieren eher auf Platzierungen von Menschen (z.B. Anordnungen von Menschen in Cliquen). Über Räume werden Ein- und Ausschlüsse organisiert. Unter Raum kann daher eine relationale (An)Ordnung sozialer Güter und Lebewesen an Orten verstanden werden (ausführlich Löw 2001a). Von „(An)Ordnung" wird gesprochen, weil Räume sowohl im Handeln entstehen (dem Prozess des Anordnens) und gleichzeitig Ordnungen zum Beispiel des Zusammengehörigen und des Ausgegrenzten oder der Platzierungen schaffen. Ein Ort ist im Unterschied zu Raum eine einmalige Stelle, meist geographisch markierbar und mit einem Namen versehen. Raum dagegen ist eine (An)Ordnung auf der Basis einer Verknüpfungsleistung („Syntheseleistung") und – mit Ausnahme virtueller Verknüpfungen – einer Platzierungspraxis („Spacing"). Raum ist demzufolge ein Hybrid aus materiellen Bedingungen und sozialer Nutzung. Räume sind nicht Untergrund oder Hintergrund des Handelns, sondern Räume sind Strukturierungen, die im gesellschaftlich geprägten Prozess der Wahrnehmung oder der Platzierung konstituiert, durch Regeln abgesichert und in Institutionen eingelagert werden.

Spacing

Ein solches Verständnis von Raum ist relativ neu für die Soziologie. Erst in den letzten zehn Jahren haben sich die „relativ" genannten (da sie sich auf die Relationenbildung und die Syntheseleistung beziehen – im Unterschied zu „absolut", welches die Behälterraumbegriffe bezeichnet) Raumverständnisse durchgesetzt (vgl. zu ähnlichen Konzeptionen und zur Entstehungsgeschichte auch Läpple 1991; Sturm 2000). In der „klassischen" Soziologie wird Raum anders entworfen.

8.1.1 Raum in den klassischen Konzepten der Chicagoer Schule und von Georg Simmel

Chigacoer Schule

Robert E. Park und Ernest W. Burgess gehören zusammen mit ihren Schülern Louis Wirth und Roderick D. McKenzie zu den berühmtesten Vertretern der in den 1920er Jahren entwickelten urban sociology. 1925 verfassen Park, Burgess und McKenzie das Buch „The city", in dem sie die sozialökologische Stadtforschung theoretisch begründen und Forschungsergebnisse präsentieren. Hier wie in anderen Publikationen entwickeln sie ihre Vorstellung, dass Überlegungen der Tier- und Pflanzenökologie auf die menschliche Gesellschaft übertragbar seien. Der zu Recht häufig kritisierte Grundgedanke ist, dass Menschen sich ihrer natürlichen Umwelt anpassen. Verschiedene Stadtviertel erscheinen in dieser Perspektive als Anpassungsformen der Menschen an ihre Umwelt (natural areas). Durch jeweils spezifische Selektionsformen versuchen die einzelnen Gruppen, basierend auf Ethnizität oder Schichtzugehörigkeit, in sich homogene Gemeinschaften aufzubauen, eben wie auch „verschiedene Pflanzenarten dazu neigen, permanente Gruppen zu bilden" (Park 1974[7], S. 90). Park zufolge betont die Sozialökologie, hier Humanökologie genannt, weniger die geographische Struktur als vielmehr den Raum. Darunter versteht er den folgenden Sachverhalt:

> „Die Humanökologie, wie sie die Soziologen verstehen, betont nicht so sehr die geographische Struktur, sondern den Raum. In der Gesellschaft lebt man nicht nur zusammen, sondern lebt gleichzeitig getrennt (...). Lokale Gemeinschaften, im folgenden Gemeinde genannt, können einmal in Bezug auf die Gebiete, die sie einnehmen, verglichen werden und zum zweiten bezüglich der relativen Dichte der Bevölkerungsverteilung innerhalb der Gebiete" (Park 1974[7], S. 91).

Sozialökologie

Wenn also Parks Sozialökologie auf der Bedeutung von Raum aufbaut, dann wendet er sich gegen den Blick auf die bloße Verteilungsstruktur (geographische Struktur) und betont die Gleichheit im Viertel und die Differenz in der Stadt. Für Park ist die Einführung des Raumbegriffs ein Versuch, sich von der bloßen Betrachtung von geographischen Grenzen und physikalischen Entfernungen zu distanzieren und stattdessen das Gesamtgebilde, welches sich durch Immigrationswellen immer neu verändern muss, in den Blick zu nehmen. Park definiert die einzelnen Gemeinden über Größe und Dichte, den Gesamtkomplex Stadt gleichzeitig über das Verhältnis von Heterogenität und Gleichheit. Raum ist für Park, wie für die anderen VertreterInnen der Sozialökologie, insofern von Bedeutung, als sie Gleichheit und Differenz auf der Basis verschieden großer und gegeneinander abgegrenzter Gebiete mit dem Raumbegriff bestimmen wollen.

Konzentrische Modelle

So geht zum Beispiel auch Ernest Burgess von den „physiologischen Tatsachen" Größe und Dichte aus, wenn er die Ausdehnung der Stadt und die damit einhergehenden räumlichen Strukturierungen untersucht. Burgess entwickelt das idealtypische Modell der konzentrischen Kreise, welches für die Bildungssoziologie, wie im Folgenden noch gezeigt werden wird, von besonderer Bedeutung ist. Diesem konzentrischen Modell zufolge teilt sich Stadt vom Zentrum ausgehend in verschiedene Zonen auf.

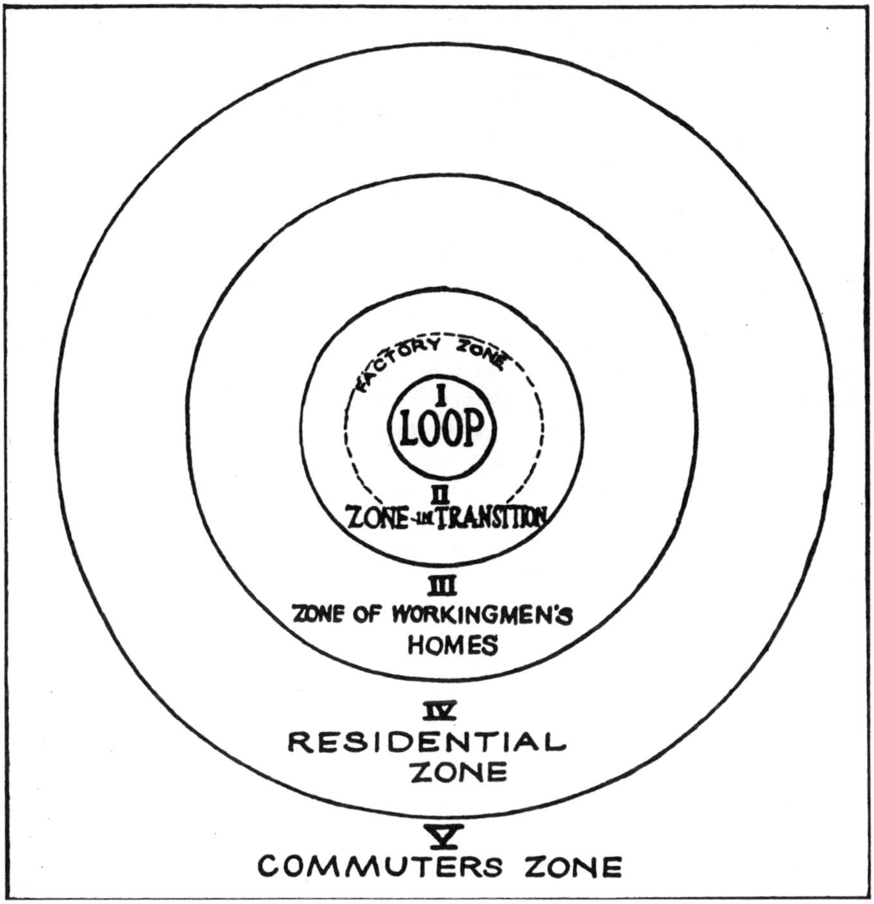

(Quelle: Burgess 1974[7], original 1925, S. 51)

Dieses Modell ist seitdem verändert, verfeinert und widerlegt worden. Es stellt sich heraus, dass es kein universelles Muster städtischer Entwicklung ist, sondern nur für nordamerikanische Städte der 1920er Jahre zutrifft (vgl. zum Modell: Park, Burgess und McKenzie 1974[7]; zur Kritik: Hamm 1982).

Es ist häufig darauf hingewiesen worden, dass die Sozialökologie der Chicgoer Schule „trotz" ihrer evolutionstheoretischen Grundlagen wesentliche Grundbegriffe der Stadtsoziologie bereitgestellt hat (Friedrichs 1977; Krämer-Badoni 1991; Häußermann/Siebel 1994). Dazu zählt vor allem: „Segregation als die Konzentration bestimmter Gruppen in spezifischen Stadtvierteln; Invasion als das Eindringen von Gruppen in ein Gebiet, in dem vormals andere Gruppen segregiert waren; Sukzession für die Beschreibung eines vollständigen Nutzungswandels in bestimmten Stadtvierteln" (Krämer-Badoni 1991, S. 20). Den Raum jedoch behandeln die Sozialökologen der Chicagoer Schule mehrheitlich als natürlich gegebene Gebiete, die zu unterschiedlich großen Behältern abgesteckt werden. Dies führte dazu, dass über die „Inhalte" der Behälter viel gestritten wurde, der Prozess der Konstitution von Raum jedoch aus den Augen verloren wird (vgl. dazu auch Löw 2001b).

Georg Simmel

Ganz anders entwickelt Georg Simmel sein soziologisch-philosophisches Raumverständnis. In einer 1905 an der Berliner Universität gehaltenen Kant-Vorlesung entwickelt Simmel die Kantsche Raumvorstellung, die auch für sein eigenes Werk richtungsweisend ist: „Was bedeutet dieses unendliche Gefäß um uns herum, in dem wir als verlorne Pünktchen schwimmen und das wir doch samt seinem Inhalt *vorstellen*, das also ebenso in uns ist, wie wir in ihm sind?" (Simmel 1905, S. 52; Hervorhebung im Original). Nicht ohne Ironie beginnt er seine sechste Vorlesung mit der Vorstellung von der Welt als Gefäß für verlorene Pünktchen, um kurz darauf diese Idee heftig zu kritisieren. „Angesichts jener Gewöhnung, uns und die Dinge innerhalb eines vor allem einzelnen bestehenden Raumes vorzustellen, ist es ein schwieriger Gedanke, dass – mit etwas paradoxer Kürze ausgedrückt – der Raum selbst nichts Räumliches ist: grade so wenig wie die Vorstellung des Roten selbst etwas Rotes ist" (Simmel 1905, S. 55).

Simmel betont, dass der unendliche, leere Raum, die Vorstellung eines Behälters eine bloße Abstraktion sei, die möglich, aber für die Soziologie nicht relevant ist. Räumlichkeit entstehe vor allem dadurch, dass Menschen sie in ihrer Vorstellung generieren. Außerhalb menschlicher Empfindungen habe der Raum keine Realität, allerdings bedeute dies eben nicht, dass er nur subjektiv oder irreal sei, da die menschlichen Empfindungen die Welt bilden: „... der Raum hat alle Realität, von der innerhalb unsrer Erkenntnis überhaupt die Rede sein kann, eben dadurch, dass er die Form und Bedingung unsrer empirischen Vorstellung ist. Die räumlichen Dinge sind dadurch und insoweit real, als sie unsere Erfahrung bilden" (Simmel 1905, S. 57).

Form

Raum ist demnach für Simmel eine an sich wirkungslose Form, vergleichbar der Form, durch die Holz zu einem Schrank wird, die aber außerhalb des Materials keine eigenständige Existenz führt (vgl. Simmel 1905, S. 55). Es sind die Dinge, deren Räumlichkeit im Prozess des Anschauens geschaffen wird, und die, so der Soziologe Simmel, von Menschen in ihren Handlungen angeordnet und emotional besetzt werden. Dabei nimmt er, entsprechend seiner euklidischen Grundannahmen, einen einheitlichen Raum an, der von „der Seele her" gegliedert werde. Aufbauend auf dieser absolutistischen Grundannahme definiert er als Grundqualitäten des Raums die Ausschließlichkeit, die Zerlegbarkeit, die Fixierung, die Nachbarschaft und die Bewegungsmöglichkeiten von Ort zu Ort. Die Bedeutung, die Simmel der Euklidik zuweist, muss nach heutigem Kenntnisstand relativiert werden. Durch die Entwicklung nichteuklidischer Geometrien und durch das Wissen, dass die Euklidik eine Idealisierung der Anschauung ist, wird deutlich, dass diese Prinzipien nicht die einzigen sind, mit deren Hilfe das Umgebende in eine Ordnung gebracht wird, sondern dass soziale Faktoren bei der Syntheseleistung berücksichtigt werden müssen.

Ein neues Verständnis von Raum, sowohl erstens in Bezug auf die Syntheseleistung, welche eben nicht vorgegeben und damit erlernbar ist, sowie zweitens in Bezug auf die Raumvorstellung vom einen, unendlichen, starren Raum, der jenseits des menschlichen Handelns existiert, wird durch viele Entwicklungen evoziert: neue Erkenntnisse in der Physik, neue Theorien in der Philosophie, andere bildliche Darstellungen in der Kunst, aber auch ein Wandel der sozialen Welt. Dieser wird im folgenden Abschnitt nachgezeichnet.

8.2 Bildungs- bzw. erziehungssoziologische Raumstudien

Straßenkindheit

Mit der Urbanisierung und Industrialisierung entsteht vor allem zwischen 1870 und 1920 die so genannte *Straßenkindheit*. Kinder zwischen drei und vierzehn Jahren finden sich nach Klassen getrennt, aber geschlechtlich durchmischt zu Cliquen zusammen und entdecken die soziale Welt über die Erkundung (städtischer) Räume (Behnken/Zinnecker 1987). Der Begriff „Straße" umfasst also nicht nur die konkrete Verkehrsstraße, sondern auch öffentliche Gebäude, angrenzende Parks und Plätze etc. Die Straße wird zu einem Lernort (Zinnecker 1979). Der öffentliche Raum, in dem Kinder sich spielend bewegen, ist der Ort, so Jürgen Zinnecker, an dem sie gesellschaftliche Prozesse studieren können. Über die Straße erhalten sie die Chance, sich selbst in die Gesellschaft einzusozialisieren, indem sie deren Regeln und Normen, ihre Strukturen und Gepflogenheit beobachten und austesten. Sie üben die Erwachsenenanforderungen der KäuferInnen, KonsumentInnen oder VerkehrsteilnehmerInnen ein. So entstehen *Kinderöffentlichkeiten* ohne Erwachsene.

Kinder-öffentlichkeiten

Die Erwachsenen tauchen zwar als *Raumwärter* mit verschiedenen Interessen auf, die Kinder können sich jedoch kollektiv mit diesen Erwachseneninteressen am Raum auseinandersetzen (Harms/Preissing 1988; Negt 1983; Ahrend 2002). Insofern ist eine Straßensozialisation explizit auch eine politische Sozialisation (Nissen 1998). Von einer Straßenkindheit kann für Kinder aus der Arbeiterklasse und dem Kleinbürgertum – nach den Zerstörungen des 2. Weltkriegs für alle Kinder – bis in die 1960er Jahre hinein die Rede sein.

Aktionsräume

Kinder, die wesentliche Anteile des Tages im Straßenraum leben, so findet 1934 Marta Muchow in einer Studie zur kindlichen Raumaneignung heraus, entdecken Raum ringförmig um den Wohnbezirk (Muchow/Muchow 1935, vgl. das Modell von Burgess im vorhergehenden Abschnitt). In den 1960er Jahren kommt Elisabeth Pfeil zu dem gleichen Ergebnis einer allmählichen Ausdehnung des Aktionsraums in konzentrischen Kreisen (Pfeil 1965). Auch sozialökologische Ansätze, die u.a. aus dem geistigen Umfeld der Chicagoer Schule entstehen und Sozialisation als Prozess des Aufwachsens in aktiver Auseinandersetzung mit der stofflich-dinglichen und der sozialen Umwelt verstehen, gehen von einer kontinuierlichen Expansion des kindlichen Erfahrungs- und Vorstellungsbereichs aus (Bronfenbrenner 1981; Baacke 1993[2]). Uri Bronfenbrenner zum Beispiel vertritt die Annahme, dass Kinder kontinuierlich ihre Erfahrungen in verschiedenen Räumen bzw. vermittelte Erfahrungen zu einem Gesamtraum verknüpfen. Die Entwicklung verlaufe von der Mikro- zur Meso- über die Exo- zur Makroebene. Dieter Baackes Ansatz der „Lebensweltanalyse" lehnt sich an Bronfenbrenner an und unterscheidet ebenfalls vier expandierende Zonen, die Kinder und Jugendliche in kontinuierlicher Reihenfolge durchlaufen. Auf diese Weise entsteht auch im alltäglichen Handeln der Eindruck eines einheitlich größer werdenden Raumes.

Funktionale Raumorganisation

Während in den zwanziger Jahren des letzten Jahrhunderts und in der Wiederaufbauphase nach dem zweiten Weltkrieg eine starke Bindung der Menschen zu einzelnen Stadtteilen oder Orten existiert, die sich zum Beispiel darin ausdrückt, dass sich Kinder einer Straße zu „unserer Straße" zusammenschließen und sich gegen Kinder anderer Straßen abgrenzen (Pfeil 1965), setzt sich in den sechziger und frühen siebziger Jahren eine funktionale Raumorganisation durch. Damit entstehen neue Bedingungen der räumlichen Sozialisation.

Die Organisation des Nebeneinanders ändert sich in den 1970er Jahren in vielen Details. Die Straßen werden dem Autoverkehr vorbehalten. Man kauft in Supermärkten am Ortsrand und nicht in „Tante-Emma-Läden". Aus Schulen werden Schulzentren; Freiflächen werden funktionsgebundene Parkplätze, Spielplätze etc. Monofunktionale Wohnsiedlungen entstehen, in die vor allem junge Familien mit kleinen Kindern ziehen. Von allen Großstadtkindern leben 1972 vier Fünftel in den Außenbezirken der jeweiligen Stadt (Zeiher/Zeiher 1994).

Parallel zu der Unterbringung von Kindern in den anregungsarmen, monofunktionalen Wohngebieten entsteht ein spezialisiertes Angebot für Kinder: Getrennte Spielplätze für jüngere und ältere Kinder, Bolzplätze und Waldspielplätze für Jungen und einige „wilde" Mädchen, Jugendzentren etc. Fragen der Bildung und Erziehung werden zu einem wichtigen gesellschaftlichen Thema, in dessen Mittelpunkt immer auch die Frage nach Umsetzungsmöglichkeiten des Chancengleichheitspostulats steht. Die gezielte Förderung möglichst unterschiedlicher kindlicher Fähigkeiten wird zu einem wesentlichen pädagogischen Ziel (vgl. Kapitel 7.1). Das Besuchen spezieller Kurse im Jugendzentrum, in der Musikschule, im Sportverein oder in der Volkshochschule wird zum normalen Alltag vieler Kinder.

Verinselte Vergesellschaftung

Die räumliche Vergesellschaftung ändert sich nun dahingehend, dass Kinder Raum nicht als etwas sie einheitlich Umgebendes kennen lernen, den sie mit zunehmendem Alter mehr und mehr entdecken, sondern dass Kinder einzelne Räume kennen, die wie Inseln über die Stadt verteilt liegen und die nur durch die eigene biographische Erfahrung einen Zusammenhang erfahren. Es entsteht ein neuer Vergesellschaftungstypus, der als *verinselte Vergesellschaftung* bezeichnet wird.

Lernorte

Helga Zeiher und Hartmut J. Zeiher zeigen auf der Basis vergleichender Fallstudien mit Kindern verschiedener Stadtviertel, dass Raum derzeit von Kindern erfahren wird als „aus einzelnen separaten Stücken (bestehend, M.L.), die wie Inseln in einem größer gewordenen Gesamtraum verstreut sind, der als ganzer bedeutungslos und weitgehend unbekannt bleibt" (Zeiher/Zeiher 1994, S. 27). Helga und Hartmut J. Zeiher können in ihrer Untersuchung zu Orten und Zeiten der Kinder überzeugend darlegen, dass Kinder heute in funktionsgebundenen, über die Stadt verteilten Kinderorten aufwachsen. Eltern transportieren die Kinder von einer Insel zur anderen. Zwischen diesen Orten dürfen sie auswählen, das Wohnumfeld jedoch bleibt vielen Kindern fremd. Erst im Alter von neun oder zehn Jahren beginnen sie, eigenständig ihre Orte aufzusuchen und ihre Umgebung zu erkunden (Ahrend 1997). Die Straße als „Lernort" verliert damit an Bedeutung (sie ist jedoch gerade für proletarische Jungen, auch für junge Männer aus verschiedenen anderen Kulturen nicht bedeutungslos). Dies trifft in besonderem Maße auf Kinder aus ländlichen Gebieten und weiblichen Geschlechts zu (Nissen 1992). Ebenfalls geschlechtsspezifisch different ist die Menge der Institutionen, die aufgesucht werden. Jungen suchen mehrmals in der Woche immer wieder den gleichen institutionalisierten Freizeitort, meistens den Sportverein, auf, wohingegen Mädchen unterschiedliche Angebote in Anspruch nehmen und sich folglich immer wieder auf neue soziale Situationen einstellen müssen (Nissen 1998, S. 186). Durch Urlaubsreisen und Besuche bei Verwandten und Bekannten, insbesondere dann, wenn diese Reisen mit Flugzeugen unternommen werden, verstärkt sich dieser Prozess für beide Geschlechter. Die soziologische Analyse der Raumkonstitution in der Kindheit kann nicht mehr, das ist eine der wesentlichen Erkenntnisse, in Kategorien von starren Hintergrundräumen analysiert werden. Raum selbst rückt ein in den Handlungsprozess.

Verhäuslichung

Eine andere soziale Entwicklung, die ebenfalls die städtische „Straßenkindheit" abgelöst hat, wird als *Verhäuslichung* bezeichnet (Zinnecker 1990b). Mit den Bildungsreformbemühungen der 1960er Jahre (vgl. dazu Kapitel 2.2 und 4.2.1 dieses Buches) ändern sich nicht nur die Bildungsinstitutionen, sondern das gesellschaftliche Interesse an Kindern allgemein wächst. Der prognostizierte Bildungsbedarf und das Chancengleichheitspostulat bringen die Aufgabe mit sich, Kindern eine umfassende Bildung zukommen zu lassen. Dies betrifft auch außerschulische Aktivitäten der Kinder. So entstehen zunehmend Sportplätze, Kindergärten, Spielplätze, Bolzplätze, Freizeithäuser (Zeiher/Zeiher 1994, S. 21). Diese spezialisierten Orte, die die verinselte Vergesellschaftung ermöglichen, lösen gleichzeitig eine Phase der Verhäuslichung aus. Waren es bis dato nur die bürgerlichen Kinder, die sich zunehmend in geschlossenen, geschützten Räumen aufhalten, so wird die pädagogische Betreuung in spezialisierter und separierter Form zur gesellschaftlichen Normalität.

Soziale Herkunft	**Zeitraum** 1800	1900	1990
Bürgerliche Familie	**Verhäusl. Kindheit**	**Verhäusl. Kindheit**	**Verhäusl. Kindheit**
Kleinbürgerl. Familie	**Straßen-kindheit**	**Mischform Straße/Haus**	**Verhäusl. Kindheit**
Arbeiterfamilie (Städt. Unter-schichten)	**Straßen-kindheit Arbeits-kindheit**	**Straßen-kindheit**	**Verhäusl. Kindheit**

(Quelle: Zinnecker 1995, S. 48)

Zinnecker (1990b, S. 156f.) zufolge entstehen dadurch einerseits eine Trennung von fein- und grobmotorischen Tätigkeiten, verlängerte, abstrakte Handlungsketten in städtischen Gebieten sowie viel wache Zeit in abgeschlossenen Räumen (auch Fahrzeuge), andererseits aber auch eine relative Autonomie gegenüber Umwelteinflüssen, eine Intensivierung des Einzelspiels und mediatisierte Kompetenzen. Die Funktionalisierung der öffentlichen Räume für Verkehr und Warentausch lässt die Straße zu einem gefährlichen Ort für Kinder werden.

Mobilitätsstrategien

Christine Ahrend (2002) stellt in einer empirischen Untersuchung fest, dass diese Entwicklungen von den Kindern biographisch different verarbeitet werden. Die *Mobilitätsstrategien*, welche sie entwickeln, unterscheiden sich darin, ob sie die Vorherrschaft des motorisierten Verkehrs erkennen und experimentell eigene Orte entdecken (experimentelle Mobilitätsstrategie), ob sie ihre Bewegungsbedürfnisse an kurzzeitig besuchte spezielle Orte wie zum Beispiel Ferienorte verschieben (verlagerte Mobilitätsstrategie) oder ob sie die Verhäuslichung akzeptieren und auf Bewegung weitgehend verzichten (rudimentäre Mobilitätsstrategie). In der Forschungslandschaft gibt es eine Reihe von Forschungsergebnissen, die die Annahme nahe legen, dass ein expansives Bewegungsverhalten im öffentlichen Raum eher von Jungen ausgeübt wird als von Mädchen sowie dass sich die Muster der Raumkonstitution heute geschlechtsspezifisch unterscheiden.

8.2.1 Geschlechtsspezifische Aspekte der Raumkonstitution

Mädchenschulhof

Ein Schulzentrum im Hamburger Stadtteil Dulsberg gibt Mädchen in der Pause jungenfrei. Ein zweiter Schulhof, eine so genannte Mädchenarena, 3500 qm groß, mit Rasenstufen, einer Wiese, Bäumen und Sportfeldern ermöglicht den Mädchen, sich in Pausen auszutoben. Dulsberg ist eine Ausnahme unter den großstädtischen Schulhöfen, welche häufig von Jungen dominierten Asphaltwüsten gleichen (vgl. Pfister 1993). Das Konzept des Hamburger Mädchenschulhofs, auf dem Jungen auch spielen dürfen, so lange sie sich den Spielstilen der Mädchen anpassen, ist ein Versuch, das expansive Raumverhalten der Jungen zu bremsen, das defensive Verhalten der Mädchen jedoch zu mehr Aktivität zu führen.

Phantasiewelten

Es ist ein bekanntes Phänomen und zuletzt in einer ethnographischen Feldstudie (vgl. Gordon 1996) belegt worden: Jungen dominieren den Schulhof, rennen wilder durch die Schulkorridore und nutzen mehr Platz im Klassenzimmer. Mit dem Körper breiten sie sich mehr aus und lassen den Mädchen neben ihnen weniger Platz bzw. umgekehrt, die Mädchen begeben sich eher in Nischen, laufen langsamer und lassen sich den Platz wegnehmen. Selbstverständlich gibt es Ausnahmen, so genannte wilde Mädchen, die bezeichnender Weise häufig als „halbe Jungen“ tituliert werden und schüchterne Jungen, die jedoch nicht als „halbe Mädchen“ bezeichnet werden (Klingt „halber Junge“ nicht wie eine Aufwertung und „halbes Mädchen“ wie eine Abwertung?). Empirisch auffällig sind die platzergreifenderen, aggressiveren Jungen und die braveren, defensiveren Mädchen. Jungen, so bringt es die britische Schulforscherin Tuula Gordon auf den Punkt, versuchen den Erwartungen der Erwachsenengesellschaft an Körperkontrolle zum Beispiel in Form des ruhigen Sitzens auf Schulstühlen durch Bewegung einzelner Körperteile zu entkommen, Mädchen, indem sie Phantasiewelten aufsuchen (vgl. Gordon 1996, S. 42).

Freizeit

Tatsächlich zeigen Studien über die Aufteilung öffentlicher Plätze, dass für Freizeitgestaltungen, die sich mehrheitlich mit den Interessen von Jungen dekken, mehr großflächige, öffentliche Freizeitorte wie zum Beispiel Fußballfelder oder Spielplätze zur Verfügung stehen und für Interessenbereiche von Mädchen wie beispielsweise Rollschuhfahren oder Reiten kaum städtische Orte zur Verfügung gestellt werden (Massey 1994, S. 185). Abenteuerspielplätze, die handwerkliche Fähigkeiten und körperliche Risikobereitschaft schulen sollen, sind ebenfalls an Jungeninteressen orientiert (Flade 1993). Während Jungen zu wenig lernen, sich vor Verletzungen zu schützen und auf den eigenen Körper zu achten, lernen Mädchen, dass ihr Körper permanent potentiell bedroht ist. Aus Angst vor gewaltsamen Übergriffen werden Mädchen häufiger von den Eltern zum Freizeitort begleitet (Nissen 1998); gehen sie allein, dann verweilen sie seltener unterwegs, sondern bewegen sich ziel- und zweckorientiert (Spitthöver

1989). Gabriele Geiger sieht daher auch die Ursachen der Angst von Frauen in öffentlichen Räumen in den „Erziehungspraktiken, nach denen kleine Mädchen immer noch vor dem ‚bösen Onkel' gewarnt werden und der ihnen zugestandene Raum grundsätzlich enger bemessen, verregelter und stärker kontrolliert ist als der von Knaben" (Geiger 1989, S. 397).

Straßenkultur

Während Imke Behnken und Jürgen Zinnecker bis 1920 noch eine Straßenkultur beobachten, in der Mädchen und Jungen gemeinsam agieren, findet Martha Muchow heraus, dass Mädchen selten nur herumtreiben und spielen, sondern sich zweckgerichtet zum Einkaufen, Aufträge erledigen und Geschwister-Beaufsichtigen im öffentlichen Raum aufhalten. Helmut Becker und Michael May (1987) schildern die kulturellen Praktiken von „Arbeiterjugendlichen", sich auf öffentlichen Plätzen und an Straßenecken zu treffen. Es werden dort Informationen ausgetauscht, Witze gemacht und Episoden berichtet. Sie bieten Möglichkeiten des Erfahrungsaustausches und der Informationsweitergabe, also Bildungsräume, wie sie die Schule nur am Rande ermöglicht. Wenn Becker/May die Rituale schildern, mit denen sich die Jugendlichen zum Beispiel begrüßen – „einen Schlag mit der flachen Hand auf den Rücken oder einen mittelschweren Faustschlag auf den Oberarm" – und diese als „maskuline Elemente von Kraft und Mut" (Becker/May 1987, S. 41) interpretieren, wird deutlich, dass es sich nicht nur um eine proletarische, sondern auch um eine männliche Kultur handelt.

Eine groß angelegte Studie des Deutschen Jugendinstituts München (1992) zeigt, dass in jeder Region und jeder Schicht der untersuchten Altersgruppe, nämlich 8- bis 12jährige Jugendliche, sich die Jungen mehr auf der Straße und öffentlichen Plätzen aufhalten als Mädchen (Deutsches Jugendinstitut 1992, insbesondere Nissen 1992). Dieses Ergebnis bestätigt auch Heinz-Hermann Krüger (1996) in einer empirischen Untersuchung, in der das Aufwachsen in Ost- und Westdeutschland verglichen wird. Die Studie des Münchner Jugendinstituts verdeutlicht aber auch, dass Kinder und Jugendliche in der Stadt eine schichtspezifische Straßennutzung aufweisen. Demzufolge halten sich Arbeiterkinder und -jugendliche mehr als Kinder anderer Schichten im Straßenraum auf (Nissen 1992; vgl. Herlyn 1990).

Sportsozialisation

Zusammenfassend bedeutet dies: Über Sozialisationsprozesse lernen die Mehrzahl der Mädchen ein sich reduzierendes räumliches Handeln, Jungen ein expandierendes. Diese Prozesse werden in der Sportsozialisation von Mädchen und Jungen verstärkt (vgl. Kröner/Pfister 1992; Rose 1992a/b; Pfister/Valentin 1993; Löw 1994; Nissen 1998). Jungen pflegen in der Freizeit prinzipiell mehr Hobbies, die auf sportlichen Aktivitäten basieren, wohingegen Mädchen mehr künstlerisch-ästhetische und kulturelle Aktivitäten aufweisen. In den Sportarten, die Mädchen wählen, überwiegen dann wieder solche, die standortgebunden sind und in denen die ästhetische Modellierung des eigenen Körpers im Vordergrund steht und nicht ein expansives Handeln wie zum Beispiel im Laufsport. Auf die Widersprüche, in die sich Mädchen begeben, wenn sie dennoch als männlich assoziierte Sportarten über die Pubertät hinaus betreiben, hat Palzkill (1990) ausführlich hingewiesen. Sie verweigern die als raumreduziert verstandene Weiblichkeit zum Beispiel durch eine erst sehr spät einsetzende Menstruation.

9. Tipps für Studierende

9.1 Institutionen, Studienorte

Bei den im Folgenden aufgelisteten Institutionen handelt es sich um solche, die sich wesentlich oder teilweise mit Fragen der Soziologie der Bildung und Erziehung beschäftigen. Es sind keine rein bildungssoziologischen Institute.

- Max-Planck-Institut für Bildungsforschung (www.mpib-berlin.mpg.de)
  Das Institut für Bildungsforschung ist eines von insgesamt 80 Instituten der Max-Planck-Gesellschaft zur Förderung der Wissenschaften e.V.. 115 MitarbeiterInnen (darunter 37 WissenschaftlerInnen) sind dort fest angestellt. Das Max-Planck-Institut für Bildungsforschung widmet sich der sozialwissenschaftlichen Grundlagenforschung. Zu den Forschungsthemen zählen Bildungs- und Entwicklungsprozesse von der Kindheit bis ins hohe Alter, Bildungssysteme sowie Bildungsinstitutionen. Die vier Forschungsbereiche des Instituts (Adaptives Verhalten und Kognition; Bildung, Arbeit und gesellschaftliche Entwicklung; Erziehungswissenschaft und Bildungssysteme; Entwicklungspsychologie) kooperieren in interdisziplinären Forschungsvorhaben miteinander sowie mit externen Forschungseinrichtungen.
- Forum Bildung (www.forum-bildung.de)
  Das Forum Bildung wurde 1999 von Bund und Ländern ins Leben gerufen, um die Qualität und Zukunftsfähigkeit des deutschen Bildungssystems sicherzustellen. Es beauftragte fünf Expertengruppen, die jeweils einen Themenschwerpunkt behandelten und einen Bericht vorlegten. Die erarbeiteten Empfehlungen wurden in einer Materialienreihe veröffentlicht, die auf der Internetseite nachzulesen sind. Die Themenschwerpunkte sind:
  - Bildungs- und Qualifikationsziele von morgen
  - Förderung von Chancengleichheit
  - Qualitätssicherung im internationalen Wettbewerb
  - Lernen, ein Leben lang
  - Neue Lern- und Lehrkultur
- Bundesinstitut für Berufsbildung (www.bibb.de)
  Die gesetzlich festgelegten Forschungs- und Dienstleistungsarbeiten des BIBB beziehen sich auf die Berufsbildung der FacharbeiterInnen, Fachangestellten, GesellInnen und MeisterInnen. Es beobachtet und untersucht die Aus- und Weiterbildungspraxis in den Betrieben, erprobt neue Wege in der Aus- und Weiterbildung, entwickelt Konzepte für die Qualifizierung der betrieblichen AusbilderInnen, fördert moderne Ausbildungszentren und vieles

mehr. Hinzu kommen verschiedenste Forschungsprojekte, die sich mit den umfassenden Themen der Berufsbildung beschäftigen.

- Deutsches Jugendinstitut (www.dji.de)
  Das DJI wurde 1963 gegründet. Es hat seinen Sitz in München und verfügt über eine Arbeitsstelle in Leipzig. Es ist ein gemeinnütziger Verein mit Mitgliedern aus Politik, Wissenschaft, Verbänden und Institutionen der Kinder-, Jugend- und Familienhilfe. Mit derzeit 125 wissenschaftlichen MitarbeiterInnen ist das DJI das größte außeruniversitäre Forschungsinstitut im Bereich Kinder, Jugendliche, Frauen und Familien. Es gliedert sich in sechs Fachabteilungen, die sich unter anderem mit folgenden Projekten beschäftigen:
  - Familienunterstützende Kinderbetreuungsangebote
  - Schule und soziale Netzwerke
  - Konfliktverhalten von Kindern in Kindertagesstätten
  - Privatschulen und Gruppenerziehung zwischen Schule, Kinder- und Jugendhilfe und Justiz
- Deutsches Institut für Erwachsenenbildung (www.die-bonn.de)
  Das DIE ist ein Institut der Leibnizgemeinschaft (WGL). Es vermittelt zwischen Forschung und Praxis der Erwachsenenbildung, liefert Grundlagen für eine praxisorientierte Forschung und entwickelt innovative Konzepte. Dabei richtet sich das Angebot des DIE an professionelle WeiterbilderInnen und WissenschaftlerInnen. Zusätzlich ist das Institut in vielfältige internationale Zusammenhänge eingebunden, wie z.B. das Engagement in der europäischen Erwachsenenbildungspolitik. Es bringt dabei die deutschen Entwicklungen in Wissenschaft und Praxis auf internationaler Ebene in die Diskussion ein. Zu den bearbeitenden Themen des DIE zählen Alters-, Grundbildung, interkulturelle Bildung, Umweltbildung und vieles mehr.
- Deutsches Institut für Internationale Pädagogische Forschung (www.dipf.de)
  Das DIPF wurde 1951 gegründet und hat seinen Sitz in Frankfurt/Main sowie Standorte in Berlin und Leipzig. Es gehört der Wissenschaftsgemeinschaft Gottfried Wilhelm Leibniz (WGL) an. Das Institut dokumentiert, beschreibt und analysiert die Bildungsentwicklung in Deutschland und anderen Ländern. Zu den weiteren Aufgaben gehören Beratung in Bildungspolitik, Bildungsverwaltung und Bildungspraxis sowie die Förderung des wissenschaftlichen Nachwuchses. Das Institut stellt bibliothekarische und archivarische Dienstleistungen zur Verfügung und führt Forschungen zu den unterschiedlichsten Bereichen der Bildung durch.
- Institut für Hochschulfoschung e.V. (www.hof.uni-halle.de)
  Das Institut wurde 1996 gegründet. Es ist das erste und bisher einzige Institut für Hochschulforschung in den neuen Bundesländern und führt Untersuchungen zur Hochschule im gesellschaftlichen Wandel durch. Erforscht wird dabei hauptsächlich die Hochschulentwicklung in Deutschland im internationalen Kontext. Der besondere regionale Schwerpunkt des Instituts liegt in den neuen Bundesländern. Laufende Projekte sind z.B.:
  - Hochschule und lebenslanges Lernen
  - Organisationsformen an Hochschulen
  - Internationalisierung, Europäisierung und nationale Steuerungsmöglichkeiten der Hochschulsysteme

Eine ständige Aufgabe des HoF besteht in der Dokumentation von Quellen und Daten zur Hochschulentwicklung in Deutschland und Europa.

- Wissenschaftliches Zentrum für Berufs- und Hochschulforschung (www.uni-kassel.de/wzi)
  Gegründet wurde das WZI 1978 als Einrichtung der Universität Gesamthochschule Kassel. Es beschäftigt sich unter anderem mit Fragen des Zusammenhangs von Hochschulausbildung und Beruf sowie von Hochschule, Staat und Gesellschaft. Dabei geht es um Themen wie die Entwicklung des Hochschulsystems, Hochschulen in Entwicklungsländern, Studienbedingungen und Lehr-/Lernprozesse, Berufswege von HochschulabsolventInnen.

Die Soziologie der Bildung und Erziehung ist ein Fach, dass an einigen Universitäten mit ein oder zwei Fachleuten vertreten ist, die nicht selten auch an erziehungswissenschaftlichen Fachbereichen beheimatet sind. Deshalb ist es schwierig, Tipps bezüglich der Studienorte zu geben. Von Interesse für bildungssoziologisch interessierte Studierende sind jedoch mit Sicherheit die folgenden Universitäten (die Liste erhebt keinen Anspruch auf Vollständigkeit):

- Am Institut für Soziologie der Martin-Luther-Universität Halle/Wittenberg arbeitet Reinhard Kreckel, der gleichzeitig Direktor des Instituts für Hochschulforschung ist. Am Fachbereich Erziehungswissenschaften der Universität Halle gibt es darüber hinaus einen Arbeitsbereich Bildungssoziologie, der durch Ursula Rabe-Kleberg vertreten wird.
- In Kassel kann man am Fachbereich Berufsbildungs-, Sozial- und Rechtswissenschaft bildungssoziologische Veranstaltungen im Rahmen des Magister-Studienangebots Soziologie (Haupt- und Nebenfach) bei Ulrich Teichler belegen.
- In Münster kann Bildungssoziologie bei Mathias Grundmann studiert werden.
- Schließlich hat das Institut für Soziologie der TU Darmstadt einen bildungssoziologischen Schwerpunkt, der primär durch Beate Krais vertreten wird.

9.2 Berufsperspektiven

Leider existiert keine Untersuchung speziell zu Berufsperspektiven von BildungssoziologInnen, es können jedoch Schlussfolgerungen aus den Arbeitsmarktchancen von SoziologInnen allgemein gezogen werden. Die Soziologie besitzt im Vergleich zu anderen Disziplinen eine geringe Berufsfeldprägnanz. Die MedizinerIn wird ÄrztIn, die JuristIn wird AnwältIn, aber was wird die SoziologIn? Wenn es überhaupt ein typisches Arbeitsfeld für ihn/sie gibt, dann ist es Forschung und Lehre.

Seit den 70er Jahren hat sich allerdings der Anteil der SozialwissenschaftlerInnen, die eine Stellung an der Hochschule haben, kontinuierlich verringert. So befinden sich 1993 nur noch 22% in Hochschule und Forschung, 22% im Dienstleistungsbereich und der größte Anteil (54%) im öffentlichen Dienst (Brüderl und Reimer 2002, S. 208f.). Viele BildungssoziologInnen arbeiten heute in Verbänden und Vereinen, die sich speziell mit Erziehungs- und Bildungsfragen beschäftigen. Das Berufsfeld für SoziologInnen weitet sich jedoch auch stark auf die Sektoren Marketing, Personal und PR ausgeweitet. 3 Jahre nach dem Examen befinden sich bereits 56% der AbsolventInnen in der Privatwirtschaft. Dieser Anteil ist seit den 70er Jahren kontinuierlich angestiegen. Beratungstätigkeiten sind daher auch für BildungssoziologInnen typisch. In Absolventenstudien geben 23% an, dass sie eine betriebswirtschaftliche Tätigkeit ausüben. Ein etwas geringerer Teil nehmen Verwaltungs- und Referententätigkeiten wahr, kulturelle und journalistische Aufgaben haben 18% und nur noch 13% betreffen Forschung und Wissenschaft (Brüderl und Reimer 2002, S. 208).

SozialwissenschaftlerInnen benötigen viel Zeit für den Übergang auf den Arbeitsmarkt. Studien darüber zeigen, dass sie im Vergleich zu Natur- und WirtschaftswissenschaftlerInnen deutlich langsamer sind (Brüderl und Reimer 2002, S. 202f.). Gleich nach dem Examen gehen nur 22% der AbsolventInnen einer regulären Erwerbstätigkeit nach. In den Folgemonaten steigt der Anteil allerdings schnell an, so dass nach 60 Monaten fast alle (92,5%) in den Arbeitsmarkt integriert sind.

Für einen problemlosen Übergang ins Berufsleben ist es wichtig, dass die Studierenden ihre theoretischen Fähigkeiten durch Praktika ergänzen. Befragungen bereits erfolgreicher Bewerber zeigen, dass praktische Berufserfahrung als wichtigstes Kriterium für den Erhalt der Stelle angegeben werde, aber auch persönliches Engagement bei der Stellensuche (Behrendt, Kallweit und Kromrey 2002, S. 189). Daneben sind EDV-Kenntnisse und der sichere Umgang mit den „neuen Medien“ (Internet) ebenfalls von großer Bedeutung. Während des Studiums vermittelte Qualifikationen wie Statistik/Methodenlehre und analytisches Denken sind Fähigkeiten, die auf dem Arbeitsmarkt sehr gefragt sind. Langfristig von Arbeitslosigkeit sind somit (Bildungs-)SoziologInnen selten betroffen.

9.3 Thematisch sortierte Literaturempfehlungen

9.3.1 Einführungen

Arbeitsgruppe Bildungsbericht am Max-Planck-Institut für Bildungsforschung : Das Bildungswesen der Bundesrepublik Deutschland. Strukturen und Entwicklungen im Überblick. Reinbek 1994.

Böhnisch, L.: Pädagogische Soziologie. Eine Einführung. München 1996.

Büchner, P.: Einführung in die Soziologie des Erziehungs- und Bildungswesens. Darmstadt 1985.

Friedeburg, L. v.: Bildungsreform in Deutschland. Geschichte und gesellschaftlicher Widerspruch. Frankfurt am Main 1992.

Grimm, Susanne: Soziologie der Bildung und Erziehung. München 1987

Krais, Beate: Erziehungs- und Bildungssoziologie. In: Kerber, H./Schmieder, A. (Hg.): Spezielle Soziologien. Reinbek 1994, S. 556-576.

Parelius, R. J./Parelius, A. P.: The sociology of education. London u.a. 1987[2].

9.3.2 Grundbegriffe

Büchner, P./Krüger, H.-H.: Soziale Ungleichheit beim Bildungserwerb innerhalb und außerhalb der Schule. Ergebnisse einer empirischen Untersuchung in Hessen und Sachsen-Anhalt. In: Aus Politik und Zeitgeschichte. Beilage zur Wochenzeitung „Das Parlament". (1996) H. B. 11, S. 21-30.

Klemm, K.: Bildung. In: Allmendinger, J./Ludwig-Mayerhofer, W. (Hrsg.): Soziologie des Sozialstaates. Weinheim/München 2000.

Korte, H./Schäfers, B. (Hrsg.): Einführung in die Hauptbegriffe der Soziologie. Opladen 2000[5].

9.3.3 Bildungs- und Erziehungstheorien

Fend, H.: Sozialisation und Erziehung. Weinberg/Berlin/Basel 1969.

Geulen, D.: Das vergesellschaftete Subjekt. Zur Grundlegung der Sozialisationstheorie. Frankfurt am Main 1989.

Hagemann-White, C.: Sozialisation: weiblich – männlich. Opladen 1984.

Humboldt, W. v.: Theorie der Bildung des Menschen. Bruchstück. In: Humboldt, W. v. (Hrsg.): Werke in fünf Bänden, Bd. III. Schriften zur Sprachphilosophie. Darmstadt 1960, S. 234-240.

Honig, M.-S.: Entwurf zu einer Theorie der Kindheit. Frankfurt am Main 1999.

Grundmann, M.: Konstruktivistische Sozialisationsforschung. Lebensweltliche Erfahrungskontexte, individuelle Handlungskompetenzen und die Konstruktion sozialer Strukturen. Frankfurt am Main 1999.

Luhmann, N.: Das Erziehungssystem der Gesellschaft. Frankfurt a.M. 2002.

Miller, A.: Am Anfang war Erziehung. Frankfurt am Main 1983.

9.3.4 Soziologie der (Vor)Schule

Arbeitsgruppe Bildungsbericht am Max-Planck-Institut für Bildungsforschung: Das Bildungswesen in der Bundesrepublik Deutschland. Strukturen und Entwicklungen im Überblick. Reinbek 1994.

Fölling-Albers, M.: Schulkinder heute. Auswirkungen veränderter Kindheit auf Unterricht und Schulleben. Weinheim/Basel 1992.

Händle, C.: Lehrerinnen in System und Umwelt: Erkundungen ihrer doppelten Sozialisation. Opladen 1998.

Helsper, W. u. a.: Schulkultur und Schulmythos. Rekonstruktionen zur Schulkultur I Opladen 2001.

Illich, I.: Entschulung der Gesellschaft. Eine Streitschrift. München 1995.

9.3.5 Soziologie der Universität

Bourdieu, P.: Homo academicus. Frankfurt am Main 1998[2]

Bourdieu, P.: Vom Gebrauch der Wissenschaft. Eine klinische Soziologie des wissenschaftlichen Feldes, Konstanz 1998.

Engler, S.: In Einsamkeit und Freiheit? Zur Konstruktion der wissenschaftlichen Persönlichkeit auf dem Weg zur Professur. Konstanz 2001.

Goldschmidt, D.: Die gesellschaftliche Herausforderung der Universität. Weinheim 1991.

Parsons, T.: Die amerikanische Universität: ein Beitrag zur Soziologie der Erkenntnis. Frankfurt am Main 1990.

9.3.6 Bildung und soziale Ungleichheit

Gildemeister, R./Wetterer, A.: Wie Geschlechter gemacht werden. Die soziale Konstruktion der Zweigeschlechtlichkeit und ihre Reifizierung in der Frauenforschung. In: Knapp, G.-A./Wetterer, A. (Hrsg.): Traditionen – Brüche. Entwicklungen feministischer Theorie. Freiburg im Br. 1992, S. 201-254.

Hartmann, M.: Der Mythos von den Leistungseliten. Spitzenkarrieren und soziale Herkunft in Wirtschaft, Politik, Justiz und Wissenschaft. Frankfurt a.M./New York 2002.

Krais, B.: Bildungsexpansion und soziale Ungleichheit in der Bundesrepublik Deutschland. In: Bolder, A.et al.. (Hrsg.): Jahrbuch Bildung und Arbeit: Die Wiederentdeckung der Ungleichheit. Aktuelle Tendenzen in Bildung und Arbeit. Opladen 1996b, S. 118-146.

Krappmann, L.: Die Reproduktion des Systems gesellschaftlicher Ungleichheit in der Kinderwelt. In: Grundmann, M. (Hrsg.): Konstruktivistische Sozialforschung: lebensweltliche Erfahrungskontexte, individuelle Handlungskompetenzen und dieKonstruktion sozialer Strukturen. Frankfurt am Main 1999, S. 228-239.

9.3.7 Periodika

Bundesminister für Bildung und Wissenschaft (Hg.): Berufsbildungsbericht. Bad Honnef 1977ff.

Bericht über die menschliche Entwicklung. (Hrsg. von der Deutschen Gesellschaft für die Vereinten Nationen e.V.) Bonn 1990ff.

Jahrbuch der Schulentwicklung. (Hrsg. von Rolff, H.-G./Klemm, K./Tillmann, K.-J.) Weinheim 1980ff.

Jahrbuch Bildung und Arbeit. (Hrsg. von Bolder, A./Heinz, W. R./Rodax, K.) Opladen 1996ff.

Zeitschrift für Soziologie der Erziehung und Sozialisation.

Literatur

Abbott, A.: The System of Profession. An Essay on the Division of Expert Labour. Chicago/London 1988.

Abels, H.: Jugend vor der Moderne. Soziologische und psychologische Theorien des 20. Jahrhunderts. Opladen 1993.

Adorno, T. W.: Erziehung nach Auschwitz. In: Adorno, T. W. (Hrsg.): Kulturkritik und Gesellschaft 2. Gesammelte Schriften. Frankfurt am Main 1977, S. 674-690.

Ahrend, C.: Lehren der Straße. Über Kinderöffentlichkeit. Zeiten und Zwischenräume. In: Ecarius, J./Löw, M. (Hrsg.): Raumbildung – Bildungsräume. Über die Verräumlichung sozialer Prozesse. Opladen 1997, S. 197-212.

Ahrend, C.: Mobilitätsstrategien zehnjähriger Jungen und Mädchen als Grundlage städtischer Verkehrsplanung. Münster/New York 2002.

Alexander, J. C.: Soziale Differenzierung und kultureller Wandel. Essays zur neofunktionalistischen Gesellschaftstheorie. Frankfurt am Main 1993.

Allmendinger, J.: Bildungsarmut: Zur Verschränkung von Bildungs- und Sozialpolitik. In: Soziale Welt (1999) H. 50, S. 35-50.

Allmendinger, J./Brückner, E. et al.: The Limitations of Individual Level Analyses. In: Heinz, W. R. (Hrsg.): Institutions and Gatekeeping in the Life Course. Weinheim 1992, S. 187-215.

Allmendinger, J./Brückner, H. et al.: Berufliche Werdegänge von Frauen in der Max-Planck-Gesellschaft: Eine Echternacher Springprozession. In: Rehberg, K.-S. (Hrsg.): Differenz und Intergration. 28. Deutscher Soziologentag. Dresden/Opladen 1997, S. 537-544.

Amendt, G.: Genderaspekte im Schüler-Lehrer-Verhältnis. In: Leviathan (1996) H. 3, S. 372-386.

Ariès, P.: Geschichte der Kindheit. München 1975.

Arnold, K.: Kind und Gesellschaft in Mittelalter und Renaissance. Paderborn 1980.

Arbeitsgruppe Bildungsbericht am Max-Planck-Institut für Bildungsforschung: Das Bildungswesen in der Bundesrepublik Deutschland. Strukturen und Entwicklungen im Überblick. Reinbek 1994.

Ashenden, S.: Foucault contra Habermas: Recasting the Dialogue between Genealogy and Critical Theory. London 1999.

Baacke, D.: Sozialökologische Ansätze in der Jugendforschung. In: Krüger, H.-H. (Hrsg.): Handbuch der Jugendforschung. Opladen 1993², S. 135-157.

Baacke, D./Ferchhoff, W.: Jugend, Kultur und Freizeit. In: Krüger, H.-H. (Hrsg.): Handbuch der Jugendforschung. Opladen 1992², S. 291-325.

Baecker, D.: Die Universität als Algorithmus. In: Laske, S./Scheytt, T. et al. (Hrsg.): Universität im 21. Jahrhundert. Zur Interdependenz von Begriff und Organisation der Wissenschaft. München 2000, S. 47-75.

Baier, H. (Hrsg.): Helmut Schelsky – ein Soziologe in der Bundesrepublik: Eine Gedächtnisschrift von Freunden, Kollegen und Schülern. Stuttgart 1986.

Balibar, E.: Die Nation-Form: Geschichte und Ideologie. In: Balibar, E./Wallerstein, I. (Hrsg.): Rasse, Klasse, Nation. Hamburg/Berlin 1990, S. 107-130.

Baraldi, C./Corsi, G. et al.: GLU. Glossar Niklas Luhmann. Frankfurt am Main 1997.

Barber, B./Gerhardt, U. (Hrsg.): Agenda for Sociology: Classic Sources and Current Uses of Talcott Parsons's Work. Baden-Baden 1999.

Bauer, A./Gröning, K. et al.: Die Regel der Ausnahme: Hochschulfrauen. Frankfurt am Main 1993.

Baumert, J./Lehmann, R.: TIMSS – mathematisch-naturwissenschftlicher Unterricht im internationalen Vergleich. Opladen 1997.

Baumert, J./Schümer, G.: Familiäre Verhältnisse, Bildungsbeteiligung und Kompetenzerwerb. In: Deutsches PISA-Konsortium (Hrsg.): PISA 2000. Basiskompetenzen von Schülerinnen und Schülern im internationalen Vergleich. Opladen 2001, S. 323-407.

Beck, U.: Risikogesellschaft. Auf dem Weg in eine andere Moderne. Frankfurt am Main 1986.

Becker, H./May, M.: „Die lungern eh' nur da ‚rum" – Raumbezogene Interessenorientierungen von Unterschichtsjugendlichen und ihre Realisierung in öffentlichen Räumen. In: Specht, W. (Hrsg.): Die gefährliche Straße. Jugendkonflikte und Stadtteilarbeit. Bielefeld 1987, S. 35-46.

Beckmann, S.: Professionalization: borderline authority and autonomy in work. In: Burrage, M./Torstendahl, R. (Hrsg.): Professions in Theory and History. Rethinking the Study of the Profession. London 1991, S. 115-138.

Behnken, I./Zinnecker, J.: Vom Straßenkind zum verhäuslichten Kind. Zur Modernisierung städtischer Kindheit 1900-1980. In: Sozialwissenschaftliche Informationen (1987) H. 2, S. 87-96.

Behrendt, E./Kallweit, H. et al.: Primat der Theorie? Arbeitsmarkt, Qualifikation und das Image der Soziologie. In: Stockmann, R./Meyer, W. et al. (Hrsg.): Soziologie im Wandel. Universitäre Ausbildung und Arbeitsmarktchancen in Deutschland. Opladen 2002, S. 187-197.

Bell, D.: Die nachindustrielle Gesellschaft. Frankfurt am Main/New York 1975.

Bellenberg, G./Böttcher, W. et al.: Schule und Unterricht. In: Böttcher, W./Klemm, K. et al. (Hrsg.): Bildung und Soziales in Zahlen. Statistisches Handbuch zu Daten und Trends im Bildungsbereich. Weinheim u.a. 2001, S. 93-126.

Benninghaus, C.: Die anderen Jugendlichen. Arbeitermädchen in der Weimarer Republik. Frankfurt am Main u.a. 1997.

Berger, P. L./Luckmann, T.: Die gesellschaftliche Konstruktion von Wirklichkeit. Eine Theorie der Wissenssoziologie. Frankfurt am Main 1972[3].

Berking, H.: Local Frames and Global Images – Nation State and New Urban Underclass: Über die Globalisierung lokaler Wissensbestände. In: Löw, M. (Hrsg.): Differenzierungen des Städtischen. Opladen 2002, S. 107-123.

Bernsdorf, W.: Wörterbuch der Soziologie. Stuttgart 1969.

Bernstein, B.: Studien zur sprachlichen Sozialisation. Düsseldorf 1973.

Blankertz, H.: Bildung im Zeitalter der großen Industrie. Hannover 1969.

Blankertz, H.: Die Geschichte der Pädagogik. Von der Aufklärung bis zur Gegenwart. Wetzlar 1982.

Blomert, R.: Intellektuelle im Aufbruch: Karl Mannheim, Alfred Weber, Norbert Elias und die Heidelberger Sozialwissenschaften der Zwischenkriegszeit. München u.a. 1999.

Blosfeld, H.-P.: Changes in Educational Opportunities in the Federal Republic of Germany. A Lonitudinal Study of Cohorts Born between 1916 and 1965. In: Shavit, Y./Blosfeld, H.-P. (Hrsg.): Persist Inequality: Changing Educational Stratification in Thirteen Countries. Boulder Colorado 1993, S. 51-74.

Bochow, M./Joas, H.: Wissenschaft und Karriere – Der berufliche Verbleib des akademischen Mittelbaus. Frankfurt 1987.

Böhme, J.: Schulmythen und ihre imaginäre Verbürgung durch oppositionelle Schüler. Bad Heilbrunn 2000.

Böhnisch, L.: Pädagogische Soziologie. Eine Einführung. München 1996.

Bolder, A./Hendrich, W.: Fremde Bildungswelten. Alternative Strategien lebenslangen Lernens. Opladen 2000.

Boll, F.: Jugend im Umbruch vom Nationalsozialismus zur Nachkriegsdemokratie. In: Archiv für Sozialgeschichte (1997) H. 5, S. 482ff.

Bollenbeck, G.: Bildung und Kultur. Glanz und Elend eines deutschen Deutungsmusters. Frankfurt am Main/Leipzig 1994.

Boris, D.: Krise und Planung: Die Politische Soziologie im Spätwerk Karl Mannheims. Stuttgart 1971.

Bourdieu, P.: Die feinen Unterschiede. Frankfurt am Main 1983².

Bourdieu, P.: Sozialer Raum und „Klassen". Frankfurt am Main 1991².

Bourdieu, P.: Die männliche Herrschaft. In: Dölling, I./Krais, B. (Hrsg.): Ein alltägliches Spiel. Geschlechterkonstruktion in der sozialen Praxis. Frankfurt am Main 1997, S. 153-217.

Bourdieu, P.: Vom Gebrauch der Wissenschaft. Eine klinische Soziologie des wissenschaftlichen Feldes. Konstanz 1998.

Bourdieu, P.: Homo academicus. Frankfurt am Main 1998².

Bourdieu, P./Boltanski, L. et al.: Titel und Stelle. Über die Reproduktion sozialer Macht. Frankfurt am Main 1981.

Bourdieu, P./Champagne, P.: Die intern Ausgegrenzten. In: Bourdieu et al. (Hrsg.): Das Elend der Welt. Zeugnisse und Diagnosen alltäglichen Leidens an der Gesellschaft. Konstanz 1997, S. 527-533.

Bourdieu, P./Wacquant, L. J. D.: Reflexive Anthropologie. Frankfurt am Main 1996.

Brehmer, I.: Sexismus in der Schule. Weinheim/Basel 1982.

Breidenstein, G.: Der Gebrauch der Geschlechterunterscheidung in der Schulklasse. In: Zeitschrift für Soziologie 5 (1997), S. 337-351.

Breidenstein, G./Kelle, H.: Geschlechteralltag in der Schulklasse. Ethnographische Studien zur Gleichaltrigenkultur. Weinheim 1998.

Briggs, A./Burke, P.: A Social History of the Media: from Gutenberg to the Internet. Cambridge 2002.

Brock, D./Suckow, J.: Nationale Bildungssysteme im Globalisierungsprozeß. In: Bolder, A./Heinz, W. R. et al. (Hrsg.): Deregulierung der Arbeit – Pluralisierung der Bildung? Opladen 2001.

Bronfenbrenner, U.: Die Ökologie der menschlichen Entwicklung: Natürliche und geplante Experimente. Stuttgart 1981.

Brüderl, J./Reimer, D.: Soziologinnen und Soziologen im Beruf. Ergebnisse ausgewählter Absolventenstudien der 90er Jahre. In: Stockmann, R./Meyer, W. et al. (Hrsg.): Soziologie im Wandel. Universitäre Ausbildung und Arbeitsmarktchancen in Deutschland. Opladen 2002, S. 199-214.

Bublitz, H.: Ich gehörte irgendwie so nirgends hin ... Arbeitertöchter an der Hochschule. Gießen 1980.

Büchner, P.: Einführung in die Soziologie des Erziehungs- und Bildungswesens. Darmstadt 1985.

Büchner, P./Fuhs, B. et al. (Hrsg.): Kinderkulturelle Praxis. Kindliche Handlungskontexte und Aktivitätsprofile im außerschulischen Lebensalltag. Marburger Beiträge zur Kindheits- und Jugendforschung. Opladen 1994.

Büchner, P./Krüger, H.-H.: Soziale Ungleichheit beim Bildungserwerb innerhalb und außerhalb der Schule. Ergebnisse einer empirischen Untersuchung in Hessen und Sachsen-Anhalt. In: Aus Politik und Zeitgeschichte. Beilage zur Wochenzeitung „Das Parlament" (1996) H. 11, S. 21-30.

Bude, H.: Das nervöse Selbst in der geschlossenen Welt des Sinns. Niklas Luhmann und Pierre Bourdieu im Vergleich. In: Merkur (1990) H. 5, S. 429-433.

Burgess, E. W.: The Growth of the City. An Introduction to a Research Project. In: Park, R. E. et al. (Hrsg.): The City. Chicago 1974⁷.

Butler, J.: Das Unbehagen der Geschlechter. Frankfurt am Main 1991.

Clarke, S. H./Campbell, F. A.: Can Intervention Early prevent Crime later? – The Abecedarian Project Compared with Other Programms. In: Early Childhood Research Quarterly (1998) H. 2, S. 319-343.
Colberg-Schrader, H./Krug, M.: Arbeitsfeld Kindergarten. Pädagogische Wege, Zukunftsperspektiven und berufliche Perspektiven. Weinheim/München 1999.
Coleman, J. S.: Die asymmetrische Gesellschaft. Vom Aufwachsen mit unpersönlichen Systemen. Weinheim/Basel 1986.
Collège de France: Vorschläge für das Bildungswesen der Zukunft. In: Müller-Rolli, S. (Hrsg.): Das Bildungswesen der Zukunft. Stuttgart 1987, S. 253-282.
Combe, A.: Pädagogische Professionalität, Hermeneutik und Lehrerbildung. Am Beispiel der Berufsbelastung von Grundschullehrkräften. In: Combe, A./Helsper, W. (Hrsg.): Pädagogische Professionalität. Untersuchungen zum Typus pädagogischen Handelns. Frankfurt am Main 1999³, S. 501-520.
Dahrendorf, R.: Bildung ist Bürgerrecht. Hamburg 1965.
Dane, G. et al. (Hrsg.): Anschlüsse. Versuche nach Michel Foucault. Tübingen 1985.
Daschner, P./Rolff, H. G. et al. (Hrsg.): Schulautonomie – Chancen und Grenzen. Weinheim/München 1995.
Davies, B.: Frösche, Schlangen und feministische Märchen. Hamburg 1992.
Daxner, M.: Ein bedrohter Stand? Selbst- und Fremdbilder des Hochschullehrerberufs. In: Das Hochschulwesen: Forum für Hochschulforschung, -praxis und -politik 1 (1999) H. 47, S. 26-29.
DeMause, L. (Hrsg.): Hört ihr die Kinder weinen. Eine psychogenetische Geschichte der Kindheit. Frankfurt am Main 1989⁶.
Deutsches Jugendinstitut (Hrsg.): Was tun Kinder am Nachmittag? Ergebnisse einer empirischen Untersuchung zur mittleren Kindheit. München 1992.
Dewe, B./Ferchhoff, W. et al. (Hrsg.): Erziehen als Profession. Opladen 1992.
Drewek, P./Harney, K.: Beteiligung und Ausschluß. Zur Sozialgeschichte von Bildungssystem und Karriere. In: Tenorth, H. E. (Hrsg.): Allgemeine Bildung. Weinheim/München 1986, S. 138-153.
Du Bois-Reymond, M.: Zum Wandel der Beziehungen zwischen Eltern und Heranwachsenden. In: Bücher, P./Krüger, H.-H. (Hrsg.): Aufwachsen hüben und drüben. Opladen 1991, S. 297-306.
Dubiel, H.: Kritische Theorie der Gesellschaft. Weinheim/München 1997.
Dudek, P.: Geschichte der Jugend. In: Krüger, H.-H. (Hrsg.): Handbuch der Jugendforschung. Opladen 1992², S. 305-332.
Durkheim, E.: Die Regeln der soziologischen Methode. Neuwied 1961.
Durkheim, E.: Erziehung und Soziologie. Düsseldorf 1972.
Durkheim, E.: Die Entwicklung der Pädagogik. Zur Geschichte und Soziologie des gelehrten Unterrichts in Frankreich. Weinheim/Basel 1977.
Durkheim, E.: Erziehung, Moral und Gesellschaft. Vorlesung an der Sorbonne 1902/1903. Frankfurt am Main 1999³a.
Durkheim, E.: Über soziale Arbeitsteilung. Studie über die Organisation höherer Gesellschaften. Frankfurt am Main 1999³b.
Ecarius, J.: Individualisierung und soziale Reproduktion im Lebensverlauf. Konzepte der Lebenslaufforschung. Opladen 1996.
Ecarius, J.: Lebenslanges Lernen und Disparitäten in sozialen Räumen. In: Ecarius, J./Löw, M. (Hrsg.): Raumbildung – Bildungsräume. Über die Verräumlichung sozialer Prozesse. Opladen 1997, S. 33-62.
Eigler, H./Hansen, R. et al.: Quantitative Entwicklungen: Wem hat die Bildungsexpansion genutzt? In: Rolff, H.-G. et al. (Hrsg.): Jahrbuch der Schulentwicklung. Weinheim 1980, S. 45-71.
Elias, N.: Über den Prozeß der Zivilisation. Soziogenetische und psychogenetische Untersuchungen. Frankfurt am Main 1976.
Engelbrech, G.: Vom Arbeitskräftemangel zum gegenwärtigen Arbeitskräfteüberschuß: Frauen und Erwerbstätigkeit in den neuen Bundesländern. In: MittAB 4 (1991) S. 648-657.

Engler, S.: Fachkultur, Geschlecht und soziale Reproduktion. Eine Untersuchung über Studentinnen und Studenten der Erziehungswissenschaft, Elektrotechnik und des Maschinenbaus. Weinheim 1993.
Engler, S.: Studentische Lebensstile und Geschlecht. In: Dölling, I./Krais, B. (Hrsg.): Ein alltägliches Spiel. Geschlechterkonstruktion in der sozialen Praxis. Frankfurt am Main 1997, S. 309-329.
Engler, S.: In Einsamkeit und Freiheit? Zur Konstruktion der wissenschaftlichen Persönlichkeit auf dem Weg zur Professur. Konstanz 2001.
Faulstich-Wieland, H.: Koedukation. Enttäuschte Hoffnungen? Darmstadt 1991.
Faulstich-Wieland, H./Horstkemper, M.: „Trennt uns bitte, bitte nicht!" Koedukation aus Mädchen- und Jungensicht. Opladen 1995.
Fend, H.: Sozialisation und Erziehung. Weinberg/Berlin/Basel 1969.
Fend, H.: Gesellschaftliche Bedingungen schulischer Sozialisation. Soziologie der Schule I. Weinheim/Basel 1974.
Fischer, A. et al.: Jugend 2000. Shell Studie. Opladen 2000.
Flade, A.: Kann der Rückzug der Mädchen aus dem öffentlichen Raum verhindert werden? Empirische Forschungsergebnisse und Schlußfolgerungen. In: Flade, A./Kustor-Hüttl, B. (Hrsg.): Mädchen in der Stadtplanung. Weinheim 1993, S. 23-40.
Fölling-Albers, M.: Schulkinder heute. Auswirkungen veränderter Kindheit auf Unterricht und Schulleben. Weinheim/Basel 1992.
Foucault, M.: Wahnsinn und Gesellschaft. Eine Geschichte des Wahns im Zeitalter der Vernunft. Frankfurt am Main 1969.
Foucault, M.: Die Geburt der Klinik. Eine Archäologie des ärztlichen Blicks. Frankfurt am Main 1973.
Foucault, M.: Überwachen und Strafen. Die Geburt des Gefängnisses. Frankfurt am Main 1977.
Foucault, M.: Sexualität und Wahrheit. Der Wille zum Wissen. Frankfurt am Main 1983.
Frasch, H./Wagner, A.: „Auf Jungen achtet man einfach mehr...". In: Bremer, I. (Hrsg.): Sexismus in der Schule. Weinheim/Basel 1982, S. 260-279 .
Frerichs, P.: Klasse und Geschlecht 1. Arbeit, Macht, Anerkennung und Interessen. Opladen 1997.
Frerichs, P./Steinrücke, M.: Kochen – ein männliches Spiel? Die Küche als geschlechts- und klassenstrukturierter Raum. In: Dölling, I./Krais, B. (Hrsg.): Ein alltägliches Spiel. Geschlechterkonstruktion in der sozialen Praxis. Frankfurt am Main 1997, S. 231-255.
Friebertshäuser, B.: Übergangsphase Studienbeginn: eine Feldstudie über Riten der Initiation in eine studentische Fachkultur. Weinheim/München 1992.
Friedeburg, L. von: Probleme der Hochschulreform. In: Die Zukunft. Sozialistische Zeitschrift für Politik, Wirtschaft und Kultur (1969) H. 15&16.
Friedeburg, L. von: Bildungsreform in Deutschland. Geschichte und gesellschaftlicher Widerspruch. Frankfurt am Main 1992.
Friedrichs, J.: Stadtanalyse. Reinbek 1977.
Fuchs, W.: Jugendliche Statuspassage oder individualisierte Jugendbiographie? In: Soziale Welt (1983) H. 3, S. 341-371.
Fuchs-Heinritz, W./Lautmann, R. et al.: Lexikon zur Soziologie. Opladen 1994.
Fuhrmann, M.: Der europäische Bildungskanon des bürgerlichen Zeitalters. Frankfurt am Main/Leipzig 1999.
Funken, C./Löw, M.: Ego-Shooters Container. Raumkonstruktionen im elektronischen Netz. In: Maresch, R./Werber, N. (Hrsg.): Raum-Wissen-Macht. Frankfurt am Main 2002, S. 69-91.
Geiger, G.: Der enteignete, der ungeeignete Ort. Psychosoziale Faktoren für das Verhalten von Frauen im städtischen Raum. In: Verhaltenstherapie und psychosoziale Praxis (1989) H. 2, S. 389-403.
Geiger, T.: Erziehung als Gegenstand der Soziologie. In: Hurrelmann, K. (Hrsg.): Soziologie der Erziehung. Weinheim/Basel 1974, S. 85-105.
Geulen, D.: Das vergesellschaftete Subjekt. Zur Grundlegung der Sozialisationstheorie. Frankfurt am Main 1989.

Giddens, A.: Durkheim. Hassocks 1978.
Giddens, A.: Interpretative Soziologie. Eine kritische Einführung. Frankfurt am Main/New York 1984.
Giddens, A.: Die Konstitution der Gesellschaft. Grundzüge einer Theorie der Strukturierung. Frankfurt am Main/New York 1988.
Giel, K.: Friedrich Fröbel. In: Scheuerl, H. (Hrsg.): Klassiker der Pädagogik I. München 1979, S. 249-269.
Gildemeister, R./Wetterer, A.: Wie Geschlechter gemacht werden. Die soziale Konstruktion der Zweigeschlechtlichkeit und ihre Reifizierung in der Frauenforschung. In: Knapp, G.-A./Wetterer, A. (Hrsg.): Traditionen – Brüche. Entwicklungen feministischer Theorie. Freiburg im Br. 1992, S. 201-254.
Glotz, P.: Bildungsziele für die Informationsgesellschaft. Lernen und Bildung in der Wissensgesellschaft. Virtuelle Konferenz 2. – 13. November 1998 (http://www.bildung2001.de/beitragforum1_1.html).
Glumper, E.: Integration von Frauenforschung in die LehrerInnenbildung. Über Möglichkeiten und Grenzen des Forschungstransfers zwischen alten und neuen Bundesländern. In: Hempel, M. (Hrsg.): Verschieden und doch gleich. Schule und Geschlechterverhältnis in Ost und West. Bad Heilbrunn 1995, S. 27-47.
Gogolin, I./Lenzen, D.: Medien-Generation. Opladen 1999.
Goldschmidt, D.: Die gesellschaftliche Herausforderung der Universität. Weinheim 1991.
Goldschmidt, D./Sommerkorn, I.: Deprivation and Disadvantage: Federal Republic of Germany. In: Passow, A. H. (Hrsg.): Deprivation and Disadvantage: Nature and Manifestations. Hamburg 1970, S. 119-169.
Gordon, T.: Citizenship, Difference and Marginality in Schools: Spatial and Embodies Aspects of Gender Construction. In: Gipps, C./Murphy, P. (Hrsg.): Equity in the Classroom: Towards Effective Pedagogy for Girls and Boys. Washington D.C. 1996, S. 34-45.
Gottschall, K.: Soziale Ungleichheit und Geschlecht. Kontinuitäten und Brüche, Sackgassen und Erkenntnispotentiale im deutschen soziologischen Diskurs. Opladen 2000.
Grimm, S.: Die Bildungsabstinenz der Arbeiter. München 1966.
Grimm, S.: Soziologie der Bildung und Erziehung. Eine Einführung und kritische Bilanz. Ehrenwirth 1987.
Grossmann, A.: Die „Neue Frau“ und die Rationalisierung der Sexualität in der Weimarer Republik. In: Snitow, A. et al. (Hrsg.): Die Politik des Begehrens. Berlin 1985, S. 38-62.
Großmann, W.: KinderGarten. Weinheim 1990.
Grundmann, H.: Vom Ursprung der Universität im Mittelalter. Darmstadt 1976².
Grundmann, M.: Konstruktivistische Sozialisationsforschung. Lebensweltliche Erfahrungskontexte, individuelle Handlungskompetenzen und die Konstruktion sozialer Strukturen. Frankfurt am Main 1999.
Habermas, J.: Protestbewegung und Hochschulreform. Frankfurt am Main 1969.
Habermas, J.: Technik und Wissenschaft als „Ideologie“. Frankfurt am Main 1973.
Hagemann-White, C.: Sozialisation: weiblich – männlich. Opladen 1984.
Hamm, B.: Einführung in die Siedlungssoziologie. München 1982.
Händle, C.: Lehrerinnen in System und Umwelt: Erkundungen ihrer doppelten Sozialisation. Opladen 1998.
Hardach-Pinke, I./Hardach, G. (Hrsg.): Kinderalltag. Reinbek 1981.
Harms, G./Preissing, C. (Hrsg.): Kinderalltag. Beiträge zur Analyse der Veränderung von Kindheit. Berlin 1988.
Hartmann, J.: Heterosexismus in der pädagogischen Ausbildung. In: Vogel, A. et al. (Hrsg.): „Wie ‚normal‘ ist es, verschieden zu sein?“ Vermittlung der Norm Heterosexualität in der Schule. Herausgegeben von der Zentralen Frauenbeauftragten der TU Berlin. Berlin 1996, S. 12-17.
Hartmann, M.: Der Mythos von den Leistungseliten. Spitzenkarrieren und soziale Herkunft in Wirtschaft, Politik, Justiz und Wissenschaft. Frankfurt am Main/New York 2002.
Hausen, K.: Die Polarisierung der „Geschlechtscharaktere“. Eine Spiegelung der Dissoziation von Erwerbs- und Familienleben. In: Conze, W. (Hrsg.): Sozialgeschichte der Familie in der Neuzeit Europas. Stuttgart 1976, S. 363-393.

Hausen, K.: Warum Männer Frauen zur Wissenschaft nicht zulassen wollen. In: Hausen, K./Nowotny, H. (Hrsg.): Wie männlich ist die Wissenschaft? Frankfurt am Main 1986, S. 31-40.
Häußermann, H./Siebel, W.: Gemeinde- und Stadtsoziologie. In: Kerber, H./Schmieder, A. (Hrsg.): Spezielle Soziologien. Problemfelder, Forschungsbereiche, Anwendungsorientierungen. Reinbek 1994, S. 363-387.
Heinsohn, G./Knieper, B. M. C.: Theorie des Kindergartens und der Spielpädagogik. Frankfurt am Main 1975.
Heintz, B. et al.: Ungleich unter Gleichen. Studien zur geschlechtsspezifischen Segregation des Arbeitsmarktes. Frankfurt am Main 1997.
Heinzel, F.: Methoden der Kindheitsforschung. Weinheim 2000.
Hekman, S. J.: Feminist Interpretations of Michel Foucault. University Park, Pennsylvania 1996.
Helsper, W.: Schule in den Antinomien der Moderne. In: Krüger, H. H. (Hrsg.): Abschied von der Aufklärung? Perspektiven der Erziehungswissenschaft. Opladen 1990, S. 175-194.
Helsper, W.: Antinomien des Lehrerhandelns in modernisierten pädagogischen Kulturen. Paradoxe Verwendungsweisen von Autonomie und Selbstverantwortung. In: Combe, A./ Helsper, W. (Hrsg.): Pädagogische Professionalität. Untersuchungen zum Typus pädagogischen Handelns. Frankfurt am Main 1999³, S. 521-569.
Helsper, W. et al.: Schulkultur und Schulmythos. Rekonstruktionen zur Schulkultur I. Opladen 2001.
Helsper, W.: Sozialisation. In: Krüger, H.H./Helsper, W. (Hrsg.): Einführung in die Grundbegriffe und Grundfragen der Erziehungswissenschaft. Opladen 2002⁵, S. 71-79.
Hempel, M.: Grundschulreform und Koedukation. In: Hempel, M. (Hrsg.): Grundschulreform und Koedukation. München 1996, S. 107-124.
Henz, U./Maas, I.: Chancengleichheit durch Bildungsexpansion? In: Kölner Zeitschrift für Soziologie und Sozialpsychologie (1995) H. 47, S. 605-633.
Herlyn, U.: Zur Aneignung von Raum im Lebensverlauf. In: Bertels, L./Herlyn, U. (Hrsg.): Spezielle Soziologien. Problemfelder, Forschungsbereiche, Anwendungsorientierungen. Opladen 1990, S. 7-34.
Herrlitz, H.-G.: Deutsche Schulgeschichte von 1800 bis zur Gegenwart. Eine Einführung. Königstein 1981.
Herzog, R.: Bildung und Elite: Wissen ist die wichtigste Ressource (1997). Rede auf dem Berliner Bildungsforum.
http://www.bundespraesident.de/top/dokumente/Rede/ix_12065.htm
Heydorn, H. J.: Über den Widerspruch von Bildung und Herrschaft. Frankfurt am Main 1970.
Hilbig, N.: Mit Adorno Schule machen: Beiträge zu einer Pädagogik der kritischen Theorie. Bad Heilbrunn 1997.
Hildebrand, S./Musiol, M.: „Es müßte eben wieder ein Plan sein“. Ambivalenz zwischen Befreiung und Angst vor Freiheit. Pädagogische Berufsbiographien im Umbruch. In: Löw, M. et al. (Hrsg.): Pädagogik im Umbruch. Kontinuität und Wandel in den neuen Bundesländern. Opladen 1995, S. 209-228.
Hitzler, R.: Perspektivwechsel – Über künstliche Dummheit, Lebensweltanalyse und Allgemeine Soziologie. In: Soziologie (1997) H. 4, S. 5-18.
Hoerning, E./Krais, B.: Der Ausbruch aus der Normalbiographie. Milieu-untypische Lebensläufe von Arbeitertöchtern. In: Bolder, A./Rodax, K. (Hrsg.): Das Prinzip der aufge(sc)hobenen Belohnung. Die Sozialisation von Arbeiterkindern für den Beruf. Bonn 1987, S. 201-231.
Holzkamp, C.: Heterosexismus – (k)ein Thema für Heterosexuelle? In: Vogel, A. et al. (Hrsg.): „Wie ‚normal‘ ist es, verschieden zu sein?“ Vermittlung der Norm Heterosexualität in der Schule. Herausgegeben von der Zentralen Frauenbeauftragten der TU Berlin. Berlin 1996, S. 9-11.
Honig, M.-S.: Entwurf zu einer Theorie der Kindheit. Frankfurt am Main 1999.
Horkheimer, M./Adorno, T. W.: Dialektik der Aufklärung. Philosophische Fragmente. Frankfurt am Main 1988.

Hornstein, W.: Jugend in ihrer Zeit. Geschichte und Lebensformen des jungen Menschen in der europäischen Welt. Hamburg 1966.
Horster, D.: Niklas Luhmann. München 1997.
Horstkemper, M.: Schule, Geschlecht und Selbstvertrauen. Weinheim 1987.
Hradil, S.: Sozialstrukturanalyse in einer fortgeschrittenen Gesellschaft. Opladen 1987.
Hradil, S.: Soziale Ungleichheiten, Milieus und Leben in den Ländern der Europäischen Gemeinschaft. In: Hradil, S./Immerfall, S. (Hrsg.): Die westeuropäischen Gesellschaften im Vergleich. Opladen 1997, S. 475-519.
Huber, L.: Fachkulturen. Über die Mühen der Verständigung zwischen den Disziplinen. In: Neue Sammlung 31 (1991) S. 3-24.
Huke-Dedier, E.: Die Wissenssoziologie Karl Mannheims in der Interpretation durch die kritische Theorie: Kritik einer Kritik. Frankfurt am Main u.a. 1995.
Humboldt, W. von: Theorie der Bildung des Menschen. Bruchstück. In: Humboldt, W. von (Hrsg.): Werke in fünf Bänden, Bd. III. Schriften zur Sprachphilosophie. Darmstadt 1960, S. 234-240.
Humboldt, W. von: Der Königsberger und der Litauische Schulplan. In: Humboldt, W. von (Hrsg.): Werke in fünf Bänden. Bd. IV. Schriften zur Politik und zum Bildungswesen. Darmstadt 1964, S. 168-195.
Hurrelmann, K.: Lebensphase Jugend. Weinheim/München 1993.
Hurrelmann, K./Albert, M.: Jugend 2002. 14. Shell Jugendstudie. Frankfurt am Main 2002.
Illich, I.: Entschulung der Gesellschaft. Eine Streitschrift. München 1995.
Institut für Schulentwicklung (IFS): Die Schule im Spiegel der öffentlichen Meinung. Ergebnisse der elften IFS-Repräsentativbefragung der bundesdeutschen Bevölkerung. In: Rolff, H-G. et al. (Hrsg.): Jahrbuch der Schulentwicklung. Bd. 11. Weinheim/München 2000, S. 13-50.
Jaeggi, U. et al.: Geist und Katastrophe. Studien zur Soziologie im Nationalsozialismus. Berlin 1983.
Janshen, D. (Hrsg.): Frauen über Wissenschaften. Die widerspenstigen Erbinnen der Männeruniversität. Mannheim/München 1991.
Joas, H.: Die Kreativität des Handelns. Frankfurt am Main 1992.
Joppke, C.: The Cultural Dimension of Class Formation und Class Struggle: On the Social Theory of Pierre Bourdieu. In: Berkley Journal of Sociology (1986) H. 31, S. 53-78.
Kaiser, A.: Zur Interaktion von Mädchen und Jungen im Grundschulalter. In: Pädagogik und Schulalltag 4 (1994) S. 558-567.
Keim, K.-D.: Milieu und Moderne. Zum Gebrauch und Gehalt eines nachtraditionalen sozialräumlichen Milieubegriffs. In: Berliner Zeitschrift für Soziologie (1997) H. 7, S. 387-399.
Kettler, D./Meja, V. et al.: Politisches Wissen. Studien zu Karl Mannheim. Frankfurt am Main 1989.
Keuffer, J./Oelkers, J. (Hrsg.): Reform der Lehrerbildung in Hamburg. Weinheim 2001.
Klatt, R. et al.: Nutzung elektronischer wissenschaftlicher Information in der Hochschulausbildung. Barrieren und Potenziale in der innovativen Mediennutzung im Lernalltag der Hochschulen. Endbericht. Dortmund 2001.
Kleinau, E./Opitz, C. (Hrsg.): Geschichte der Mädchenbildung und Frauenbildung. Frankfurt am Main 1996.
Klemm, K.: Jugendliche ohne Ausbildung. Die „Kellerkinder" der Bildungsexpansion. In: Zeitschrift für Pädagogik 37 (1991), S. 887-898.
Klemm, K.: Bildung. In: Allmendinger, J./Ludwig-Mayerhofer, W. (Hrsg.): Soziologie des Sozialstaates. Weinheim/München 2000, S. 145-165.
Kneer, G.: Rationalisierung, Disziplinierung und Differenzierung: zum Zusammenhang von Sozialtheorie und Zeitdiagnose bei Jürgen Habermas, Michel Foucault und Niklas Luhmann. Opladen 1996.
Kneer, G./Nassehi, A.: Niklas Luhmanns Theorie sozialer Systeme. München 1997[3].
Knorr-Cetina, K.: Spielarten des Konstruktivismus. Einige Notizen und Anmerkungen. In: Soziale Welt (1989) H. 40, S. 86-96.

Kocka, J.: „Bürgertum“ and professions in the nineteenth century: two alternative approaches. In: Burrage, M./Torstendahl, R. (Hrsg.): Professions in Theory and History. Rethinking the Study of the Profession. London 1991, S. 62-74.
Koebner, T./Janz, R.-P. et al. (Hrsg.): „Mit uns zieht die neue Zeit“. Der Mythos Jugend. Frankfurt am Main 1985.
Köhler, H.: Bildungsbeteiligung und Sozialstruktur in der Bundesrepublik. Zu Stabilität und Wandel der Ungleichheit von Bildungschancen. Max-Planck-Institut für Bildungsforschung. Studien und Berichte 53. Berlin 1992.
Kohli, M.: Die Institutionalisierung des Lebenslaufs. Historische Befunde und theoretische Argumente. In: Kölner Zeitschrift für Soziologie und Sozialpsychologie (1985) H. 37, S. 1-29.
Kohli, M.: Gesellschaftszeit und Lebenszeit. Der Lebenslauf im Strukturwandel der Moderne. In: Berger, J. (Hrsg.): Die Moderne. Kontinuitäten und Zäsuren. Soziale Welt, Sonderband 4, 1986, S. 183-207.
Kolbe, F.-U./Sünker, H. et al.: Neue bildungssoziologische Beiträge zur Theorie institutionalisierter Bildung – Markierungen zur Theorieentwicklung. In: Kolbe, F.-U./Sünker, H. et al. (Hrsg.): Bildung, Gesellschaft, soziale Ungleichheit. Frankfurt am Main 1994, S. 11-33.
König, R.: Emile Durkheim zur Diskussion. München/Wien 1978.
Korte, H./Schäfers, B.: Einführung in Hauptbegriffe der Soziologie. Opladen 2000[5].
Krais, B.: Soziales Feld, Macht und kulturelle Praxis. Die Untersuchungen Bourdieus über die verschiedenen Fraktionen der „herrschenden Klasse“ in Frankreich. In: Eder, K. (Hrsg.): Klassenlage, Lebensstil und kulturelle Praxis. Frankfurt am Main 1989, S. 47-70.
Krais, B.: Erziehungs- und Bildungssoziologie. In: Kerber, H./Schmieder, A. (Hrsg.): Spezielle Soziologien. Reinbek 1994, S. 556-576.
Krais, B.: The Academic Disciplines: Social Field and Culture. In: Comparative Social Research Supplement 2 (1996a) S. 93-111.
Krais, B.: Bildungsexpansion und soziale Ungleichheit in der Bundesrepublik Deutschland. In: Bolder, A. et al. (Hrsg.): Jahrbuch Bildung und Arbeit: Die Wiederentdeckung der Ungleichheit. Aktuelle Tendenzen in Bildung und Arbeit. Opladen 1996b, S. 118-146.
Krais, B.: Einleitung: Die Wissenschaft und die Frauen. In: Krais, B. (Hrsg.): Wissenschaftskultur und Geschlechterordnung. Frankfurt am Main/New York 2000.
Krämer-Badoni, T.: Die Stadt als sozialwissenschaftlicher Gegenstand. In: Häußermann, H. et al. (Hrsg.): Stadt und Raum: soziologische Analysen. Pfaffenweiler 1991, S. 1-29.
Krappmann, L.: Die Reproduktion des Systems gesellschaftlicher Ungleichheit in der Kinderwelt. In: Grundmann, M. (Hrsg.): Konstruktivistische Sozialforschung: lebensweltliche Erfahrungskontexte, individuelle Handlungskompetenzen und die Konstruktion sozialer Strukturen. Frankfurt am Main 1999, S. 228-239.
Krappmann, L.: Bildung als Ressource der Lebensbewältigung. Manuskript. 2001.
Krause, D.: Luhmann-Lexikon: Eine Einführung in das Gesamtwerk von Niklas Luhmann mit 27 Abbildungen und über 500 Stichwörtern. Stuttgart 1999[2].
Kreckel, R.: Class, Status and Power? Begriffliche Grundlagen für eine politische Sozialisation der sozialen Ungleichheit. In: Kölner Zeitschrift für Soziologie und Sozialpsychologie (1982) H. 34, S. 617-648.
Kreckel, R.: Theorien sozialer Ungleichheit im Übergang. Einleitung. In: Kreckel, R. (Hrsg.): Soziale Ungleichheiten. Soziale Welt. Sonderband 2. Göttingen 1983, S. 3-15.
Kreckel, R.: Politische Soziologie der Sozialen Ungleichheit. Frankfurt am Main/New York 1992.
Kröner, S./Pfister, G.: Frauen – Räume. Körper und Identität im Sport. Pfaffenweiler 1992.
Krüger, H. (Hrsg.): Frauen und Bildung. Wege der Aneignung und Verwertung von Qualifikationen in weiblichen Erwerbsbiographien. Bielefeld 1992.
Krüger, H.-H.: Wege aus der Kindheit in Ost- und Westdeutschland. Bilanz und Perspektiven. In: Büchner, P./Krüger, H.-H. (Hrsg.): Vom Teddybär zum ersten Kuß. Wege aus der Kindheit in Ost- und Westdeutschland. Opladen 1996, S. 225-237.

Krüger, H.-H./Ecarius, J. et al.: Kinderbiographien: Verselbständigungsschritte und Lebensentwürfe. In: Du Bois-Reymond, M./Büchner, P. et al. (Hrsg.): Kinderleben. Opladen 1994, S. 221-271.

Krüger, H.-H./Haak, G. et al.: Kindheit im Umbruch. Biographien ostdeutscher Kinder. In: Aus Politik und Zeitgeschichte. Beilage zur Wochenzeitung „Das Parlament" (1993) H. 24, S. 32-39.

Krüger, H.-H./Olbertz, J.-H. (Hrsg.): Bildung zwischen Staat und Markt. Opladen 1997.

Krüger, H.-H./Rauschenbach, T.: Erziehungswissenschaft – eine ganz normale Disziplin? Eine einführende Problemskizze. In: Krüger, H.-H./Rauschenbach, T. (Hrsg.): Erziehungswissenschaft. Die Disziplin am Beginn einer neuen Epoche. Weinheim/München 1994, S. 7-16.

Kunnemann, H./de Vries, H. (Hrsg.): Die Aktualität der „Dialektik der Aufklärung". Frankfurt am Main/New York 1989.

Lampert, M. D.: Studying Gender Differencies in the Conversational Humor of Adults and Children. In: Slobin, D. I. et al. (Hrsg.): Social Interaction, Social Context, and Language. Mahwah, New Jersey 1996, S. 579-596.

Läpple, D.: Essay über den Raum. Für ein gesellschaftswissenschaftliches Raumkonzept. In: Häußermann, H. et al. (Hrsg.): Stadt und Raum. Pfaffenweiler 1991, S. 157-207.

Lenzen, D.: Zwischen Stabilisierung und Differenzierung. Paradoxien im Erziehungssystem. In: Luhmann, N./Schorr, K. E. (Hrsg.): Zwischen System und Umwelt. Fragen an die Pädagogik. Frankfurt am Main 1996, S. 256-278.

Lichtenstein, E.: Der Ursprung der Pädagogik im griechischen Denken. Hannover 1970.

Lindemann, G.: Das paradoxe Geschlecht. Transsexualität im Spannungsfeld von Körper, Leib und Gefühl. Frankfurt am Main 1993.

Longhurst, B.: Karl Mannheim and the Contemporary Sociology on Knowledge. Basingstoke u.a. 1989.

Loviscach, J.: Die elektronische Uni. Neue Medien in der Lehre. In: c't 4 (2001) S. 108-115.

Löw, M.: Raum ergreifen. Alleinwohnende Frauen zwischen Arbeit, sozialen Beziehungen und der Kultur des Selbst. Bielefeld 1994.

Löw, M.: Was wird aus Lippenstift und Puderdose? Geschlechtertheorie und Frauenpolitik im Wandel. In: Löw, M. et al. (Hrsg.): Pädagogik im Umbruch. Kontinuität und Wandel in den neuen Bundesländern. Opladen 1995, S. 59-67.

Löw, M.: Raumsoziologie. Frankfurt am Main 2001a.

Löw, M.: Gemeindestudien heute: Sozialforschung in der Tradition der Chicagoer Schule? In: Zeitschrift für qualitative Bildungs-, Beratungs- und Sozialforschung 1 (2001b) S. 111-131.

Löw, M. (Hrsg.): Differenzierungen des Städtischen. Opladen 2002.

Luhmann, N.: Soziologische Aufklärung 1. Opladen 1970.

Luhmann, N.: Gesellschaftsstruktur und Semantik. Studien zur Wissenssoziologie der modernen Gesellschaft. Frankfurt am Main 1980.

Luhmann, N.: Zum Begriff der sozialen Klasse. In: Luhmann, N. (Hrsg.): Soziale Differenzierung: Zur Geschichte einer Idee. Opladen 1985, S. 119-162.

Luhmann, N.: Ökologische Kommunikation. Kann die moderne Gesellschaft sich auf ökologische Gefährdungen einstellen? Opladen 1986.

Luhmann, N.: Die Wissenschaft der Gesellschaft. Frankfurt am Main 1990a.

Luhmann, N.: Soziologische Aufklärung 5. Konstruktivistische Perspektiven. Opladen 1990b.

Luhmann, N.: Soziologische Aufklärung 2. Aufsätze zur Theorie der Gesellschaft. Opladen 1991[4].

Luhmann, N.: Soziale Systeme. Grundriß einer allgemeinen Theorie. Frankfurt am Main 1993[4].

Luhmann, N.: Inklusion und Exklusion. In: Berding, H. (Hrsg.): Nationales Bewußtsein und kollektive Identität. Studien zur Entwicklung des kollektiven Bewußtseins in der Neuzeit 2. Frankfurt am Main 1994, S. 15-45.

Luhmann, N.: Das Erziehungssystem und die Systeme seiner Umwelt. In: Luhmann, N./Schorr, E. (Hrsg.): Zwischen System und Umwelt. Fragen an die Pädagogik. Frankfurt am Main 1996, S. 14-52.

Luhmann, N.: Arbeitsteilung und Moral. Durkheims Theorie. In: Durkheim, E. (Hrsg.): Über soziale Arbeitsteilung. Frankfurt am Main 1999[3], S. 19-38.
Luhmann, N.: Das Erziehungssystem der Gesellschaft. Frankfurt am Main 2002.
Luhmann, N./Schorr, K. E.: Reflexionsprobleme im Erziehungssystem. Frankfurt am Main 1988.
Mann, M.: A Crisis of Stratification Theory? Persons, Households/Families/Lineages, Genders, Classes and Nations. In: Crampton, R./Mann, M. (Hrsg.): Gender and Gratification. Cambridge 1986, S. 40-56.
Mannheim, K.: Die Bedeutung der Konkurrenz im Gebiete des Geistigen. Verhandlungen des 6. Deutschen Soziologentages vom 17.-19.9.1928 in Zürich. Tübingen 1929b, S. 35-83.
Mannheim, K.: Ideologie und Utopie. Bonn 1929a.
Mannheim, K.: Diagnose unserer Zeit. Zürich/Wien/Konstanz 1951.
Mannheim, K.: Mensch und Gesellschaft. Bad Homburg v.d.H. u.a. 1967.
Mannheim, K.: Konservatismus. Ein Beitrag zur Soziologie des Wissens. (Hrsg. von Kettler, D. et al.). Frankfurt am Main 1984.
Mannheim, K./Stewart: Einführung in die Soziologie der Erziehung. Düsseldorf 1973.
Maresch, R.: Cyberhypes: Möglichkeiten und Grenzen des Internet. Frankfurt am Main 2001.
Marotzki, W.: Bildungsprozesse in lebensgeschichtlichen Horizonten. In: Hoerning, E. et al. (Hrsg.): Biographieforschung. Bad Heilbrunn 1991.
Marotzki, W.: Digitalisierte Biographien? Sozialisations- und bildungstheoretische Perspektiven virtueller Welten. In: Lenzen, D./Luhmann, N. (Hrsg.): Bildung und Weiterbildung im Erziehungssystem. Lebenslauf und Humanontogenese als Medium und Form. Frankfurt am Main 1997, S. 175-198.
Maroztki, W./Meister, D. M. et al.: Zum Bildungswert des Internet. Opladen 2000.
Marshall, J.: Michel Foucault: Personal Autonomy and Education. Dordrecht u.a. 1996.
Marx, K./Engels, F.: Marx-Engels-Werke. Band 4. Berlin 1964.
Marx, K./Engels, F.: Manifest der Kommunistischen Partei. Berlin 1984.
Massey, D.: Space, Place and Gender. Oxford 1994.
Maturana, H. R.: Erkennen. Die Organisation und Verkörperung von Wirklichkeit. Braunschweig/Wiesbaden 1982.
Mayer, S.: Eia und Popeia ist nicht genug. In: Die Zeit 8. März 2001, www.zeit.de/2001/11/Kultur/200111_sm-kind_liste.html, 5 Seiten (8. März 2001).
McWhorter, L.: Bodies and Pleasures: Foucault and the Politics of Sexual Normalization. Bloomington, Indiana 1999.
Merton, R. K.: Science and the Economy of 17th Century in England. In: Science and Society (1939) H. 3, S. 3-27.
Merton, R. K.: Entwicklung und Wandel von Forschungsinteressen. Frankfurt am Main 1985.
Meulemann, H.: Expansion ohne Folgen? Bildungschancen und sozialer Wandel in der Bundesrepublik. In: Glatzer, W. (Hrsg.): Entwicklungstendenzen der Sozialstruktur. Frankfurt am Main/New York 1992, S. 123-157.
Miller, A.: Am Anfang war Erziehung. Frankfurt am Main 1983.
Mitterauer, M.: Sozialgeschichte der Jugend. Frankfurt am Main 1986.
Muchow, M./Muchow, H.: Der Lebensraum des Großstadtkindes. Hamburg 1935.
Müller, H.-P.: Wertkrise und Gesellschaftsreform. Emile Durkheims Schriften zur Politik. Stuttgart 1983.
Müller, U.: Soziologie und Geschlechtergerechtigkeit am Beispiel der Forschung zu Frauen an Hochschulen. In: Dausien, B. et al. (Hrsg.): Erkenntnisprojekt Geschlecht. Opladen 1999, S. 141-166.
Müller, U.: Asymmetrische Geschlechterkultur in Organisationen und Frauenförderung als Prozeß – mit Beispielen aus Betrieben und der Universität. In: Lenz, I. et al. (Hrsg.): Geschlecht – Arbeit – Zukunft. Münster 2000, S. 126-149.
Müller, W.: Erwartete und unerwartete Folgen der Bildungsexpansion. In: Friedrichs, J. et al. (Hrsg.): Die Diagnosefähigkeit der Soziologie. Sonderheft 38 der Kölner Zeitschrift für Soziologie und Sozialpsychologie. Opladen 1998, S. 81-112.
Müller, W./Haun, D.: Bildungsungleichheit im sozialen Wandel. In: Kölner Zeitschrift für Soziologie und Sozialpsychologie (1994) H. 46, S. 1-42.

Münch, R.: Theorie des Handelns. Zur Rekonstruktion der Beiträge von Talcott Parsons, Emile Durkheim und Max Weber. Frankfurt am Main 1982.
Münch, R.: Code, Structure, and Action: Building a Theory of Structuration from a Parsonian Point of View. In: Turner, J. H. (Hrsg.): Theory Building in Sociology: Assessing Theory Cumulation. Newbury Park 1989, S. 101-117.
Münch, R.: Dialektik der Kommunikationsgesellschaft. Frankfurt am Main 1991.
Nassehi, A.: Differenzierungsfolgen. Beiträge zur Soziologie der Moderne. Opladen 1999.
Nefiodow, L. A.: Der sechste Kondratieff. Wege zur Produktivität und Vollbeschäftigung im Zeitalter der Information. Stankt Augustin 1997.
Negt, O.: Kindheit und Kinderöffentlichkeit. In: Neue Rundschau (1983) H. 3, S. 40-55.
Nissen, U. (Hrsg.): Raum und Zeit in der Nachmittagsgestaltung von Kindern. München 1992.
Nissen, U.: Kindheit, Geschlecht und Raum. Sozialisationstheoretische Zusammenhänge geschlechtsspezifischer Raumaneignung. Weinheim 1998.
Oevermann, U.: Programmatische Überlegungen zu einer Theorie der Bildungsprozesse und zur Strategie der Sozialisationsforschung. In: Hurrelmann, K. (Hrsg.): Sozialisation und Lebenslauf. Reinbek 1976, S. 34-52.
Offe, C.: Bildungssystem, Beschäftigungssystem und Bildungspolitik. Aufsätze zu einer gesamtgesellschaftlichen Funktionsbestimmung des Bildungswesens. In: Roth, H./Friederich, D. (Hrsg.): Bildungsforschung. Probleme – Perspektiven – Prioritäten. Stuttgart 1975.
Olk, T.: Jugend an den Grenzen der Moderne. Ein gescheitertes Projekt? In: Breyvogel, W. (Hrsg.): Pädagogische Jugendforschung. Erkenntnisse und Perspektiven. Opladen 1989, S. 31-48.
Ortmann, H.: Arbeiterfamilie und sozialer Aufstieg. Weinheim/München 1971.
Palzkill, B.: Zwischen Turnschuh und Stöckelschuh. Die Entwicklung lesbischer Identität im Sport. Bielefeld 1990.
Parelius, R. J./Parelius, A. P.: The Sociology of Education. London u.a. 1987[2].
Park, R. E.: Die Stadt als räumliche Struktur und als sittliche Ordnung. In: Atteslander, P./ Hamm, B. (Hrsg.): Materialien zur Siedlungssoziologie. Köln 1974[7], S. 90-100.
Park, R. E. et al.: The City. Chicago. 1974[7]
Parsons, T.: The Structure of Social Action. New York 1968, orig. 1949.
Parsons, T.: Die amerikanische Universität: ein Beitrag zur Soziologie der Erkenntnis. Frankfurt am Main 1990.
Parsons, T.: Sozialstruktur und Persönlichkeit. Eschborn bei Frankfurt am Main 1997[5].
Parsons, T. et al.: Family, Socialization and Interaction Process. Glencoe Illinois 1955.
Paschen, H./Wigger, L. (Hrsg.): Schulautonomie als Entscheidungsproblem. Weinheim 1996.
Paulsen, F.: Bildung. In: Rein, W. (Hrsg.): Enzyklopädisches Lexikon der Pädagogik. Langensalza 1903, S. 658-670.
Paulsen, J.: Zur Geschichte der Soziologie im Nationalsozialismus. Oldenburg 1988.
Peisert, H.: Soziale Lage und Bildungschancen in Deutschland. München 1967.
Perry, T./Delpit, L.: The Real Ebonics Debatte: Power, Language and the Education of African-American Children. Boston 1998.
Pfeil, E.: Das Großstadtkind. München/Basel 1965.
Pfister, G.: Der Widerspenstigen Zähmung. Raumaneignung, Körperlichkeit und Interaktion. In: Pfister, G./Valtin, R. (Hrsg.): MädchenStärken. Probleme der Koedukation in der Grundschule. Frankfurt am Main 1993, S. 67-83.
Pfister, G./Valtin, R. (Hrsg.): MädchenStärken. Probleme der Koedukation in der Grundschule. Frankfurt am Main 1993.
Plake, K. (Hrsg.): Klassiker der Erziehungssoziologie. Düsseldorf 1987.
Pollock, L.: Forgotten Children. Parent-Child-Relations since 1500-1900. Cambridge 1983.
Preuss-Lausitz, U.: Geschlechtersozialisation und Schulpädagogik in der Nachmoderne. In: Die Deutsche Schule (1997) H. 4, S. 429-445.
Quaiser-Pohl, C. (1997). Geschlechtsspezifische Raumaneignung von Mädchen. (Unveröffentlichtes Manuskript.) 10 S.
Qvortrup, J.: Placing Children in the Division of Labour. In: Close, P./Collins, R. (Hrsg.): Family and Economy in Modern Society. London 1985, S. 129-145.

Qvortrup, J.: Die soziale Definition von Kindheit. In: Markefka, M./Nauk, B. (Hrsg.): Handbuch der Kindheitsforschung. Neuwied/Kriftel/Berlin 1993, S. 109-124.
Rabe-Kleberg, U.: Frauenberufe. Zur Segmentierung der Berufswelt. Bielefeld 1987.
Rabe-Kleberg, U. (Hrsg.): Besser gebildet und doch nicht gleich! Frauen und Bildung in der Arbeitsgesellschaft. Bielefeld 1990.
Rabe-Kleberg, U.: Verantwortung und Macht. Ein Beitrag zum Verhältnis von Geschlecht und Beruf angesichts der Krise traditioneller Frauenberufe. Bielefeld 1993.
Rabe-Kleberg, U.: Wird die Frauensache zur Privatsache? Geschlechterpolitik in den neuen Bundesländern. In: Krüger, H. H. et al. (Hrsg.): Transformationsprobleme in Ostdeutschland. Arbeit. Bildung. Sozialpolitik. Opladen 1995, S. 117-126.
Rabe-Kleberg, U.: Frauen in sozialen Berufen – (k)eine Chance auf Professionalisierung? In: Friebertshäuser, B. et al. (Hrsg.): Sozialpädagogik im Blick der Frauenforschung. Weinheim 1997, S. 59-69.
Rabe-Kleberg, U.: Wie aus Berufen für Frauen Frauenberufe werden. Ein Beitrag zur Transformation des Geschlechterverhältnisses. In: Nickel, H. M. et al. (Hrsg.): Transformation – Unternehmensreorganisation – Geschlechterforschung. Opladen 1999³, S. 93-107.
Rabe-Kleberg, U./Löw, M.: Geschlechtsspezifische Sozialisation und deren Auswirkungen auf die Schulentwicklung. In: Braun, K.-H. et al. (Hrsg.): Schule mit Zukunft. Bildungspolitische Empfehlungen und Expertisen der Enquete-Kommission des Landtages von Sachsen-Anhalt. Opladen 1998, S. 323-339.
Reinhardt, S.: Etappen und Perspektiven der Bildungspolitik. In: Ellwein, T./Holtmann, E. (Hrsg.): 50 Jahre Bundesrepublik. Rahmenbedingungen – Entwicklungen – Perspektiven. Opladen/Wiesbaden 1999, S. 310-326.
Rettke, U.: Berufswünsche von Mädchen unter dem Diktat des Arbeitsmarktes. Die schrittweise „Verweiblichung" der Bildungs- und Berufsbiographien von Hauptschülerinnen. In: Bolder, A./Rodax, K. (Hrsg.): Das Prinzip der aufge(sc)hobenen Belohnung. Die Sozialisation von Arbeiterkindern für den Beruf. Bonn 1987, S. 127-141.
Ritsert, J.: Die Rationalität Adornos. Frankfurt am Main 1995.
Ritter, M.: Computer oder Stöckelschuh? Eine empirische Untersuchung über Mädchen am Computer. Frankfurt am Main/New York 1994.
Rolff, H.-G.: Sozialisation und Auslese durch die Schule. Heidelberg 1967.
Rolff, H.-G./Bauer, K.-O. et al. (Hrsg.): Jahrbuch der Schulentwicklung. Weinheim und München 1994.
Roloff, C.: Informatik und Karriere. In: GI Jahrestagung 2 (1990) S. 307-317.
Roloff, C. et al.: Nicht nur ein gutes Examen. Dortmund 1987.
Rose, L.: Das Drama des begabten Mädchens. Lebensgeschichte junger Kunstturnerinnen. Weinheim 1992a.
Rose, L.: Körper ohne Raum. Zur Vernachlässigung weiblicher Bewegungs- und Sportwelten in der feministischen Körper-Debatte. In: Feministische Studien (1992b) S. 113-120.
Rosenbaum, H.: Formen der Familie. Frankfurt am Main 1990⁵.
Roth, L.: Die Erfindung des Jugendlichen. München 1983.
Rötzer, F.: Die Achillesferse des Internet. Frankfurt am Main 2000.
Ruhloff, J.: Pädagogik und anderes. Transzendentalkritische Bemerkungen zu Niklas Luhmann, „Das Erziehungssystem und die Systeme seiner Umwelt". In: Luhmann, N./Schorr, K. E. (Hrsg.): Zwischen System und Umwelt. Fragen an die Pädagogik. Frankfurt am Main 1996, S. 53-74.
Rutschky, K. (Hrsg.): Schwarze Pädagogik. Frankfurt am Main 1977.
Schäfer, A.: Systemtheorie und Pädagogik: Konstitutionsproblem von Erziehungstheorien. Königstein 1983.
Schelsky, H.: Soziologische Bemerkungen zur Rolle der Schule in unserer Gesellschaftsverfassung. In: Schelsky, H. (Hrsg.): Auf der Suche nach Wirklichkeit – Gesammelte Aufsätze. Düsseldorf/Köln 1965, S. 131-159.
Schimank, U.: Theorien gesellschaftlicher Differenzierung. Opladen 1996.
Schimank, U./Volkmann, U.: Gesellschaftliche Differenzierung. Bielefeld 1999.
Schimpl-Neimanns, B.: Hat die Bildungsexpansion zum Abbau der sozialen Ungleichheit in der Bildungsbeteiligung geführt? Methodische Überlegungen zum Analyseverfahren und

Ergebnisse multinominaler Logik-Modelle für den Zeitraum 1950-1989. Mannheim 2000.
Schlüter, A./Metz-Göckel, S.: „Die Gymnasiallehrer, sie haben uns nie begriffen" – Bildungsbarrieren und Bildungskarrieren von Arbeitertöchtern. In: Kreienbaum, M. A. (Hrsg.): Frauen bilden Macht. Dokumentation des 7. Fachkongresses Frauen und Schule. Dortmund 1989, S. 192-203.
Schmid, P.: Das Allgemeine, die Bildung und das Weib. Zur verborgenen Konzipierung von Allgemeinbildung als allgemeiner Bildung für Männer. In: Tenorth, H.-E. (Hrsg.): Allgemeine Bildung. Analysen zu ihrer Wirklichkeit, Versuche über ihre Zukunft. Weinheim 1986, S. 202-214.
Schuller, M.: Vergabe des Wissens. Notizen zum Verhältnis von „weiblicher Intellektualität" und Macht. In: Konkursbuch 12: Frauen Macht. Zeitschrift für Vernunftkritik (o.J.), S. 13-21.
Schulze, G.: Die Erlebnisgesellschaft. Kultursoziologie der Gegenwart. Frankfurt am Main/New York 1997[7].
Schütze, F.: Organisationszwänge und hoheitsstaatliche Rahmenbedingungen im Sozialwesen: Ihre Auswirkungen auf die Paradoxien professionellen Handelns. In: Combe, A./Helsper, W. (Hrsg.): Pädagogische Professionalität. Untersuchungen zum Typus pädagogischen Handelns. Frankfurt am Main 1999[3], S. 183-275.
Schütze, F.: Schwierigkeiten bei der Arbeit und Paradoxien des professionellen Handelns. In: Zeitschrift für Bildungs-, Beratungs- und Sozialforschung (2000) H. 1, S. 49-97.
Schwinn, T.: Jenseits von Subjektivismus und Objektivismus: Max Weber, Alfred Schütz, und Talcott Parsons. Berlin 1993.
Schwinn, T.: Soziale Ungleichheit und funktionale Differenzierung. Wiederaufnahme einer Diskussion. In: Zeitschrift für Soziologie (1998) H. 1, S. 3-17.
Seidenfaden, F.: Studien zur Geschichte der Universität im Deutschsprachigen Raum. In: Körner, A./Seidenfaden, F. (Hrsg.): Universitäten und Hochschulen im Wandel. Teil 2. Historischer Rückblick und Bundesrepublik, Großbritannien und Niederlande. Gießen 1988, S. 11-133.
Shahar, S.: Kindheit im Mittelalter. München/Zürich 1991.
Shavit, Y./Blossfeld, H.-P.: Persistent Inequalities: a Comparative Study of Educational Attainment in Thirteen Countries. Boulder Colorado 1993.
Shields, R.: Cultures of Internet: Virtual Spaces, Real Histories, Living Bodies. London 1996.
Siemering, H.: Deutschlands Jugend in Bevölkerung und Wirtschaft. Berlin 1937.
Simmel, G.: Kant. Sechzehn Vorlesungen gehalten an der Berliner Universität. Leipzig 1905.
Sobiech, G.: „Ich hatte das Gefühl, irgend etwas ist jetzt vorbei." Die Pubertät – Brüche und Ambivalenzen in der Körper- und Bewegungsentwicklung von Mädchen. In: Palzkill, B. et al. (Hrsg.): Bewegungs(t)räume. Frauen Körper Sport. München 1991, S. 47-61.
Spitthöver, M.: Frauen in städtischen Freiräumen. Köln 1989.
Stanat, P./Kunter, M.: Kooperation und Kommunikation. In: Deutsches PISA-Konsortium (Hrsg.): PISA 2000. Basiskompetenzen von Schülerinnen und Schülern im internationalen Vergleich. Opladen 2001, S. 299-321.
Statistisches Jahrbuch 2002 für die Bundesrepublik Deutschland. Stuttgart 2002.
Stehr, N.: Die Zerbrechlichkeit moderner Gesellschaften. Die Stagnation der Macht und die Chancen des Individuums. Weilerswist 2000.
Steinkamp, G.: Sozialstruktur und Sozialisation. In: Hurrelmann, K./Ulich, D. (Hrsg.): Neues Handbuch der Sozialisationsforschung. Weinheim 1991, S. 251-277.
Stichweh, R.: Akademische Freiheit, Professionalisierung der Hochschullehre und Politik. In: Oelkers, J. et al. (Hrsg.): Pädagogik, Erziehungswissenschaft und Systemtheorie. Weinheim/Basel 1987, S. 125-145.
Stichweh, R.: Globalisierung der Wissenschaft und die Rolle der Universität. In: Rusterholz, P./Liechti, A. (Hrsg.): Universität am Scheideweg. Herausforderungen, Probleme, Strategien. Zürich 1988, S. 63-74.
Stichweh, R.: Der frühmoderne Staat und die europäische Universität. Zur Interaktion von Politik und Erziehungssystem im Prozess ihrer Ausdifferenzierung (16.-18. Jahrhundert). Frankfurt am Main 1991.

Sturm, G.: Wege zum Raum. Methodologische Annäherung an ein Basiskonzept raumbezogener Wissenschaften. Opladen 2000.
Terlinden, U.: Gebrauchswirtschaft und Raumstruktur. Ein feministischer Ansatz in der soziologischen Stadtforschung. Stuttgart 1990.
Theling, G.: „Vielleicht wäre ich als Verkäuferin glücklicher geworden". Arbeitertöchter & Hochschule. Münster 1986.
Tietze, W. (Hrsg.): Wie gut sind unsere Kindergärten? Eine Untersuchung zur pädagogischen Qualität in deutschen Kindergärten. Neuwied 1998.
Timmermann, D.: Bildungsökonomisches Abwägen heterogener Argumente zur Schulautonomie. In: Paschen, H./Wigger, L. (Hrsg.): Schulautonomie als Entscheidungsproblem. Weinheim 1996, S. 59-88.
Tizard, B./Hughes, M.: Young Children Learning. Talking and Thinking at Home and at School. London 1984.
Tough, J.: Listening to Children Talking. London 1976.
Turkle, S.: Leben im Netz. Identität in Zeiten des Internet. Reinbek bei Hamburg 1998.
Varela, F.: Die Biologie der Freiheit. In: Psychologie heute (1982) H. 11, S. 82-93.
Vester, M.: Soziale Milieus im gesellschaftlichen Strukturwandel: zwischen Integration und Ausgrenzung. Köln 2001.
Wacquant, L. J. D.: The Pugilistic Point of View: How Boxers Think and Feel about their Trade. In: Theory and Society (1995a) H. 24, S. 489-535.
Wacquant, L. J. D.: Pugs at Work: Bodily Capital and Bodily Labour Among Professional Boxers. In: Body & Society (1995b) H. 1, S. 65-93.
Wacquant, L. J. D.: Auf dem Wege zu einer Sozialpraxeologie. In: Bourdieu, P./Wacquant, L. J. D. (Hrsg.): Reflexive Anthropologie. Frankfurt am Main 1996, S. 17-93.
Wagner, H. J.: Eine Theorie pädagogischer Professionalität. Weinheim 1998.
Walford, G. (Hrsg.): Durkheim and Modern Education. London 1998.
Weber, M.: Wirtschaft und Gesellschaft. Grundriss der Verstehenden Soziologie. Tübingen 1980[5], orig. 1921.
Weber, M.: Wissenschaft als Beruf. Ditzingen 1992.
Weedon, C.: Wissen und Erfahrung. Feministische Praxis und poststrukturalistische Theorie. Zürich 1990.
Wenneras, C./Wold, A.: Vetternwirtschaft und Sexismus im Gutachterwesen. In: Krais, B. (Hrsg.): Wissenschaftskultur und Geschlechterordnung. Frankfurt am Main 2000.
Wenzel, H.: Die Ordnung des Handelns: Talcott Parsons' Theorie des allgemeinen Handlungssystems. Frankfurt am Main 1990.
Wetterer, A.: Profession und Geschlecht. Über die Marginalität von Frauen in hochqualifizierten Berufen. Frankfurt am Main 1992.
Wetterer, A.: Professionalisierung und Geschlechterhierarchie. Vom kollektiven Frauenausschluß zur Integration mit beschränkten Möglichkeiten. Kassel 1993.
Wetterer, A. (Hrsg.): Die soziale Konstruktion von Geschlecht in Professionalisierungsprozessen. Frankfurt am Main/New York 1995.
Weymann, A. (Hrsg.): Bildung und Beschäftigung. Grundzüge und Perspektiven des Strukturwandels. Soziale Welt. Sonderband 5. Göttingen 1987.
Wiley, N.: The History and Politics of Recent Sociological Theory. In: Ritzer, G. (Hrsg.): Frontiers of Social Theory. The New Syntheses. New York 1990, S. 392-415.
Willis, P.: Spaß am Widerstand. Gegenkultur in der Arbeiterschule. Frankfurt am Main 1982[2].
Willis, P.: Jugend-Stile. Zur Ästhetik der gemeinsamen Kultur. Hamburg/Berlin 1991.
Willms-Herget, A.: Frauenarbeit. Zur Integration der Frauen in den Arbeitsmarkt. Frankfurt am Main/New York 1985.
Wingens, M.: Wissensgesellschaft und gesellschaftliche Wissensproduktion. In: Honegger, C./Hradil, S. et al. (Hrsg.): Grenzenlose Gesellschaft? Opladen 1999, S. 433-446.
Zeiher, H. J./Zeiher, H.: Orte und Zeiten der Kinder. Soziales Leben im Alltag von Großstadtkindern. Weinheim/München 1994.
Zinnecker, J.: Straßensozialisation. In: Zeitschrift für Pädagogik (1979) H. 25, S. 727-746.

Zinnecker, J.: Kindheit, Jugend und soziokultureller Wandel in der Bundesrepublik Deutschland. In: Büchner, P./Krüger, H.-H. et al. (Hrsg.): Kindheit und Jugend im interkulturellen Vergleich. Opladen 1990a, S. 17-36.

Zinnecker, J.: Vom Straßenkind zum verhäuslichten Kind. Kindheitsgeschichte im Prozeß der Zivilisation. In: Behnken, I. (Hrsg.): Stadtgesellschaft und Kindheit im Prozeß der Zivilisation. Opladen 1990b, S. 142-162.

Zinnecker, J.: Jugend als Bildungsmoratorium. Zur Theorie des Wandels der Jugendphase in west- und osteuropäischen Gesellschaften. In: Melzer, W. et al. (Hrsg.): Osteuropäische Jugend im Wandel. Weinheim/München 1991, S. 9-24.

Zinnecker, J.: Kindheitsort Schule – Kindheitsort Straße. In: Reiß, G. (Hrsg.): Schule und Stadt. Lernorte, Spielräume, Schauplätze für Kinder und Jugendliche. Weinheim/ München 1995, S. 45-67.

Zuckermann, H./Cole, J. R. et al.: The Outer Circle. Women in Scientific Community. New York/London 1991.

Anmerkungen zur Autorin

Dr. Martina Löw ist Professorin für Soziologie an der TU Darmstadt. Ihre Arbeitsgebiete sind Stadt- und Raumsoziologie, Bildungssoziologie und Geschlechterforschung. Ihre wichtigsten Veröffentlichungen der letzten Jahren sind:

Raumsoziologie. Frankfurt am Main: Suhrkamp Taschenbuch Wissenschaft. 2001.

Geschlechtsspezifische Sozialisation. Modul im Rahmen eines multimedialen Lehrprogramms. Erscheint auf CD und im elektronischen Netz. 2003, 129 Seiten (zusammen mit Regina Siemers).

Differenzierungen des Städtischen. Opladen: Leske und Budrich. 2002, (Herausgabeschrift).